Core Physics

Core Physics

Geoff Cackett
Portobello High School, Edinburgh

Ron Kennedy
Formerly at Holy Rood High School, Edinburgh

Alastair Steven
Wester Hailes Education Centre, Edinburgh

Oxford University Press

Oxford University Press, Walton Street,
Oxford OX2 6DP

Oxford New York Toronto
Delhi Bombay Calcutta Madras Karachi
Petaling Jaya Singapore Hong Kong Tokyo
Nairobi Dar es Salaam Cape Town
Melbourne Auckland

and associated companies in
Berlin Ibadan

Oxford is a trade mark of Oxford University Press

ISBN 0 19 914051 0

First published 1979
Reprinted 1979, 1981 (twice), 1982, 1983, 1985,
1986, 1987, 1989, 1991

Colour illustration by
Colin Ratray, photographs
by Paul Brierley

Typeset in Monophoto Photina by
Tradespools Ltd., Frome, Somerset
Printed in Hong Kong

Preface

This book is written specifically for students preparing for Scottish 'O' grade Physics but it will also be useful for GCSE courses. Core Physics will also provide a firm theoretical foundation for the new Standard Grade (applications oriented) syllabus. The book is designed to equip the students with the understanding and basic knowledge which should provide the foundation for greater appreciation and enjoyment of the subject.

The book follows closely the order of the topics as presented in the Scottish Certificate of Education syllabus. Each chapter covers a particular topic of the syllabus with a full explanation. The practical nature of the course is emphasized by reference to many experiments with diagrams and photographs. Sample results are used to derive relationships and worked examples have been used to illustrate particular points and to help with understanding. Each chapter ends with a summary and problems, many of the problems being from past 'O' grade papers. Numerical answers have been provided to all problems; SI units are used throughout, other units being referred to only when they are in everyday use. The negative index notation for units has been used, (e.g. $m\ s^{-1}$ rather than m/s), in accordance with examination requirements. Vector notation is introduced and used where essential. 'Electron flow' current convention is used and conventional current is not referred to at all, apart from a warning to students that they may encounter it in other texts.

We should like to express our appreciation to the various establishments listed on page 277 which gave permission to reproduce their drawings and photographs, and to The Scottish Certificate of Education Examinations Board for allowing us to include questions from past 'O' grade examination papers. Finally we should like to thank our families for their tolerance.

Geoff Cackett
Ron Kennedy
Alastair Steven

Contents

1 Wave motion

1.1 Introduction

There are many different kinds of waves. One that we all recognize is the **water wave**. When a stone is thrown into a calm pond, waves spread out over the surface, Figure 1.1.

What is travelling out from the centre of the disturbance? After the wave has passed, the water settles back to its original level, so it is not the water that is travelling out. To answer the question, let us imagine a small piece of wood floating on the pond surface. The piece of wood is seen to vibrate as the wave passes. Since the wave causes the wood to move, it must be supplying energy.

This appears to be the answer to the question. It is **energy** that is travelling out from the centre. It is even more obvious that waves carry energy if we watch waves on the sea, Figure 1.2.

During stormy weather, the energy carried by the waves has often resulted in great damage, Figure 1.3.

This energy comes from winds moving over the surface of the sea. Figure 1.4 is a photograph of a device which can convert wave energy to useful electrical energy.

When we considered the effect of the water wave on the floating piece of wood, we saw that the actual vibrations of the water were in a **vertical** direction, while the energy was transmitted in a **horizontal** direction along the water surface.

A wave in which the vibrations are at right angles to the direction in which the wave is moving is called a **transverse** wave.

Figure 1.1 A circular wave on a pond

Figure 1.2 Waves smashing against a breakwater

Figure 1.3 A damaged pier

Figure 1.4 A device for converting wave energy to electricity

1.2 Transverse waves and pulses

Transverse waves can be investigated using a long 'slinky' spring stretched out on a bench. One end is held in a fixed position and the other end is moved with your hand, Figure 1.5.

By moving your hand once from side to side, you would generate what is called a **wave pulse** which would travel down the spring. It can be seen that the actual disturbance is **at right angles** to the direction of the wave pulse.

Another method of showing transverse waves is to use a wave machine, Figure 1.6.

The beads can be made to move up and down and give the impression that a wave is travelling at right angles to that direction.

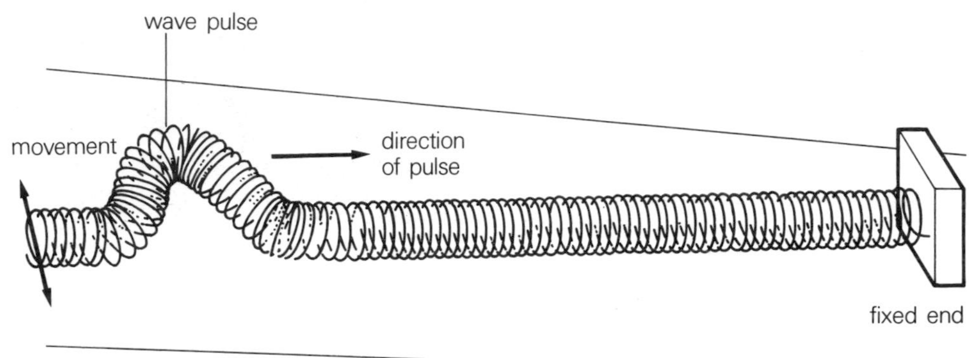

Figure 1.5 A wave on a 'slinky'

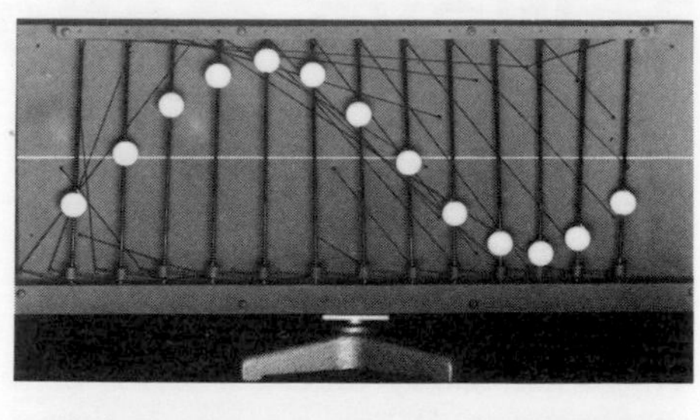

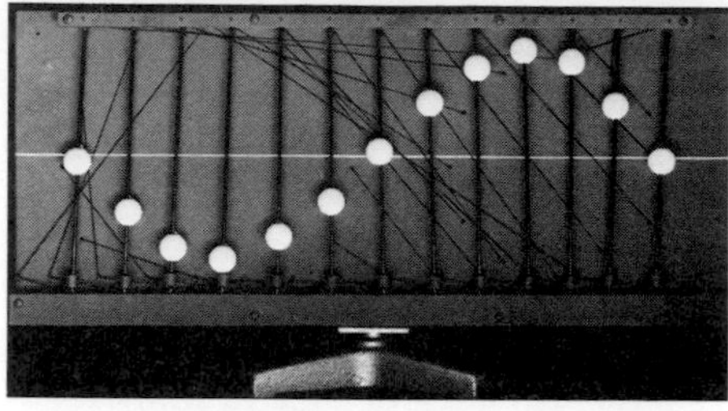

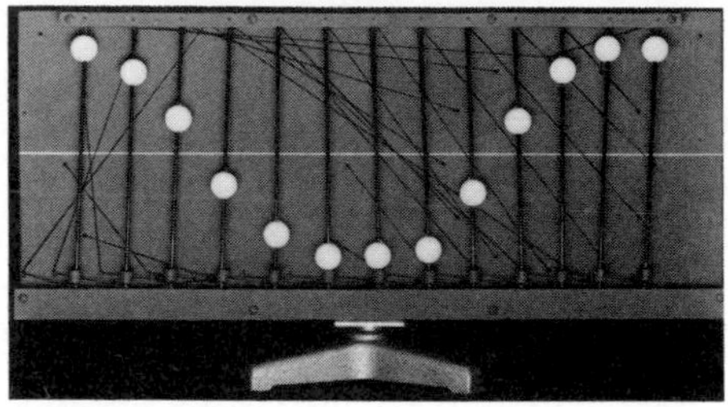

Figure 1.6 A wave machine

1.3 Quantities to describe waves

Various quantities are required to describe waves: they can be investigated using a length of rubber tubing on a bench. A travelling wave (often called a **continuous** wave) can be made by continuously moving your hand from side to side, Figure 1.7.

Amplitude

The greater the movement of your hand, the greater is the size of the disturbance which travels down the tubing.

The **amplitude** of a wave is the size of the maximum disturbance measured from the 'zero' position, Figure 1.8.

Figure 1.7 A wave travelling along rubber tubing

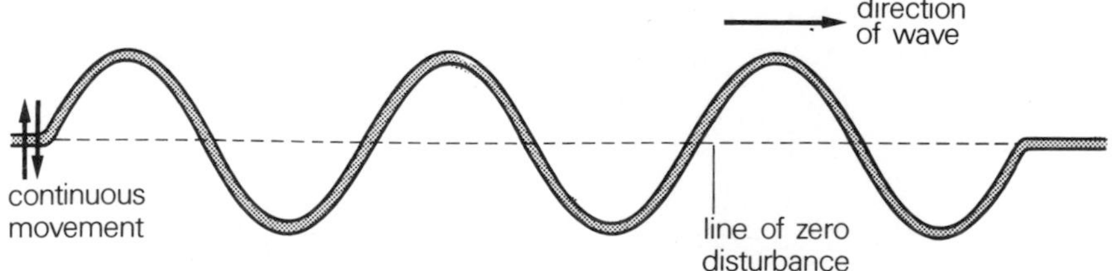

Figure 1.8 Amplitude

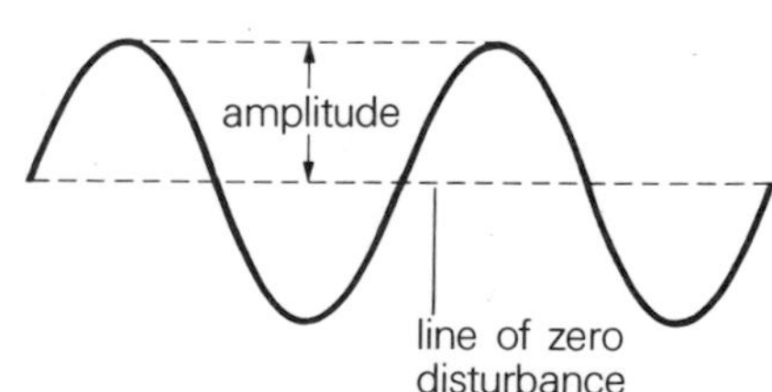

Wavelength (symbol λ)

The wave can be seen to repeat itself after a certain distance.

The **wavelength** of a wave is the minimum distance in which a wave repeats itself.

The SI unit for measuring wavelength is the metre (m). Another unit often used is the centimetre (cm). The wavelength may be measured between any two points at which the wave repeats itself, Figure 1.9.

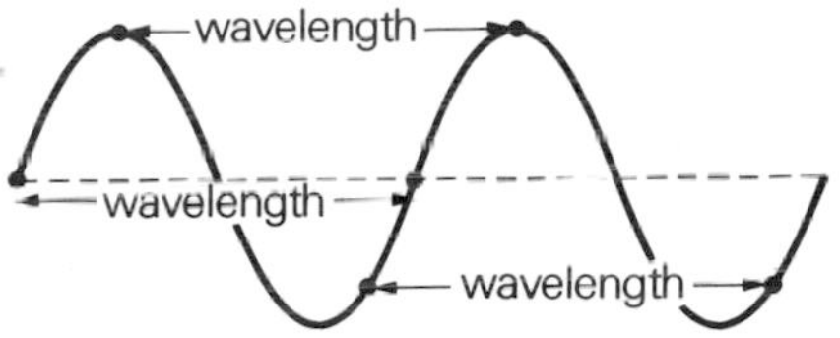

Figure 1.9 Wavelength

Frequency (symbol f)

The faster the movement of your hand, the greater is the number of waves produced in a given time.

The **frequency** of a wave is the number of complete wavelengths produced in one second.

The SI unit for measuring frequency is the hertz (Hz). The word hertz can be thought of as meaning 'per second'. For example if a wave has a frequency of 2 Hz, then two complete wavelengths are produced per second. Hertz was a German physicist who discovered radio waves in 1888. Our unit of frequency is named after him, Figure 1.10.

Figure 1.10 Heinrich Hertz

Wave speed (symbol v)

The wave travelled along the tubing at a constant speed. The **wave speed** is the distance travelled by the wave in one second.

The SI unit for measuring wave speed is the metre per second ($m\ s^{-1}$). Another unit often used is the centimetre per second ($cm\ s^{-1}$).

1.4 A wave equation

The diagrams in Figure 1.11 show how three waves of different frequencies but the same wavelength appear 1 s (one second) after they are started.

When the frequency is 1 Hz, one wavelength (λ) leaves the starting point in 1 s. Hence the distance travelled by this wave in 1 s is $1 \times \lambda$.

When the frequency is 2 Hz, two wavelengths leave the starting point in 1 s. Hence the distance travelled by this wave in 1 s is $2 \times \lambda$.

When the frequency is 3 Hz, three wavelengths leave the starting point in 1 s. Hence the distance travelled by this wave in 1 s is $3 \times \lambda$.

If the frequency is f Hz, f wavelengths leave the starting point in 1 s. And so the distance travelled by the wave in 1 s is $f \times \lambda$.

But the distance travelled in 1 s is the wave speed (v). Hence,

$$v = f \times \lambda$$

wave speed = frequency × wavelength
($m\ s^{-1}$) (Hz or s^{-1}) (m)

Figure 1.11

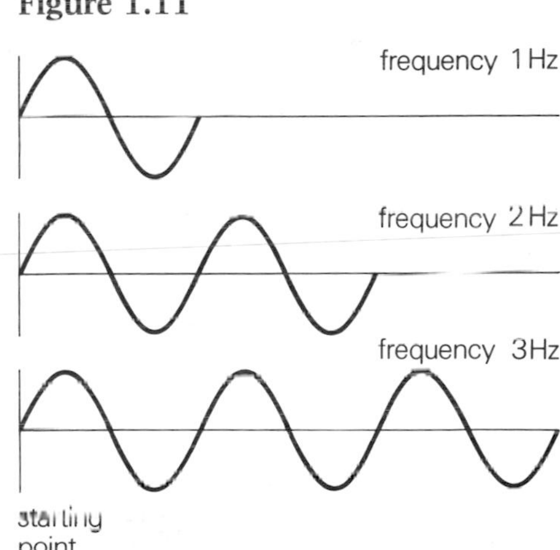

Example 1

Waves on a spring are produced at the rate of 20 wavelengths every 5 s.
a) Calculate the frequency of the wave motion.
b) If the wavelength of the waves is 0·01 m, find the speed of the waves.

a) $\text{frequency} = \dfrac{\text{number of complete wavelengths}}{\text{time taken}}$

$\Rightarrow \quad f = \dfrac{20}{5}$

$\Rightarrow \quad f = 4$ **The frequency of the wave motion is 4 Hz**

b) wave speed = frequency × wavelength

$\Rightarrow \quad v = 4 \times 0{\cdot}01$

$\Rightarrow \quad v = 0{\cdot}04$ **The speed of the waves is $0{\cdot}04\ m\ s^{-1}$.**

Example 2

A water wave travels 12 m in 4 s.
a) Calculate the speed of the wave.
b) If the vibrations producing the wave have a frequency of 2 Hz, find the wavelength of the wave.

a) $\text{speed} = \dfrac{\text{distance travelled}}{\text{time taken}}$

$\Rightarrow \quad v = \dfrac{12}{4}$

$\Rightarrow \quad v = 3$

The speed of the wave is 3 m s^{-1}

b) speed = frequency × wavelength

$\Rightarrow \quad 3 = 2 \times \lambda$ multiply both sides of the equation by $\frac{1}{2}$

$\Rightarrow \dfrac{1}{2} \times 3 = \dfrac{1}{2} \times 2 \times \lambda$

$\dfrac{3}{2} = \lambda$

The wavelength of the wave is 1·5 m.

1.5 Phase

Let us think about the way the particles of water move in a water wave. First we need two new words, Figure 1.12. A **crest** is the part of the wave above the line of zero disturbance. A **trough** is the part of the wave below the line of zero disturbance.

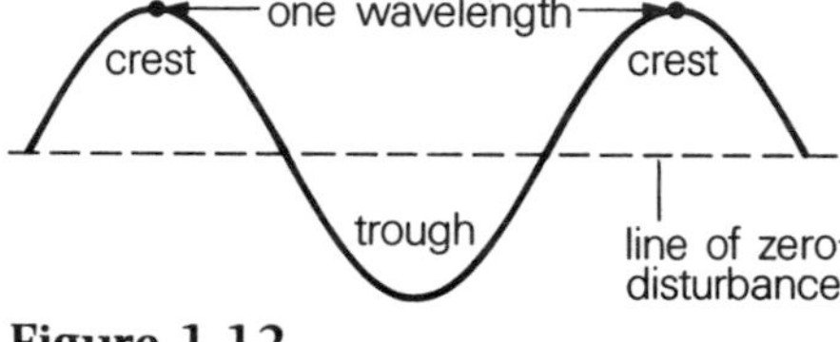

Figure 1.12

Figure 1.13 represents a wave that has travelled from the position shown by the dotted line to the position shown by the continuous line. The vertical arrows show how the particles at A, B, C, etc. have moved.

Particle A has moved down. Particle F, exactly one wavelength along, has also moved down by exactly the same amount. Particles A and F are said to be **in phase**. Particles B and G are one wavelength apart and have moved up by exactly the same amount so they are also in phase with each other.

Points in a wave separated by a whole wavelength are said to be in phase.

Particle C has moved up, while particle E (exactly one half wavelength along) has moved down, but by exactly the same amount. Particles C and E are said to be exactly out of phase. Particles A and D are half a wavelength apart and are also exactly out of phase since particle A has moved down while particle D has moved up by the same amount. Points in a wave separated by half a wavelength are said to be exactly out of phase.

Figure 1.13 Phase

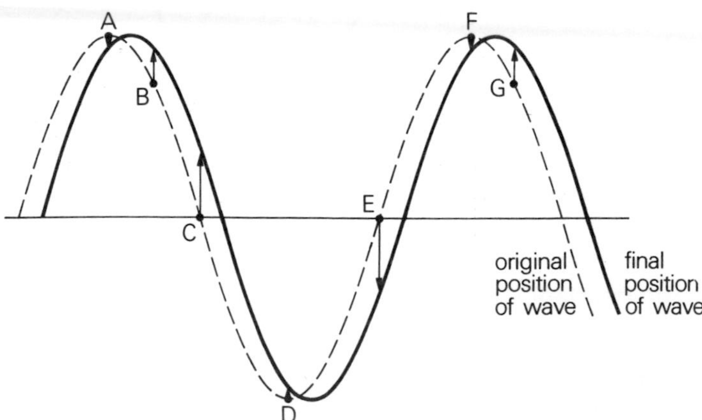

1.6 Energy changes in a wave

The transmission of waves involves energy changes. When a particle is displaced, it is forced back towards the line of zero disturbance. For example, in water waves the water tends to level out. In waves on rubber tubing, the displaced tubing tends to spring back to its original position.

The existence of a force on the particle trying to move it back to the line of zero disturbance means that it has **potential energy** (stored energy). When a particle is moving, it has **kinetic energy** (movement energy).

The greater the displacement of the particle, the greater is the force and the greater is its potential energy.

The faster the particle is moving, the greater is its kinetic energy.

As the wave travels along, the particles vibrate about the line of zero disturbance and there are continuous energy changes which are represented in Figure 1.14.

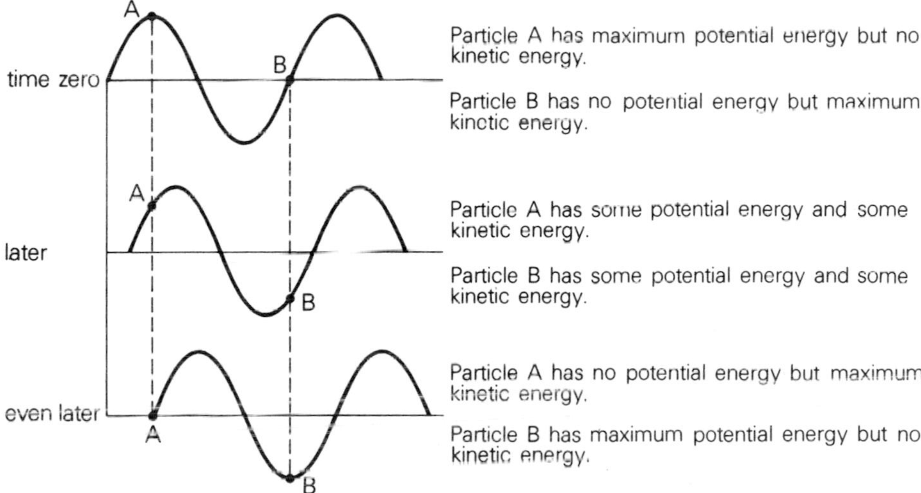

Figure 1.14

Finally let us see how displacement and speed are related to potential energy and kinetic energy, Figure 1.15.

maximum displacement | zero speed | maximum potential energy, zero kinetic energy

some potential and kinetic energies

zero displacement | maximum speed | zero potential energy, maximum kinetic energy

some potential and kinetic energies

maximum displacement | zero speed | maximum potential energy, zero kinetic energy

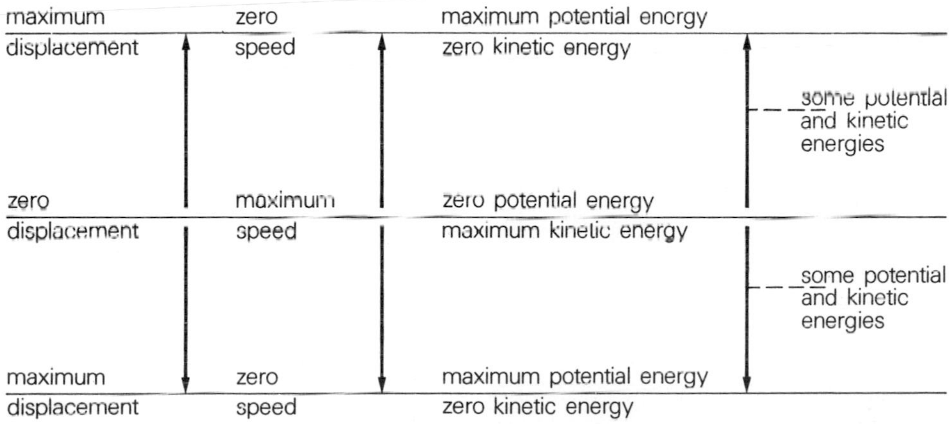

Figure 1.15

The total amount of energy in the vibration (potential energy + kinetic energy) stays constant.

1.7 Stationary waves

Consider a wave pulse travelling along a spring. If the far end of the spring is firmly held, the incident pulse is **reflected** back along the spring, Figure 1.16. Before reflection we call the pulse an **incident** pulse. After reflection we call it a **reflected** pulse.

It is seen that, if the incident pulse is a crest, then the reflected pulse is a trough. This is called a **change of phase**.

If a continuous wave is sent along the spring instead of a pulse, the incident wave meets the reflected wave coming back in the opposite direction. The result of the combination of these two waves is called a **standing** wave (or a **stationary** wave). It is called a stationary wave because, although there are still vertical vibrations, there is no obvious motion of a wave along the spring. The displacement of any part of the spring is the result of the combination of the incident and reflected waves.

A typical stationary wave could appear as shown in Figure 1.17.

There are parts of the wave where there is no displacement. These are called **nodes** and are labelled N in Figure 1.17.

At points midway between nodes the vibrations are of maximum displacement. These are called **antinodes** and are labelled A in Figure 1.17.

The distance between nodes is **half** a wavelength, Figure 1.18.

At an antinode there is a continuous conversion of potential energy to kinetic energy to potential energy etc., because that point of the spring vibrates from one side to the other, Figure 1.19.

Figure 1.20 shows the production of stationary waves on an elastic string by means of a signal generator and vibrator.
The frequency of the vibrations can be varied by changing the frequency at which the signal generator operates. Stationary waves are found to be set up only at certain frequencies, Figure 1.21.

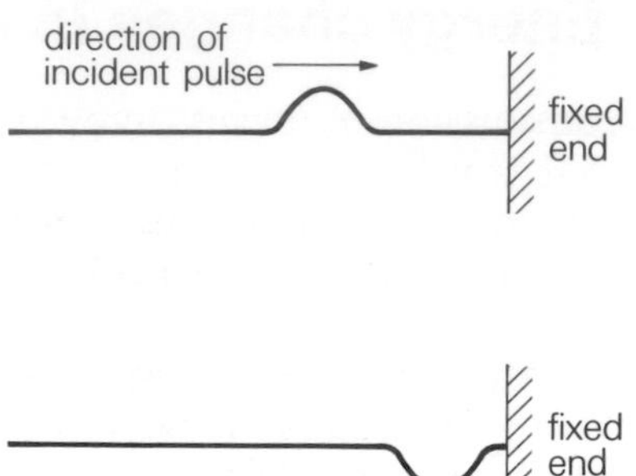

Figure 1.16 Reflection of a pulse

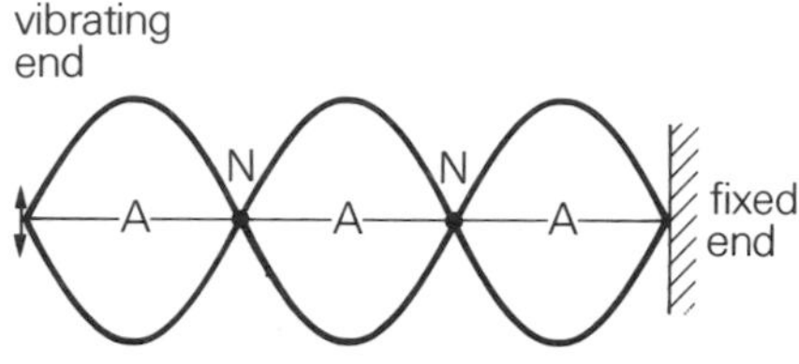

Figure 1.17 A stationary wave

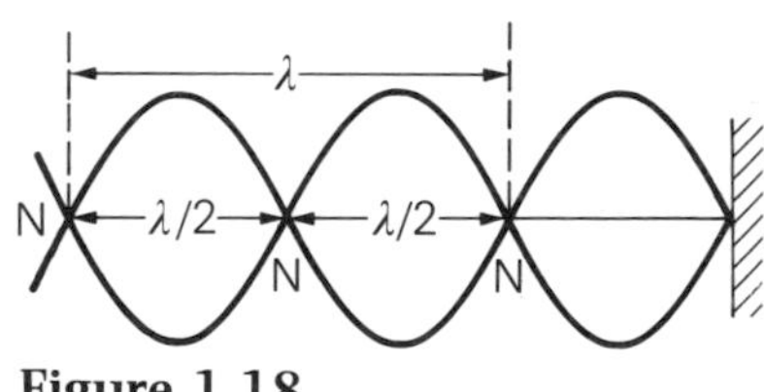

Figure 1.18

Figure 1.20

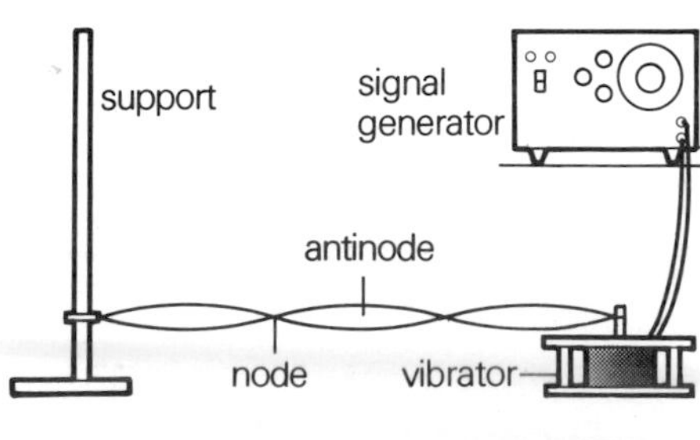

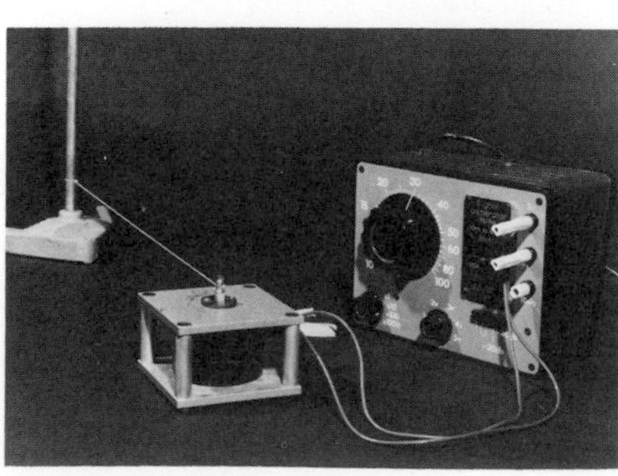

Figure 1.21 Frequency f

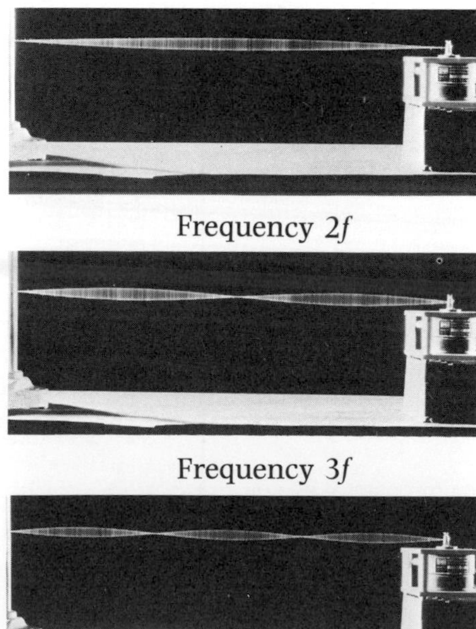

Figure 1.19

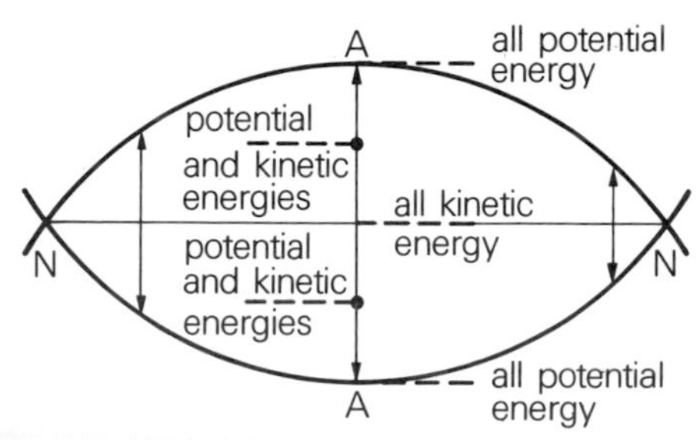

Summary

A travelling wave transmits energy.

The vibrations in a transverse wave are at right angles to the direction of travel.

The amplitude of a wave is the size of the maximum disturbance, measured from the line of zero disturbance.

The wavelength of a wave is the minimum distance in which a wave repeats itself. The unit of wavelength is the metre (m).

The frequency of a wave is the number of complete wavelengths produced in one second. The unit of frequency is the hertz (Hz).

The wave speed is the distance travelled by the wave in one second. The unit of wave speed is the metre per second (m s^{-1}).

$$v = f \times \lambda$$

wave speed = frequency × wavelength

Points in a wave separated by a whole wavelength are in phase. Points in a wave separated by half a wavelength are exactly out of phase.

Waves involve energy changes of the form

potential energy → potential energy and kinetic energy → kinetic energy → potential energy and kinetic energy → potential energy etc.

but the total amount of energy stays constant.

A stationary wave is produced when an incident wave combines with its reflected wave.

Problems

1 The diagram represents a wave 0·2 s after it has started.

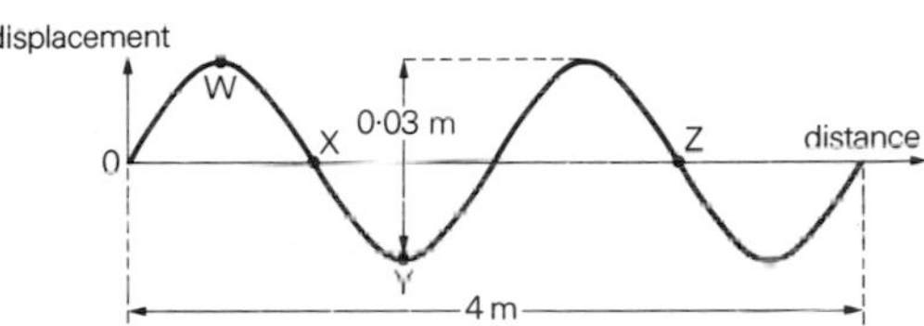

1·1 Calculate
a) the amplitude of the wave.
b) the wavelength of the wave.
c) the wave speed.
d) the frequency of the wave.

1·2 State
a) two points which are in phase.
b) two points which are exactly out of phase.
c) which point is on a wave crest.
d) which point is in a wave trough.

2 A water wave of frequency 4 Hz has a wavelength of 0·02 m. What is its speed?

3 If a wave travels with a speed of 0·1 m s^{-1} and has a wavelength of 0·05 m, what is its frequency?

4 A wave of frequency 5 Hz has a speed of 10 cm s^{-1}. What is its wavelength?

5 In this diagram of part of a transverse wave, the arrow shows the direction in which the energy is being transferred.

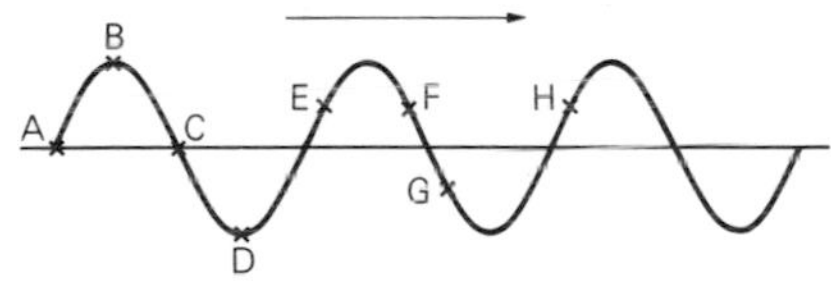

5·1 Use the letters A to H to indicate:
a) a wavecrest,
b) a wave trough,
c) two points which are in phase.

5·2 Measure:
a) the wavelength in millimetres,
b) the amplitude in millimetres.

5·3 Draw a sketch of the wave and mark the directions of motion of E. *SCEEB*

6 The diagram shows a side-view of water waves in a glass-walled tank.

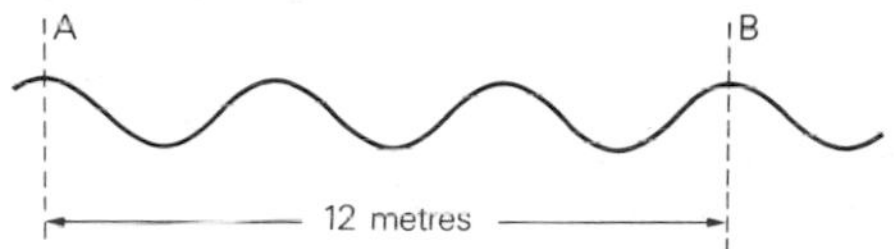

6·1 What is their wavelength?

6·2 If the crest at A takes 2 seconds to reach the point B, what is their frequency? *SCEEB*

2 Water waves

2.1 The ripple tank

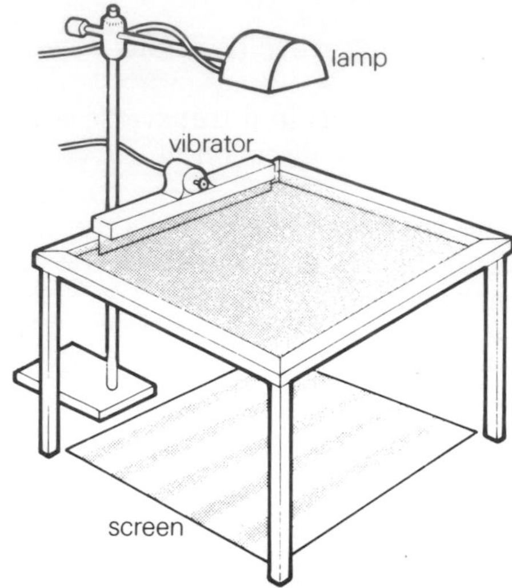

Figure 2.1 The ripple tank

We shall see that energy is transmitted by waves in many aspects of physics. In the first chapter we discussed water waves in a pond and in the sea. In this chapter we look more closely at the properties of water waves because they are easily observed and their wavelength and frequency are easily measured. We can then use this knowledge to study other forms of wave motion and to explain their behaviour.

The properties of water waves are most easily studied in a ripple tank, Figure 2.1.

An image of the water waves is produced on the screen underneath the ripple tank. This image is formed by the light shining through the water on to the screen, and consists of a series of bright lines (corresponding to crests), with darker spaces in between (corresponding to troughs).

We can understand how this comes about if we consider how a magnifying glass is used to concentrate the Sun's rays, Figure 2.2. The curved lens of the magnifying glass bends the rays and brings them to a point. This process is called **focusing.**

Similarly in the ripple tank when the light from the lamp passes through the water, the curve of the water surface through which a wave is moving acts like a series of lenses and focuses the light to give a series of bright lines, Figure 2.3.

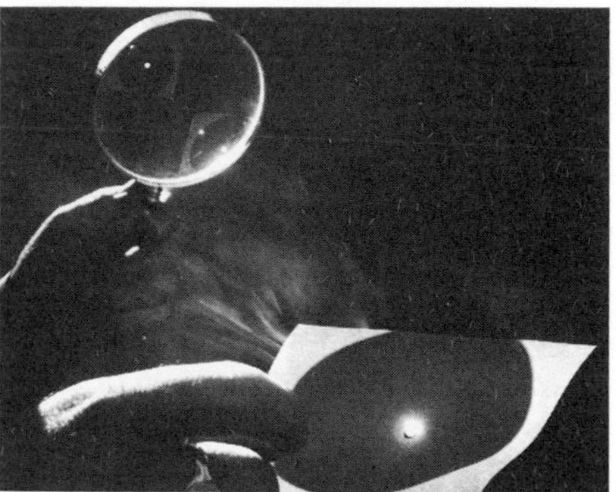
Figure 2.2

Figure 2.3

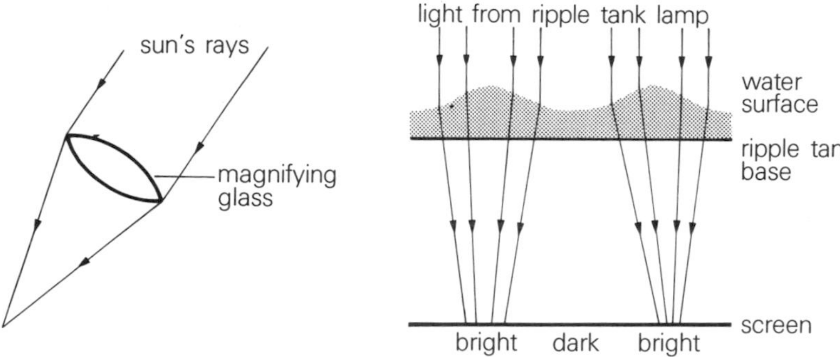

2.2 Generation of pulses

Figure 2.4

If a disturbance is created at a point on the surface of the water, a circular pulse travels out from that point. One way of creating such a disturbance is to allow a drop of water to fall on to the surface of the water, Figure 2.4. As the pulse travels out, it retains its circular shape; the boundary of this shape is called the **wavefront.**

The direction of travel of a pulse is always at right angles to the wavefront. Figure 2.5 shows a circular wavefront travelling out from O, and the arrows show the direction in which the different parts of the wavefront are travelling.

Figure 2.5

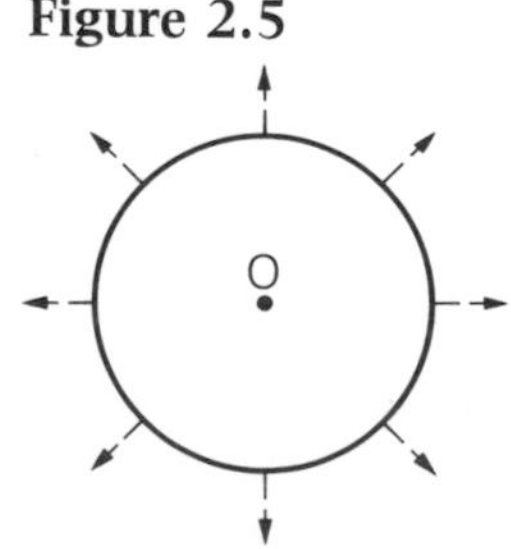

As a circular pulse spreads out, the circumference of the circle increases, and the amplitude of the wave decreases.

If a disturbance is created along a straight line on the surface of the water, a straight pulse is generated. This can be done by dipping a straight object into the water. The straight pulse that is generated continues to move across the surface of the water as a straight line. Again, the direction of travel is at right angles to the wave front, Figure 2.6.

Figure 2.6

2.3 Reflection of pulses

When a barrier is placed in the path of a pulse, the pulse is reflected. Before it strikes the barrier the pulse is called the **incident pulse**; after it has been reflected the pulse is called the **reflected pulse**.

Figure 2.7 shows a circular pulse being reflected at a straight barrier. The incident pulse is circular, with a centre at O. The reflected pulse is also circular, with a centre at I. The points I and O are an equal distance from the barrier.

When a straight pulse is reflected by a straight barrier, the reflected pulse is also straight. Such a reflection is illustrated in Figure 2.8.

The **normal** is the line at right angles to the reflecting surface.

The angle between the path of the incident pulse and the normal is called the **angle of incidence** ($\angle i$).

The angle between the path of the reflected pulse and the normal is called the **angle of reflection** ($\angle r$).

Whenever a pulse is reflected, it is observed that the angle of incidence is always equal to the angle of reflection. This is called the **Law of Reflection** $\angle i = \angle r$.

Figure 2.9 shows two examples of straight pulses being reflected by curved barriers.

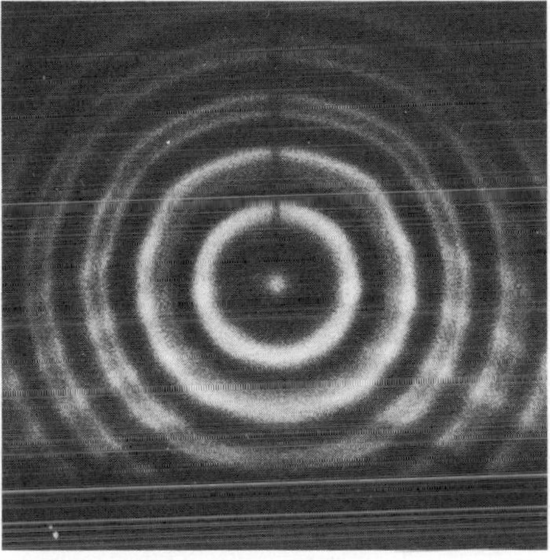
Figure 2.7

Figure 2.9

a) Reflection at a convex barrier

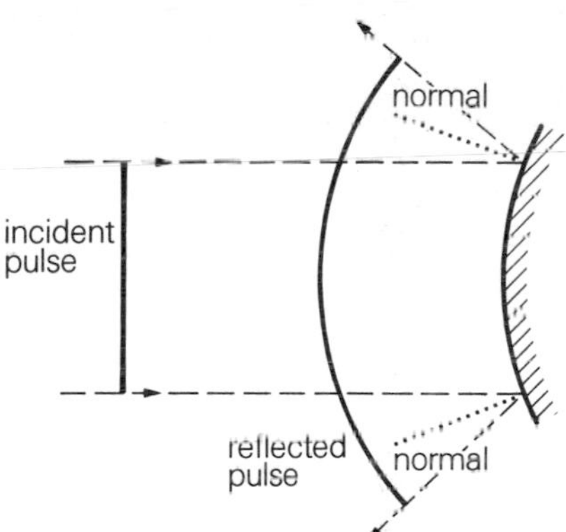

b) Reflection at a concave barrier

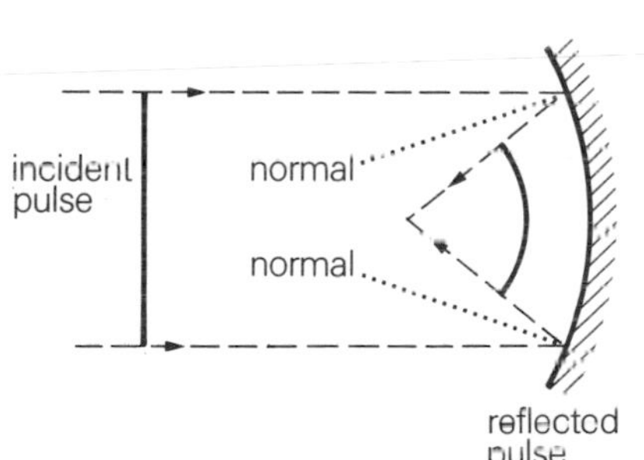

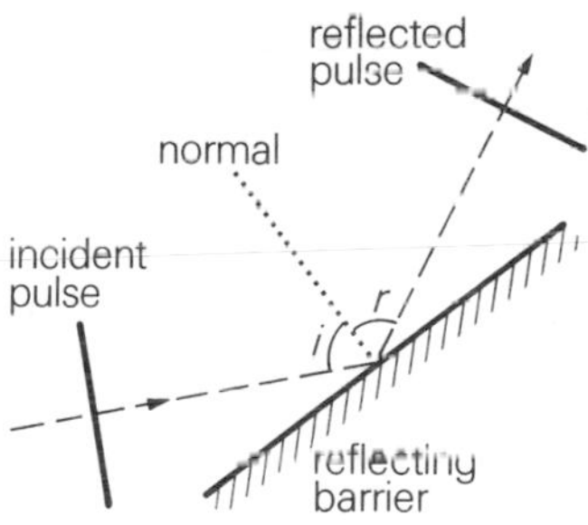

Figure 2.8

2.4 Generation of continuous waves

Figure 2.10 shows an apparatus that can be used to generate continuous circular waves. As the electric motor rotates, it causes the bar to vibrate; these vibrations cause the dipper to move up and down in the water. The vibrations at a single point on the surface of the water generate a series of pulses which make up a continuous circular wave.

Figure 2.10

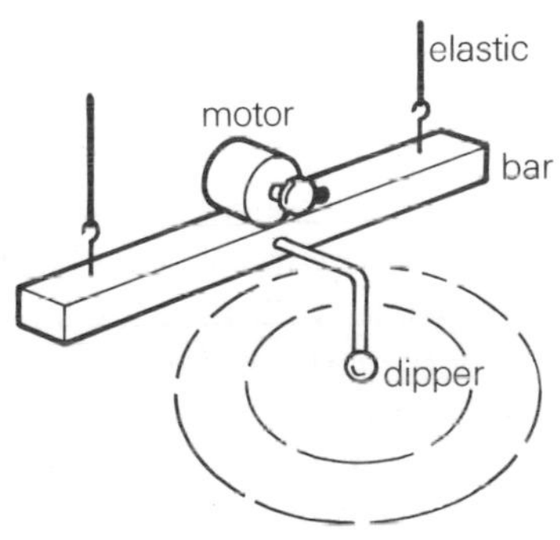

If the whole length of the bar is dipping in the water (Figure 2.11), the energy from the rotation of the motor is transmitted to the water along the length of the bar. This generates a series of straight pulses that make up a continuous straight wave.

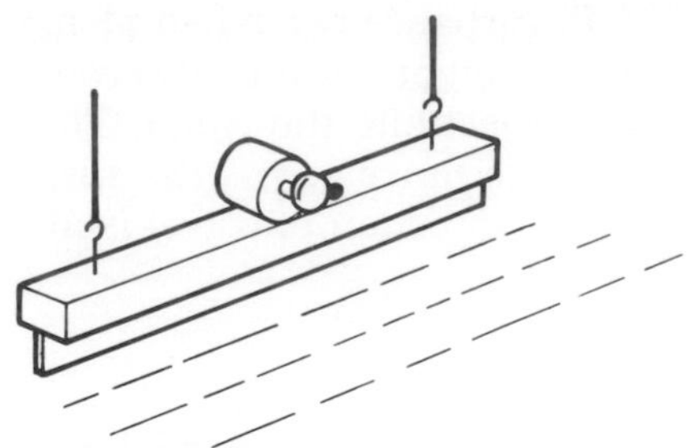

Figure 2.11

2.5 'Stopping' the wave pattern

It is difficult to observe and measure waves that are continuously on the move. One way of 'stopping' or 'freezing' the wave, when we wish to make observations, is to take a photograph and study it. In the laboratory we use a hand stroboscope to 'stop' the wave pattern. The hand stroboscope (Figure 2.12) consists of a disc with slits in it at regularly spaced intervals. If the disc is rotated in front of your eye, the scene appears whenever a slit is passing the eye but it is blocked for the rest of the time. If the stroboscope is rotated at a steady rate, you see the scene as a series of pictures separated by fixed time intervals.

Figure 2.13 shows how the stroboscope can 'stop' the wave pattern. Of course the wave does not actually stop, but to the observer it does appear to be stationary.

Figure 2.13

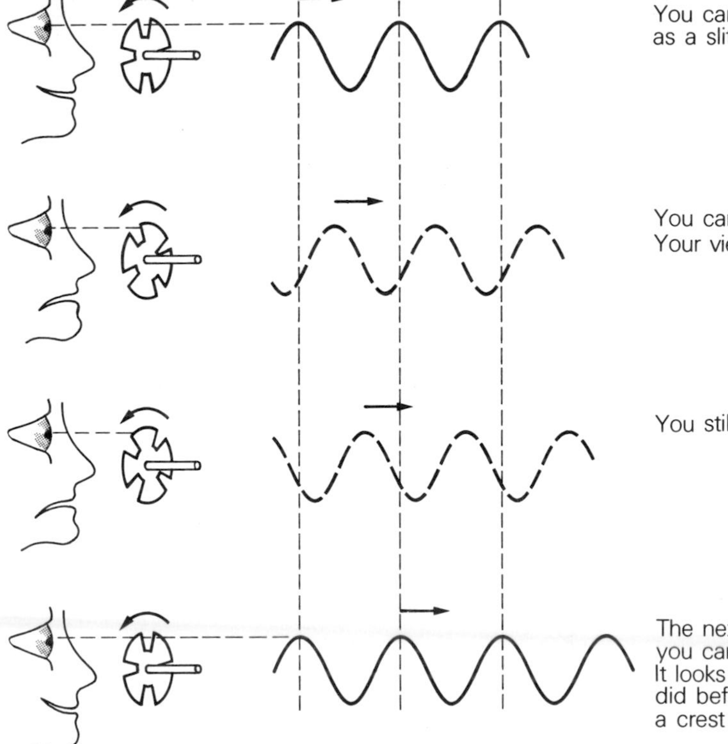

Figure 2.12 Hand stroboscope

If the frequency of the series of pictures seen by the observer is such that, whenever he sees the wave, there is a crest in the same position, the observer will see the wave standing still.

Using the ripple tank and the wave generator, we can now generate continuous waves and investigate their behaviour. In some cases it will be easier to make observations if we 'stop' the wave pattern by viewing it through a hand stroboscope. The word 'stroboscope' is often abbreviated to 'strobe'.

2.6 Properties of continuous waves

Reflection

A continuous wave consists of a series of pulses. Whenever a pulse is reflected, it obeys the Law of Reflection (Figure 2.8 on page 9):

the angle of incidence = the angle of reflection

$$\angle i = \angle r$$

It can be observed that the wavelength does not change when the waves are reflected, Figure 2.14. Any stroboscope frequency which will 'stop' the incident waves will also stop the reflected waves. This indicates that both the incident waves and the reflected waves have the same frequency.

The wave speed v is also unchanged by reflection. This must be true if both the frequency f and the wavelength λ are unchanged, since the wave equation tells us that $v = f \times \lambda$

Figure 2.15 shows a straight pulse being reflected by a curved barrier. Again, v, f and λ are unchanged on reflection.

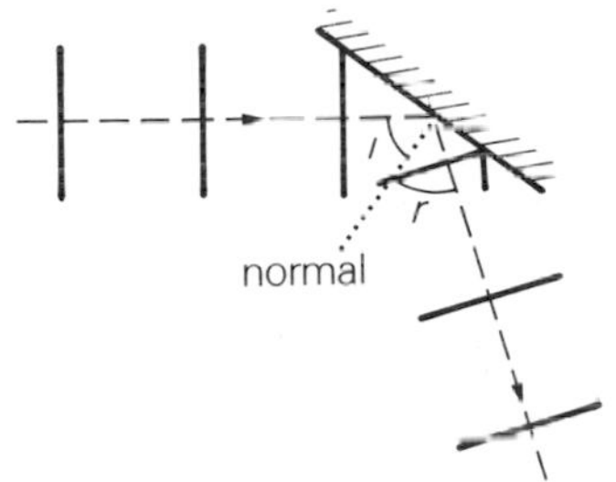

Figure 2.14

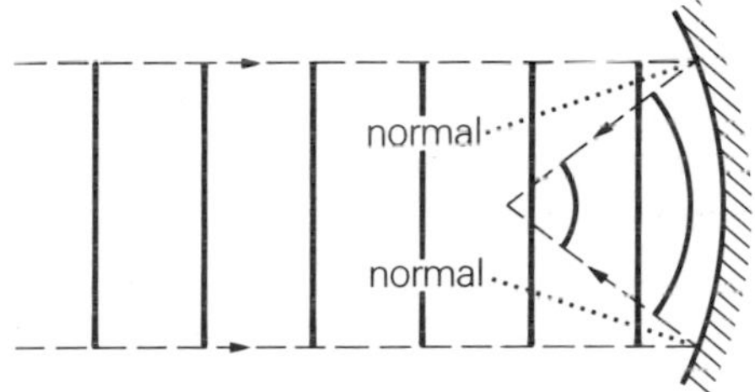

Figure 2.15

Diffraction

When water waves pass the edge of an obstacle, they bend around the edge. This bending around corners is called **diffraction**, Figure 2.16.

When a water wave passes an obstacle of less than one wavelength in width, the wavefronts are not affected by the obstacle, Figure 2.17.

When the width of the obstacle is greater than one wavelength, the wavefronts do bend around each edge, but do not immediately join up to form the original wavefronts Figure 2.18.

A gap has two edges, and we have just noted that a wave will bend around an edge. As you can see from Figure 2.19, waves spread out after passing through a gap. The narrower the gap, the greater is the spreading out of the wave. The narrowest gap results in a circular wave pattern. In fact, whenever the width of the gap is less than one wavelength, the diffracted wavefronts are circular.

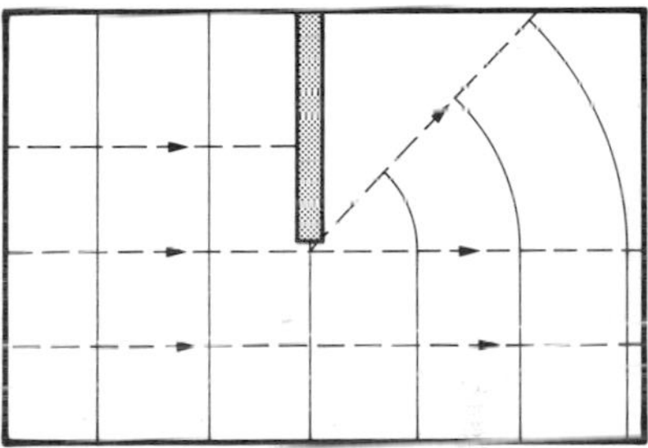

Figure 2.16

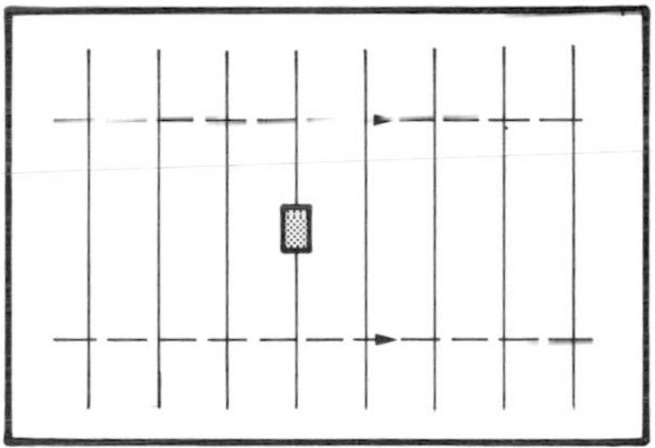

Figure 2.17

Figure 2.18

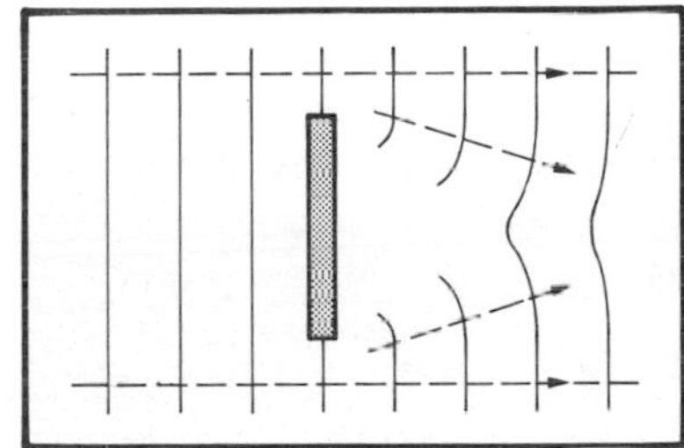

Figure 2.19

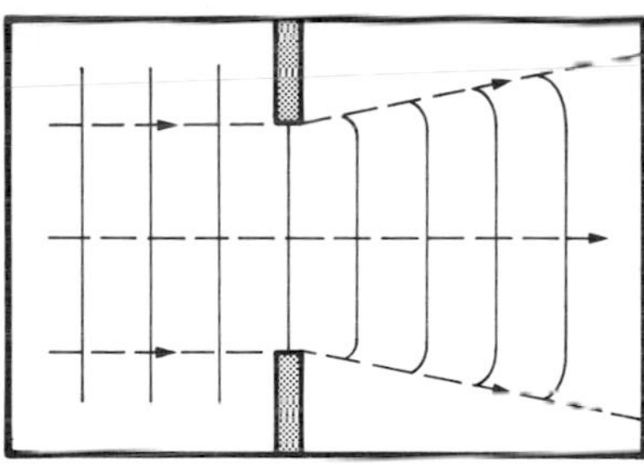

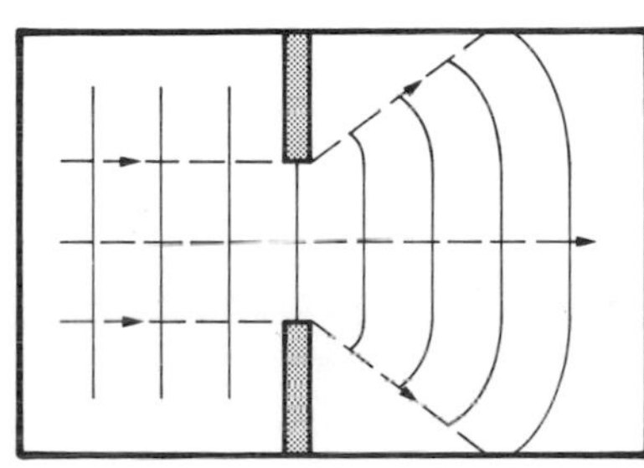

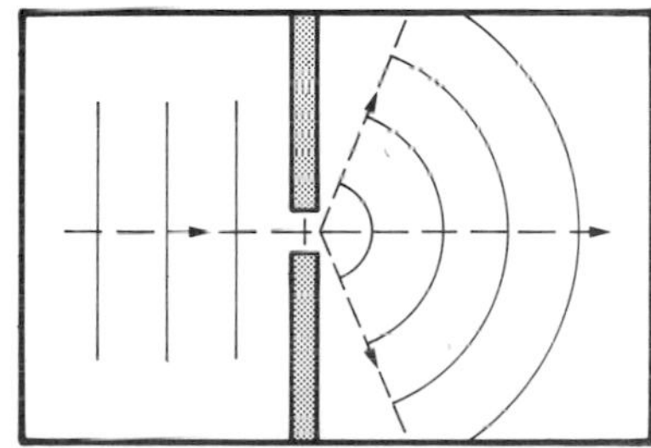

A low frequency wave (longer wavelength) is bent more than a high frequency wave (shorter wavelength) at the edge of an obstacle. This is shown in Figure 2.20 in which the low frequency wave (a) is diffracted more than the high frequency wave (b).

We may also note that diffraction causes

a) no change in wavelength (an observed fact),

b) no change in frequency (a strobe frequency that 'stops' the incident wave also 'stops' the diffracted wave),

c) no change in wave speed (because $v = f\lambda$ and neither wavelength nor frequency change).

Figure 2.20

(a) Low frequency wave

(b) High frequency wave

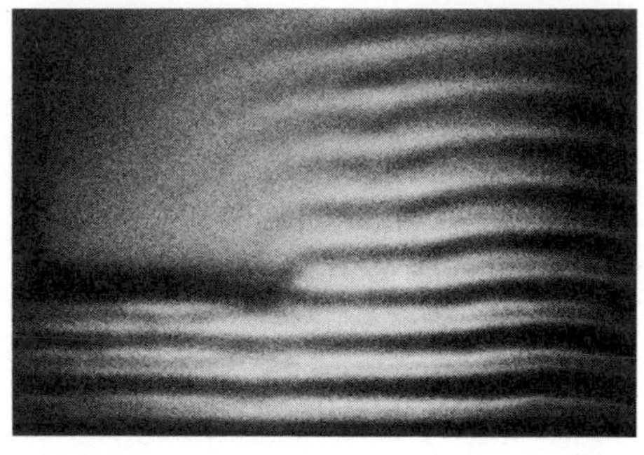

Interference

When two circular waves of the same frequency and wavelength overlap, they form an **interference** pattern, Figure 2.21.

One way of producing two such waves is to attach two dippers to the same vibrating bar above the ripple tank, Figure 2.22.

In order to understand how the interference pattern is formed, we must consider what happens when two waves meet each other. How the waves combine will depend on whether they are in phase or out of phase. We shall look at the two extreme cases where the waves are exactly in phase or exactly out of phase.

In Figure 2.23 the black line represents the total displacement resulting from the combination of the two waves represented by the grey line and the broken line:

When the two waves are in phase Figure 2.23a, the two waves reinforce each other to produce a wave of greater amplitude, but of the same wavelength. This is called **constructive interference**.

When the two waves are out of phase, they tend to cancel each other out to produce a wave of smaller amplitude. In the extreme case shown in Figure 2.23b, they are exactly out of phase and cancel each other out completely to give an area of calm; this is called **destructive interference**.

We can now consider how the two separate circular waves combine to form an interference pattern.

Figure 2.23

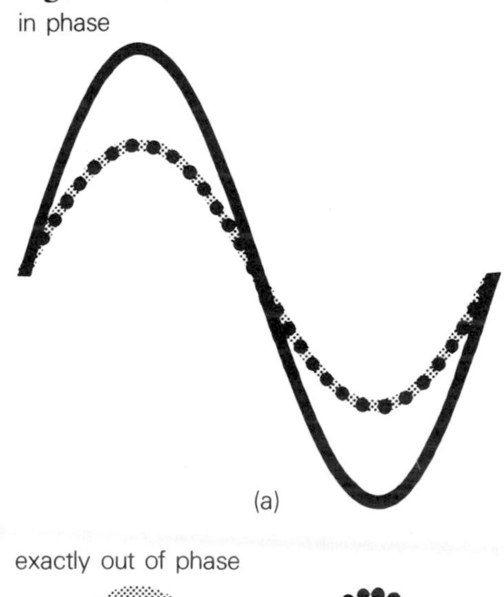

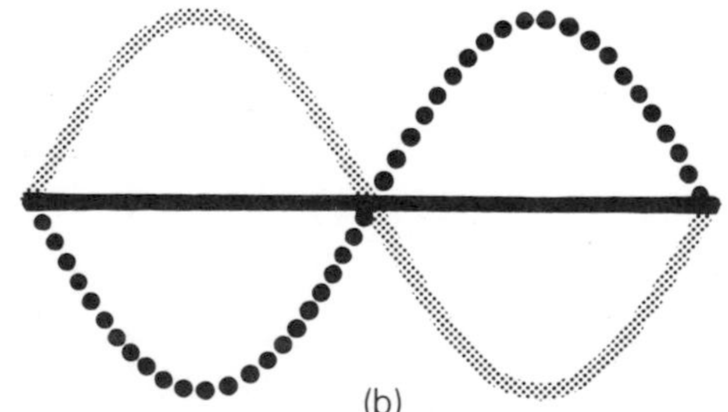

Figure 2.21

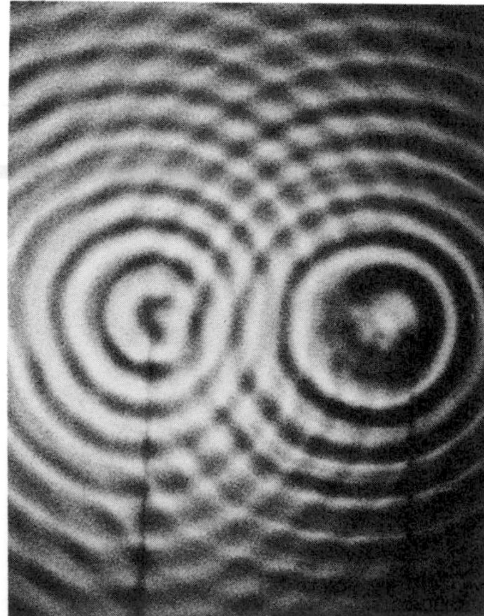

Figure 2.22

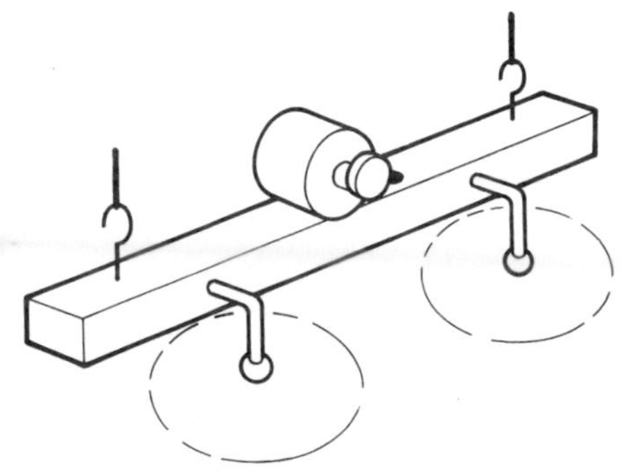

Example 1

Figure 2.24 shows the wave pattern produced in a ripple tank one second after the two sources were set vibrating. If the lines represent crests, state what is represented at this instant at:

a) point A,
b) point B,
c) point C,

Give reasons for each answer.

a) Point A is between two wave crests produced by source 1, and is, therefore, a trough.
b) At point B there is a wave crest produced by source 1 and a wave crest produced by source 2. Hence a large crest is produced at B by constructive interference.
c) At point C there is a wave crest produced by source 1 and a wave trough produced by source 2. Hence an area of calm water has been produced at C by destructive interference.

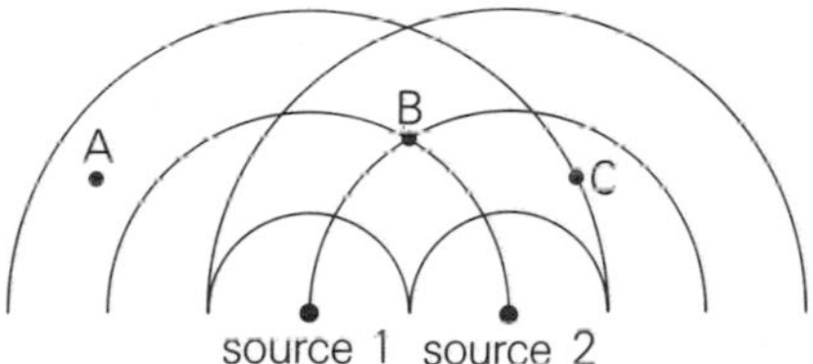

Figure 2.24

Figure 2.25 shows an interference pattern. If you look along the lines labelled *c* you will see that, wherever there is a crest from one source, there is a crest from the other source. Similarly, along these lines, troughs from both sources always coincide. Consequently, all the way along these lines the two waves are in phase and interfere constructively to give a wave of larger amplitude.

If you check along the lines labelled *d* you will see that the waves are always exactly out of phase and the two waves interfere destructively along these lines to form small areas of calm water.

It is important to note that, from two wave sources, a series of maxima (greater amplitude) and minima (zero or near zero amplitude) are formed. As we have seen, this is explained by the ways in which overlapping waves combine. The importance of this observation is that the production of an interference pattern is definite evidence of wave motion.

In order to produce an interference pattern, we used two point sources vibrating together to produce two circular wave patterns. When we investigated diffraction (page 11), we saw that a circular wave pattern was produced when a wave passed through a gap of less than one wavelength in width. Thus, if we pass a plane wave through two such slits, we obtain two circular wave patterns. The two slits act as two sources of circular waves which are generated exactly in step, since wavefronts from each slit are generated by the same plane wavefront.

Figure 2.25

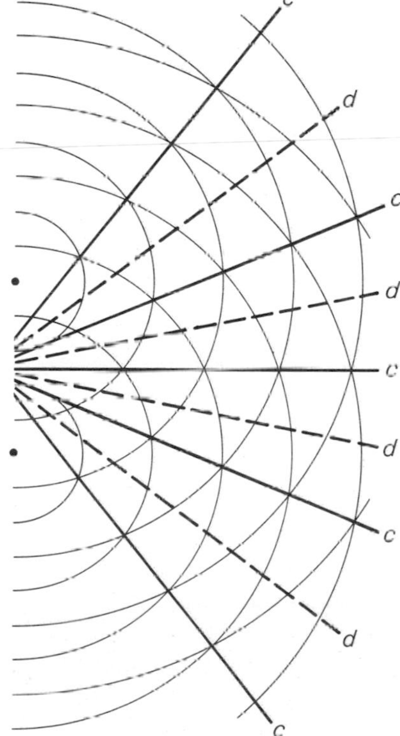

Figure 2.26 shows an interference pattern produced by passing a plane wave through two narrow slits. The fact that this occurs reinforces our previous statement that the production of an interference pattern provides definite evidence of wave nature. We might expect two slits to give rise to two maxima opposite the slits, and a minimum at the centre

Figure 2.26

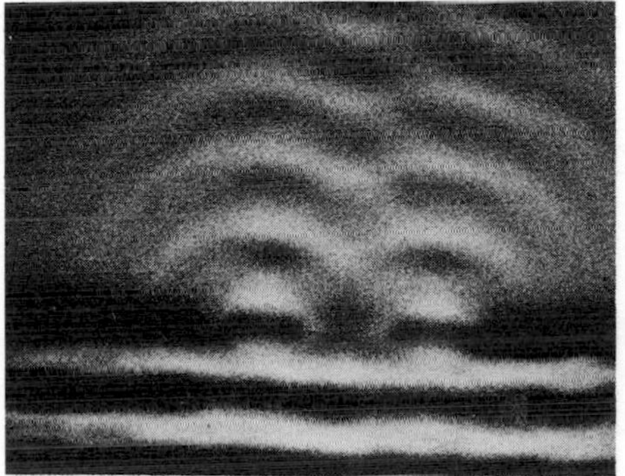

opposite the barrier between the two slits. The facts are that we obtain a series of maxima and minima and that the centre is a maximum. This can only be explained by considering the wave nature of the motion.

The effect of varying the distance between the two sources Figure 2.27 shows interference patterns formed by wavefronts from sources of different separations. All other quantities are the same in both cases. It can be seen that, when the sources are closer together, the interference pattern is more spread out.

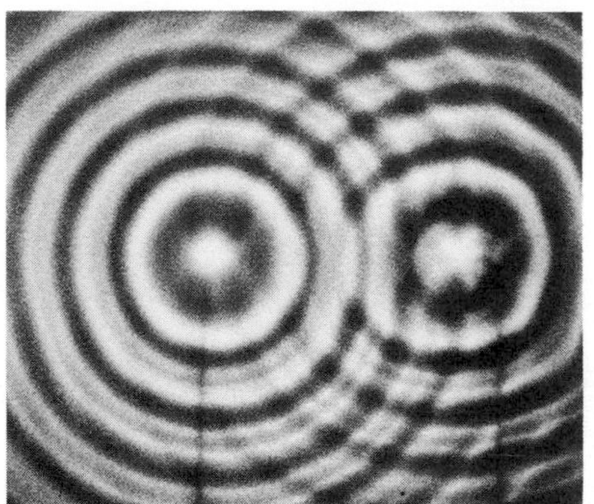
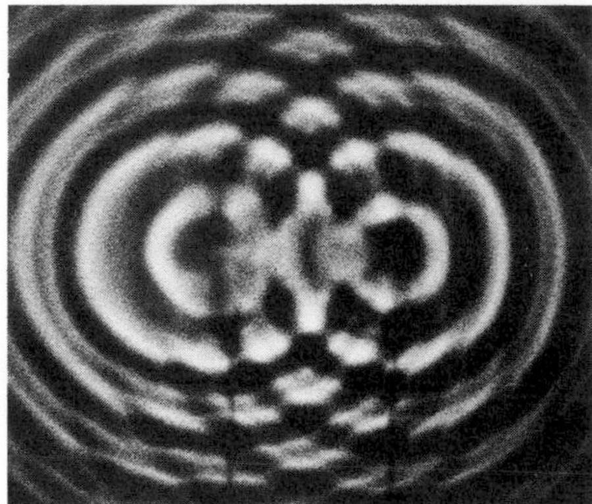

Figure 2.27

The effect of varying the wavelength Figure 2.28 shows two interference patterns produced by waves of different wavelengths. In the second pattern formed by the waves of greater wavelength, the lines of maxima and minima are more widely spaced. When the waves have a longer wavelength, the interference pattern is more spread out.

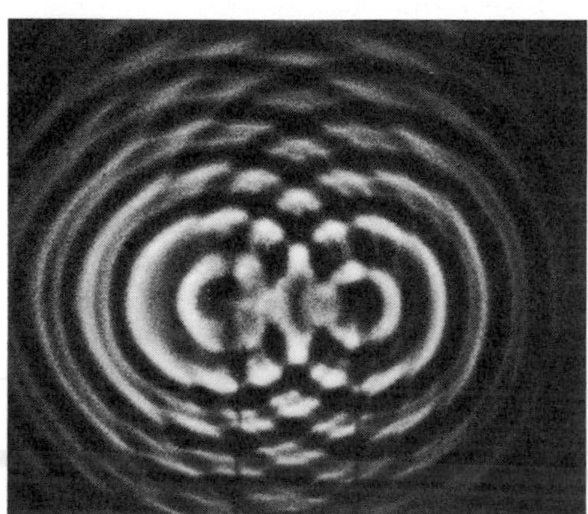
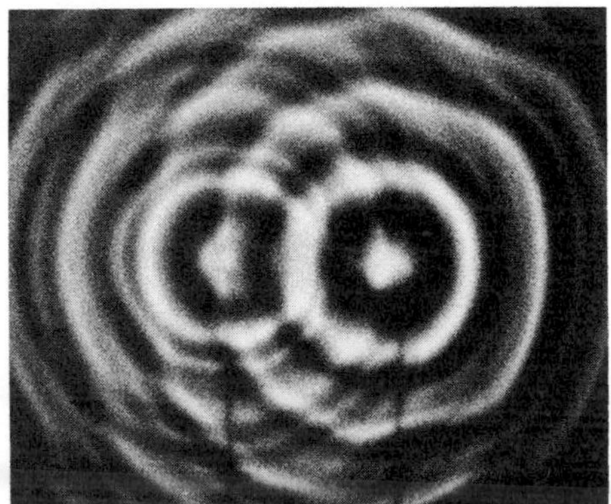

Figure 2.28

Refraction

Refraction of water waves occurs when waves move from water of one depth to another.

When we use the ripple tank to investigate refraction, we create a shallow region by placing a perspex or glass plate in the water. The water level is adjusted so that the water just covers the plate. The plate must be transparent so that the light from the ripple tank lamp can pass through this region and project an image of the waves on to the screen below.

Effect on speed If a single pulse is sent towards the shallow region it is seen to slow down on entering the shallow water. On re-entering the deeper water, the pulse speeds up. The same changes in wave speed are observed with a continuous water wave.

Water waves travel more slowly in shallower water than in deeper water.

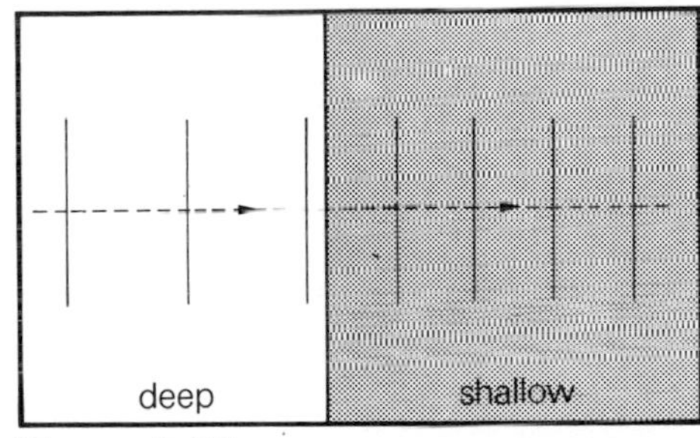

Figure 2.29

Effect on frequency When a stroboscope is used to 'stop' a continuous wave passing from deep to shallow water, any stroboscope frequency which 'stops' the motion in one region also 'stops' the motion in the other region. This indicates that the frequency is the same in both regions. Refraction produces no change in the frequency of a wave.

You might have predicted this result if you had considered that the arrival of a pulse from the deeper water is the start of a pulse in the shallower water. Thus in one second the number of pulses starting in the shallower region is equal to the number of pulses arriving from the deeper water. The frequency is the same in both depths of water.

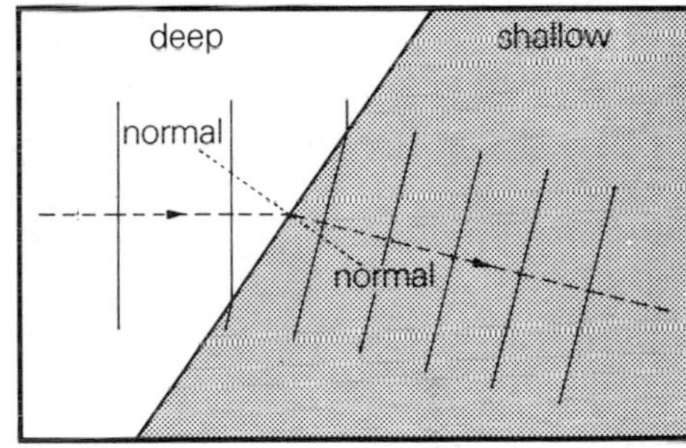

Figure 2.30

Effect on wavelength Figure 2.29 shows a continuous wave passing from deeper to shallower water. The wavelength is shorter in the shallower region. Water waves have a shorter wavelength in shallower water than in deeper water.

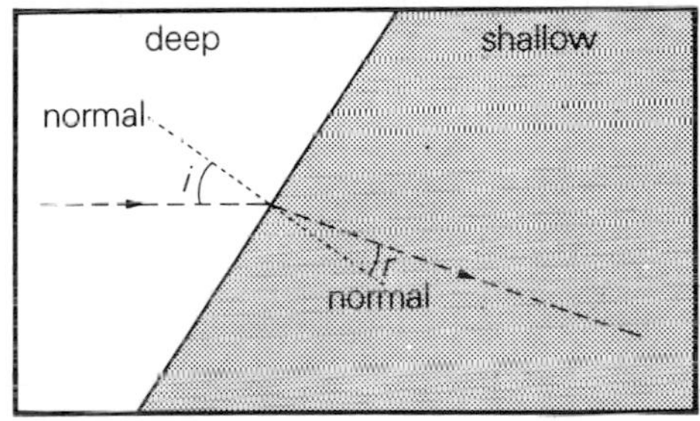

Figure 2.31

Effect on direction When the boundary between the different depths of water is at an angle to the wavefronts, a change occurs in the direction in which the wave is travelling Figure 2.30. Remember that the wave direction is always at right angles to the wavefronts.

You will see from Figure 2.31 that $\angle i > \angle r$ where:

$\angle i$ = angle of incidence = angle between the incident wave direction and the normal,
$\angle r$ = angle of refraction = angle between the refracted wave direction and the normal.

If the incident wave is travelling along the normal, it will continue to travel along the normal after entering water of a different depth. In all other cases, refraction produces a change in wave direction. On entering shallower water, the wave direction bends towards the normal. On entering deeper water, the wave direction bends away from the normal.

Figure 2.32

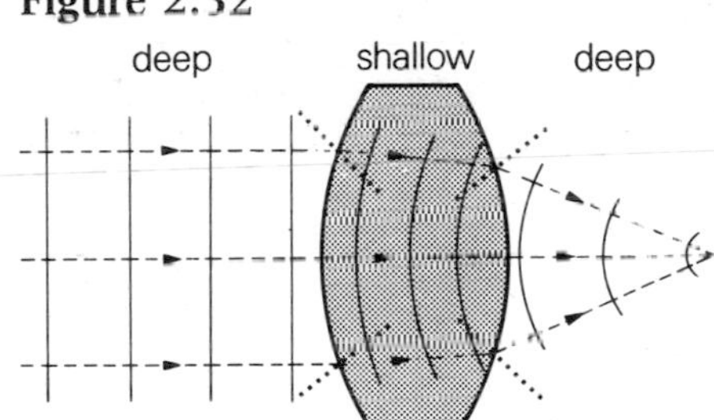

Example 2

Figure 2.32 shows wavefronts approaching a shallow region and how the wave direction and the wavefronts change as the wave travels in the shallow region and as it re-enters the deeper water.

Points to note in the diagram:

a) To find the new direction for any part of the wavefront, we draw in the normal at the point where that part of the wavefront crosses the boundary between the two depths. If the wave is entering shallower water, the wave direction bends towards the normal. If it is entering deeper water, the wave direction bends away from the normal.

b) On meeting the curved boundary, the central part of the wavefront enters the shallow water first, and is the first to be slowed down. On leaving the shallow region, the central part of the wavefront is the last to leave the shallow region and is the last to speed up. Thus the central parts of the wavefronts have a lower speed for a longer time, and lag behind the outside parts of the wavefronts after the wave has passed through this shallow region.

2.7 Estimation of frequency, wavelength and wave speed

Estimating frequency

Figure 2.33 shows two photographs of successive views through the slits of a stroboscope being used to 'stop' the wave pattern. Can you tell how far the wave has moved in between pictures?

Figure 2.33

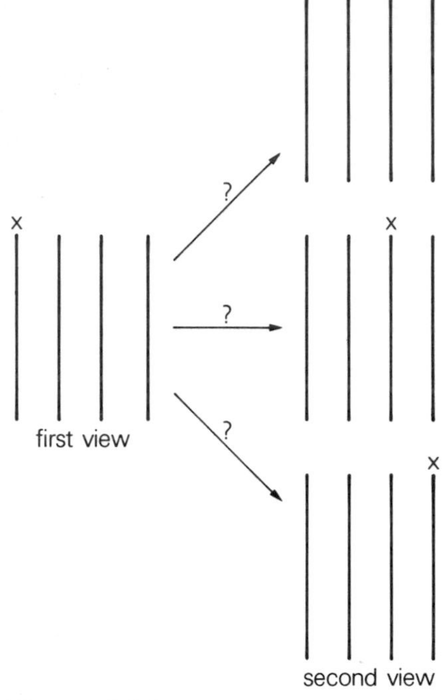

Figure 2.34

If the wave pattern is 'stopped', we know that every view through a slit of the stroboscope shows wavefronts in the same positions, but we do not know whether the wave has moved on by 1, 2, 3 or more wavelengths (see Figure 2.34 where one wave front has been labelled X). The stroboscope will 'stop' the wave pattern if the wave moves on by any whole number of wavelengths between views through the slits of the stroboscope. The number of views per second obtained with a stroboscope is known as the **strobe frequency**.

If we can find the strobe frequency which will give us a view of the wave pattern every time it has moved on by one wavelength, then that strobe frequency is the same as the wave frequency.

Let us consider a situation where a stroboscope has 'stopped' the wave pattern and the wave pattern is moving on by three wavelengths between successive views (although the observer may not realize it). If the strobe frequency is increased steadily, a time will come when the wave pattern moves on by two wavelengths in between views, and the pattern is again 'stopped'. If we increase the strobe frequency further, the wave pattern will again be 'stopped' when the observer views the pattern every time it moves on by one wavelength. At this point, the wave frequency and the strobe frequency are the same but the observer has no way of knowing this. If the strobe frequency is increased further, the successive views must catch the wavefronts travelling between positions, and the pattern will not be 'stopped' without showing a reduction in wavelength.

Figure 2.35

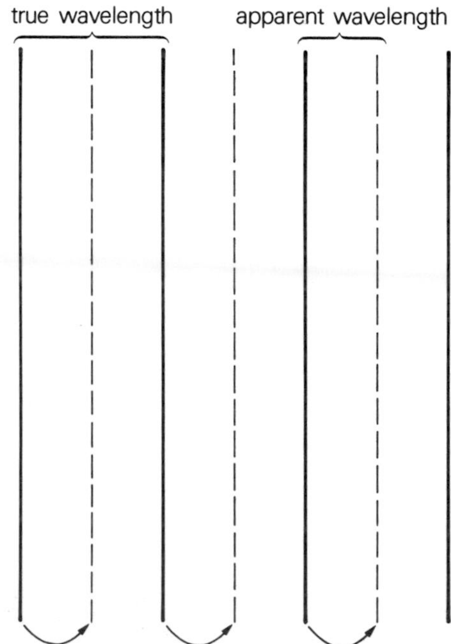

Figure 2.35 shows how the wave pattern can be 'stopped' at a higher strobe frequency, but with an apparent reduction in wavelength. The wave pattern moves forward by a half wavelength between successive views, and the observer sees wavefronts alternating between the positions shown by the continuous and the dotted lines. Thus the observer sees a wave pattern with an apparent wavelength which is half the actual wavelength.

The **wave frequency** is equal to the highest strobe frequency which will 'stop' the wave pattern without an apparent reduction in wavelength.

The frequency of the stroboscope can be calculated if we count the number of slits in the stroboscope disc, and measure the time taken for the disc to rotate a fixed number of times.

Example 3

A stroboscope with 5 slits rotates 10 times in 5 seconds. What is the strobe frequency?

In 5 seconds the disc performs 10 revolutions.
In 1 second the disc performs 2 revolutions.
For each revolution 5 slits pass the observer's eye.
For 2 revolutions 10 slits pass the observer's eye.
In 1 second the observer obtains 10 'views'.
Strobe frequency = 10 Hz

Example 4

An observer views a wave pattern through a hand stroboscope with 6 slits in it. Figure 2.36 shows the patterns that he observes when the stroboscope 'stops' the wave and, next to each diagram the time for 10 revolutions of the strobe disc is given. What is the frequency of the wave?

The higher the strobe frequency, the less is the time for 10 revolutions of the disc.

In C the observer sees an apparent reduction in wavelength.

B is the case in which the strobe frequency is the highest that will 'stop' the wave without an apparent reduction in wavelength.

In B the strobe disc rotates 10 times in 8 s.

$\Rightarrow$ in 1 s the disc rotates 1·25 times

Number of slits in the disc = 6.

$\Rightarrow$ strobe frequency = 6 × 1·25
= 7·5 Hz

Highest strobe frequency which will 'stop' the wave pattern without apparent reduction in wavelength = 7·5 Hz.

$\Rightarrow$ wave frequency = 7·5 Hz.

Frequency of wave is 7·5 Hz.

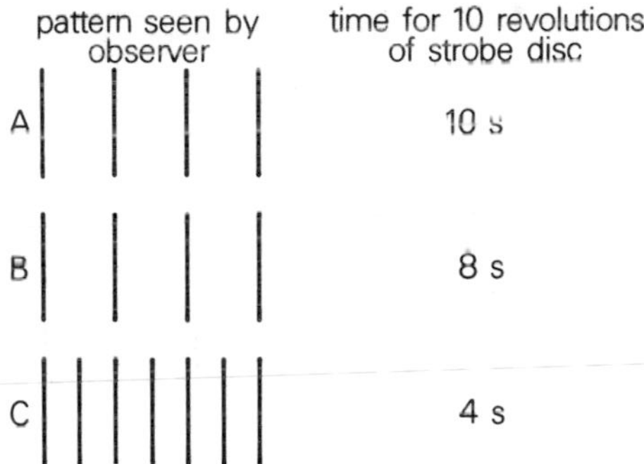

Figure 2.36

Estimating wavelength

The behaviour of the waves in a ripple tank is observed by viewing the image of the waves projected on to a screen below the ripple tank. You can see in Figure 2.37 that the image on the screen is magnified so that the apparent wavelength is greater than the wavelength of the waves in the ripple tank. This magnification can be measured by placing an object of known length in the ripple tank and measuring the length of its shadow on the screen.

$$\text{wavelength} = (\text{apparent wavelength on screen}) \times \frac{\text{object length}}{\text{shadow length}}$$

The apparent wavelength is measured by placing a metre stick on the screen, 'stopping' the pattern with a stroboscope and measuring the length of five apparent wavelengths. This length is then divided by five to give the apparent wavelength.

Figure 2.37

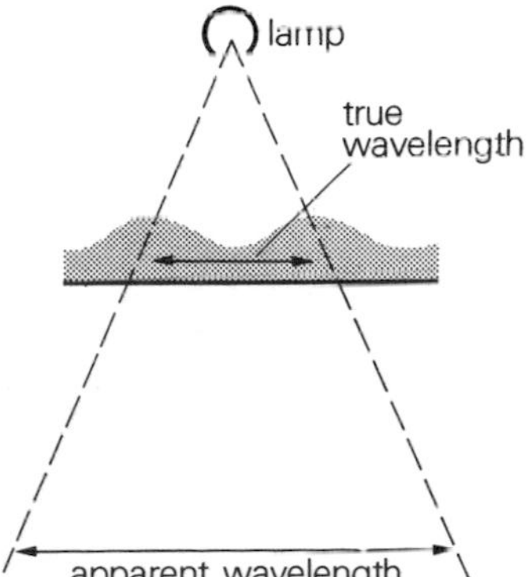

Example 5

The image of ten wavelengths of a water wave is 0·3 m in length on the screen below the ripple tank. When an object 0·1 m in length is placed in the ripple tank, it casts a shadow 0·2 m long on the screen. What is the wavelength of the water waves?

$10 \times$ apparent wavelength = 0·3 m
⇒ apparent wavelength = 0·03 m
length of object = $\frac{1}{2} \times$ length of object's shadow
⇒ wavelength = $\frac{1}{2} \times$ apparent wavelength
⇒ wavelength = $\frac{1}{2} \times 0{\cdot}03$
= 0·015

Wavelength of water waves = 0·015 m.

Estimating wave speed

If we have estimates for the frequency f and the wavelength λ, we can substitute into the wave equation $v = f \times \lambda$ and obtain an estimate for the wave speed v.

A more direct method is to measure the time it takes for a wavefront to travel a known distance. If two marks are placed on the base of the ripple tank, 40 cm apart, a stopwatch can be used to give an approximate measure of the time taken by one of the wavefronts to travel between the two marks.

Wave speed = distance travelled per second

$$\text{Wave speed} = \frac{\text{distance travelled}}{\text{time taken}}$$

Example 6

A pulse travelled a distance of 40 cm in 2 seconds.
a) What was the speed of the pulse?
b) If the image on the screen had a magnification of 2, what would be the speed of the pulse's image on the screen?

a) Distance = 40 cm = 0·40 m

$$\text{Speed} = \frac{\text{distance travelled}}{\text{time taken}}$$

$$v = \frac{0{\cdot}40}{2} = 0{\cdot}20$$

Speed of pulse = 0·20 m s^{-1}

b) The distance between the two marks on the ripple tank = 0·40 m
Magnification = 2
The distance between the images of the two marks = $2 \times 0{\cdot}40 = 0{\cdot}80$
The image of the pulse travels 0·80 m in 2 s

$$\text{Speed of the image of the pulse} = \frac{0{\cdot}80}{2} = 0{\cdot}40$$

The speed of the pulse's image = 0·40 m s^{-1}.

Summary

When waves are reflected they obey the law of reflection: $\angle i = \angle r$

Diffraction of water waves is the bending of waves around corners.

In diffraction at a slit, the narrower the slit, the more the wave spreads out.

Diffraction through a slit of width less than one wavelength produces circular wavefronts.

An interference pattern is produced when two circular waves that are generated in step overlap each other.

Constructive interference occurs when two waves that are in phase reinforce each other to give a wave of greater amplitude.

Destructive interference occurs when two waves that are out of phase cancel each other out to give an area of calm.

Production of an interference pattern provides definite evidence of wave motion.

Refraction occurs when water waves pass from water of one depth to another

Deep to shallow
Wave speed decreases
Wavelength decreases
Frequency does not change
Bends towards the normal

Shallow to deep
Wave speed increases
Wavelength increases
Frequency does not change
Bends away from the normal

The frequency of a wave is found by measuring the highest strobe frequency that will 'stop' the wave pattern without an apparent reduction in wavelength.

Problems

1 Copy and complete each diagram to show what will happen to the water wave after it hits the reflector.

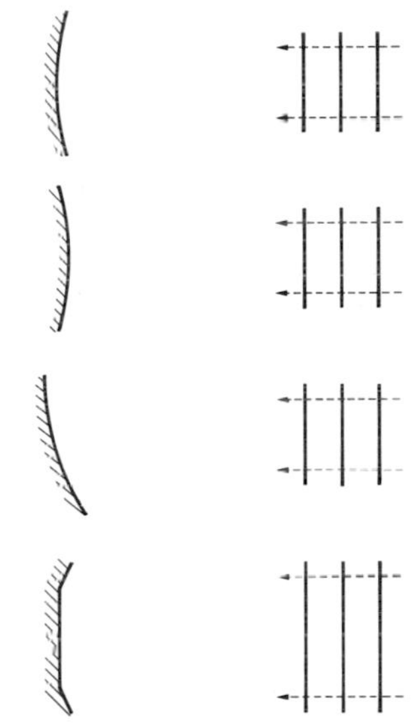

2 **2·1** Define 'angle of incidence' and 'angle of reflection'.
2·2 State the Law of Reflection.
2·3 When a water wave is reflected which (if any) of the following are changed? Frequency; wavelength; wave speed.

3 Copy and complete each diagram to show what happens to the wave after it passes through the gap.

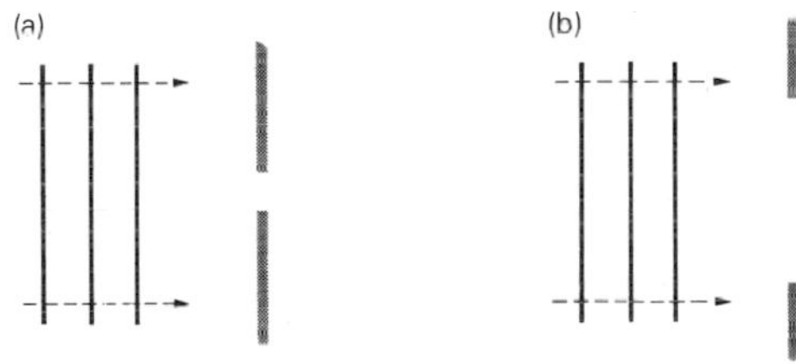

4 Water waves bend around corners and spread out after passing through gaps.
4.1 What is the name given to this property of water waves?
4.2 When this occurs, which (if any) of the following are changed? Frequency; wavelength; wave speed.

5 Copy and complete each diagram to show what happens to the wavefront and wave direction of the water wave in the shallow region and after leaving the shallow region.

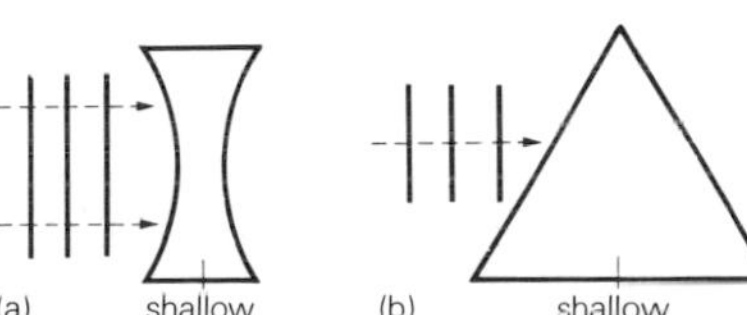

6 The diagram gives a full-scale representation of the water waves in a ripple tank one second after the vibrator was started. The dark lines represent crests.

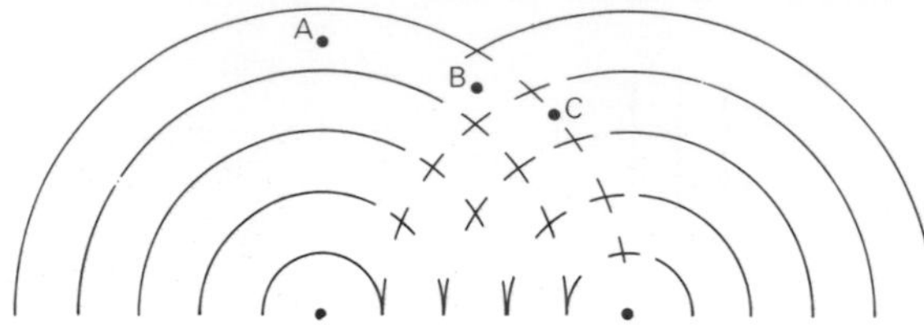

6·1 What is represented at A at this instant?
6·2 Estimate
a) the wavelength;
b) the speed of the waves;
c) the frequency of the vibrator.
6·3 Sketch a suitable attachment which could have been vibrated up and down to produce this wave pattern.
6·4 Explain how the waves combine
a) at B;
b) at C. *SCEEB*

7 Water waves are generated in a ripple tank with a frequency of 10 Hz and travel with a speed of 0·1 m s^{-1}.
7·1 What is the wavelength of the water waves?
The frequency of the waves is now increased to 20 Hz by speeding up the ripple tank motor. When this has been done:
7·2 What is the wave speed of the water waves?
7·3 What is the wavelength of the water waves?

8 A water wave has a frequency of 4 Hz. State two strobe frequencies which would 'stop' this wave pattern.

9 A stroboscope is used to 'stop' a wave pattern. The stroboscope has 6 slits in it. When the stroboscope is being rotated at:
1 revolution per second the observer sees pattern A,
2 revolutions per second the observer sees pattern B,
4 revolutions per second the observer sees pattern C.

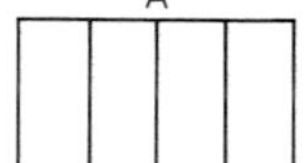

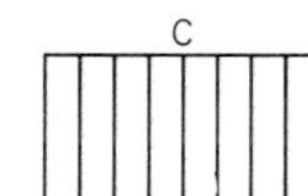

What is the frequency of the wave?

10 A pencil of length 12 cm is placed in a ripple tank. Its shadow is 18 cm long. A continuous wave is being generated in the ripple tank, with a frequency of 5 Hz. A pupil uses a stroboscope to 'stop' the wave pattern on the screen below.
If the speed of the water waves is 8 cm s^{-1}:
10·1 At what speed would the bright lines move across the screen?
10·2 What would be the distance between two bright lines on the screen?

11 **11·1** The pattern of water waves produced by a vibrator in a ripple tank is viewed on a screen by illuminating the tank using a lamp as shown in this chapter on page 8.
Describe how you would measure:
a) the frequency,
b) the wavelength of the waves.
If you knew the value of the velocity of the waves, how could you use it to check the results of your measurements in (a) and (b)?
11·2 How would you demonstrate the refraction of water waves in the ripple tank? Indicate any additional apparatus you require.
Describe what would be observed.
11·3 Name any instrument which depends on refraction and explain very briefly how the operation of the instrument depends on refraction. *SCEEB*

12 The diagram shows a side-view of water waves in a glass-walled tank.

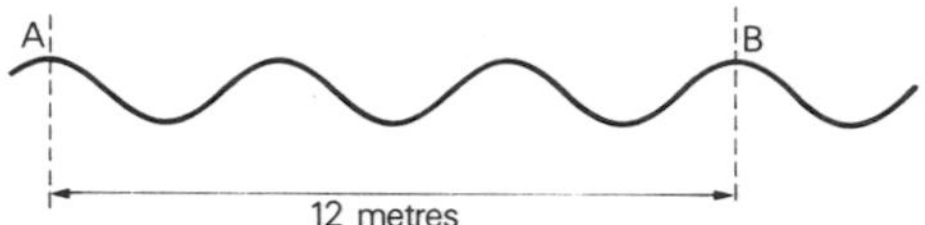

12·1 What is their wavelength?
12·2 If the crest at A takes 2 seconds to reach the point B, what is their frequency? *SCEEB*

13 **13·1** A barrier is placed in a ripple tank in which plane waves are being generated.

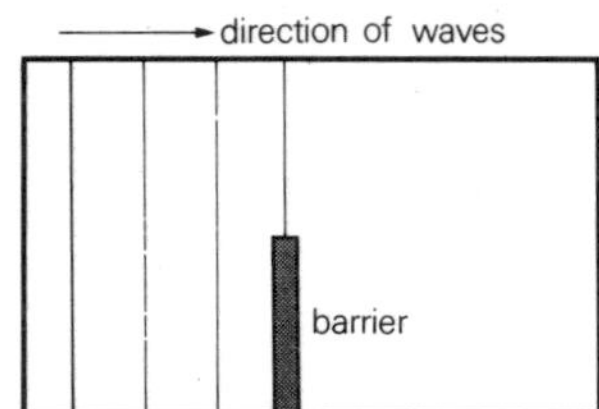

Redraw the diagram and complete it to show the pattern of waves produced in the right hand side of the tank.
13·2 The wavelength of the waves is reduced.

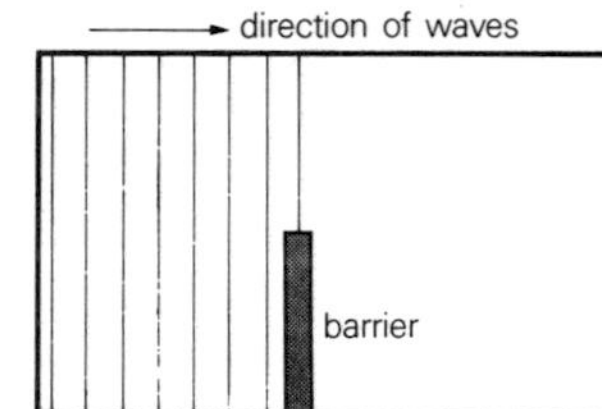

Redraw and complete this diagram. *SCEEB*

3 Light

3.1 Introduction

Figure 3.1 Solar eclipse

Light is a form of energy. The main source of light on Earth is the Sun which gives off a million, million, million, million times more energy every second than a light bulb. We cannot look directly at the Sun because this can damage our eyes. Figure 3.1 shows the Sun during an eclipse when the Moon blocks out most of its energy.

Light travels at the enormously high speed of 300 million metres per second (3×10^8 m s^{-1}). Though a rocket might take two days to reach the Moon, light takes only 1·28 seconds to travel the same distance. It would take 0·13 seconds for light to travel a distance equal to the circumference of the Earth.

Scientists continue to wonder at the nature of light. For many years it was thought that light consisted of tiny fast-moving particles, and there are still facts to support this theory. In other respects light behaves like waves in water.

In this chapter we compare the behaviour of light with that of water waves and we look at the evidence for the wave theory of light.

3.2 Interference of light

When we studied water waves, we found that one very important property of waves was interference. If light exhibits this behaviour, then we may conclude that light behaves as waves.

When we used two identical sources of water waves (which had the same frequency, wavelength and speed), we obtained a regular interference pattern (page 12). When one source produced a wave, the other source produced an identical wave at the same instant. We said that the waves were in phase or 'in step'.

These are the conditions we need to produce interference of light. We can arrange for the same frequency, wavelength and speed by using identical light sources. But if we use two separate light sources, we cannot be sure that the 'waves' are in phase as the water waves were. We can use the method devised by Thomas Young in 1801 to make sure that the light 'waves' are in phase.

Young used a barrier in front of a single light source and the two slits in the barrier effectively resulted in two identical light sources, Figure 3.2.

When each wavefront of light from the one source reaches the slits, it produces two secondary wavefronts of light which are therefore produced at the same instant. These light waves have the same frequency and wavelength since they were originally produced by the same light source. They also have the same speed and are in phase.

Now we might expect to see only **two** bright lines on a screen placed in front of the **two** slits. In fact we see many bright lines separated by dark ones. The pattern of bright and dark lines or **fringes** (Figure 3.2) on the screen can only be explained as constructive and destructive interference of light waves, similar to the constructive and destructive interference of water waves. So we find that light also behaves as waves.

Figure 3.2 Young's double slits experiment

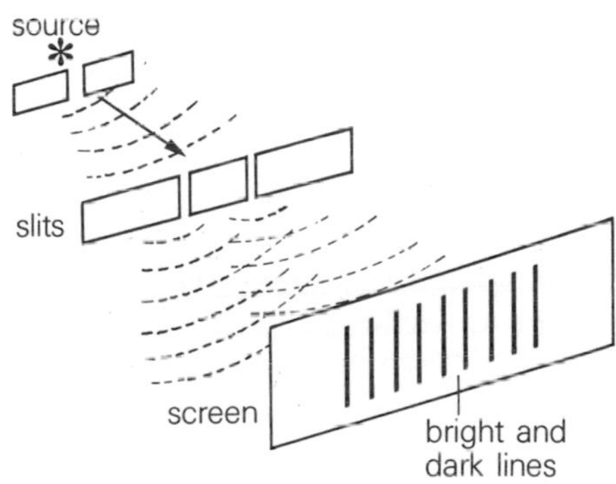

Effect of colour

We can use a different colour light source for Young's experiment. The results are shown in Figure 3.3 (and in colour on the back cover).
We see that the pattern of fringes changes with different colours, and that the spacing of the fringes also changes.

Comparing these results with those using water waves (page 14), we find that the greater the wavelength the wider apart become the lines of maxima and minima, Figure 3.4.
In the same way, red light has the greater fringe spacing and therefore the longer wavelength.

Since the wave speed, frequency and wavelength are related by the equation $v = f \times \lambda$, this means that red light has the lower frequency and blue light the higher frequency; green light is somewhere in between the two.

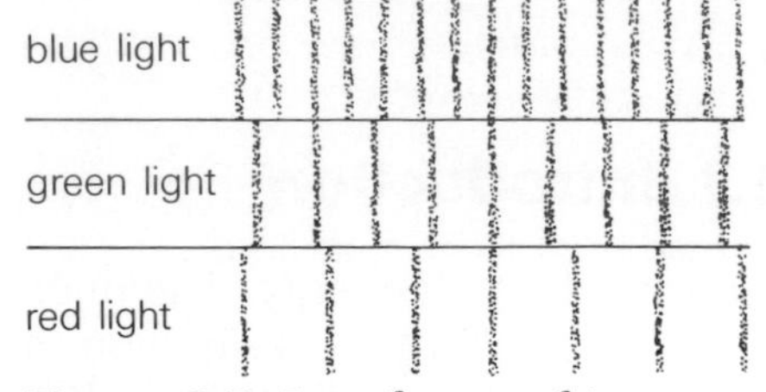

Figure 3.3 Interference fringes with different wavelengths

Figure 3.4 Effect of wavelength on spacing of maxima and minima

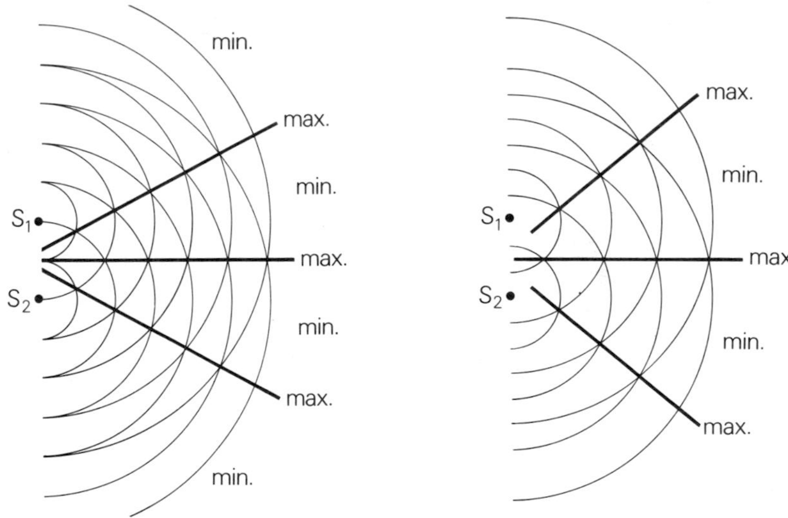

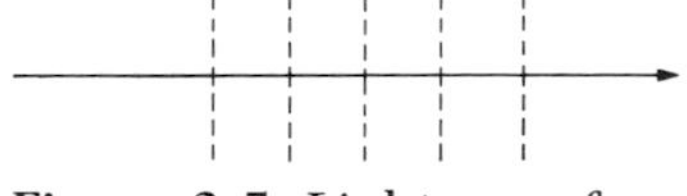

Figure 3.5 Light wavefronts

3.3 Reflection of light

Light travels in straight lines as we can see from the beam of a car headlight. We shall therefore show the direction that the light energy travels by a straight line with an arrow on it. A narrow beam of light is called a **ray**. Since light behaves as waves, we can draw wavefronts of light at right angles to the line of travel of the ray, Figure 3.5.

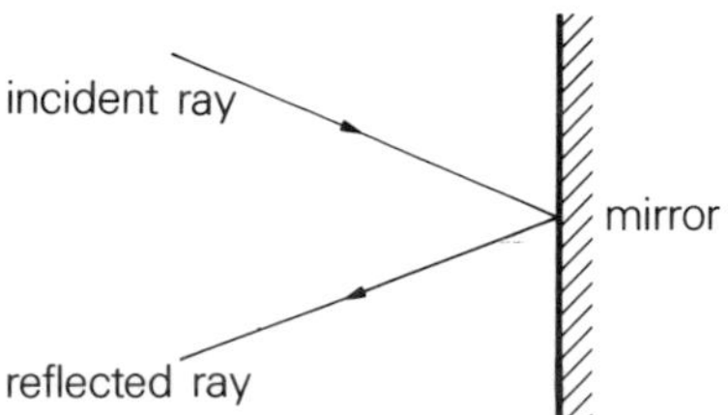

Figure 3.6 Reflection at a plane mirror

Plane reflector

A ray of light changes direction when it strikes a mirror. This is called **reflection** and is shown for a plane mirror in Figure 3.6. The narrow beam before reflection is called the incident ray; after reflection it is called the reflected ray.
As with water waves, we measure the angles that the incident ray and the reflected ray make with the **normal** (the line at right angles to the reflecting surface), Figure 3.7.
We find that the angle of incidence i always equals the angle of reflection r. This is the **Law of Reflection**.

$\angle i = \angle r$

Figure 3.7 The Law of Reflection

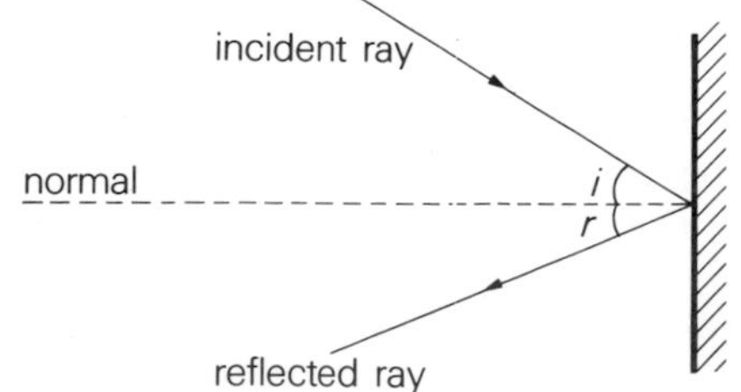

Curved reflector

If we take a simple arrangement of three plane reflectors as in Figure 3.8, we find from the Law of Reflection that three rays of light are reflected to a point.
A concave reflector is like a series of plane reflectors arranged in a curve. Using this type of reflector (Figure 3.9), we find that all rays of light are reflected to a point called the **focal point** or **focus**.

Figure 3.10 shows the solar furnace at Odeillo in France. Here many plane reflectors are arranged in a curve to reflect many solar rays to a focus. The furnace is placed at this focal point where the energy of the Sun is sufficiently concentrated to melt steel. The Law of Reflection operates for each ray striking a reflector.

Curved reflectors are also used as driving mirrors, shaving mirrors, make-up mirrors, and in car headlights.

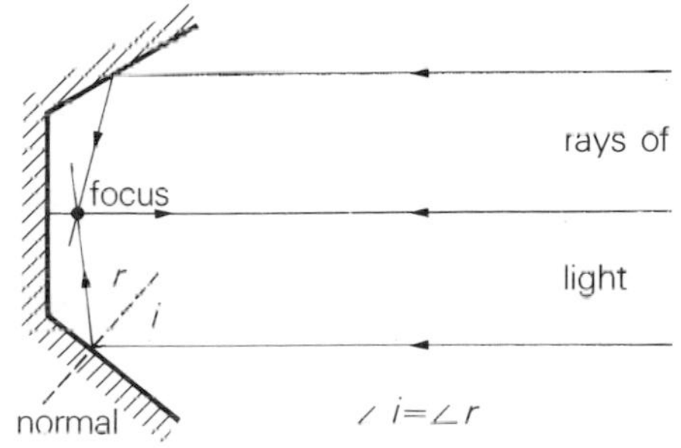

Figure 3.8 Reflection from 'curved' reflector

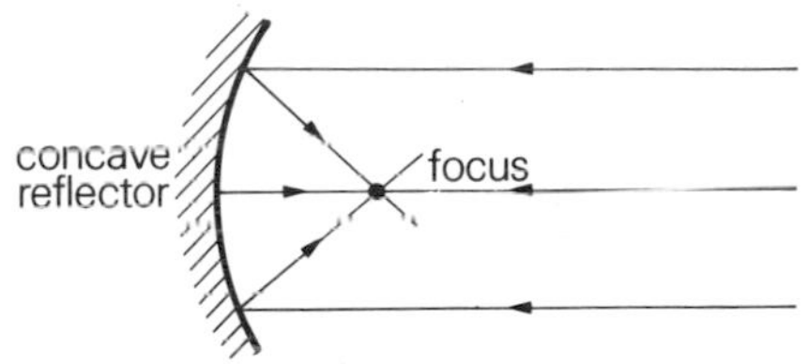

Figure 3.9 Reflection from a concave reflector

Figure 3.10 The solar furnace at Odeillo

Wave speed, frequency and wavelength

There is no effect on the wave speed, frequency or wavelength of light waves when they are reflected.

3.4 Refraction of light

Refraction by glass

When a ray of light hits the surface of a glass block at an angle, it changes direction. It is being **refracted**, as shown in Figure 3.11.
To describe more precisely this change in direction, we draw a normal. On refraction from air to glass, the ray is bent towards the normal. In this case the angle of incidence ($\angle i$) is greater than the angle of refraction ($\angle r$).

Figure 3.11 Refraction of light by glass

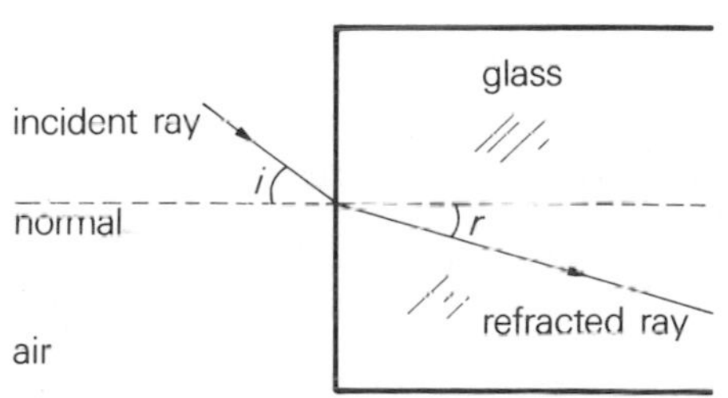

A ray of light bends **towards** the normal on entering glass, and then bends **away** from the normal on emerging from the glass into air, Figure 3.12.
Note that because the glass block has parallel sides, the ray of light emerges parallel to the direction at which it enters the glass block.

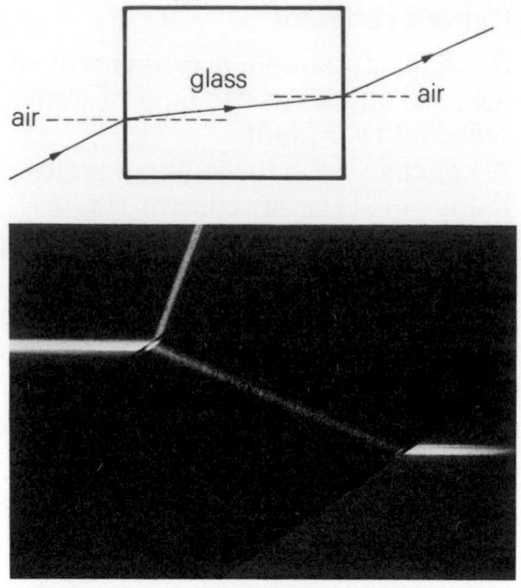

Figure 3.12 Refraction by a rectangular glass block

Wave speed, frequency and wavelength

When a water wave is refracted (page 15) we found that its speed and wavelength decreases while its frequency remains constant. This causes the water wave to bend towards the normal.

When light travels from air to another medium it also bends towards the normal, so it is reasonable to assume that its speed and wavelength decrease similarly.

Figure 3.13 shows more examples of the fact that light rays travelling from air to glass are bent towards the normal and that when they emerge from the glass they are bent away from the normal. The direction of the rays is affected by the shape of the piece of glass.

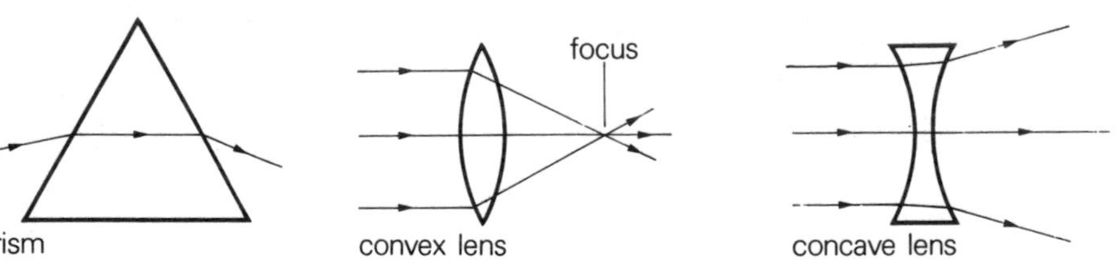

Figure 3.13 Refracting glass shapes

The prism-shaped piece of glass refracts the ray of light towards its base.

The convex lens converges the rays to a focus.

The concave lens diverges the rays of light.

Refraction of light by a specially shaped piece of glass called a **lens** is particularly useful. Figure 3.14 gives some examples of how a lens can be used.

Figure 3.14 Applications of lenses

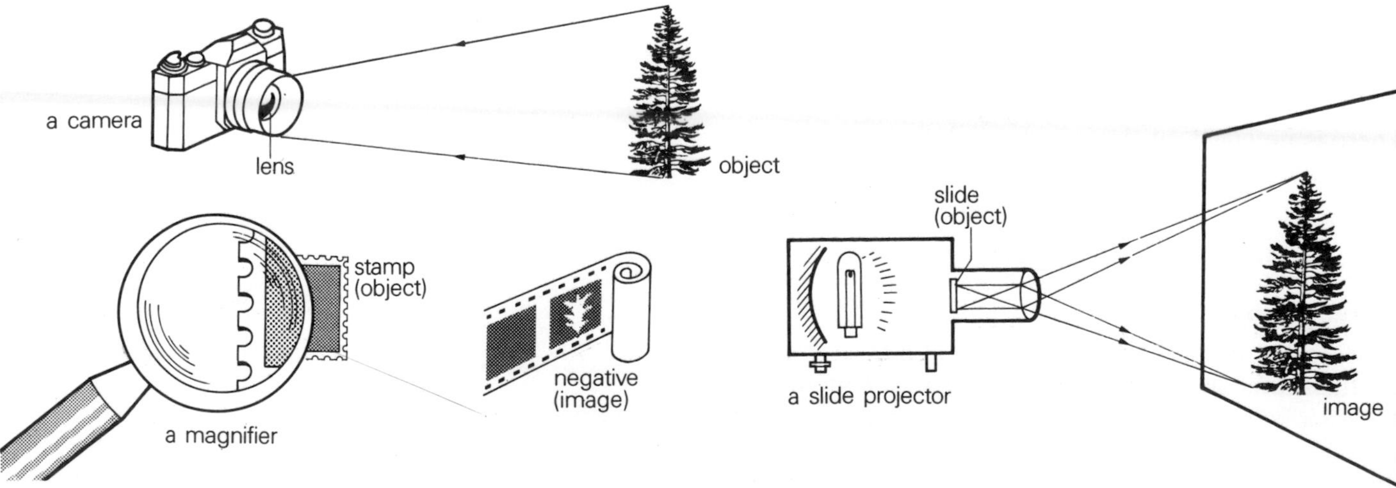

Notice that the slide projector reverses the effect of the camera.

Refraction by water

If a ray of light passes through a transparent tank containing water we find (Figure 3.15) that the ray is refracted in a similar manner to light travelling through glass.

If we try the experiment in reverse by immersing the light source in the tank of water (Figure 3.16) we see that the emerging ray is bent away from the normal as it comes out into the air.

So we have seen that light is refracted by water as well as glass. All transparent materials refract light.

We should now be able to explain why a stone under water appears nearer the surface than it really is. As shown in Figure 3.17, the light is being refracted as it leaves the water and enters our eyes.
We see that the light travelling in direction YZ enters our eyes and we imagine that it has come from X′. We call X′ the **image** of X. We see the stone at X′ even though the light has come from point X. The stone appears to be nearer the surface.

We can also understand why we can see a coin in a cup even though it is not in our direct line of vision, as in Figure 3.18.

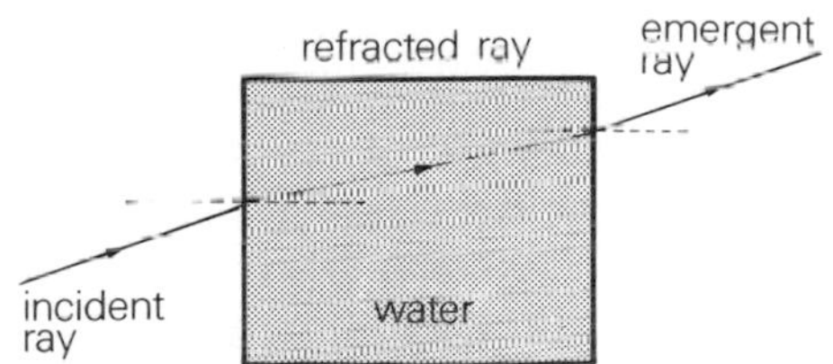

Figure 3.15 Refraction of light by water

Figure 3.16 Refraction from water to air

Figure 3.17 Viewing a stone in water

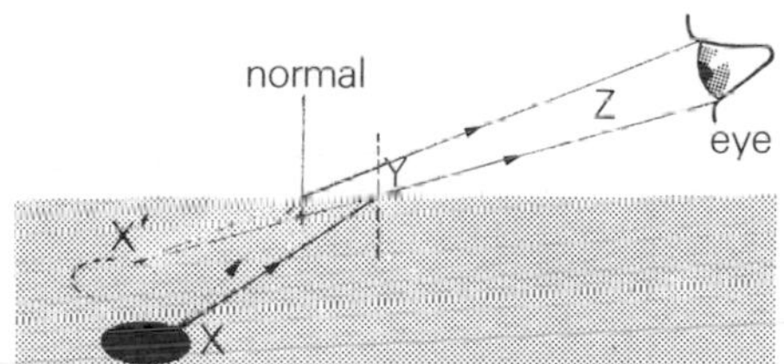

Figure 3.18 The appearing coin

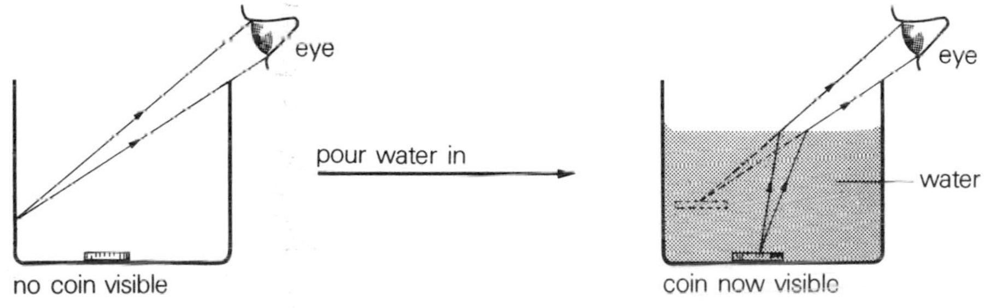

3.5 Visible spectrum

Sir Isaac Newton (1642–1727) showed experimentally that white light is composed of a series of colours, called the **visible spectrum**. These colours in order of increasing frequency (decreasing wavelength) are red, orange, yellow, green, blue, and violet. Now we have seen that light is refracted when it enters glass and Newton, when he placed a glass prism in the path of a beam of white light, found that for larger angles of incidence the light was spread out or **dispersed** into coloured bands (spectrum), Figure 3.19.

Figure 3.19 Dispersion of light by a prism

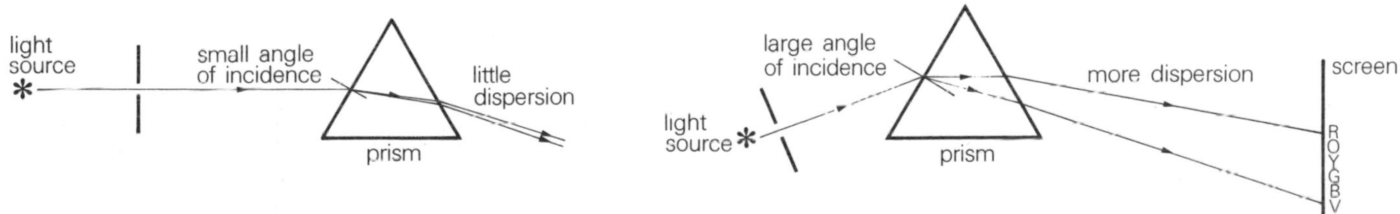

In effect, the frequency or colour of the light determines the amount by which it is refracted: blue light is refracted more than red light. When white light from the Sun encounters small droplets of water in the atmosphere it is refracted, reflected and hence dispersed to form a rainbow in the sky. The water droplets act like many small prisms.

We can produce interesting colour effects if we use two or three light sources, each of one colour only. Before we do this, we must understand first how the colours are produced. When we place a red filter in front of a light source we are effectively allowing only red light through and stopping (or absorbing) the other colours of the spectrum. A blue filter allows through blue, and a green filter allows through only green. Obviously, then, if we use three light sources with these three filters in front of them and allow all three beams to coincide, then we should obtain red, blue and green superimposed, which results in **white** light, Figure 3.20.

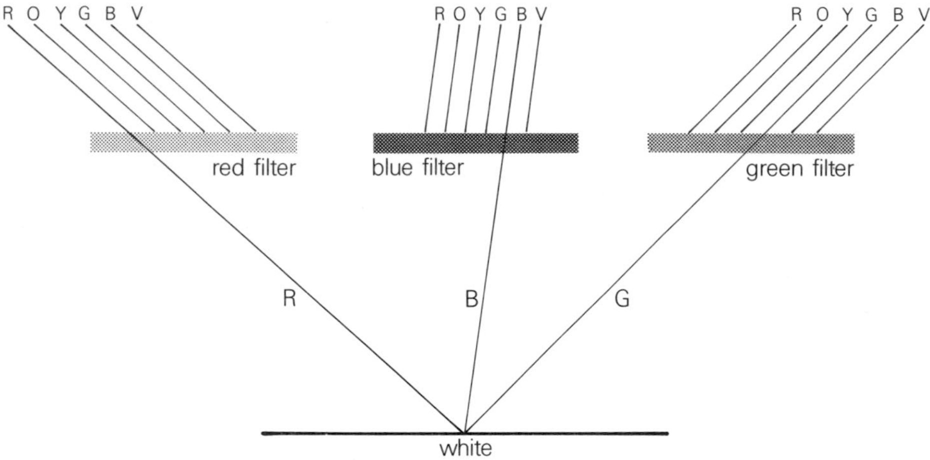

Figure 3.20 Colour mixing to give white light

Perhaps more surprising, when we mix pairs of colours we obtain some unexpected results, Figure 3.21.

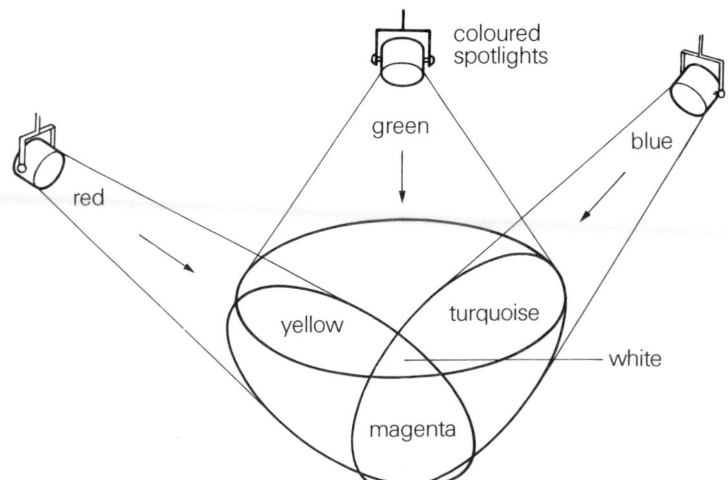

Figure 3.21 Colour mixing in pairs

The colours red, green and blue are known as the primary colours. The colours yellow, turquoise and magenta are the secondary colours.

Summary

The speed of light in air is $3{\cdot}00 \times 10^8$ m s^{-1}.

Light behaves as a wave and produces an interference pattern from two sources in step.

The greater the wavelength of light, the greater is the interference fringe spacing.

When light waves are reflected, they obey the Law of Reflection; $\angle i = \angle r$.

The wavelength of red light is greater than the wavelength of blue light.

Refraction occurs when light passes from air to glass (or to water).

air to glass
wave speed decreases
wavelength decreases
frequency does not change
ray bends towards normal

glass to air
wave speed increases
wavelength increases
frequency does not change
ray bends away from normal

White light can be dispersed into its constituent colours of red, orange, yellow, green blue, violet.

A prism will disperse white light to give the visible spectrum.

Problems

1 A periscope has two straight mirrors parallel to each other and at 45° to the horizontal. Draw ray diagrams and try to explain how the periscope works.

2 Here is a curved mirror. Copy the drawing and add the reflected rays.

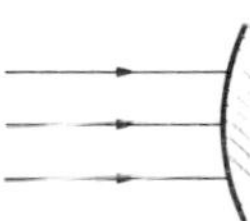

3 Draw a diagram to show how a ray of light is refracted when it emerges from a rectangular glass block.

4 Copy and complete these refraction drawings.

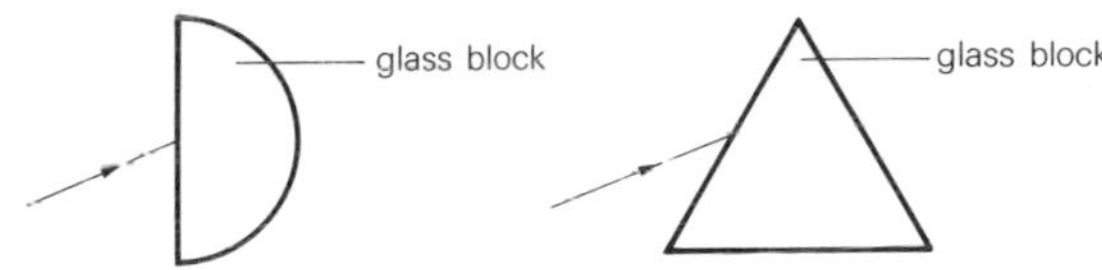

5 Try to explain how a fishbowl of water could start a fire.

6 Light travels at 3×10^8 m s^{-1}. How long would it take light to travel one kilometre?

7 Hold the paper level with your eye, and look along the diagram from S_1S_2 towards the side AB.

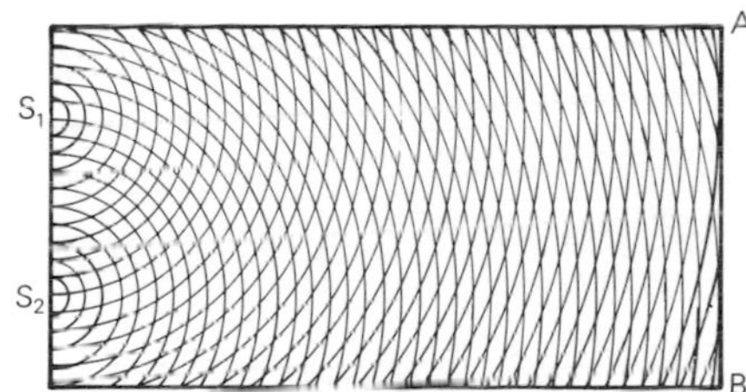

7·1
a) Describe what you see.
b) What wave phenomenon does this illustrate?

7·2
a) Suppose you wish to demonstrate to a friend that this effect occurs with light waves. How would you do so, assuming that you have any apparatus required?
b) What would your friend see?
c) Use the above diagram S_1S_2 BA to explain what she would see. *SCEEB*

8 How can you show the following with red light?
a) Reflection
b) Refraction
c) Interference *SCEEB*

9 A ray X of red light is directed towards a glass prism.

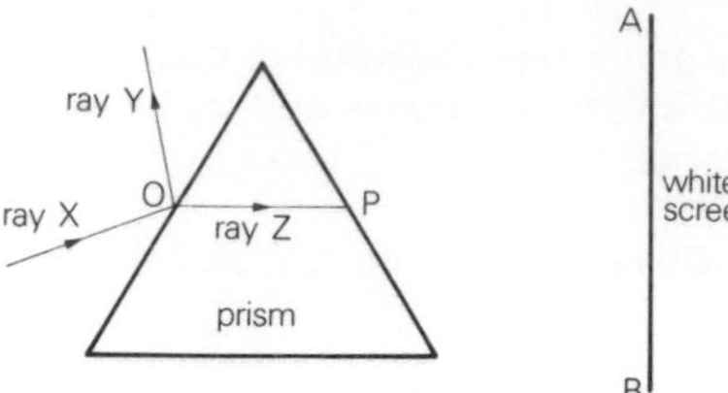

9·1 State what happens to the ray X after it strikes the glass at O.
9·2 Comment on the brightness of ray Z compared to the brightness of ray X.
9·3 Copy the above diagram and show how the light leaves the prism at P.
9·4 Describe the changes which take place in
a) the wavelength,
b) the speed of the red light as it enters the prism at O.
9·5 The ray X of red light is now replaced by a ray of white light. Describe in detail what is now seen on the screen AB. Explain how this happens. *SCEEB*

10 The diagrams below show rays of light approaching a glass obstacle. In each case the wavefronts are shown.

(i)

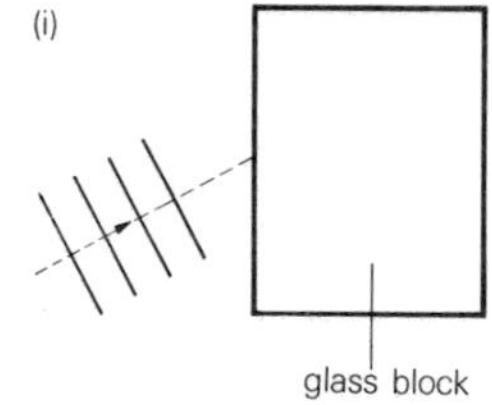

(ii)

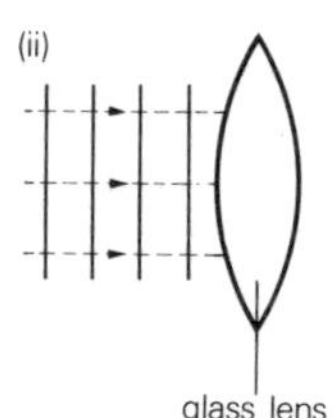

Copy these diagrams and draw the path of the rays
10·1 inside the glass block
10·2 after passing through the lens.
Show also the wavefront in each case. *SCEEB*

11 **11·1** A ray of light strikes a rectangular glass block as shown.

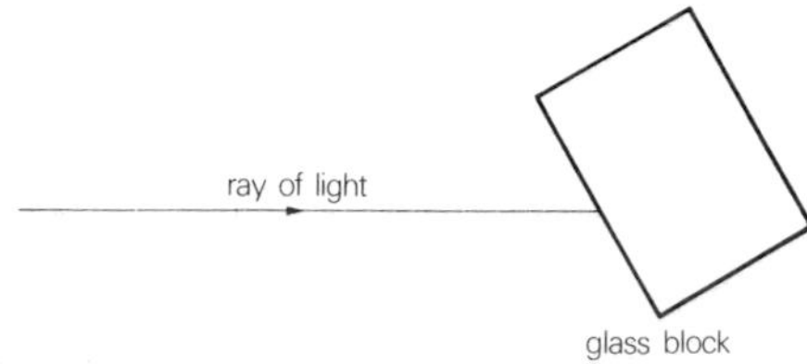

It passes through the glass block and emerges from the opposite side.
a) Copy the diagram and show the complete path of the ray.
b) What changes take place in the speed and wavelength of light when it passes from air into glass?
11·2 A ray of red light strikes a triangular glass prism as shown.

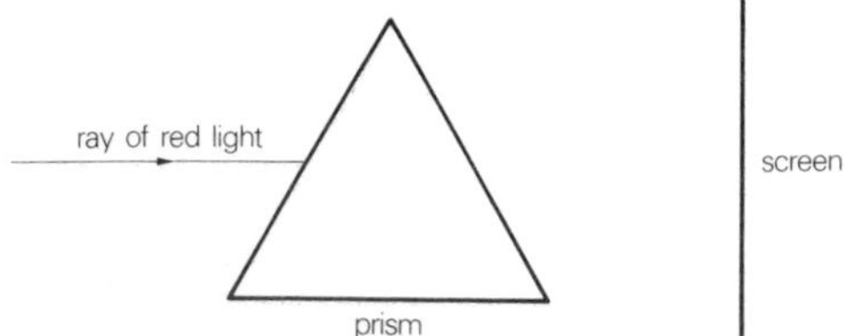

It passes through the prism and strikes the screen.
a) Copy this diagram and show the complete path of the ray.
b) The ray of red light is replaced by a ray of blue light. Mark on your diagram where the blue ray meets the screen.
c) The ray of blue light is replaced by a ray of white light. Describe what will be seen on the screen.
11·3 Red light has a smaller frequency than blue light. Compare the wavelengths of red and blue light in air. *SCEEB*

4 Electromagnetic waves

4.1 Introduction

We have already shown that light is a wave because it can produce interference patterns, but it is only one of a set of waves known as the **electromagnetic spectrum.**

The different waves in the spectrum are shown in Figure 4.1. They are so named because of their different properties. The wavelengths of light which we can see range from 4×10^{-7} m (violet) to 7×10^{-7} m (red) and this corresponds to a very narrow range of wavelengths when compared with the whole of the electromagnetic spectrum.

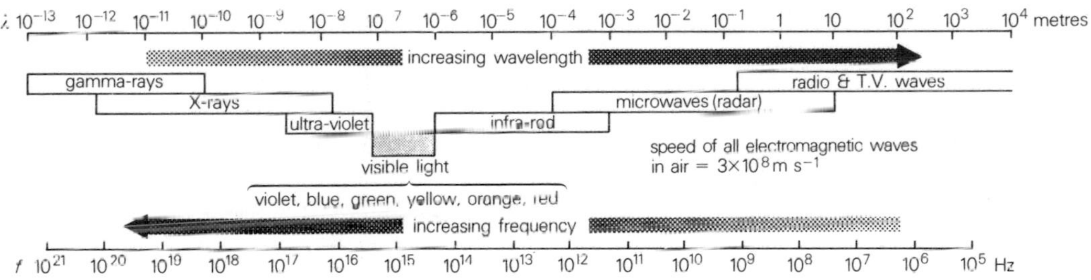

Figure 4.1 Electromagnetic spectrum

Notice that in the diagram there is an overlap in the range of wavelengths for some waves. This is because it is sometimes the source which decides what we call the waves and at other times it is the energy we supply to produce them. For example, an X-ray machine could produce X-rays of very short wavelength (high energy) which have the same wavelength as the gamma radiation emitted by a radioactive source, but they are called X-rays because of the way in which they are produced.

4.2 Light

Light reaches us from a source called the Sun, and therefore it can travel through empty space. Light is the name given to the range of waves which is visible and it is useful to consider first how it can be produced. Most of our artificial light sources are either hot objects or electrical discharges (passing electricity through a gas). They can produce what is often called 'white light' which we saw at the end of the previous chapter is a range of colours known as the 'visible spectrum'. The properties of light have been dealt with in Chapter 3.

4.3 Infra-red radiation

When an electric fire is first switched on, no visible light is produced but energy is being converted into an invisible radiation. In the laboratory if we use a radiant heater we can detect this radiation even before the heater starts emitting visible radiation, Figure 4.2.

Figure 4.2 Detection of radiation emitted by a radiant heater

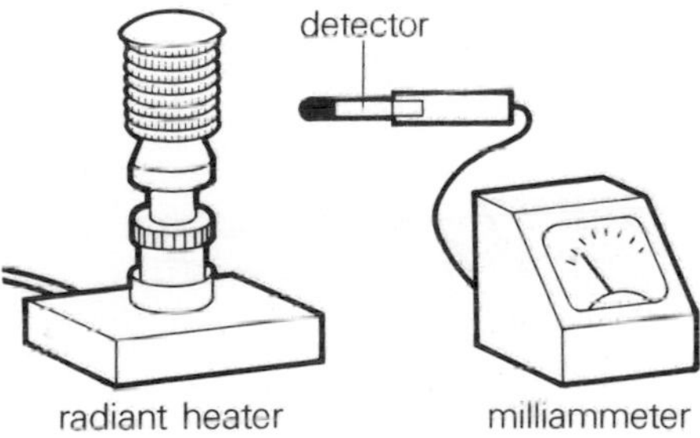

When we examine the spectrum produced by a light bulb with the same detector (Figure 4.3), we obtain an increased reading when the detector is in the region just beyond the visible red end of the spectrum.

Even when the bulb is switched on at low power (when the bulb filament isn't even red hot), we detect this radiation at the same position beyond the red. Since this radiation is detected in this region of the spectrum, it is called **infra-red radiation**. Its frequency is lower and therefore it has a longer wavelength than visible red.

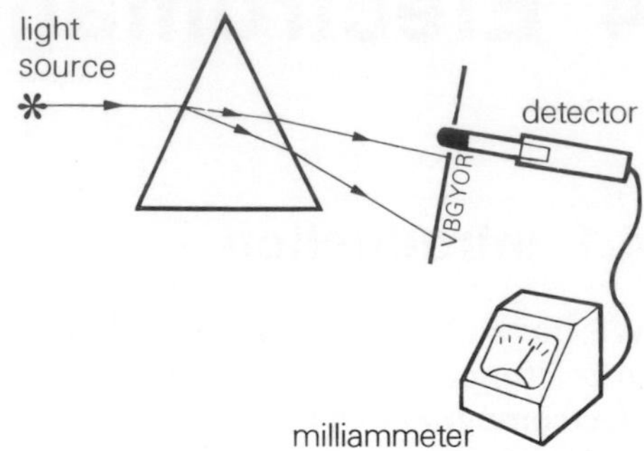

Figure 4.3 Detection of radiation 'beyond red'

Infra-red detectors

We can detect infra-red waves ourselves. When they are absorbed by our skin, their energy is converted into heat.

If we perform an experiment to discover the absorption properties of various surfaces (Figure 4.4), we find that dark surfaces absorb infra-red radiation very well, whereas silvered surfaces do not: they reflect it.

The thermometer in the blackened flask shows a greater increase in temperature in a given time.

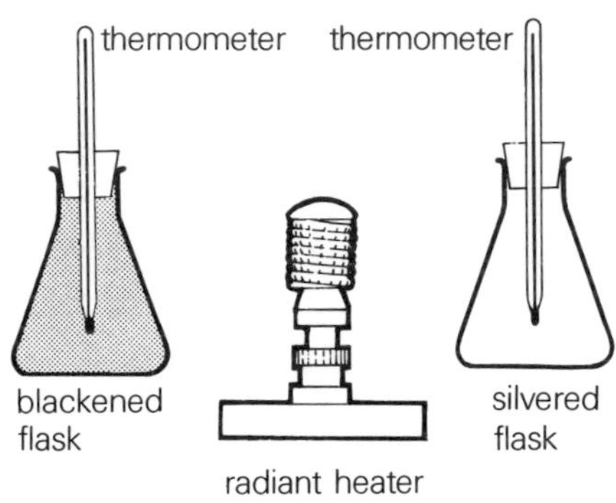

Figure 4.4 Absorption and reflection of infra-red radiation

a) Black bulb thermometer
If we blacken the bulb of a thermometer with dull black paint then, when infra-red radiation is absorbed, the heat produced raises the temperature of the bulb of the thermometer and the reading increases, Figure 4.5.

b) Thermistor
A thermistor consists of a small bead of material enclosed in glass. It is connected by two leads to an amplifier. When the thermistor absorbs infra-red radiation, the resulting increased temperature produces a greater current. This change is amplified and displayed on a meter as a temperature or temperature change, Figure 4.5.

c) Phototransistor
A blackened phototransistor connected to a battery and meter absorbs infra-red which produces a change in temperature and hence in the current, Figure 4.5.

d) Photographic film
A specially prepared photographic film which is coated with infra-red sensitive chemicals is affected by infra-red in the same way that ordinary film is affected by light, Figure 4.5.

Figure 4.5 Infra-red detectors

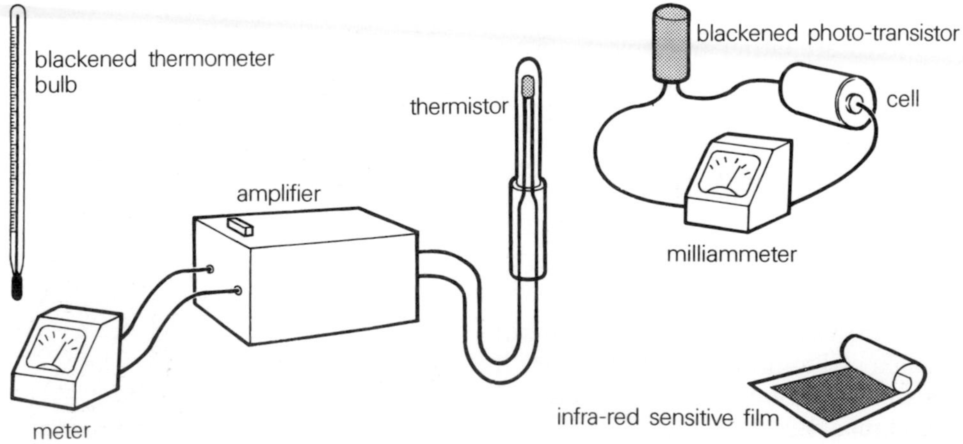

Any warm object emits infra-red radiation. The amount of infra-red emitted depends on the temperature and colour of the object. Using infra-red sensitive film, an Earth resources satellite is able to provide details of vegetation covering large areas of the globe (Figure 4.6), since different vegetations emit different amounts and frequencies of infra-red.

It is possible to produce 'heat pictures' of buildings to find where there is greatest heat loss. Medical conditions can be diagnosed by investigating the infra-red emission from the skin, Figure 4.7.

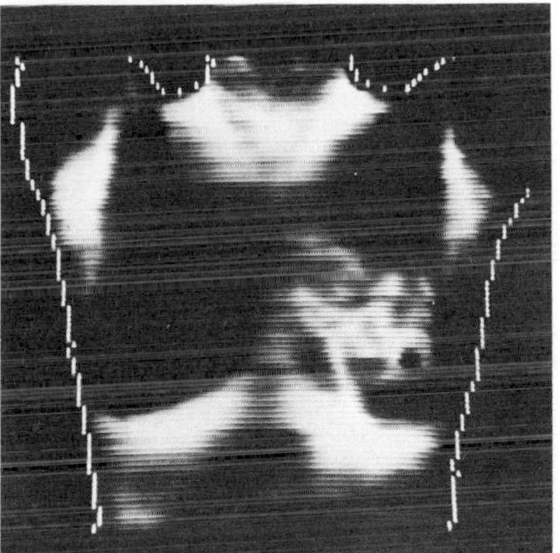

Figure 4.7 Medical infra-red photograph

Figure 4.6 Infra-red photograph from satellite

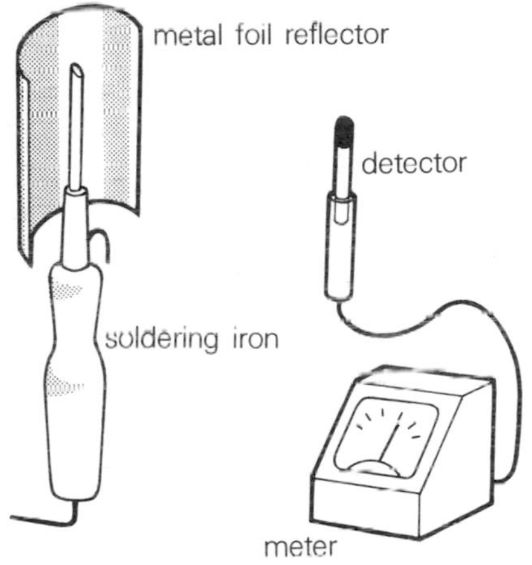

Figure 4.8 Reflection of infra-red radiation

Reflection of infra-red

A soldering iron is a convenient source of infra-red. A detector placed close to it will indicate emission of this radiation. The infra-red can be reflected on to the detector (Figure 4.8) by inserting a concave metal foil reflector behind the soldering iron. When it is inserted, there is an increase in the reading.

Reflectors of this type are used in domestic electric heaters. They are also used in the car industry to reflect infra-red from lamps to dry car paint rapidly (Figure 4.9).

Figure 4.9 Using infra-red reflection

Refraction of infra-red

Radiation from the Sun that passes through a convex lens may be focused to a point; there is usually a large increase in temperature at this point because infra-red radiation is refracted.

We can demonstrate the refraction of infra-red radiation alone by passing it through a spherical flask containing a strong solution of iodine in carbon tetrachloride (Figure 4.10). Take care: carbon tetrachloride (tetrachloromethane) is a dangerous chemical.
The detector registers a low reading before the flask is inserted in front of the infra-red source. After it is inserted, the reading increases because the infra-red radiation has been refracted on passing through the solution.

Figure 4.10 Refraction of infra-red radiation

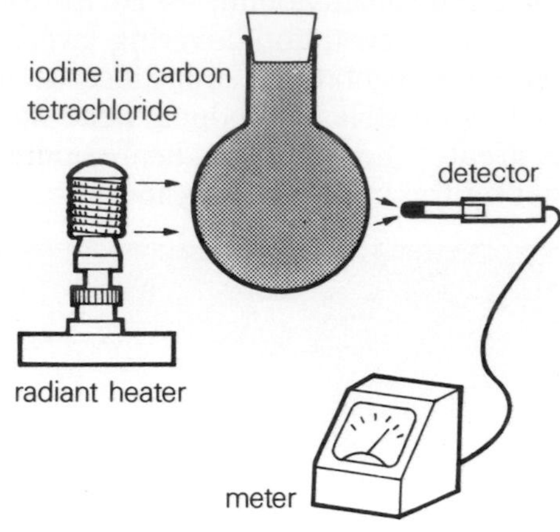

4.4 Microwave radiation

At lower frequencies than infra-red, there is a region in the electromagnetic spectrum which consists of waves called **microwaves**.

All waves produce a heating effect when absorbed. Food will absorb microwave radiation and become hot in a very short time. This is made use of in microwave ovens which cook quickly and economically.

Microwaves are used in radar, but since this involves reflection, this will be dealt with in the next section.

In the laboratory we use microwave radiation of wavelength about 3 cm, sometimes known as '3-cm waves'. A microwave source or 'transmitter' and a 'receiver' or detector are shown in Figure 4.11.

Both the transmitter and receiver have 'horns'. They help the transmitter to direct the energy and the receiver to collect the microwave energy. The receiver is often connected to an amplifier which converts the received energy into sound energy. This does not mean that microwaves can be heard: their energy is merely converted into a different form.

Figure 4.11 Microwave transmitter and receiver

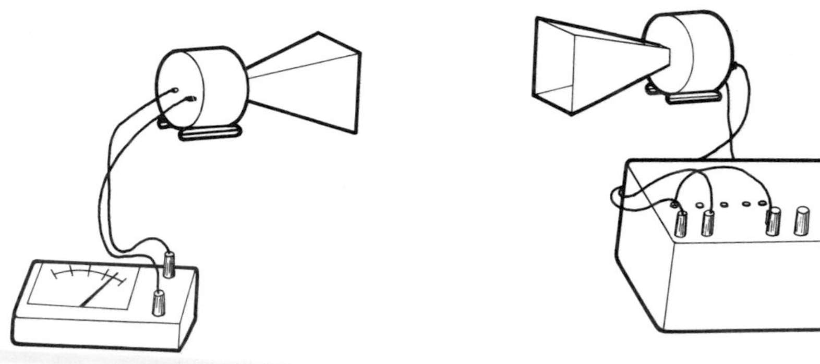

Reflection of microwaves

If we direct the microwaves at an angle to a metal reflector (Figure 4.12), we find that they obey the law of reflection and the receiver detects a maximum signal when $\angle i = \angle r$.

The reflective properties of microwaves are used in the direction finding and navigation system called **radar**.

A radar transmitter sends out a narrow beam of microwaves in short pulses. Distant objects in the path of the beam reflect some of the radiation back to a receiver alongside the transmitter. The direction in which the beam is being transmitted when the reflection is observed

Figure 4.12 Reflection of microwaves

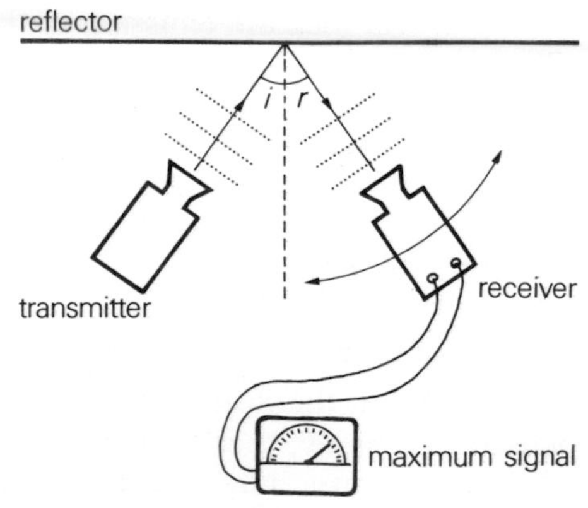

gives the direction of the object (an aeroplane for example), and the time between the transmitted and received pulse gives an indication of the distance of the object.

Sometimes the information is displayed on the screen of a cathode ray oscilloscope (Figure 4.13). P_1 is the transmitted pulse and P_2 is the echo pulse. The time taken is found from the separation on the screen between P_1 and P_2. We can use the equation

$$\text{speed } v = \frac{\text{distance travelled } s}{\text{time taken } t}$$

to find the distance of the reflecting object. The speed v is 3×10^8 m s^{-1} since all electromagnetic waves travel at this speed.

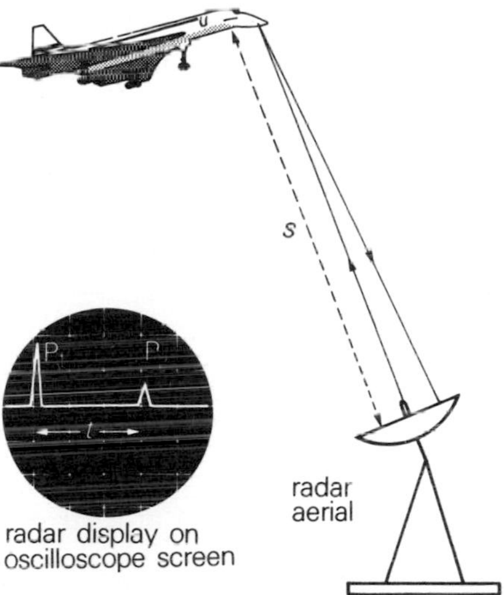

Figure 4.13 Radar

Example 1

A radar wave pulse is reflected off an object and received 6×10^{-5} s after it is transmitted. What is the distance from the radar station to the reflecting object?

The time for the pulse to travel to the object is half the time to travel there and back, i.e. 3×10^{-5} s.

Substituting in the equation

$$\text{speed} = \frac{\text{distance travelled}}{\text{time taken}}$$

$$\Rightarrow \quad 3 \times 10^8 = \frac{\text{distance travelled}}{3 \times 10^{-5}}$$

Multiplying both sides by 3×10^{-5}

$$\Rightarrow 3 \times 10^{-5} \times 3 \times 10^8 = \frac{\text{distance} \times 3 \times 10^{-5}}{3 \times 10^{-5}}$$

$$\Rightarrow \quad 9 \times 10^3 = \text{distance}$$

The aeroplane is 9 km away.

Microwaves are also used in closed-circuit television (Figure 4.14) to transmit live colour television coverage from outside broadcast vehicles to the studio.

Figure 4.14 Television microwave link

Diffraction of microwaves

If the microwave transmitter is placed behind an aluminium sheet (Figure 4.15) and the receiver is moved from X to Y, it detects microwaves which have been bent around the corner, indicating that they have been diffracted.

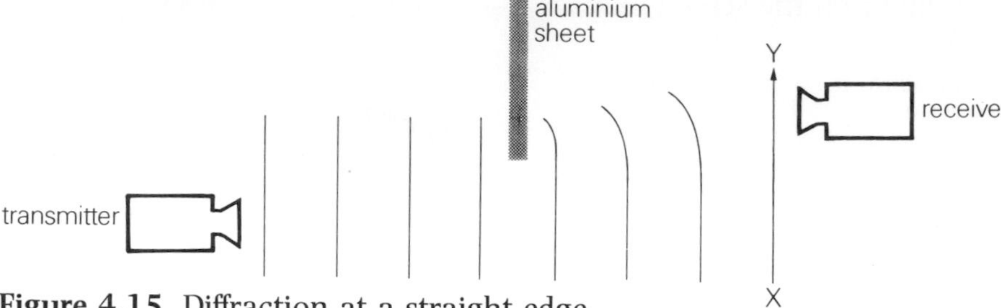

Figure 4.15 Diffraction at a straight edge

Figure 4.16 shows how microwaves diffract through a gap. The receiver detects microwave radiation on either side of the gap as it is moved from X to Y.

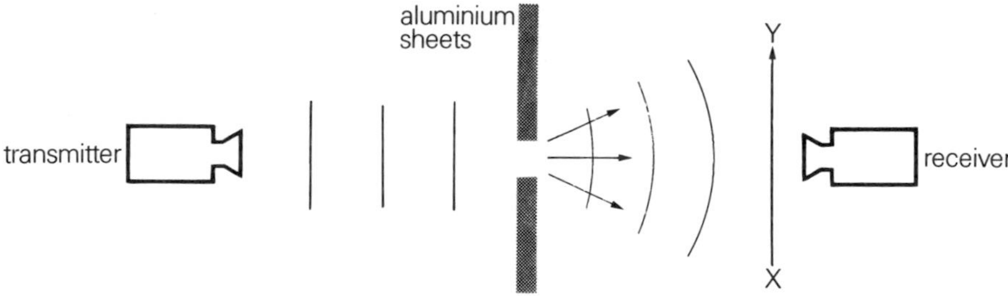

Figure 4.16 Diffraction of microwaves at a gap

Interference of microwaves

The microwave equipment used in the school laboratory is well suited to demonstrating that microwave radiation, like visible radiation, has wave properties. We can demonstrate the interference of microwaves by using three aluminium sheets to make two slits through which the waves can pass, Figure 4.17. The two gaps act as two sources of microwaves in phase. Where their waves combine, the two sources produce constructive and destructive interference, as in the Young's Slits Experiment on page 21.

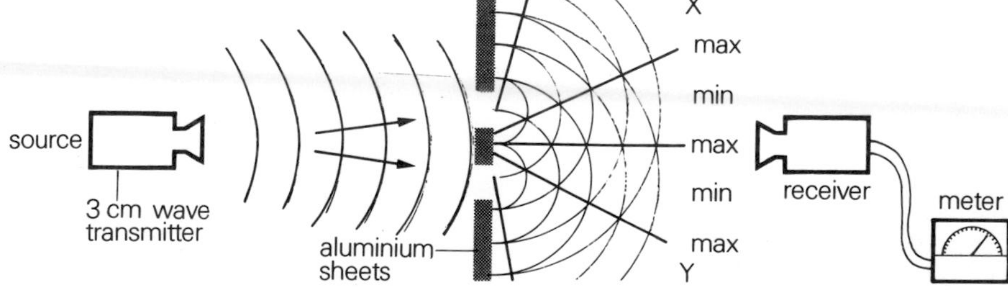

Figure 4.17 Interference of microwaves

The gap separation should be about the same as the wavelength of the microwaves. As the receiver is moved from X to Y in front of the gaps, a series of maximum and minimum readings is obtained, indicating interference of microwaves. There is a maximum signal at the centre of the interference pattern. To show convincingly that interference is really taking place, we may move the receiver to a 'minimum' (destructive interference) position and then block off one of the gaps. There is immediately an increase in the signal received, showing that there is now no destructive interference taking place with one source only, and that it is this source (gap) which we are detecting.

Measuring the wavelength of microwaves

In Chapter 1 (page 6) we saw that waves can be reflected in such a way that they produce a stationary wave. Using a microwave transmitter and detector, we can show the presence of a stationary wave, Figure 4.18.

We use a detector known as a **diode probe** which is small and will detect waves from both directions at once.

The transmitter is placed about 1 metre away from a metal reflector. Waves are reflected off the metal sheet. As the probe is moved along a line between the transmitter and reflector, it detects a varying signal and maximum and minimum readings are obtained. The maxima are antinodes and the minima are nodes of the stationary wave pattern (Figure 4.19). The separation of the nodes is found to be about 1·5 cm, giving an approximate value for the wavelength of 3·0 cm. In practice it is usual to measure the distance for several nodes and then to find the average value.

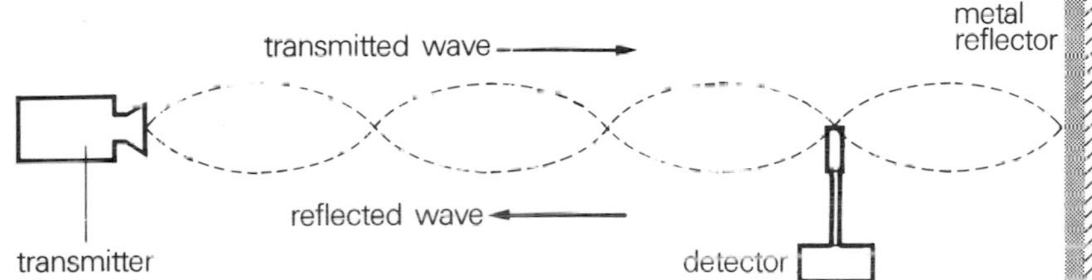

Figure 4.18 Microwave stationary wave

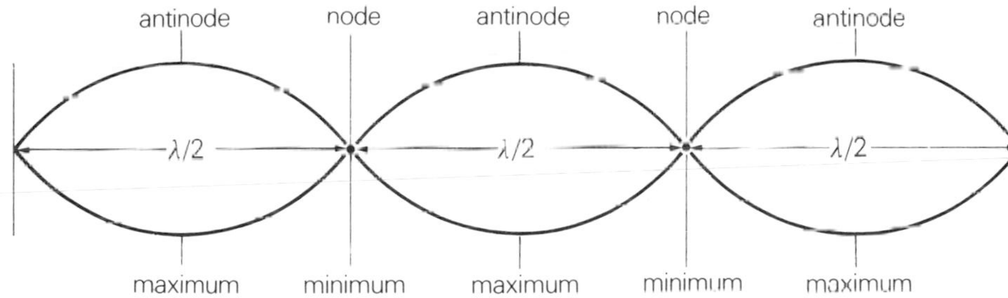

Figure 4.19 Nodes and antinodes

Refraction of microwaves

Microwaves change their speed and wavelength when they pass through different materials and this can produce a change in their direction. The refraction of microwaves may be demonstrated in the laboratory, Figure 4.20.

When the microwaves enter the perspex prism filled with paraffin, their direction is changed and they are received by the detector, Figure 4.20a.

The lens-shaped container filled with paraffin has the effect of focusing the radiation. The detector used here is the diode probe as it can locate more precisely the position of a maximum signal, Figure 4.20b.

Figure 4.20 Refraction of microwaves

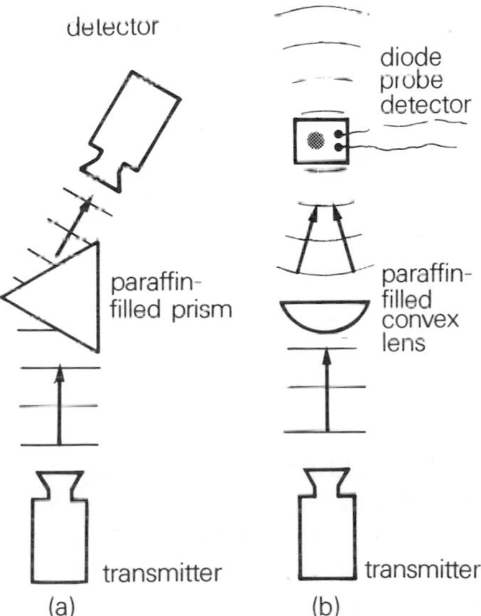

4.5 Radio waves

Waves of greater wavelength and lower frequency than microwaves are called radio waves. We will use the term 'radio waves' to include television waves. Their wavelengths range from about 10 cm to 10 million metres. This set of waves includes waves called long wave, medium wave and short wave. The short wave range contains waves known as VHF (very high frequency) and UHF (ultra high frequency).

Radio waves can be produced by an electrical circuit called an oscillator and can be accurately controlled. They are transmitted from and received by aerials. The aerial size is governed by the range required and the wavelength of the radio waves produced. We cannot hear radio waves, but a radio receiver will convert their energy to sound energy. Radio transmitters and receivers vary a great deal in size. A transmitter may be so small that it can be implanted in an animal to transmit information from measuring instruments, Figure 4.21. A receiver aerial may be very large (about 400 m long) and detect radio waves from distant stars (Figure 4.22); these radio waves can give us information about the stars.

Figure 4.21 Small radio transmitter

Figure 4.22 Large radio receiver

Reflection of radio waves

The reflective properties of radio waves depend on their frequency. Certain frequencies of radio waves will reflect off layers in the Earth's upper atmosphere called the Appleton and Heaviside layers. Owing to the curvature of the Earth, radio waves that are not reflected are lost out in space (Figure 4.23), but the reflective properties of these layers increases the range of the signal.

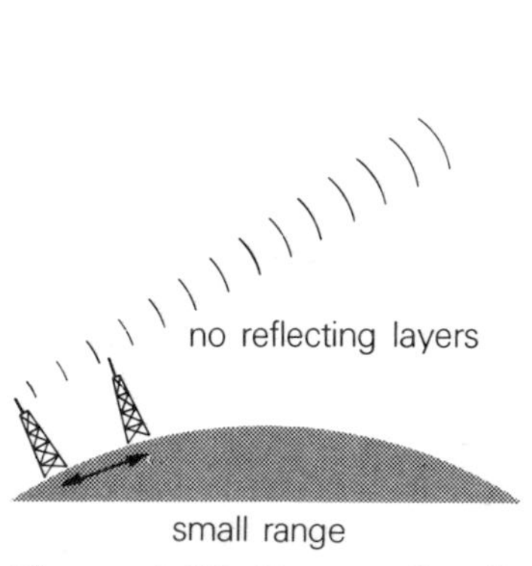

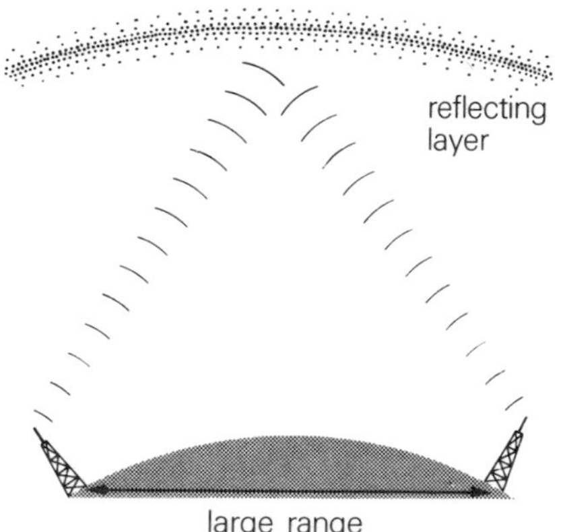

Figure 4.23 Range of radio waves

The radio waves obey the law of reflection on encountering these atmospheric layers, Figure 4.24.

TV waves and VHF radio waves are not reflected by these layers and so instead must be transmitted between repeater stations on the ground or by satellites. However TV waves are reflected by solid structures like buildings, hills and even aeroplanes. This reflection produces a 'ghosting' effect on a TV screen (Figure 4.25), because, while most of the signal is received directly by the aerial, some of it is reflected back off an obstacle and reaches the aerial slightly later since it has travelled a little further. The two signals received at two slightly different times produce two pictures side by side on the screen – a 'ghosting' effect. The ghost picture is of course fainter since it comes from the weaker reflected signal.

Figure 4.24 Reflection of radio waves

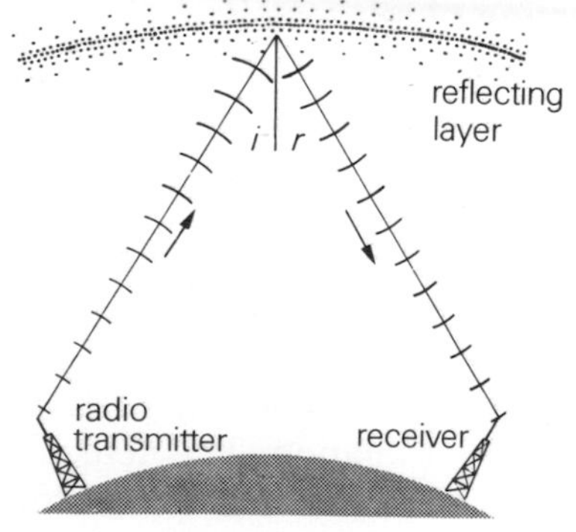

Diffraction of radio waves

With water waves we found that the amount of diffraction depended on the wavelength. This also applies to radio waves: long wavelengths (low frequency) are diffracted more than short wavelengths (high frequency), Figure 4.26.

Interference of radio waves

Two radio transmitters transmitting the same frequency of signal at the same time are illustrated in Figure 4.27.
They are broadcasting the same programme at the same frequency and in step. If a car driver travels from point A to point B, he finds that his car radio (detector) gives a series of readings which vary in intensity (loudness). He is passing through regions of constructive and destructive interference corresponding to the interference pattern produced by the two radio wave sources.

Figure 4.27 Interference of radio waves

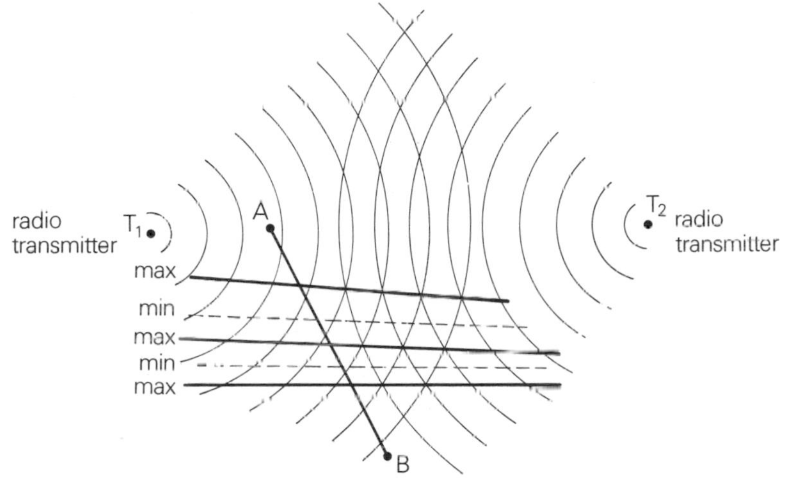

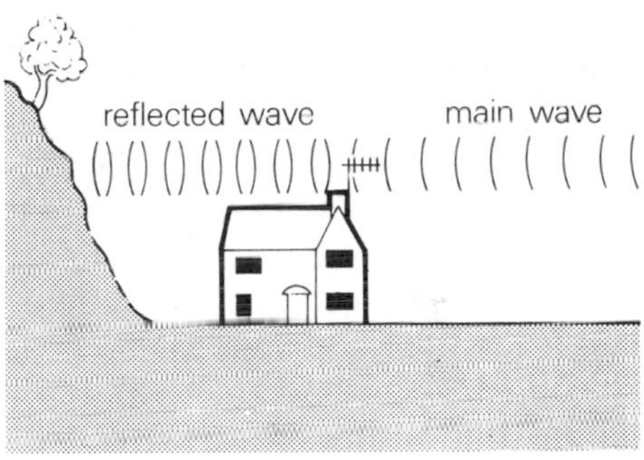

Figure 4.25 Reflection of TV waves

Figure 4.26 Diffraction of radio waves

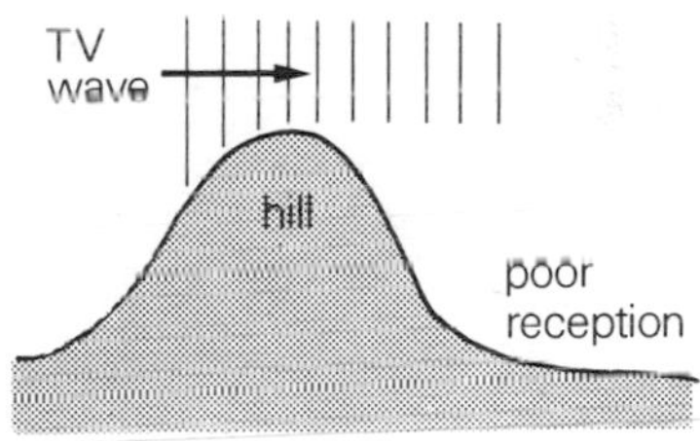

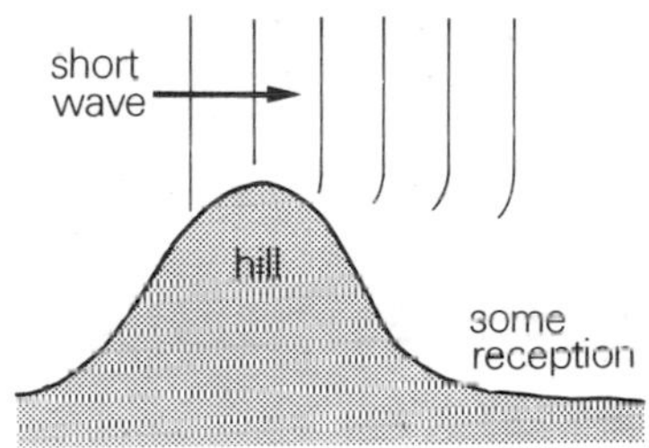

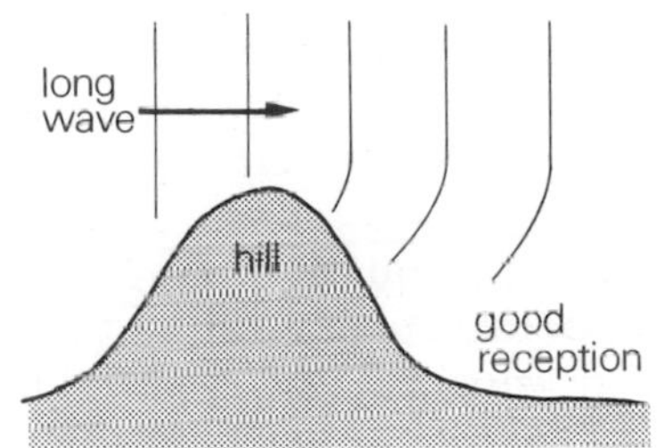

4.6 Ultra-violet radiation

Radiation of shorter wavelength (and hence higher frequency) than visible violet light is called ultra-violet. We cannot see ultra-violet radiation, but we can detect its presence using specially prepared photographic film. There are also chemicals which are able to **fluoresce**: they absorb ultra-violet and re-radiate it as visible radiation.

The Sun is a source of ultra-violet and exposure to this radiation can produce a change in the colouring of the skin (suntan).

Man-made ultra-violet lamps are used for such varied things as inducing chemical reactions, deodorizing, sterilization and accelerated weathering of timber and paints. The most useful detector of ultra-violet is photographic film, but the camera must have a lens made of quartz glass since ultra-violet radiation does not pass through ordinary glass. Ultra-violet radiation must be treated with respect because too much is dangerous.

Ultra-violet radiation exhibits the wave properties of reflection, refraction, diffraction and interference. Of these properties, only reflection can be easily demonstrated in the school laboratory in the same way in which the reflection of visible light is shown. If the experiment is performed on fluorescent paper, the path of the radiation can be observed.

4.7 X-rays

Extremely penetrating radiation of an even higher frequency is known as X-radiation or X-rays.

It was in 1895 that a scientist called Röntgen discovered what he called 'a new kind of light' which was invisible but could penetrate solid materials. He found that it affected photographic film in the same way that light does. One of his early X-ray pictures (radiographs) is shown in Figure 4.28. The X-rays pass through flesh more easily than through bone.

Human tissue is severely damaged if exposed to X-rays for too long, and yet we are all familiar with their useful and relatively safe penetrating properties when we attend the mass radiography unit for a chest X-ray or the dentist's for a dental X-ray.

X-rays can be produced by an X-ray tube, Figure 4.29.
Their penetrating properties make them very useful for 'looking inside' things without opening them up, Figure 4.30.

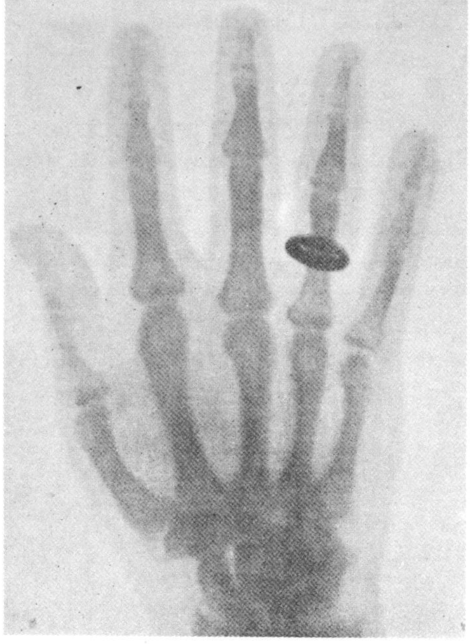

Figure 4.28 X-ray of Mrs Röntgen's hand

Figure 4.30

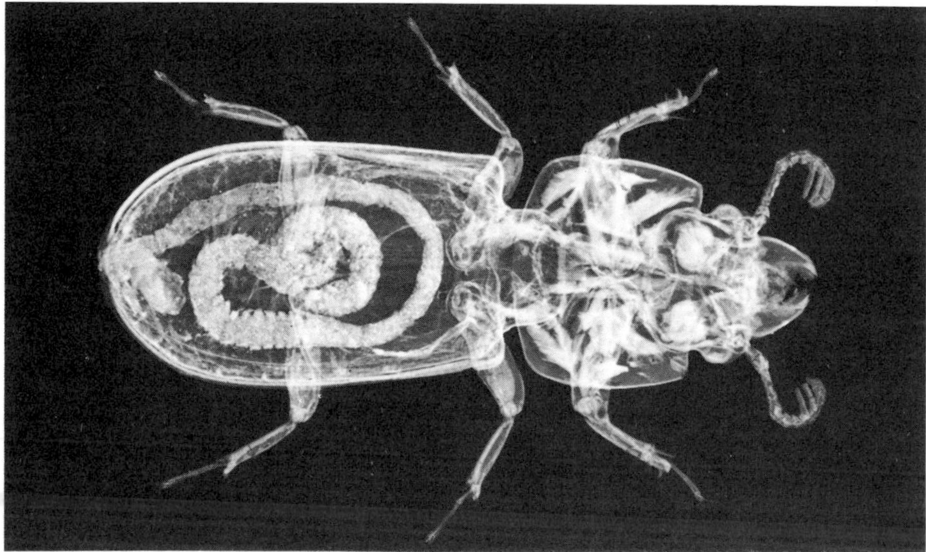

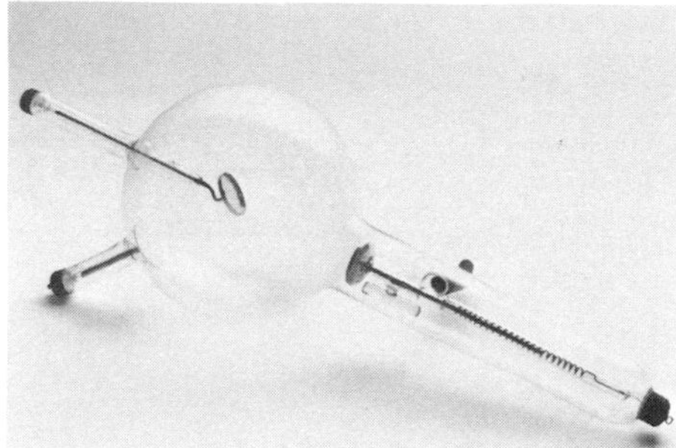

Figure 4.29 X-ray tube

Figure 4.31 Gamma radiation source and detector

4.8 Gamma radiation

Even more dangerous than X-rays is the gamma radiation emitted by some radioactive substances. These are waves of extremely short wavelength. Their properties are discussed at length in Chapter 22 on radioactivity.

Gamma radiation photographs can be produced in a similar manner to X-ray photographs. Gamma radiation can also be detected by an instrument known as a Geiger-Müller tube or 'Geiger counter' (Figure 4.31). You will find out about this instrument also in Chapter 22.

4.9 Electromagnetic wave calculations

In this chapter we have described how the waves of the electromagnetic spectrum behave. They all can show reflection, refraction, diffraction and interference. They all also obey the equation

wave speed = frequency × wavelength

$$v = f \times \lambda$$

It is important to remember that all electromagnetic waves have a speed of 300 million metres per second (3×10^8 m s^{-1}) in air.

Example 2

A VHF radio transmitter broadcasts at a frequency of 108 MHz. What is the wavelength of the radio waves it emits?

Using the equation:

wave speed = frequency × wavelength

$$v = f \times \lambda$$

$$\Rightarrow 3 \times 10^8 = 108 \times 10^6 \times \lambda$$

Multiply both sides of the equation by $\frac{1}{(108 \times 10^6)}$

$$\Rightarrow \frac{1}{(108 \times 10^6)} \times 3 \times 10^8 = \frac{1}{\cancel{(108 \times 10^6)}} \times \cancel{(108 \times 10^6)} \times \lambda$$

$$\Rightarrow \frac{3 \times 10^8}{108 \times 10^6} = \lambda$$

$$\Rightarrow \frac{300 \times \cancel{10^6}}{108 \times \cancel{10^6}} = \lambda$$

$$\Rightarrow 2{\cdot}8 = \lambda$$

The wavelength of the waves is 2·8 m.

Example 3

What is the frequency of the waves emitted by a microwave transmitter whose waves are 3 cm in wavelength?

Using the equation:

wave speed = frequency × wavelength

$$v = f \times \lambda$$

$$\Rightarrow 3 \times 10^8 = f \times 3 \times 10^{-2}$$

Multiply both sides of the equation by $\frac{1}{(3 \times 10^{-2})}$

$$\Rightarrow 3 \times 10^8 \times \frac{1}{3 \times 10^{-2}} = f \times \cancel{(3 \times 10^{-2})} \times \frac{1}{\cancel{(3 \times 10^{-2})}}$$

$$\Rightarrow \frac{3 \times 10^8}{3 \times 10^{-2}} = f$$

$$\Rightarrow 10^{10} = f$$

The frequency of the waves is 10^{10} Hz.

Example 4

At what speed do waves travel whose frequency is 7×10^{16} Hz and wavelength 3×10^{-9} m?

Using the equation:

wave speed = frequency × wavelength

$$v = f \times \lambda$$

$$\Rightarrow v = 7 \times 10^{16} \times 3 \times 10^{-9}$$

$$\Rightarrow v = 21 \times 10^7$$

$$\Rightarrow v = 2{\cdot}1 \times 10^8$$

The waves travel at $2{\cdot}1 \times 10^8$ m s^{-1}.

Summary

Electromagnetic waves can be transmitted through a vacuum.

Electromagnetic waves travel at $3{\cdot}00 \times 10^8$ m s^{-1}.

Electromagnetic radiation can show reflection, refraction, diffraction and interference.

Electromagnetic waves obey the wave equation $v = f \times \lambda$.

The electromagnetic spectrum consists of the following waves in order of increasing wavelength:
gamma rays
X-rays
ultra-violet waves
visible light
infra-red waves
microwaves
TV waves
radio waves

Problems

1 Complete the following statements about electromagnetic radiations and detectors:

1·1 When infra-red is absorbed by a detector the energy is converted to ________.
1·2 When ultra-violet radiation is detected its energy is converted to ________.
1·3 When microwaves are absorbed their energy is converted to ________.

2 During an eclipse of the Sun the visible light and infra-red are both cut off at the same time. What does this tell you about the speed of light waves and infra-red waves?

3 The Earth continually receives electromagnetic radiation from the Sun, producing a warming effect. Why doesn't the Earth eventually get as hot as the Sun?

4 An object emits infra-red and ultra-violet radiation. How could it be photographed so that only its infra-red was seen?

5 Why do people in sun-bathing lounges need to wear special glasses?

6 Why are the Appleton and Heaviside layers important for radio communication?

7 Describe briefly how a radar system works.

8 What is the frequency of the infra-red radiation of wavelength 900×10^{-9} m?

9 When choosing the wavelength of radio waves for transmission in hilly countryside what factors are important? Explain your choice.

10 A spectrum of white light is allowed to fall on a sheet of photographic paper in a darkened room. The position of the spectrum is marked (figure 1). When the photographic paper is developed, a black area appears as shown in figure 2.

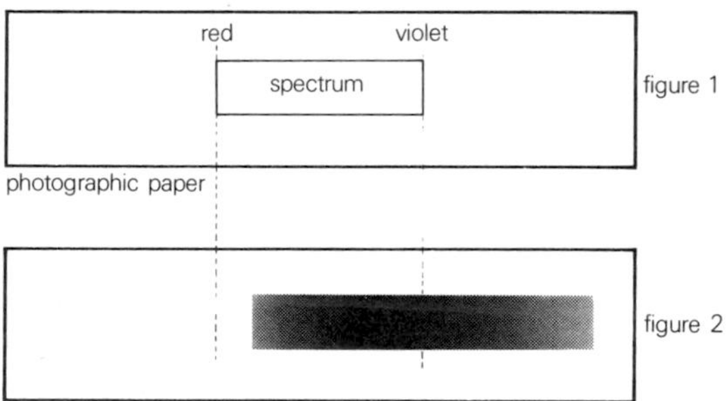

10·1 What conclusion can be drawn from the fact that the blackened area on the developed photographic paper is much longer than the area of the spectrum?
10·2 Explain how this experiment illustrates that it is quite safe to develop this photographic paper in a room illuminated only by red light. *SCEEB*

11 **11·1** Which of the following would detect X-rays? phototransistor, photographic film, thermocouple.
11·2 As light of a particular colour passes from one medium into another, which of the following does not alter? wavelength, frequency, velocity.
11·3 Which of the following waves has the shortest wavelength?
ultra-violet radiation, infra-red radiation, radio waves. *SCEEB*

12 A VHF radio transmitter broadcasts radio waves at a frequency of 300 MHz. What is their wavelength?

5 Sound waves

A crystal glass can be shattered by an opera singer singing a high note and a window can be broken by the bang from an aircraft 'breaking the sound barrier'. Sound is a form of energy and it is this energy which results in the breakage of the glass and the window.

5.1 Production of sound

If you rest your fingers gently on your throat and say 'ah', you will feel your throat vibrating. It is these vibrations which produce the sound. Our voices produce a variety of sounds and it is easier to begin our study of sound by considering a single note produced. When a tuning fork is struck (Figure 5.1), it produces a single note. If you look closely at the prongs of the tuning fork while it is emitting a note, you will see that they look slightly blurred. This is because they are vibrating backwards and forwards rapidly. We can demonstrate that a tuning fork is vibrating by some simple experiments.

When a tuning fork is struck and then dipped into water, the water is splashed by the vibrating prongs, Figure 5.2.

If a light ball hanging on a thread is brought up to touch a sounding tuning fork, the ball is kicked away by the vibrations of the prong, Figure 5.3.

Figure 5.4 shows light beads bouncing up and down in the cone of a loudspeaker which is emitting a note. This shows that the cone is vibrating.

Stringed musical instruments produce notes when the strings are made to vibrate. This can be done by plucking the string, as with a guitar, or by scraping it with a bow, as with a violin, but the end result is the same. When a string vibrates it produces a sound. Percussion instruments, such as drums, cymbals and triangles, are made to vibrate

Figure 5.1 Tuning fork

Figure 5.2

Figure 5.3

Figure 5.4

by striking them. Wind instruments such as trumpets and clarinets produce sounds as a result of vibrations of the player's lips or of a reed in the mouthpiece of the instrument.

Sounds are produced by vibrating objects.

5.2 Transmission of sound

Sounds are produced by vibrating objects and we know that sound can travel through air to our ears. The substance through which the sound travels is called the 'medium'. In 1654 Otto Von Guericke carried out an experiment to see whether sound does need a medium or whether it can travel through empty space. He placed a clockwork bell in a sealed bottle which was attached to a pump. As he pumped the air out, the sound faded and became very faint. We can carry out a similar experiment using the apparatus shown in Figure 5.5.

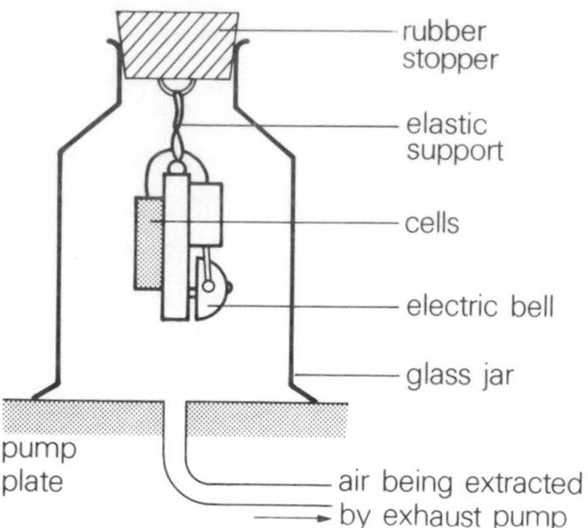

Figure 5.5

The electric bell is powered by the cells. The bell is switched on before the bell jar is placed on the pump plate. The seal between the bell jar and the pump plate is made airtight with grease. The bell can be clearly heard and the clapper can be seen to be striking the bell. When the pump is switched on, the sound dies away and finally becomes very faint, but it can still be seen that the clapper is striking the bell as before. It is not possible to remove all the air from the bell jar, but an efficient pump can remove most of it. The sound cannot travel through the space in the bell jar, from which most of the air has been removed, but it never disappears completely, because the sound travels through the elastic supports to the surrounding air.

Because sound cannot travel through space, astronauts who are 'walking in space' (Figure 5.6) are not able to talk directly to each other, even though they may be very close. They can, however, talk to each other using a radio because, like all electromagnetic waves, radio waves can travel through space.

Figure 5.6

The experiment that we have just described shows that sound travels through air but not through empty space. Air is not the only medium through which sound will travel. You have probably heard sounds which have travelled through water. When you are under water in the swimming baths, you can hear sounds made by the other bathers, and these sounds must have travelled through the water to reach you. Dolphins make mewing and whistling sounds which scientists believe are used to signal through the water to other dolphins. Some scientists hope that they will be able to understand these signals and be able to 'talk' to dolphins.

If you press your ear to a bench, cover your other ear and ask a friend to tap on the other end of the bench, you will hear the sound as it travels through the wood of the bench. In western films, you often see cowboys or Indians pressing their ears to the ground to listen for the hoofbeats of approaching horses, because the sound travels through the solid earth better than it does through air.

Sound does not travel through empty space, but it will travel through solids, liquids and gases.

5.3 Speed of sound

The speed of sound was first measured in 1640 by a French mathematician, Marin Mersenne. He measured the distance travelled by a sound which was reflected back to its source, and the time taken for the sound to travel that distance. Using the equation

$$\text{wave speed} = \frac{\text{distance travelled}}{\text{time taken}}$$

he estimated the speed of sound to be approximately 315 m s^{-1}.

You may carry out an experiment using Mersenne's idea of reflecting a sound back to its source, and measuring the time taken for the sound to travel to the reflector and back.

You may make a loud noise by clapping two pieces of wood together. This sound will reflect off a wall and you will hear the echo of the sound reflected back to you. The time taken for the sound to travel the distance to the wall and back is the time between clapping the pieces of wood together and hearing the echo.

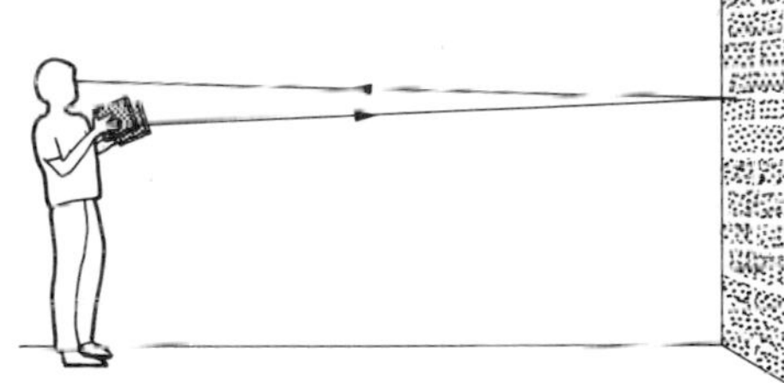

Figure 5.7

Example 1

A pupil stands a distance of 180 m from a large wall, Figure 5.7. He claps two pieces of wood together, and 1·2 s later he hears the echo of the sound that has been reflected off the wall. What value for the speed of sound is obtained from these results?

$$\text{speed} = \frac{\text{distance travelled}}{\text{time taken}}$$

the sound travels from the pupil to the wall and back to the pupil

distance between the pupil and the wall = 180 m

distance travelled by the sound = 2 × (the distance between the pupil and the wall)
= 2 × 180
= 360 m

$$\text{speed} = \frac{\text{distance travelled}}{\text{time taken}} = \frac{360}{1{\cdot}2} = 300$$

The speed of sound = 300 m s^{-1}.

When you do this experiment, it is important that you find a large wall facing an open space. You should avoid getting reflections off other buildings or objects, so as to be able to detect clearly the reflections off the wall that you are using.

The time interval that you will be measuring is very small, and you will not be able to measure it directly with a stopwatch. You and a partner can get over this difficulty by measuring ten of the time intervals and calculating the value of one time interval. To do this you should set up a steady rhythm of clapping and adjust the frequency of the claps so that the echoes of the claps are heard between the claps. When you have settled into this steady rhythm, your partner should count the claps, starting with a countdown, 3, 2, 1, 0, 1, . . . 9, 10. He should start the stopwatch on the count of 0, and stop it on the count of 10. Since the frequency of the claps is such that the echoes are heard mid-way between the claps, the time interval between the claps is equal to twice the time interval between the clap and its echo.

stopwatch reading = 10 × (time interval between the claps)
time between the claps = 2 × (time between the clap and the echo)
⇒ stopwatch reading = 20 × (time interval between the clap and the echo).

Thus from the reading on the stopwatch you can calculate the time interval between a clap and its echo. This is equal to the time taken for the sound to travel to the wall and back to you.

Example 2

A pupil stands 150 m from a large wall facing an open space. He claps two pieces of wood together with a steady rhythm so that he hears the echoes mid-way between the claps. His partner who is counting the claps, starts a stop-watch on the count of zero and stops it on the count of 10. The reading on the stopwatch is 18 s. What value do these results give for the speed of sound?

$$\text{wave speed} = \frac{\text{distance travelled}}{\text{time taken}}$$

distance between the pupil and the wall = 150 m

distance travelled by the sound = 2 × (distance between the pupil and the wall)
= 2 × 150
= 300 m.

reading on the stopwatch = 10 × (time interval between claps)

10 × (time interval between claps) = 18·0

⇒ (time interval between claps) = 1·80 s.

(time interval between claps) = 2 × (time interval between clap and echo)

2 × (time interval between clap and echo) = 1·80

⇒ (time interval between clap and echo) = $\frac{1}{2}$ × 1·80
= 0·90 s.

time taken for sound to travel to the wall and back = 0·90 s

$$\text{wave speed} = \frac{\text{distance travelled}}{\text{time taken}}$$

$$\text{wave speed} = \frac{300}{0{\cdot}90} = 333$$

The speed of sound = 333 m s^{-1}.

The speed of sound depends very much upon the nature and temperature of the medium through which it travels. The table below gives some examples of the speeds of sound in different media.

medium	*speed of sound* (m s^{-1})
air at 0 °C	331
air at 100 °C	385
water at 0 °C	1 410
metals	2 000 to 7 000
rock	1 500 to 3 500
solids	1 500 to 13 000

A knowledge of the speed of sound through various types of rock is used in the search for oil and other minerals.

Race between light and sound

Some of the earlier measurements of the speed of sound were carried out by measuring the time between seeing the flash of a distant cannon and hearing its 'boom'. It was assumed that the flash was seen at the instant the cannon was fired, and that the time lapse before the sound was heard was the time taken for the sound to travel from the cannon to the observer.

The flash of the cannon is signalled to the eye of the observer by light travelling at 300 000 000 m s^{-1} (3·0 × 10^8 m s^{-1}) while the 'boom' is signalled by sound travelling at about 340 m s^{-1}. Were these early experimenters justified in ignoring the time taken for the light to reach the observer? The following example should help us to answer this question.

Example 3

A cannon is fired at a distance of 3·0 km from an observer. If the speed of light is $3{\cdot}0 \times 10^8$ m s^{-1} and the speed of sound is $3{\cdot}4 \times 10^2$ m s^{-1}, calculate
a) the time it took for the light from the flash to reach the observer,
b) the time it took for the sound from the 'boom' to reach the observer.

a)

$$\text{wave speed} = \frac{\text{distance travelled}}{\text{time taken}}$$

speed of light = $3{\cdot}0 \times 10^8$ m s^{-1}
distance travelled by the light = 3·0 km
= $3{\cdot}0 \times 10^3$ m

Let the time taken for the light to travel from the cannon to the observer be t_1.

$$\text{wave speed} = \frac{\text{distance travelled}}{\text{time taken}}$$

$$3{\cdot}0 \times 10^8 = \frac{3{\cdot}0 \times 10^3}{t_1}$$

Multiplying both sides of the equation by t_1

$$3{\cdot}0 \times 10^8 \times t_1 = 3 \times 10^3$$

Multiplying both sides of the equation by $\frac{1}{3 \times 10^8}$

$$t_1 = \frac{3{\cdot}0 \times 10^3}{3{\cdot}0 \times 10^8} = 1{\cdot}0 \times 10^{-5}$$

The time taken for the light to reach the observer = $1{\cdot}0 \times 10^{-5}$ s.

b) $\text{wave speed} = \frac{\text{distance travelled}}{\text{time taken}}$

speed of sound = $3{\cdot}4 \times 10^2$ m s^{-1}
distance travelled by the sound = 3·0 km
= $3{\cdot}0 \times 10^3$ m

Let the time taken for the sound to travel from the cannon to the observer be t_2.

$$\text{wave speed} = \frac{\text{distance travelled}}{\text{time taken}}$$

$$3{\cdot}4 \times 10^2 = \frac{3{\cdot}0 \times 10^3}{t_2}$$

Multiplying both sides of the equation by t_2

$$3{\cdot}4 \times 10^2 \times t_2 = 3{\cdot}0 \times 10^3$$

Multiplying both sides of the equation by $\frac{1}{3{\cdot}4 \times 10^2}$

$$t_2 = \frac{3{\cdot}0 \times 10^3}{3{\cdot}4 \times 10^2} = 8{\cdot}8$$

The time taken for the sound to reach the observer = 8·8 s.

Comparison of these two answers shows that the time taken by the sound is nearly a million times greater than the time taken by the light. An error of approximately one in a million is extremely small. It is quite reasonable to assume that the time taken for the light to travel from the cannon to the observer is negligible and that the time between seeing the flash and hearing the 'boom' is equal to the time taken for the sound to reach the observer.

Figure 5.8

A thunderstorm provides us with another example of light reaching us before sound. When the storm is in the distance we see the flash of lightning and wait for the clap of thunder to follow. The lightning and the thunder are produced at the same time but, while it takes a negligible time for the light from the flash of lightning to reach us, the sound takes some seconds to reach us. By estimating the time taken for the sound of the thunder to reach us we can estimate how far away the storm is.

Example 4

If the speed of sound is approximately 340 m s^{-1}, estimate to the nearest second the time that sound takes to travel 1 km.

$$\text{wave speed} = \frac{\text{distance travelled}}{\text{time taken}}$$

speed = 340 m s^{-1}

distance travelled = 1 km

= 1 000 m

Let time taken to travel 1 000 m = t.

$$\text{speed} = \frac{\text{distance travelled}}{\text{time taken}}$$

$$340 = \frac{1\,000}{t}$$

Multiplying both sides of the equation by t

$340 \times t = 1\,000$

Multiplying both sides of the equation by $\frac{1}{340}$

$$t = \frac{1\,000}{340}$$

$= 2{\cdot}9$

Time (to nearest second) for sound to travel 1 km = 3 s.

You can get a rough estimate of the distance from you to a thunderstorm by counting off the seconds between seeing a flash of lightning and hearing the roll of thunder. Every three seconds of time represent a distance of approximately 1 km. Every five seconds of time represent a distance of approximately 1 mile.

Example 5

A girl saw a flash of lightning in the sky and estimated that 12 seconds passed before she heard the thunder. Approximately how far away was the girl from the storm?

In 3 s sound travelled 1 km

in 12 s sound travelled 4 km

Distance from the girl to the storm = 4 km.

5.4 The nature of sound

We shall see later (page 51) that sound can produce an interference pattern and this provides evidence of wave nature. So, we can think of sound energy as being transmitted by means of a wave.

Sounds are produced by vibrating objects and require a medium, but we have not yet considered how sound carries energy through the medium. Figure 5.9 shows a candle flame in front of a loudspeaker which is producing a low note.

The candle flame moves towards and away from the loudspeaker. These to-and-fro movements of the flame indicate vibrations in the air,

Figure 5.9

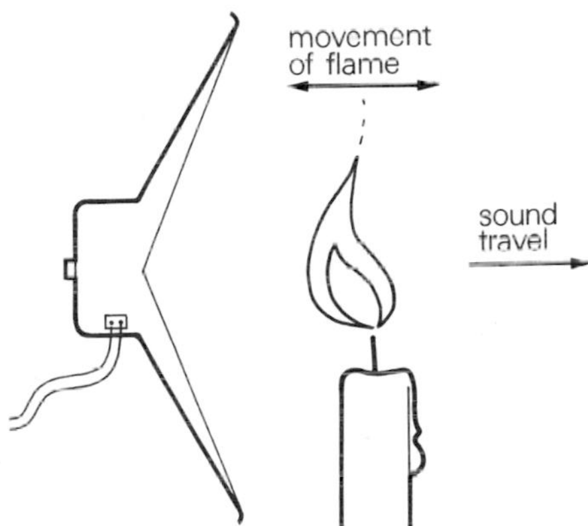

Figure 5.10

Figure 5.10. When we considered water waves, we saw that they were transmitted by vibrations of water particles at right angles to the direction of wave travel. The experiment with the candle flame in front of the loudspeaker indicates vibrations along the direction of travel of the sound, Figure 5.11.

We may use a slinky spring to investigate how energy may be transmitted by vibrations in this way. A pulse of energy may be sent along the spring by moving its end to-and-fro along the line of the spring, Figure 5.12.

This shows a **compression** formed by the left-to-right movement of the end of the spring which forces the coils closer together. The following right-to-left movement of the end of the spring pulls the coils further apart and produces a **rarefaction**. The compression and rarefaction travel along the spring.

In a spring
a compression is a section in which the coils are closer together than average;
a rarefaction is a section in which the coils are further apart than average.

You will see that a compression and a rarefaction produced by this to-and-fro movement travel along the spring. After they have passed, the spring returns to its original shape. It is energy that has travelled along the spring.

If you watch the spring closely when a compression and rarefaction move along it, you will see that each part of the spring vibrates to-and-fro. If the end of the spring is moved to-and-fro continuously, a series of compressions and rarefactions pass along the spring, and this makes up a continuous wave.

A wave in which the vibrations are along the line of travel of the wave is called a **longitudinal wave**, Figure 5.13.

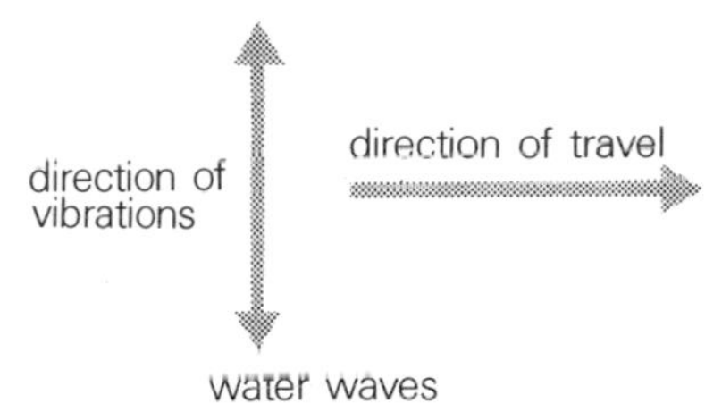

sound waves

Figure 5.11

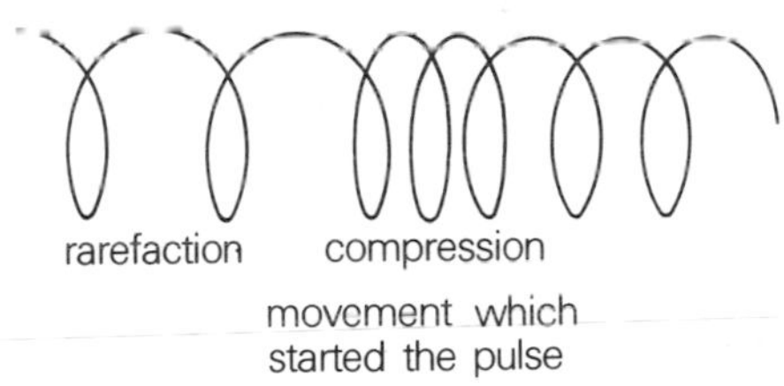

Figure 5.12

Figure 5.13

longitudinal wave vibrations are along the direction of wave travel — vibrations — direction of travel

transverse wave vibrations are at right angles to the direction of wave travel — vibrations

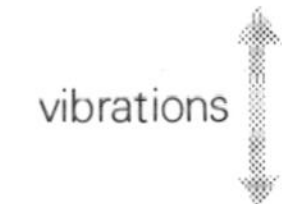

direction of travel

The experiment with the candle flame indicates that sound vibrations in air vibrate along the direction of travel of the sound; this shows that sound is a longitudinal wave.

Sound is a **longitudinal** wave and consists of a series of compressions and rarefactions.

As we have already seen, sound requires a medium. The medium is made up of a large number of particles (atoms and molecules). In a compression the particles of the medium are pushed closer together; in a rarefaction they are further apart, Figure 5.14.

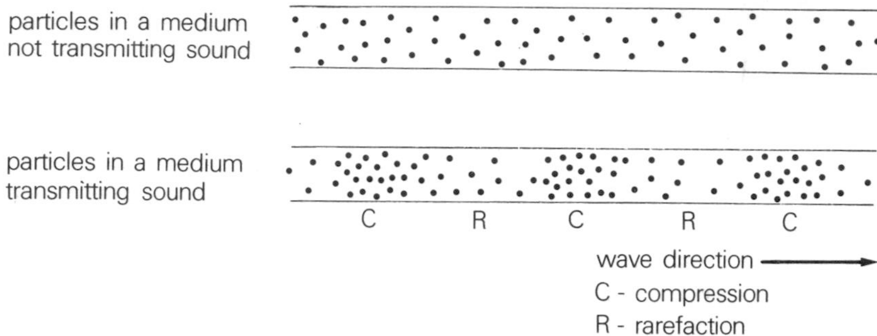

Figure 5.14

Wavelength λ

When we considered transverse waves, we defined the wavelength as 'the minimum distance in which a wave repeats itself'. This definition also applies to the wavelength of a longitudinal wave; the wavelength is the combined length of a compression and a rarefaction.

Frequency f

Like that of a transverse wave, the frequency of a longitudinal wave is the number of complete wavelengths produced in one second.

Wave speed v

The speed of sound depends on the medium through which it is travelling.

The wave equation $v = f \times \lambda$ applies to longitudinal waves.

5.5 Frequency and pitch

A loudspeaker can be made to vibrate using a signal generator, Figure 5.15. The frequency of these vibrations can be varied by adjusting the controls on the signal generator, and the frequency can be read off the scale on the signal generator.

The lowest frequency that most people can hear is about 20 Hz, and this is heard as a low rumble. Young people can usually hear sounds of frequencies up to about 20 000 Hz (20 kHz), while this upper limit tends to be lower for older people.

Figure 5.15

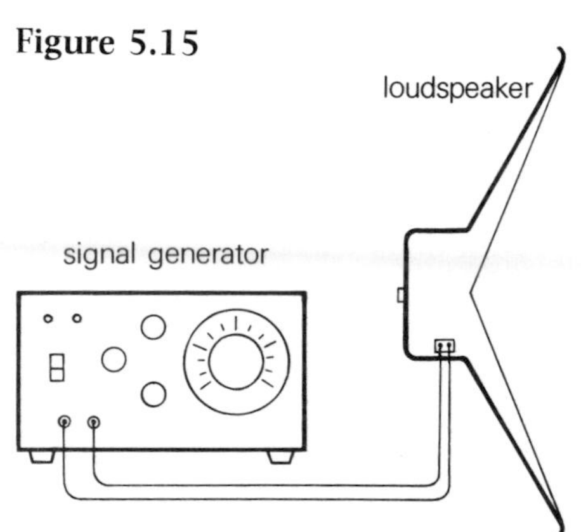

Example 6

If a person can hear sounds over a frequency range of 20 Hz to 20 kHz, what is the range of wavelengths that he can hear?
(Speed of sound = 340 m s^{-1}.)

$v = f \times \lambda$

let the wavelength of sound of frequency 20 Hz = λ_1

$340 = 20 \times \lambda_1$

multiply both sides of the equation by $\frac{1}{20}$

$$340 \times \frac{1}{20} = \lambda_1$$

$$17 = \lambda_1$$

let the wavelength of sound of frequency 20 kHz = λ_2

20 kHz = 20 000 Hz

$340 = 20\,000 \times \lambda_2$

multiply both sides of the equation by $\frac{1}{20\,000}$

$$340 \times \frac{1}{20\,000} = \lambda_2$$

$$0{\cdot}017 = \lambda_2$$

Wavelength range = 0·017 m to 17 m.

Notes of a low frequency, such as those produced by a foghorn, are said to have a low pitch, while notes of high frequencies, such as a shrill whistle or a scream, are said to have a high pitch.

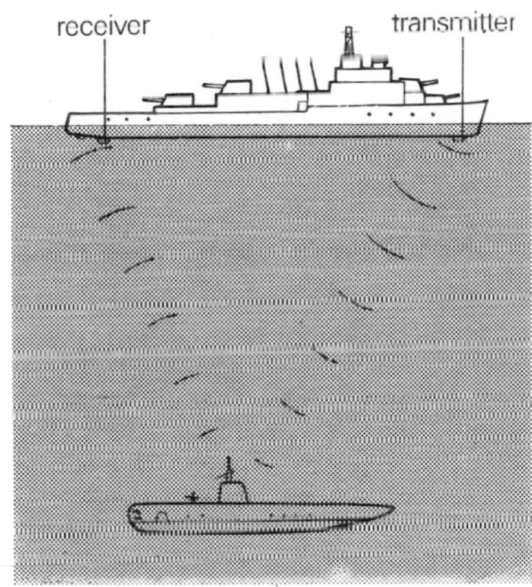

Figure 5.16 Sonar

Ultrasonics

Humans cannot hear sounds of frequencies much above 20 kHz, but other animals can. Dogs can hear frequencies of up to about 50 kHz and bats and porpoises can hear frequencies of up to about 120 kHz. Sound waves which have frequencies too high to be detected by the human ear are called **ultrasonic** waves.

There are 'silent' dog-whistles that are used to signal to dogs. They produce notes of frequencies too high to be detected by the human ear, but within the range of hearing of dogs.

A number of animals can detect objects in the dark by emitting ultrasonic waves and hearing the echoes of these notes which have been reflected off objects. Bats emit 'beeps' at frequencies ranging from 20 kHz to 120 kHz; they use the echoes to locate small insects in flight. It has been shown that bats can successfully avoid wires as small as 0·5 mm in diameter using echo-location. Porpoises use a similar method to detect the fish on which they feed.

Sonar was developed during World War II for detecting submarines. It works on the same principle as the echo-location systems of bats and porpoises, Figure 5.16. The method is now used by fishing boats to detect shoals of fish. It has been developed so that it can be used for mapping the sea bed, for locating wrecks with a view to salvaging them, and even for building a crude picture of underwater objects.

Ultrasonics have been used for some time in medicine. Ultrasonic beams may be focused on parts of the body and reflected by some of the organs and tissues. These reflections are used to form an electronic picture of the organs and to identify diseased parts. The technique is particularly useful for studying those parts of the body that cannot be shown on X-ray photographs, Figure 5.17. Ultrasonics also have many

Figure 5.17 Ultrasonic scan

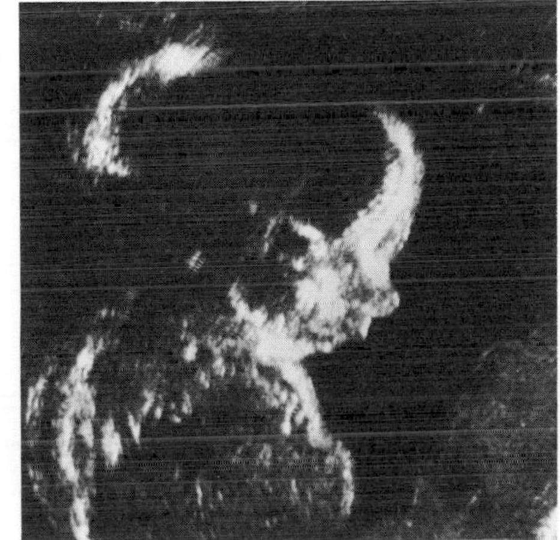

uses in industry. They are used to detect a flaw in a metal component. The ultrasonic waves are reflected off a flaw (Figure 5.18), and this will appear as an echo.

Some drills such as certain dentists' drills are made to vibrate at very high frequencies by passing ultrasonic waves through them.

Delicate instruments can be cleaned by placing them in a bath of liquid and exposing them to ultrasonic waves. The vibrations help the liquid to dissolve dirt off the surface.

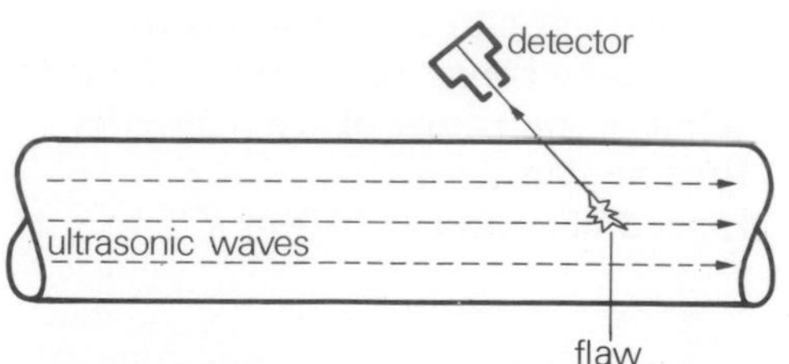

Figure 5.18

5.6 Amplitude and loudness

Sounds are produced by vibrating objects. If more energy is supplied to the vibrating object, it will vibrate through a greater distance. We say that the amplitude of the vibrations has increased. If the amplitude of the vibrations of the sound source is increased, the amplitude of the vibrations of the medium transmitting the sound is also increased. When this happens we hear an increase in volume of the sound, that is to say, the sound gets louder.

5.7 'Seeing' sounds

A sound wave is not something that we can see, but we can get a picture which tells us something about the sound wave, using a cathode ray oscilloscope (CRO). Figure 5.19 shows a pattern produced on the screen when someone whistles into a microphone connected to the CRO. The picture on the screen looks like a transverse wave, but the sound controlling it is a longitudinal wave. The trace (picture) on the screen has crests and troughs corresponding to the compressions and rarefactions in the sound wave. A steady note will produce a uniform transverse wave trace on the screen, and this trace gives information about the wave controlling it. Figure 5.20 shows how changes in the volume and frequency of the sound are shown by changes in the trace on the screen of the CRO.

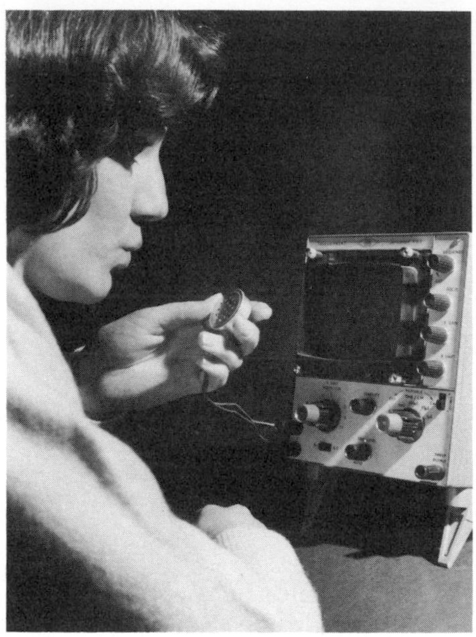

Figure 5.19

When the amplitude of the vibrations of the longitudinal sound wave increases, the volume of the sound increases, and the amplitude of the transverse wave trace on the CRO increases.

When the frequency of a sound wave increases, its pitch increases and the number of wavelengths on the trace of the CRO increases.

Figure 5.20

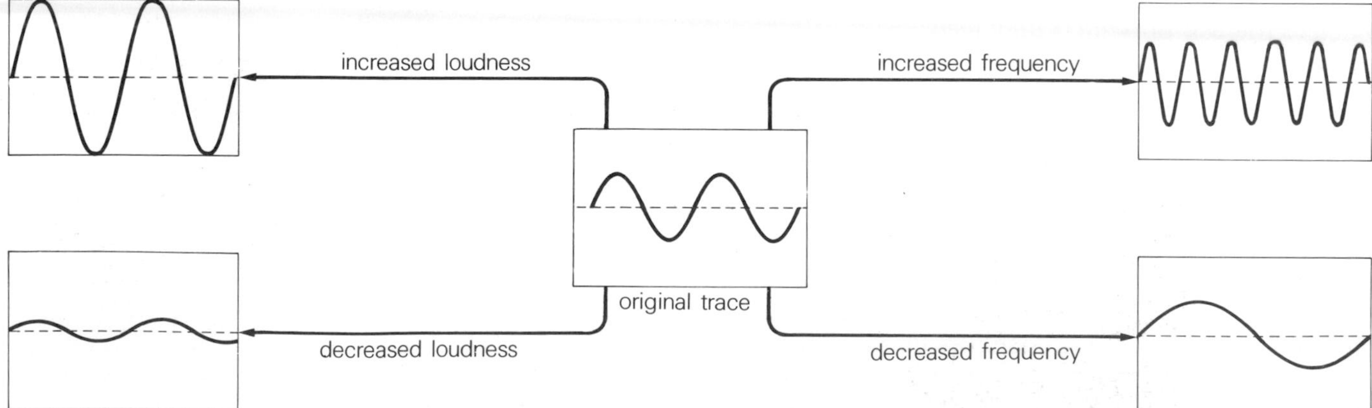

5.8 Wave properties of sound

Interference

If we can produce an interference pattern with sound, we shall have evidence of the wave nature of sound. When we produced an interference pattern with water waves, we did so by using two identical wave sources which were about one wavelength apart. We can produce two identical sources of sound by feeding the same signal into two identical loudspeakers (Figure 5.21). Before we decide how far apart we should place the loudspeakers, we should have an estimate of the wavelength of the sound that they are producing.

signal generator

soft
loud
soft
loud
soft
loud

Figure 5.21 Sound interference

Example 7

A signal of frequency 340 Hz is fed into two loudspeakers. If they are to be placed one wavelength apart, what should be the distance between the two loudspeakers?

The speed of sound in air = 340 m s^{-1}
The frequency of the sound = 340 Hz
Let the wavelength of the sound = λ
Using the wave equation $v = f \times \lambda$
$340 = 340 \times \lambda$

Multiply both sides of the equation by $\frac{1}{340}$

$$\frac{1}{340} \times 340 = \frac{1}{340} \times 340 \times \lambda$$
$$1 = \lambda$$

The wavelength of the sound = 1 m.
The spacing between the loudspeakers = 1 m.

In the experiment illustrated in Figure 5.21 the speakers should be placed approximately 1 m apart, and the frequency of the note produced by the speakers should be about 300 Hz. If you walk across the front of such an arrangement, you will hear the loudness of the sound rise and fall. This series of loud and soft sounds is the series of maxima and minima of the interference pattern.

Figure 5.22 shows two pictures in which the students were asked to stand at positions at which they heard loud notes. As you can see, they formed lines going out from the loudspeakers, and these lines indicate an interference pattern similar to that shown with water waves in a ripple tank, Figure 2.28 on page 14. In the first photograph the frequency of the sound produced by the loudspeakers was half the frequency of the sound in the second. The experiment was performed out of doors to avoid reflections of the sound.

As the photographs show, the higher frequency produces a more closely spaced interference pattern. This is because higher frequencies have smaller wavelengths.

The horseshoe bat produces an interference pattern with an ultrasonic note that it emits in its echo-location system. The bat emits a note from both of its nostrils which are half a wavelength apart, and it is the strong central maximum of the interference pattern which the bat uses for echo-location. By this method the bat is better able to locate direction than by using a note that is spreading out in all directions.

Figure 5.22

Reflection

We are already familiar with the fact that sound can be reflected. We used the reflections of sound in the 'clap-echo' experiment and we have discussed the reflections of sounds and ultrasonics used in sonar devices, and by bats and porpoises. We can find out more about the reflection of sound by using apparatus which will transmit sound in a given direction, and a detector which detects the direction from which the sound is coming.

Such a transmitter is made by attaching a cardboard tube to a small speaker connected to a signal generator, Figure 5.23.

A directional detector is made by attaching a similar tube to a microphone connected to a cathode ray oscilloscope, Figure 5.24. The amplitude of the trace on the screen of the CRO gives a measure of the strength of the sound being received by the detector.

If the transmitter is pointed at a metal plate, and the detector is moved about, the maximum strength of the sound signal is received when the directions are such that:—

the angle of incidence = the angle of reflection

Sound obeys the law of reflection:—

$\angle i = \angle r$

As with other waves, when sound is reflected there is no change in the frequency, wavelength or speed of the wave.

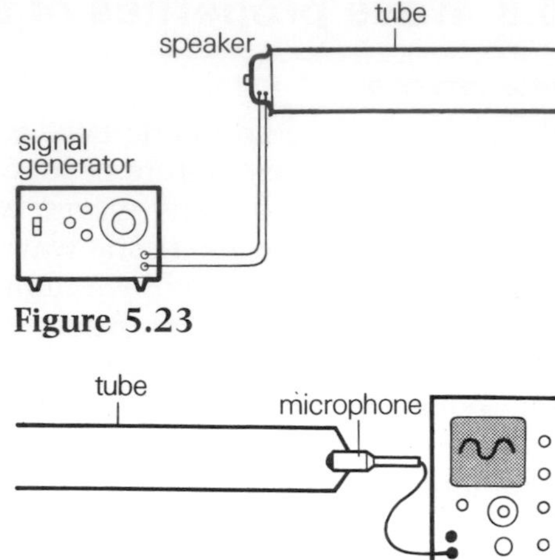

Figure 5.23

Figure 5.24

Diffraction

The fact that sound will bend round corners is one with which we are all familiar. You know that you may hide from someone by standing round a corner, but you also know that you would have to be quiet or the person might hear you. When you are in a building you cannot be sure whether sounds that reach you from an unseen source are reaching you by reflection or diffraction. Figure 5.25 shows two possible routes by which sound from a record player in one room might reach someone in the other room.

Diffraction has no effect on the frequency, wavelength or speed of the sound wave.

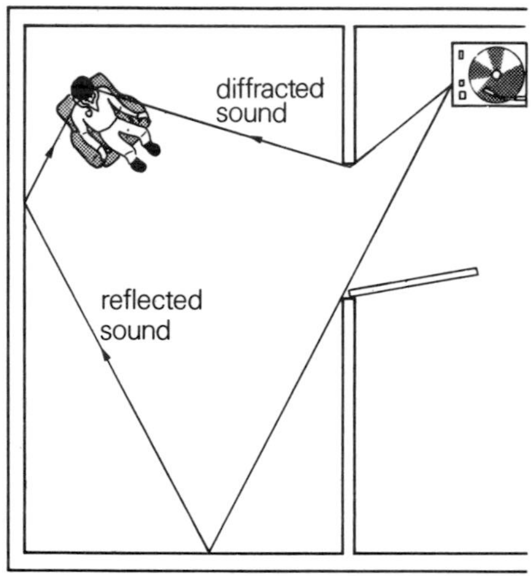

Figure 5.25

Refraction

Figure 5·26 shows an experiment in which the strength of sound reaching the microphone is increased. A balloon filled with carbon dioxide is placed between the speaker and the microphone to focus the sound. In Figure 3.13 (page 24) we saw that a convex lens focuses light by refraction. The light waves travel more slowly in glass than in air. The balloon in Figure 5.26 is also convex in shape; the fact that it focuses the sound indicates that the sound waves travel more slowly in carbon dioxide than in air.

Figure 5.26 Refraction of sound

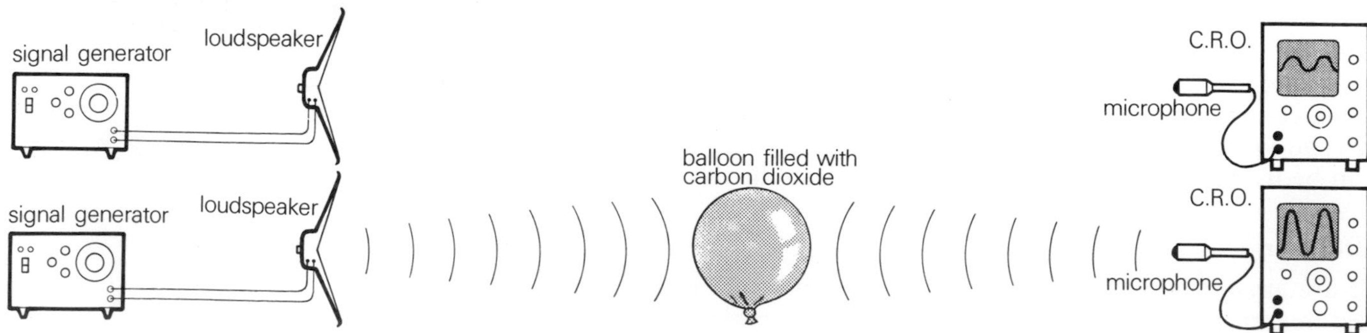

Summary

Sounds are produced by vibrating objects.

Sound will not travel through empty space, but it will travel through solids, liquids or gases.

The speed of sound depends on the medium through which it is travelling.

Sound is a longitudinal wave; i.e. one in which the vibrations are along the same line as the direction of travel of the wave. It is made up of a series of compressions and rarefactions.

An increase in the frequency of a sound wave results in an increase of pitch.

Longitudinal waves of a frequency higher than those that can be heard by humans are called ultrasonic waves.

An increase in the amplitude of vibration of sound waves produces an increase in volume (loudness).

A microphone changes sound energy into electrical energy and this can be used to control the trace on a cathode ray oscilloscope.

Sound waves exhibit the wave properties of interference, reflection, diffraction and refraction.

Problems

1 What is a longitudinal wave? Give an example.

2 A man sees the flash of a gun in the hills. Three seconds later he hears the bang. How far is he from the gun? (Take the speed of sound as 340 m s^{-1}).

3 In a clap-echo experiment, a girl stood 200 m away from a wall and clapped with a steady rhythm so that she heard the echoes in between the claps. Her partner measured ten time intervals between claps. If the time for ten time intervals was 24 seconds, what value for the speed of sound in air is given by these results?

4 When a boy whistled into a microphone connected to an oscilloscope, he observed a wave trace on the screen. What type of wave was viewed on the screen, and how did it differ from the sound wave controlling it?

5 A girl saw a flash of lightning and, being afraid of thunder, she immediately covered her ears with her hands. After ten seconds she removed her hands from her ears. If the storm was three kilometres away, would she have had her ears covered when the sound of thunder reached her? (Take the speed of sound in air as 340 m s^{-1}.)

6 The depth of the sea may be found by echo-sounding equipment on a ship. In one method, a short pulse of sound is sent down vertically from an emitter and after reflection it is detected by a receiver. The equipment measures the time interval between the emission and detection of the pulse. The sketch of the arrangement is **not** drawn to scale.

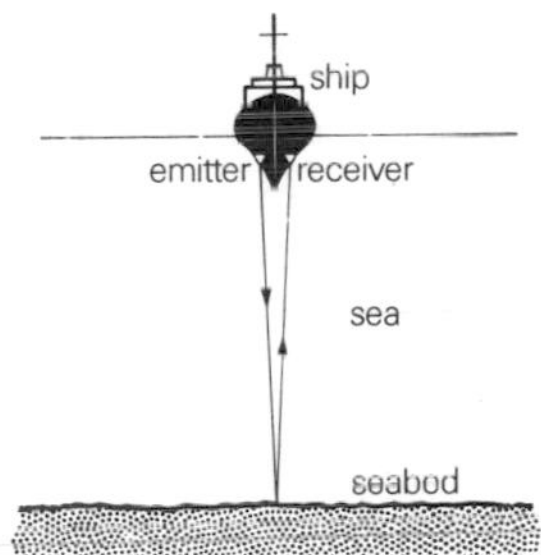

6·1 The time interval between the emission and detection of the pulse is measured when the ship is in deep water and again when the ship is in shallow water. Explain why these intervals are different.

6·2

a) If the speed of sound in water is 1 400 m s^{-1}, find the depth of the sea at a position for which the measured time interval is 0·1 s.

b) The frequency of the emitted sound is increased when the ship is at this position. How is the measured time interval affected? Explain your answer.

6·3 Suggest how the equipment could be used to detect a shoal of fish. *SCEEB*

7 In order to test whether sound travels through a vacuum, a teacher demonstrates the ringing of an electric bell in a bell jar from which air is being extracted by means of a pump. A girl sitting at the front bench sees the hammer striking the bell and hears the bell ringing very faintly.

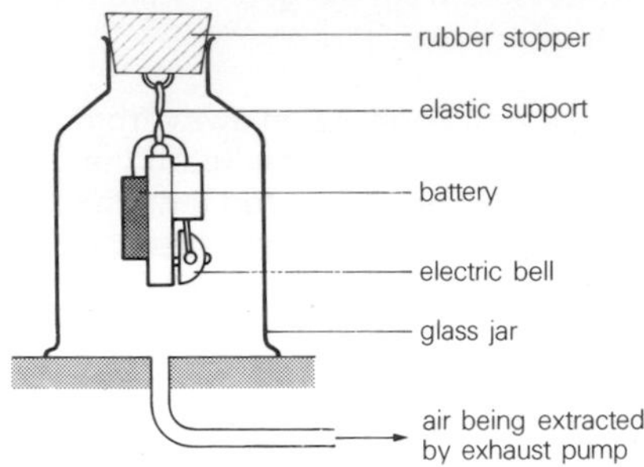

7·1 Suggest **two** possible paths for the sound to travel from the bell to the glass jar and hence through the air to the girl.
7·2 What difference would the girl hear if air were allowed slowly back into the jar?
7·3 Describe a sound wave and hence explain why sound cannot travel through a vacuum.
7·4 How does the above experiment suggest that light waves can travel through a vacuum? *SCEEB*

8 S_1 and S_2 are two identical sources of sound which have the same frequency. The point P is equidistant from the two sources of sound. A pupil set off from P and walked slowly along the line in the direction of the arrow.

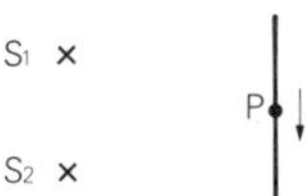

8·1 How did the intensity and frequency of the sound heard by the pupil vary on his walk (if at all)?
8·2 What is this wave effect called? Explain how it occurs. *SCEEB*

9 A new bridge is being opened by a Member of Parliament. The opening ceremony is being broadcast world-wide on radio and is also being broadcast to spectators by loudspeakers. A spectator, 1 kilometre away at the opposite end of the bridge, hears the opening announcement on his transistor radio 3 seconds before he hears the sound of the same announcement coming to him directly from the loudspeakers.

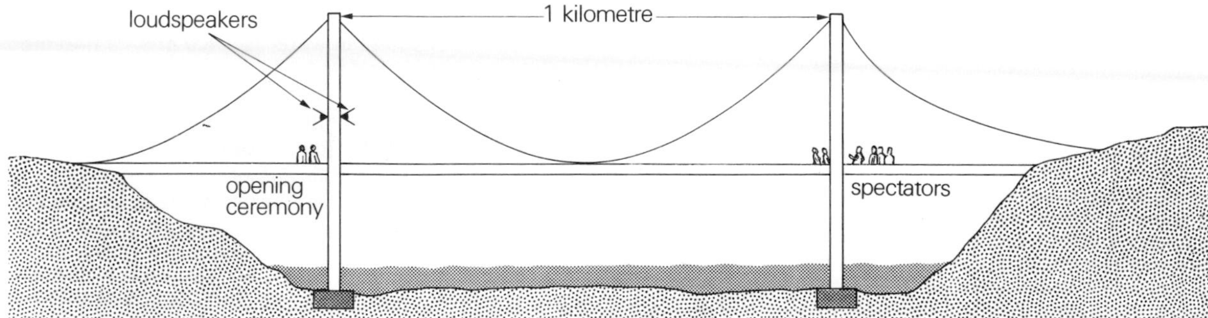

9·1 Explain why this spectator hears the announcement of the opening of the bridge on his radio before he hears the same announcement directly from the loudspeakers.
9·2 A man in Australia is also listening to the live broadcast of the opening ceremony. If the radio waves travel a distance of 18 000 kilometres to him, calculate whether he will hear the opening announcement on his radio before or after the spectator at the other end of the bridge hears the sound directly from the loudspeakers. *SCEEB*

6 Time

6.1 Introduction

What is time? The word can describe both an instant in time and a time interval. In this chapter we shall discuss time intervals. Our lives are governed by time. The regularly occurring periods of day and night form our basic time intervals. We must arrive at school on time. People are paid by the hour. Buses are scheduled to run at regular time intervals. Time is an important quantity which is vital to measurements in science.

The sundial was one of the earliest instruments used to measure time intervals. A simple sundial consists of a stick set up vertically on flat open ground, and the position of its shadow tells us the time, Figure 6.1. The shadow is long at sunrise and at sunset, and is shortest at midday.

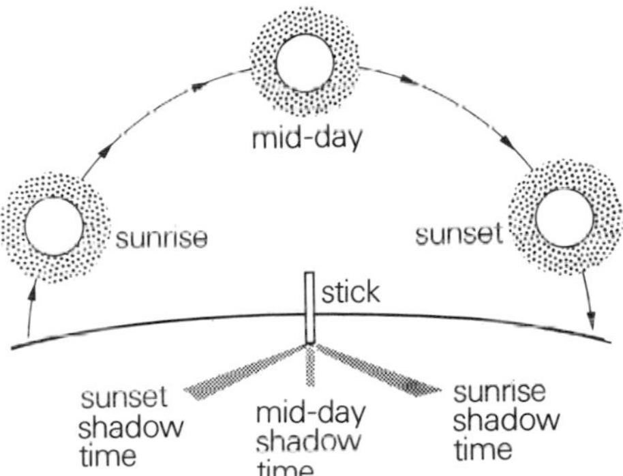

Figure 6.1 Simple sundial

The sand-glass such as that used for timing the boiling of an egg is an interval timer that has been used for centuries past. The length of time taken for the sand to pass from one bulb of the sand-glass to the other gives a constant time interval. A candle marked off in equal steps can also be used to measure time intervals. The 'water clock' is another timer that consists of a container of water with a small drainage hole in the bottom. The amount of water which has drained out indicates the time which has passed, Figure 6.2.

The next step in the development of timing devices was taken by Galileo in the sixteenth century. He wrote '. . . thousands of times I have observed vibrations, especially in churches, where lamps, suspended by long cords, had been set in motion.' He had noted that a hanging object such as a lamp, appeared to swing at its own constant frequency which did not depend on the size of the swing. It is said that Galileo used his pulse to time swinging objects in experiments when he was a medical student.

The arrangement of a small mass (or bob) swinging at the end of a long cord is called a simple pendulum, and it can be used to measure time intervals.

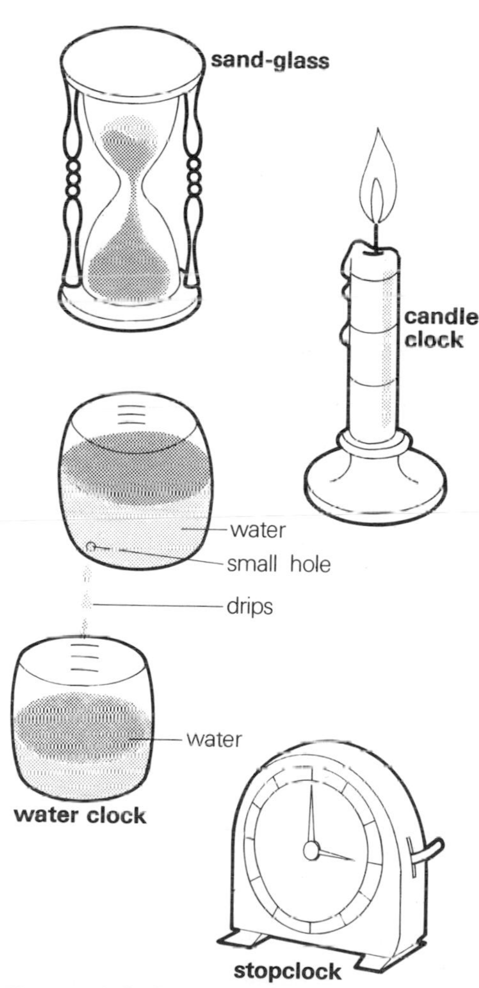

Figure 6.2 Some timers

6.2 Measuring time

Unit of time

Nature has provided us with a basic unit of time, the day, which can be further broken down into hours, minutes and then seconds. The SI unit of time is the second, which is abbreviated to s. Times, measured in seconds, have an extremely large range of values, as shown below.

Event	*Time* (s)
Age of oldest rocks	10^{17}
Human lifespan	10^{9}
One year	10^{7}
One day	10^{5}
Time between heartbeats	10^{0} (= 1)
Time for light to cross a room	10^{-8}
Time for light to cross the nucleus of an atom	10^{-23}

Table 1 A range of times in order of magnitude

The simple pendulum

A simple pendulum consists of a mass (or bob) attached to the end of a long thread. Experiments have shown that the time for one complete swing of the pendulum depends on the length of the thread and does not depend on the mass of the bob or the amplitude of the swing. The time a vibrating system takes to make one complete swing or oscillation is called the **period**, which is given the symbol T. One complete swing of a pendulum can be timed from any reference point. Two examples are given in Figure 6.3.

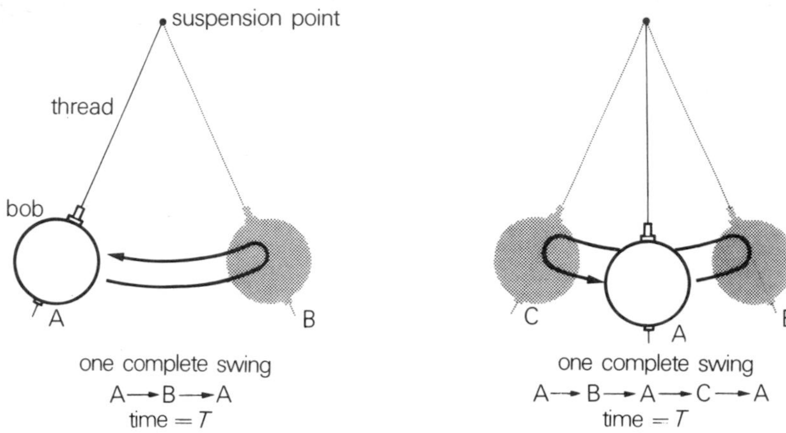

Figure 6.3 Period of a simple pendulum

Example 1

In an experiment to measure the period of a simple pendulum, the time for 20 complete swings was found to be 40 s.

a) Why are 20 swings timed?

b) What is the period of the pendulum?

c) A pupil counted 100 heartbeats during 60 swings of the pendulum. What is the period of his pulse?

a) We time a large number of pendulum swings so that a more accurate estimate of the period may be obtained. The time for one swing is too small to time accurately.

b) Time for 20 complete swings of pendulum = 40 s

$$\text{Time for 1 complete swing of pendulum} = \frac{40}{20} = 2\text{ s}$$

Period of pendulum T = 2 seconds

c) Time for 100 heartbeats = 60 complete swings

$$= 60 \times T = 60 \times 2$$

$$\text{Time for 1 heartbeat} = \frac{60 \times 2}{100} = 1{\cdot}2$$

Period of pulse = 1·2 seconds.

Stopclock

The most common type of clock which is used in the science laboratory is the stopclock or stopwatch. To measure a time interval, the clock must be started and stopped by hand. An error is therefore introduced because it takes a certain time to start or stop a clock. This error is called the **reaction time** and is on average about 0·3 s. This type of clock cannot be used to time an interval of less than about one second since the reaction time would introduce too great an error. A typical stopwatch can be read to the nearest 0·2 s and a typical stopclock to the nearest 0·5 s, Figure 6·4.

Figure 6.4 Stopwatch

Ticker-tape timer

A ticker-tape timer (or vibrator) is an electrical device in which an arm vibrates at the mains frequency of 50 Hz. Its period of vibration is therefore 0·02 s, i.e. $\frac{1}{50}$ s, Figure 6.5.

The recording of 0·02 s time intervals is achieved by drawing a piece of ticker-tape through the timer underneath a carbon-paper disc which is free to rotate. The disc is tapped at a frequency of 50 Hz by a small bolt in the end of the metal arm, Figure 6.6.

Each time the bolt strikes the carbon-paper disc it prints a dot on the tape. This tape can be used to measure short time intervals. Consider the tape in Figure 6.7.

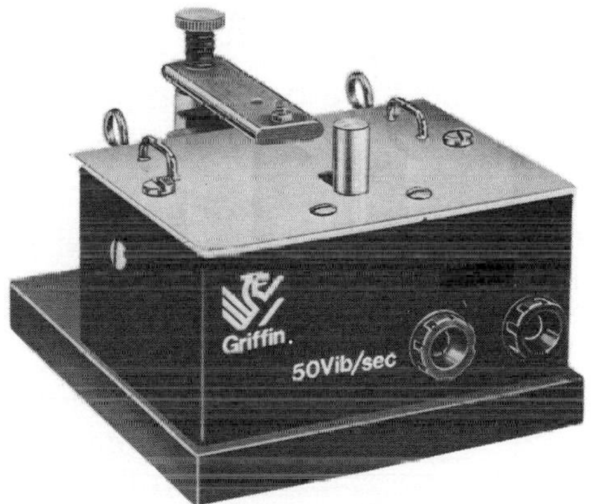

Figure 6.5 Ticker-tape timer

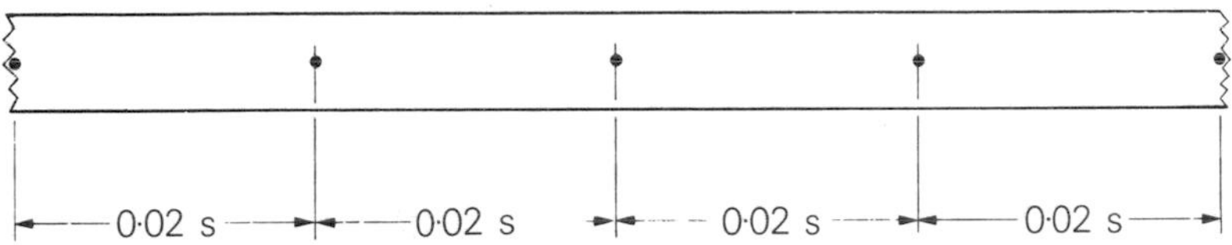

Figure 6.7 Ticker-tape timing

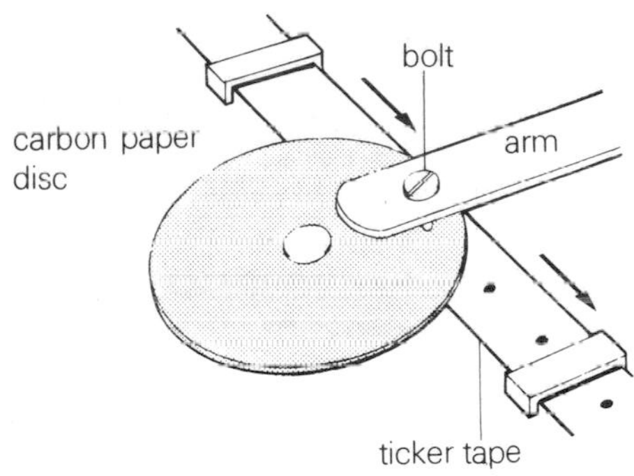

Figure 6.6 Detail of a ticker-tape timer

The timer makes a dot; 0·02 s later it makes another dot, and so on. This means that the spaces between the dots represent time intervals of 0·02 s. There are four spaces, i.e. four time intervals of 0·02 s. The time for this section of tape to pass through the timer was $4 \times 0{\cdot}02 = 0{\cdot}08$ s.

Electric stopclocks

Many timing devices suffer from the disadvantage that their accuracy is affected by the reaction time of the person using them. One method of reducing this error is to use automatic switching.

The types of electrically operated stopclocks that are commonly used are the Centisecond timer and the Venner stopclock, Figure 6.8.

These clocks will measure time intervals to the nearest 0·005 s. There is no point in measuring to this degree of accuracy if you switch on and off by hand. Figure 6.9 shows how we could accurately measure the time for a ball-bearing to roll down a grooved track. The stopclock is operated by connecting two electrical terminals.

The grooved track is lined with strips of metal foil. When the ball-bearing is in contact with both strips, it completes the circuit and the stopclock goes, Figure 6.10.

Figure 6.8 An electrical stopclock

Figure 6.9 Automatic timing

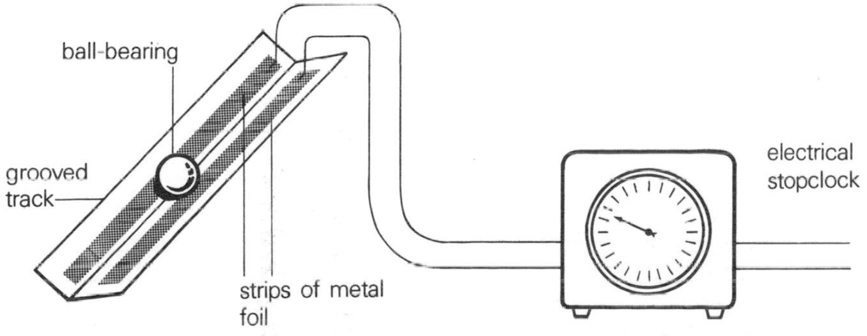

Figure 6.10 Electrical stopclock switching

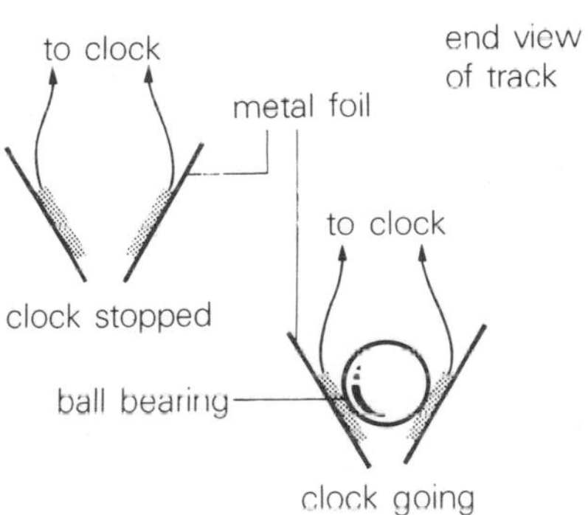

When the ball-bearing is not in contact with the metal foil the circuit is broken and the clock stops. So the clock records the time taken by the ball to roll down the foil strips.

Digital timer

A more accurate timer is the digital timer, Figure 6.11.
The digital timer can measure times to the nearest millisecond (0·001 s). The timer begins timing when a contact is closed between two terminals on the front of the instrument, and will stop when the contact is broken. If switching is carried out by hand, a reaction time is introduced and the accuracy of the measured time interval is reduced. Using a **photo-detector**, we can switch on the timer photoelectrically, Figure 6.12.
The photodetector conducts electricity when illuminated by a small light source. This can be arranged so that the timer starts when a light beam falling on the photodetector is interrupted and then stops when the light returns. In this way it is possible to time an object passing between the detector and the light source.

The timer can also be used in such a way that the breaking of one electrical connection starts the timer; on breaking another connection, it stops. You will see in Chapter 8 how the scaler can be used to measure the time for an air rifle pellet to travel a certain distance.

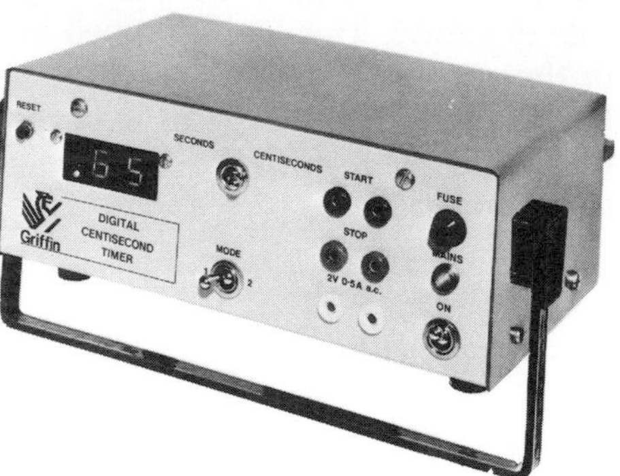

Figure 6.11 Digital timer

Figure 6.12 Photoelectric switching

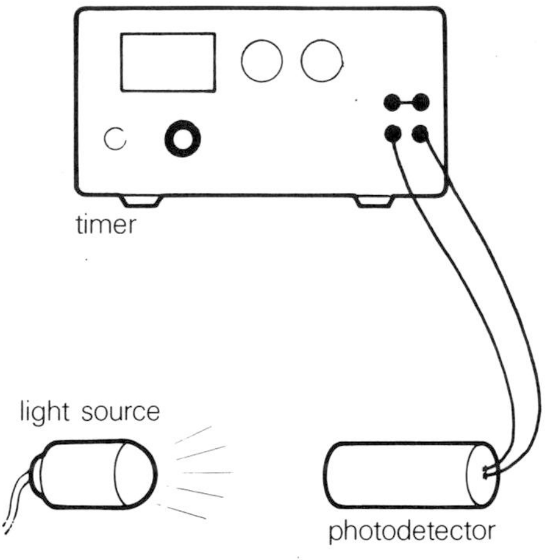

Stroboscopic timing

If we view a moving object through the slits of a rotating stroboscope (or 'strobe'), it appears to move jerkily from one position to the next. We see the object only when a slit passes in front of our eyes. If we rotate the disc by hand at a constant rate, we obtain pictures of the object in motion at equal time intervals, Figure 6.13.

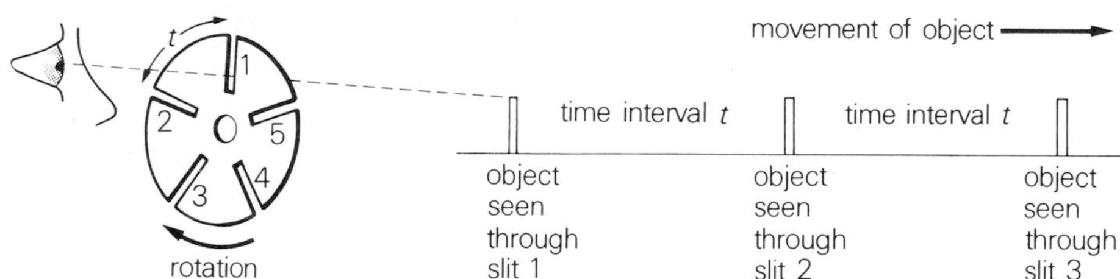

Figure 6.13 Viewing through a strobe

The difficulty in using the hand strobe as a timer is that it must itself be timed and the rate at which it is rotated cannot easily be kept constant.

We can overcome this difficulty by using a motor-driven strobe disc. The motor rotates the disc at a constant rate, usually 300 revolutions per minute, i.e. 5 revolutions per second.

When used as a timing device, the strobe disc is not 'stopping' the motion as it did with water waves. It is being used to view the position of a moving object at regular time intervals.

Example 2

A motor strobe disc rotates 5 times a second.
a) If the disc has one slit in it, what is the time between each view of an object's motion?
b) If the disc is replaced by one with 5 slits, what is the new viewing time interval?

a) Strobe disc rotates 5 times in a second.

$\Rightarrow$ Time for one revolution $= \frac{1}{5}$

$= 0{\cdot}2$ s

Number of slits in disc $= 1$

$\Rightarrow$ Number of views per rotation of disc $= 1$

$\Rightarrow$ Time between views $= 0{\cdot}2$

Time between each view = 0·2 s.

b) Strobe disc rotates 5 times in a second

$\Rightarrow$ Time for one rotation $= \frac{1}{5}$

$= 0{\cdot}2$ s

Number of slits in disc $= 5$

$\Rightarrow$ Number of views per rotation $= 5$

Time for 5 views $= 0{\cdot}2$ s

$\Rightarrow$ Time between views $= \frac{0{\cdot}2}{5} = 0{\cdot}04$

Time between each view = 0·04 s.

Another method of observing motion at regular time intervals is to use a **flashing stroboscope**. A flashing stroboscope may contain a tube like an electronic flash gun tube, which produces a flash of light at regular time intervals, Figure 6.14.
This instrument is usually used in conditions where the background is poorly illuminated. The flash rate of this stroboscope can be varied over a wide range of frequencies and so is suitable for very short as well as longer time intervals.

Both the flashing stroboscope and motor strobe can provide us with very short time intervals between viewing, but it soon becomes impossible to count the time intervals. However, if a camera is directed at the moving object through the strobe disc, then we can obtain on one exposure a record of how the motion proceeded during successive time intervals, and also the total time of motion, Figure 6.15.
The strobe photograph provides us with a record of the object's position each time the camera 'sees' through a strobe disc slit. The camera shutter must be kept open during the whole of the motion so that all these positions can be recorded on film.

Figure 6.16 shows a strobe photo of a ball rolling down a slope, obtained using the equipment.

Figure 6.14 Flashing stroboscope

Figure 6.15 Strobe photography

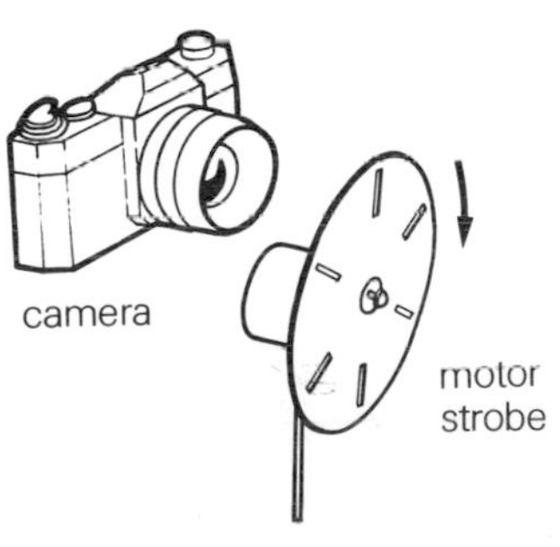

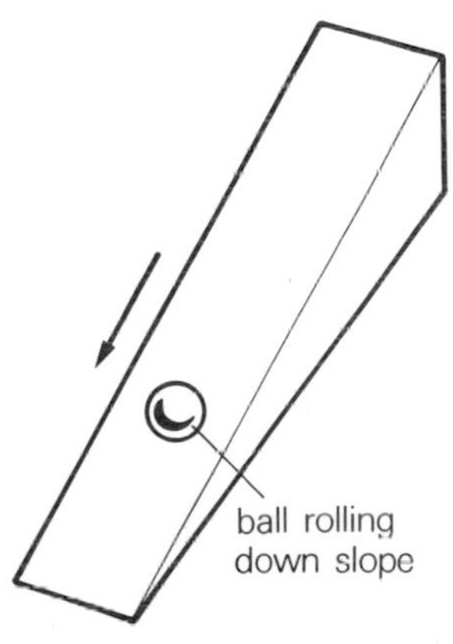

Figure 6.16 Strobe photograph

Example 3

The photograph in Figure 6.16 was taken using a camera and a motor strobe disc with 4 slits in it. If the disc rotates 5 times a second, how long did it take the object to travel from its first to its last photographed position?

Disc rotates 5 times in a second
$\Rightarrow$ Time for one revolution of disc $= \frac{1}{5} = 0{\cdot}2$ s

Number of slits on disc $= 4$
$\Rightarrow$ Time interval between slits $= \frac{0{\cdot}2}{4} = 0{\cdot}05$ s
$\Rightarrow$ Time interval between each image on photo $= 0{\cdot}05$ s

Number of spaces between images $= 5$
$\Rightarrow$ Total time between first and last image $= 5 \times 0{\cdot}05$
$= \mathbf{0{\cdot}25\ s}$

Alternatively the motion could be illuminated using a flashing stroboscope lamp and then photographed. Once again the camera shutter must be left open for the duration of the motion, Figure 6.17.

Figure 6.17 Strobe illuminated photography

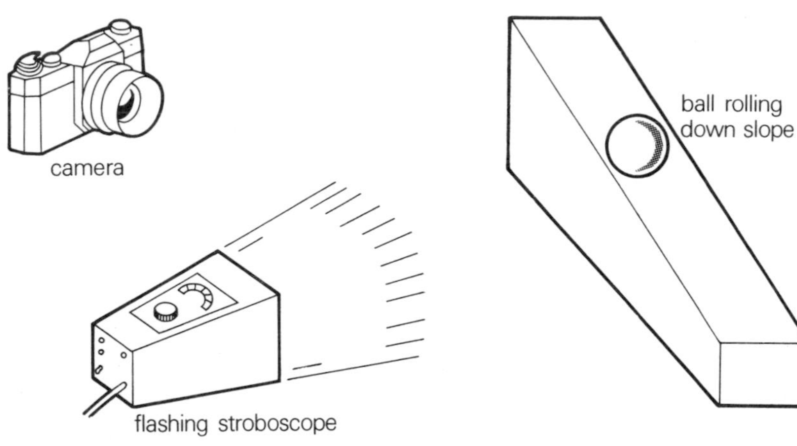

Use of strobes to 'stop' motion

If a flashing strobe lamp is used to 'stop' the motion of a vibrating or rotating object, then the maximum strobe frequency which will 'stop' the motion without double viewing is equal to its frequency.

Example 4

A white disc with a black arrow painted on it rotates clockwise. It is illuminated by a flashing strobe lamp and the arrow appears stationary. This is the maximum strobe frequency to do this without double viewing.

a) If the strobe frequency is 50 Hz, what is the rotation rate of the disc?

b) What would be seen if the strobe frequency was 25 Hz, 100 Hz, 250 Hz?

a) strobe frequency $= 50$ Hz
$\Rightarrow$ time interval between flashes $= 0{\cdot}02$ s
$\Rightarrow$ time interval between views of arrow $= 0{\cdot}02$ s
time for arrow to make one revolution $= 0{\cdot}02$ s
$\Rightarrow$ rotation rate of disc $= \frac{1}{0{\cdot}02}$
$=$ **50 revolutions per second**

b) strobe frequency $= 25$ Hz
$\Rightarrow$ time interval between flashes $= 0{\cdot}04$ s
time for arrow to make one revolution $= 0{\cdot}02$ s

$\Rightarrow$ number of times arrow rotates between flashes $= \dfrac{0{\cdot}04}{0{\cdot}02} = 2$

The arrow makes exactly 2 revolutions between flashes and is always seen in the same position, therefore appearing stationary as shown below.

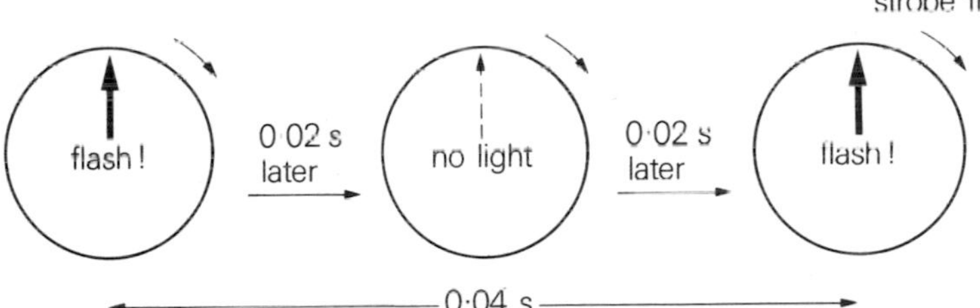

strobe frequency $= 100$ Hz
$\Rightarrow$ time interval between flashes $= 0{\cdot}01$ s
time for arrow to make one revolution $= 0{\cdot}02$ s

$\Rightarrow$ number of times arrow rotates between flashes $= \dfrac{0{\cdot}01}{0{\cdot}02} = \dfrac{1}{2}$

Therefore the arrow appears as shown below.

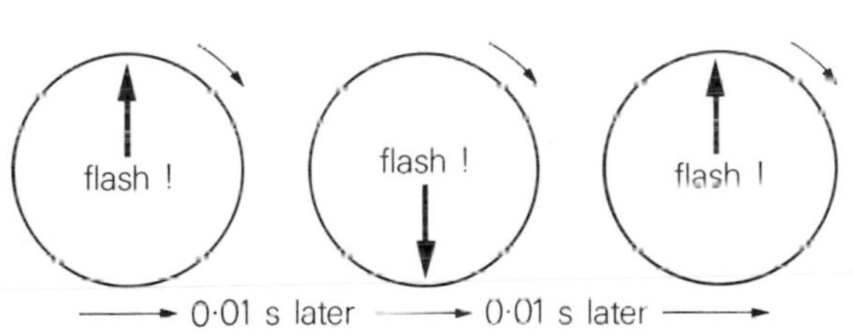

strobe frequency $= 250$ Hz
$\Rightarrow$ time interval between flashes $= 0{\cdot}004$ s
time for arrow to make one revolution $= 0{\cdot}02$ s

$\Rightarrow$ number of times arrow rotates between flashes $= \dfrac{0{\cdot}004}{0{\cdot}02} = \dfrac{1}{5}$

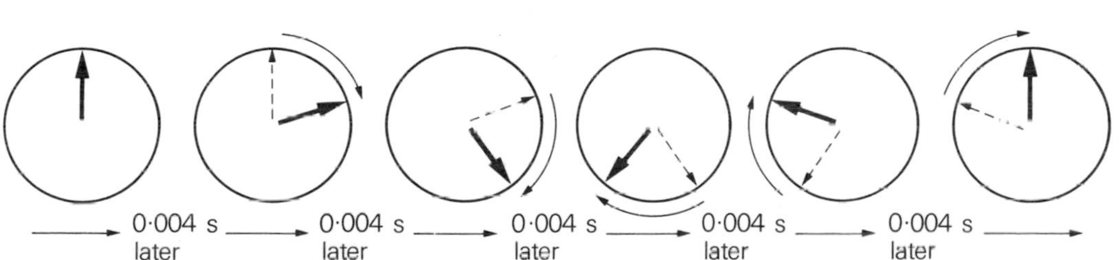

A strobe lamp can be used to 'stop' the motion of a high-speed drill to enable the operator to check for wear. A frequency check can be made on a grinding wheel, which should be rotating at a certain rate, by illuminating with a flashing strobe at the required frequency.

When the strobe frequency is not quite at the correct frequency to 'stop' the motion, interesting effects are seen – either a slow forwards or backwards rotation, as in the next example.

Example 5

A white disc with an arrow painted on it rotates clockwise. The maximum strobe frequency which 'stops' the motion is 10 Hz. How will the disc appear if

a) the frequency of the strobe is slightly less than 10 Hz?

b) the frequency of the strobe is slightly more than 10 Hz?

strobe frequency = 10 Hz
$\Rightarrow$ time for one revolution of disc = 0·1 s
frequency 10 Hz $\Rightarrow$ time interval between flashes = 0·1 s
frequency <10 Hz $\Rightarrow$ time interval between flashes = $>0{\cdot}1$ s
frequency >10 Hz $\Rightarrow$ time interval between flashes = $<0{\cdot}1$ s

strobe frequency	positions of arrow when lamp flashes			
	1st flash	2nd flash	3rd flash	
10 Hz				stationary
slightly less than 10 Hz	A	A_1	A_2	appears to rotate slowly clockwise
slightly more than 10 Hz	B	B_1	B_2	appears to rotate slowly anticlockwise

---dotted line represents rotation between flashes

a) Since the time interval between flashes is slightly greater than 0·1 s, the disc makes slightly more than one revolution and is in position A_1 as the second flash occurs, and so on, through points A_2, A_3, etc.

b) Since the time interval between flashes is slightly less than 0·1 s, the disc makes slightly less than one revolution and is in position B_1 as the second flash occurs, and so on, through points B_2, B_3, etc.

A cine camera takes a series of pictures at fixed time intervals and this can produce an effect similar to that above. Stagecoach wheels in a film can appear to rotate backwards.

Summary

The unit of time is the second (s).

The time for a vibrating system to make one oscillation is called the period *T*.

Timer	*Measurement to nearest*
stopclock	0·5 s
stopwatch	0·2 s
ticker-tape timer	0·02 s
electric stopclock	0·005 s
digital timer	0·001 s

Timing devices operated by hand have an error introduced called the reaction time.

Automatic switching of timers can be done electrically or photoelectrically.

Times may be found from strobe photos by counting the number of spaces between images and multiplying by the time interval between them.

Problems

1 Choose your answers to this question from the following.

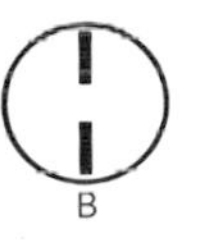

1·1 (a)
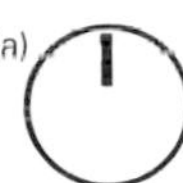
This disc is rotated 6 times a second. What do we see when we view it
a) twice every second?
b) 3 times every second?
c) 6 times every second?
d) 12 times every second?
e) 36 times every second?

1·2 (b)
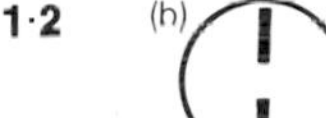
This disc is rotated 12 times a second. What do we see when we view it
a) 2 times every second?
b) 4 times every second?
c) 12 times every second?
d) 36 times every second?

2 A boy observes his rotating bicycle wheel through a strobe disc which rotates 5 times a second. If the tyre valve appears stationary at 6 positions around the wheel, at what rate was the wheel rotating? The strobe disc has only one slit.

3 The following tapes were produced by a 50 Hz ticker-tape timer. What time does each represent?

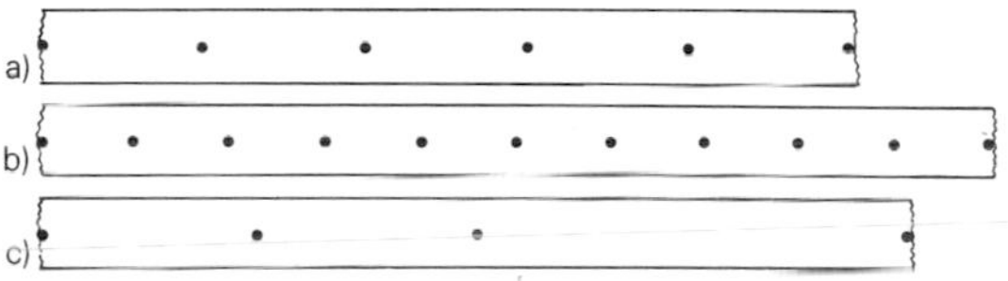

4 What is the smallest unit of time which each of the following timers will measure?
4·1 digital timer,
4·2 ticker timer,
4·3 electrical stopclock,
4·4 stopwatch.

5 A vibrator operates at 40 Hz and produces the following tape. How long did it take to do this?

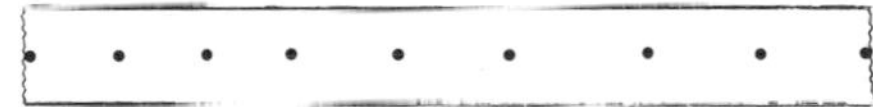

6 A pendulum took 30 seconds to make 20 complete swings. A pupil timed 120 heartbeats with the pendulum and found that they took 80 swings of the pendulum.
6·1 What is the period of the pendulum?
6·2 What is the period of his pulse?

7 The vibrating arm of a ticker timer operating at 50 Hz is 'stopped' by a flashing stroboscope at the highest possible frequency. At what frequency below this will it next appear stationary?

7 Distance and displacement

7.1 Introduction

In order to give a definition of distance or length, we must first refer to some other length for comparison. Many assorted reference lengths have been taken through the ages. These include a pace, the distance from a king's nose to the end of his thumb, an average thumbwidth, and a foot! These references, of course, were not very long lasting or accurate. More recently, the dimensions of the Earth were considered to be reasonably constant to be used as a reference. A unit called the **metre** was defined to be one ten-millionth of a quadrant of the earth, Figure 7.1.

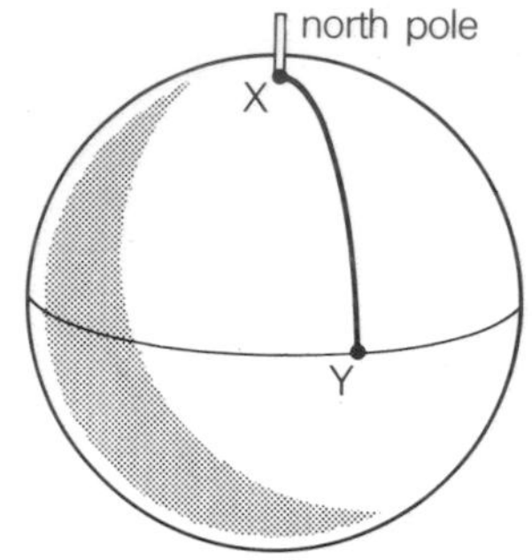

Figure 7.1 Original definition of a metre

A metal bar of this length was produced and kept safely at Sevres near Paris. Until 1960 this was the standard metre from which others were copied, Figure 7.2.

But this was not considered satisfactory for scientists, who wanted greater accuracy than that from copies of copies. The dimensions of these copies were unlikely to remain precisely constant over a period of years. For example, if the metal bar sagged in the middle or its temperature changed, its length would change. And so it was proposed that the metre should be compared with a certain number of wavelengths of light emitted by krypton-86 atoms, under carefully controlled conditions. In fact 1 650 763·73 wavelengths of this radiation correspond to one metre.

The SI unit of length is the **metre** (m). All lengths are multiples or sub-multiples of this basic unit.

1 km	$= 10^3$ m	= 1 kilometre
1 m	= 1 m	= 1 metre
1 cm	$= 10^{-2}$ m	= 1 centimetre
1 mm	$= 10^{-3}$ m	= 1 millimetre
1 μm	$= 10^{-6}$ m	= 1 micrometre
1 nm	$= 10^{-9}$ m	= 1 nanometre

Figure 7.2 Standard metre

Distance	*Order of magnitude in metres*
Distance from Sun to Earth	10^{11}
Diameter of Sun	10^{9}
Distance from Earth to Moon	10^{8}
Diameter of Earth	10^{7}
One mile	10^{3}
Thickness of one page	10^{-4}
Diameter of an atom	10^{-10}
Diameter of a nucleus	10^{-14}

Table 1 A range of distances

7.2 Measuring distances

The instruments we use to measure distance depend very much on the magnitude of the distance. We don't use a metre stick to measure the diameter of an atom or the distance to the moon!

On the other hand, it is quite possible to estimate the thickness of a page in this book with a metre stick by first measuring the thickness of many sheets and then dividing by the total number of sheets.

We can also use a metre stick to measure the height of a flight of steps if we assume that each step is of approximately the same height. The height of one step is measured and then multiplied by the number of steps.

We have already discussed in Chapters 4 and 5 how measurement of distance and time give information about speed. We can use our knowledge of speed, along with timing techniques, to find the distance indirectly. Examples of these indirect distance-measuring techniques include sonar (using sound waves) and radar (using radio waves).

When we are instructing a car driver how to reach his destination along a motorway, we need only tell him to travel so many miles along that particular motorway. We specify the **distance** and, providing the driver stays on the route, he arrives at his destination. It is a different matter, however, if we attempt to tell an air pilot or a ship's captain how to arrive somewhere. We need to specify **distance** and also **direction** in the form of a bearing.

Displacement

In physics and mathematics we describe quantities which have magnitude but no direction by the term **scalar**. Distance is a scalar. A quantity with direction as well as magnitude is a **vector**. When we are dealing with distance in a particular direction, we use a vector quantity which is called **displacement**.

Displacement is distance in a straight line in a certain direction from the starting point. Table 2 shows examples of the distinction between distance and displacement.

Table 2 Distance and displacement

Distance (scalar)	*Displacement (vector)*
10 m	10 m due east
300 km	300 km south west

Example 1

A cyclist travels once around a circular track of length 400 m.
a) What distance does he travel?
b) What is his total displacement?

a) Distance is the actual ground covered.
distance travelled = length of track = 400 m
Distance travelled = 400 metres
b) Because the cyclist finishes where he started, he has zero displacement.
Total displacement = 0 metres

The above example is a special case where the final, or **resultant** displacement is zero. Notice that the displacement does not depend on the path taken.

7.3 Representing displacement

A vector quantity has magnitude and direction, and we can draw a line to scale to represent it in magnitude and in direction. The line is one of a set which must be drawn in a specific direction, Figure 7.3.
Any one of these lines describes the vector quantity since they are all of the same length and in the same direction.

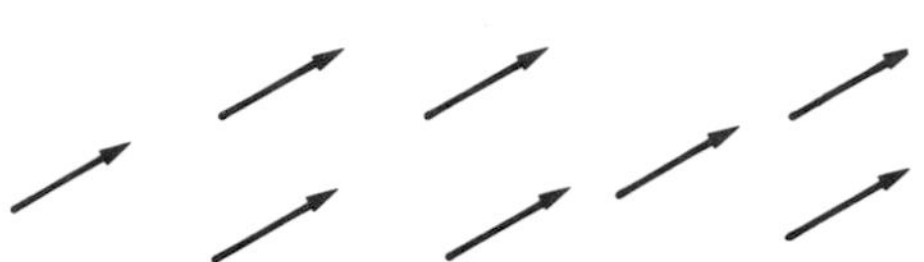

Figure 7.3 A set of vectors

The scale used for the drawing must be defined e.g. '1 cm represents 20 m'. The direction is given with reference to some given direction (often North, as shown in Example 2).
When a number of displacements are taken in sequence, the vectors are 'added' by joining them head-to-tail.

The diagram in Figure 7.4 shows the addition of two displacement vectors **a** and **b** to give a resultant displacement **c**.
This can be written $\mathbf{a} \oplus \mathbf{b} = \mathbf{c}$
where the symbol $\oplus$ means that we are adding 'vectorially', or it can be written $\vec{AB} \oplus \vec{BC} = \vec{AC}$
where $\vec{AC}$ (or **c**) is the resultant displacement.

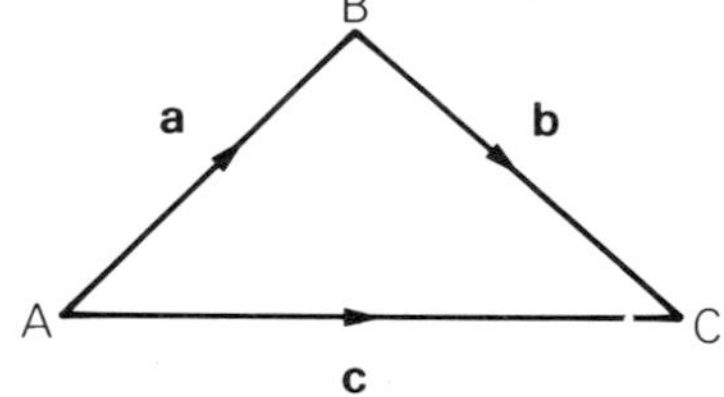

Figure 7.4

The resultant displacement is obtained by joining the starting point A to the finishing point C.

It is normally clear that we are 'adding' vectors, and from here on we shall leave out the circle round the + sign to indicate vector addition.
A distance is represented by a symbol such as s and the corresponding vector (displacement) is given by **s**.

If the resultant displacement in Figure 7.4 was to be expressed in terms of magnitude and direction, then the displacement length would be the length of the vector $\vec{AC}$; this is written $|AC|$ where $|\quad|$ means the magnitude of the vector and is called the **modulus.** The angle would be given as a 'three-figure bearing' with reference to North. North is described as 000° and all other directions are measured by the angle in a clockwise direction from North, Figure 7.5.

Figure 7.5 Three-figure bearings

direction	three-figure bearing	
N	(000°)	
E	(090°)	90°
S	(180°)	180°
W	(270°)	270°
SW	(225°)	225°

Example 2

Draw a vector diagram to represent the progress of a soldier who marches 100 m to the North, then 100 m to the West, followed by 100 m to the South.

a) What is the total distance marched?

b) What is his final displacement?

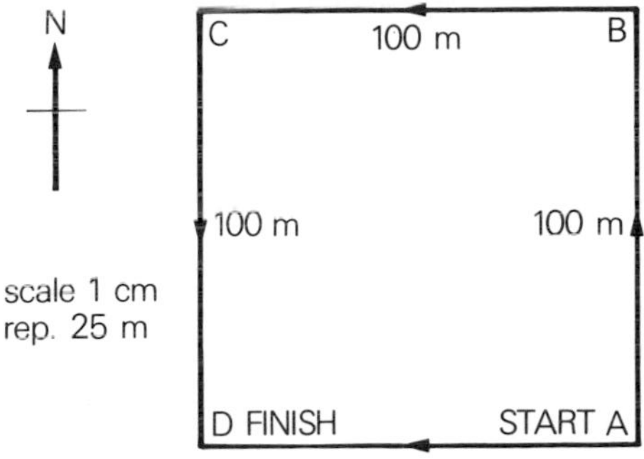

a) Total distance marched $= |AB| + |BC| + |CD|$
$= 100 + 100 + 100$
$=$ **300 m**

b) Final displacement is $\vec{AD}$ (resultant displacement)

$\vec{AD} = \vec{AB} + \vec{BC} + \vec{CD}$

Displacement $\vec{AD}$ = 100 m due West of point A or **100 m (270°)**

Notice that the distance travelled and the displacement have completely different values.

Example 3

Find the resultant displacement of a car which travels along the road shown below:

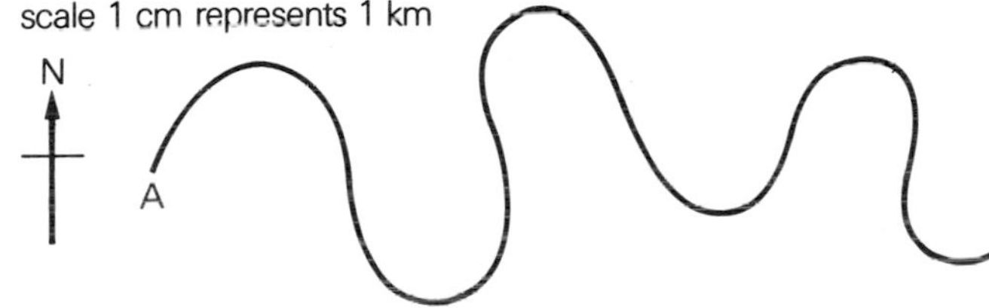

If the length is measured using a piece of thread it is found to be 16 cm which represents a distance of 16 km. The actual final displacement of the car is found by drawing a displacement vector from the start to the finish.

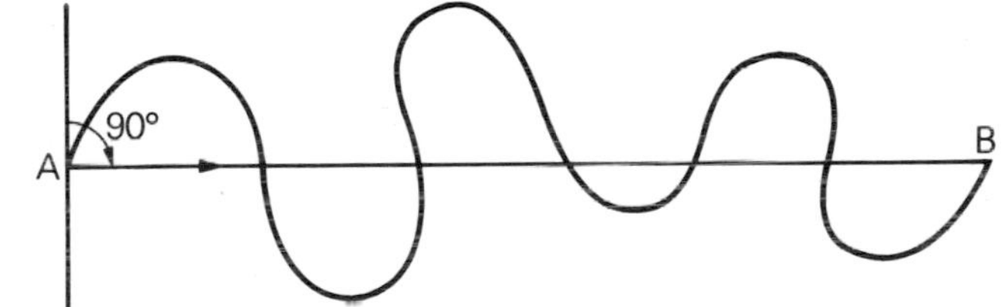

The length of AB is 8 cm, which represents a distance of 8 km. The resultant displacement $\vec{AB}$ is **8 km (090°)**.

Example 4

In an orienteering competition, a competitor's instructions are as follows: Travel 1 km (045°), 6 km (315°), 2 km (180°) and finally 2 km (270°). Find the resultant displacement from the start.

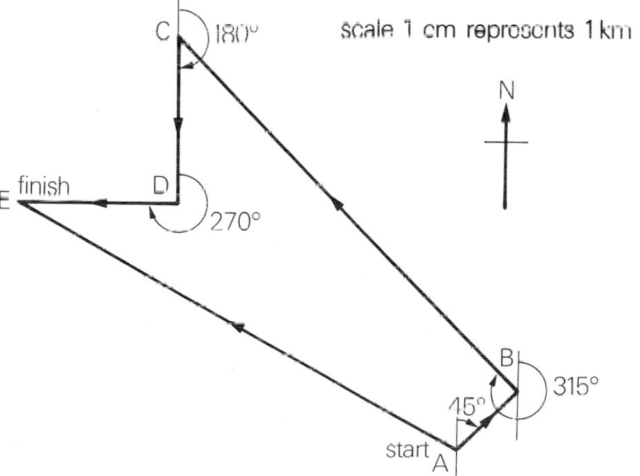

$\vec{AB} + \vec{BC} + \vec{CD} + \vec{DE} = \vec{AE}$

$|AE| = 6{\cdot}25$ cm; $\vec{AE}$ represents the resultant displacement.

By measurement, the resultant displacement of E from A is approximately **6·25 km (298°)**

Summary

Distance is a scalar quantity. The unit of distance is the metre (m).

Displacement is a vector quantity which describes distance in a specific direction.

A scalar quantity has magnitude only.

A vector quantity has magnitude and direction.

Problems

1 What is the difference between a vector and a scalar?

2 Is displacement a vector or a scalar?

3 Is distance a vector or a scalar?

4 Draw diagrams to illustrate the following displacements:
4·1 3 cm east followed by 5 cm north followed by 6 cm west.
4·2 3 cm east followed by 6 cm west followed by 5 cm north.
4·3 6 cm west followed by 5 cm north followed by 3 cm east.
What is the resultant displacement in each case?
What other conclusion can you draw from **4·1**, **4·2** and **4·3**?

5 If a man walks 500 m North and then 300 m South,
5·1 What is his total displacement?
5·2 What is the total distance he has walked?

6 A group of hikers followed this schedule:
7·5 km (045°)
followed by 6·0 km (315°)
followed by 4·0 km (270°)
followed by 2·0 km (180°)
followed by 5·0 km (060°)
6·1 What is the total distance travelled?
6·2 What is their resultant displacement from the start?
6·3 What displacement will take them 'home'?

8 Speed, velocity and acceleration

8.1 Introduction

We all have some idea of what is meant by speed. We can tell when one car is moving faster than another. We know that a car moving at high speed covers a certain distance in less time than one moving more slowly. We can say that the fast car will go further in a fixed time interval than the slower one.

Figure 8.1

However, to talk about fast and slow is not very precise. What is slow for an Olympic runner will be very fast for a snail. The best way to describe a quantity is to measure it. Before we can measure speed, we must define it. We have already said that speed is concerned with the two quantities distance and time, and we have used the following equation to measure the speed of a wave.

$$\text{speed} = \frac{\text{distance travelled}}{\text{time taken}}$$

In these cases we were dealing with movement at a steady speed, but very often movement involves changing speeds.

If you go on a car journey, you will know that the speedometer reading changes continually during the journey. To get some idea of the speed of the car over the whole journey, we can measure its **average speed.** We define average speed by the following equation.

$$\text{average speed} = \frac{\text{total distance travelled}}{\text{total time taken}}$$

or $\bar{v} = \frac{s}{t}$ where $\bar{v}$ = average speed
t = total time taken
s = total distance travelled

In fact, we have used this equation in previous chapters because the average speed is equal to the actual speed when the speed is constant.

We find the average speed by dividing the total distance travelled by the total time taken. So the unit for speed is the unit for distance divided by the unit for time.

In the SI system of units:
unit for distance = metre (m);
unit for time = second (s);
unit for speed = metre per second ($m\ s^{-1}$)

Other common units for speed are:
kilometre per hour abbreviated to km h^{-1} (or sometimes km p h);
centimetre per second, abbreviated to cm s^{-1};
miles per hour, commonly abbreviated to m p h.
Table 1 gives some examples of approximate speeds in the different units.

Movement	m s^{-1}	km h^{-1}	m p h
walking	2·2	8	5
fast car	31	113	70
sound in air	340	1 224	760
jet aircraft	600	2 160	1 340
light in vacuuo	3×10^8	$1{\cdot}08 \times 10^9$	$6{\cdot}7 \times 10^8$

Table 1 Some typical speeds

8.2 Speed measurement using a ticker-tape timer

In Chapter 6 we saw how a ticker-tape timer could be used as a device for measuring time. When the tape is pulled through the timer operating at mains frequency (50 Hz), a dot is produced every 0·02 s. The distance between two dots is, therefore, the distance through which the tape moves during that particular time interval.

Here then we have a method of measuring the distance moved by the tape in a certain time, and we can calculate the average speed of the tape during that time interval. Greater spacing between the dots indicates that the tape was moving at a higher speed. If the tape is pulled through the timer by a moving object, the speed of the tape is equal to the speed of the object.

Let us consider a section of tape that has been pulled through a 50 Hz timer:

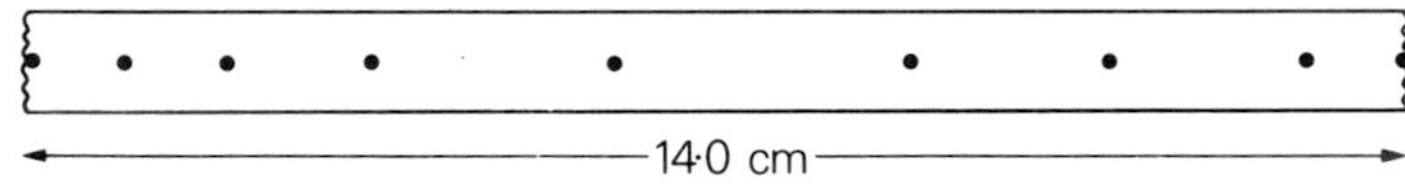

For the section of tape shown:

total distance = 14·0 cm
= 0·140 m

number of time intervals = number of spaces between the dots
= 8

each time interval = 0·02 s

total time for complete section of tape = 8 × 0·02
= 0·16 s

$$\bar{v} = \frac{s}{t}$$

$$\Rightarrow \bar{v} = \frac{0{\cdot}14}{0{\cdot}16}$$

$$\Rightarrow \bar{v} = 0{\cdot}875$$

Average speed = 0·875 m s^{-1}

By this method we have found the average speed for the whole of this section of tape. Let us now divide the tape into two sections, each representing four time intervals:

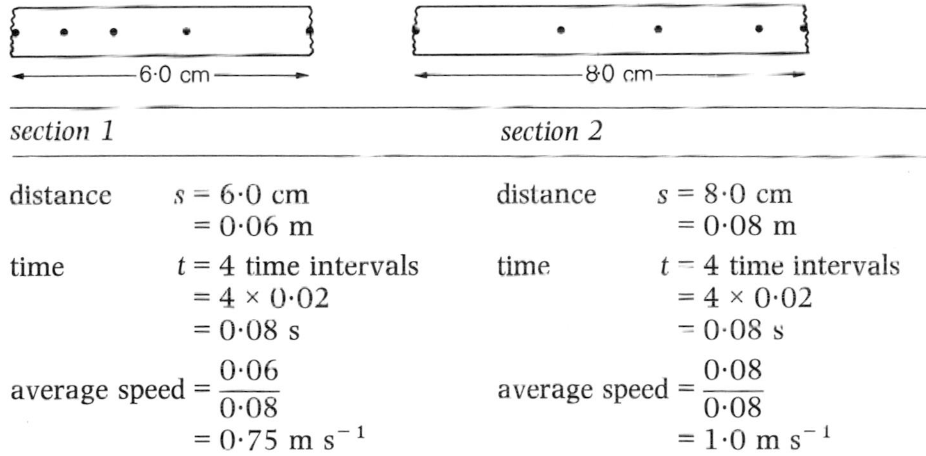

section 1		*section 2*	
distance	$s = 6{\cdot}0$ cm $= 0{\cdot}06$ m	distance	$s = 8{\cdot}0$ cm $= 0{\cdot}08$ m
time	$t = 4$ time intervals $= 4 \times 0{\cdot}02$ $= 0{\cdot}08$ s	time	$t = 4$ time intervals $= 4 \times 0{\cdot}02$ $= 0{\cdot}08$ s
average speed	$= \dfrac{0{\cdot}06}{0{\cdot}08}$ $= 0{\cdot}75$ m s^{-1}	average speed	$= \dfrac{0{\cdot}08}{0{\cdot}08}$ $= 1{\cdot}0$ m s^{-1}

Breaking down the tape into four sections, each representing two equal time intervals:

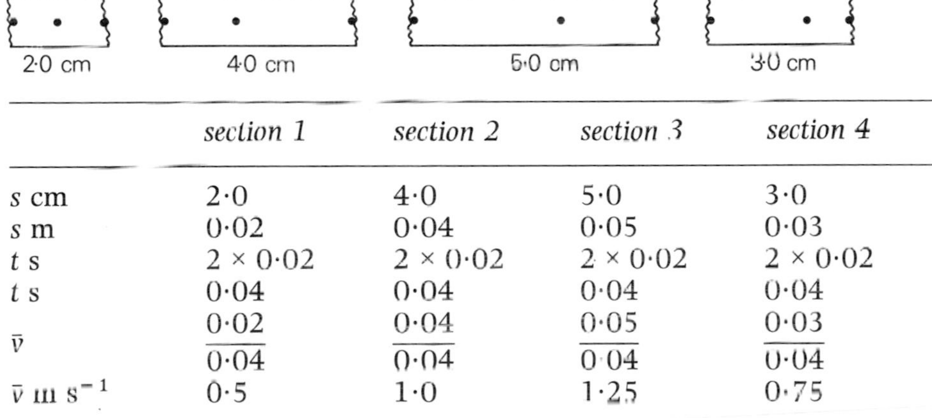

	section 1	*section 2*	*section 3*	*section 4*
s cm	2·0	4·0	5·0	3·0
s m	0·02	0·04	0·05	0·03
t s	2 × 0·02	2 × 0·02	2 × 0·02	2 × 0·02
t s	0·04	0·04	0·04	0·04
$\bar{v}$	$\dfrac{0{\cdot}02}{0{\cdot}04}$	$\dfrac{0{\cdot}04}{0{\cdot}04}$	$\dfrac{0{\cdot}05}{0{\cdot}04}$	$\dfrac{0{\cdot}03}{0{\cdot}04}$
$\bar{v}$ m s^{-1}	0·5	1·0	1·25	0·75

Breaking the tape down even further, into eight sections, each representing one time interval:

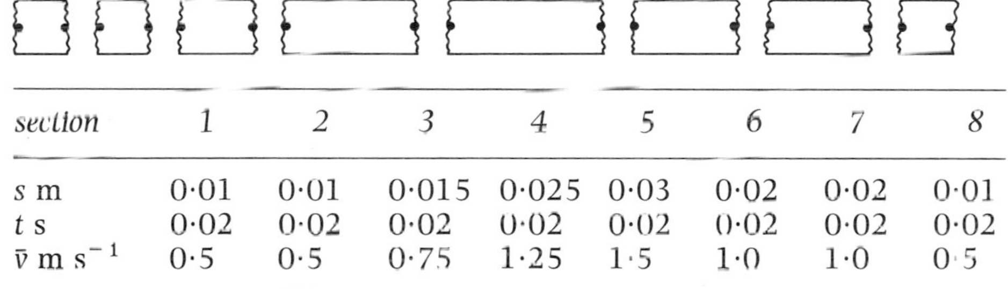

section	*1*	*2*	*3*	*4*	*5*	*6*	*7*	*8*
s m	0·01	0·01	0·015	0·025	0·03	0·02	0·02	0·01
t s	0·02	0·02	0·02	0·02	0·02	0·02	0·02	0·02
$\bar{v}$ m s^{-1}	0·5	0·5	0·75	1·25	1·5	1·0	1·0	0·5

By considering smaller and smaller time intervals, we get more and more information about how the speed varies throughout the journey. If we were able to continue considering even smaller time intervals, we would get to the stage where we would know the speed at any particular instant.

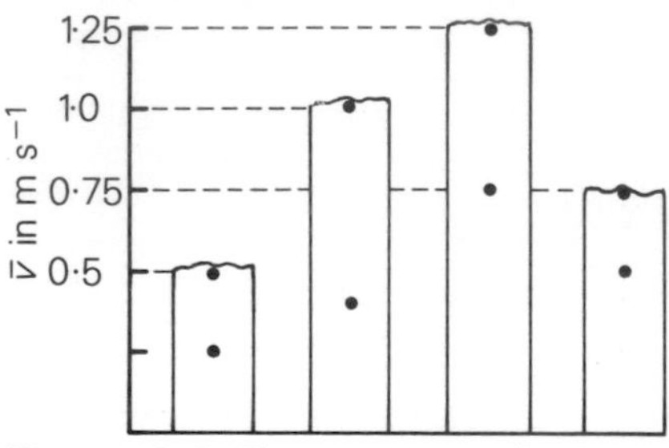

Figure 8.2 Rearranging the tapes

8.3 Speed-time graphs

Let us consider, again, the tape that we had divided into sections representing two time intervals. We have already calculated the average speeds for the separate sections of tape and we know that each section took 0·04 s to pass through the ticker-timer.

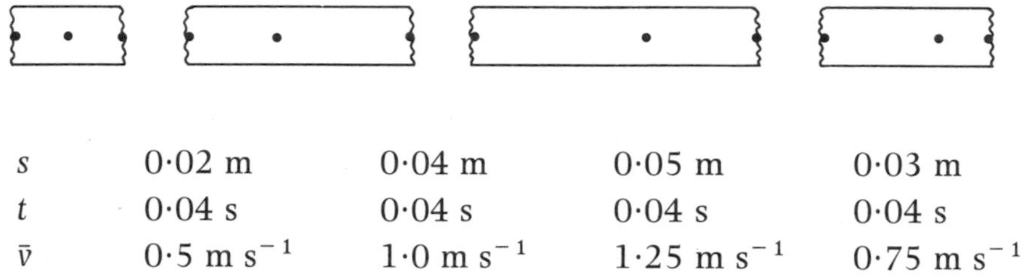

s	0·02 m	0·04 m	0·05 m	0·03 m
t	0·04 s	0·04 s	0·04 s	0·04 s
$\bar{v}$	0·5 m s^{-1}	1·0 m s^{-1}	1·25 m s^{-1}	0·75 m s^{-1}

s = length of the section of tape
t = time for the section of tape to pass through the timer
$\bar{v}$ = average speed of the section of tape as it passes through the timer.

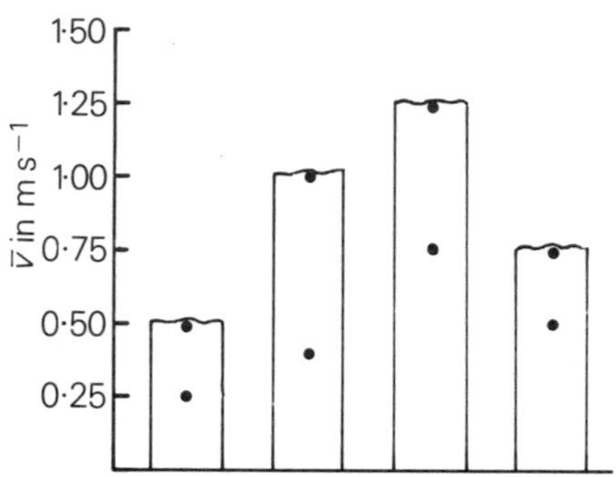

Figure 8.3 Adding a speed scale

If these sections of tape are now turned so that they are vertical (Figure 8.2), we see that the greater the height of a section, the greater is the average speed for that section.

By arranging the four tapes in this way, we have produced a vertical scale with four speeds on it. In order to produce a complete scale, we can calculate what speeds would be represented by tape sections of lengths 1 cm, 2 cm, 3 cm, etc. (Table 2).

s m	0·01	0·02	0·03	0·04	0·05	0·06
t s	0·04	0·04	0·04	0·04	0·04	0·04
$\bar{v}$ m s^{-1}	0·25	0·50	0·75	1·00	1·25	1·50

Table 2

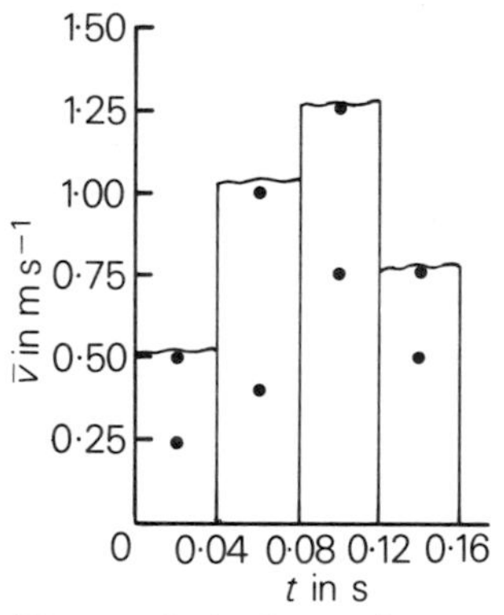

Figure 8.4 Tape chart

You will see from Table 2 that each increase of 1 cm in section length represents an increase in speed of 0·25 m s^{-1}. A scale can be placed on the vertical axis by marking off 1 cm lengths, each centimetre representing an average speed of 0·25 m s^{-1}, Figure 8.3.

By arranging the tapes in this way, we have produced a vertical scale for the average speeds of the sections of the tape.

Since we divided the tape into sections each representing two time intervals of 0·02 s (2 spaces between the dots), we know that it took 0·04 s for each section of tape to pass through the timer.

We can use this to add a horizontal time scale to the chart, with the sections of tape placed vertically next to each other, Figure 8.4.

We have produced a tape chart by considering the average speeds over sections of tape. In fact, the speed is likely to change steadily rather than in a series of steps as shown by the tape chart. A more realistic

Figure 8.5 Producing a graph

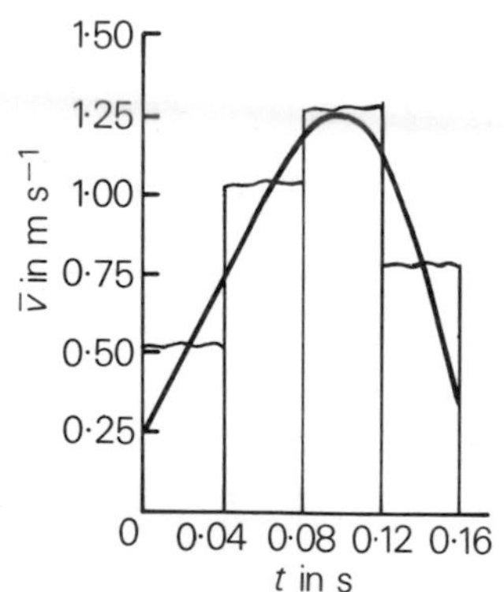

picture of how the speed changes with time is obtained if we join up the centres of the tops of the tape sections, Figure 8.5.

By joining the points together, we have produced a **speed-time graph** showing the variation of speed with time, Figure 8.6.

An upward slope on a speed-time graph indicates an increasing speed. A horizontal section represents a constant speed. A downward slope represents a decreasing speed. Also we can read off the graph the speed at any particular instant, or the times at which the speed has certain values.

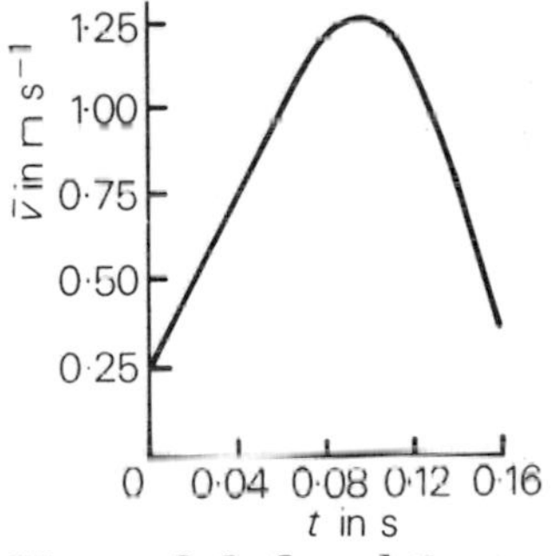

Figure 8.6 Speed-time graph

Example 1

Figure 8.7 shows the speed-time graph for part of the journey of a sledge.

a) During which time intervals of the journey was the sledge moving with:

i) increasing speed?
ii) constant speed?
iii) decreasing speed?

b) When was the speed of the sledge $2{\cdot}5\ \text{m s}^{-1}$?

c) What was the speed of the sledge after 5 s?

d) What was the speed of the sledge at the beginning of the time interval over which the measurements were taken?

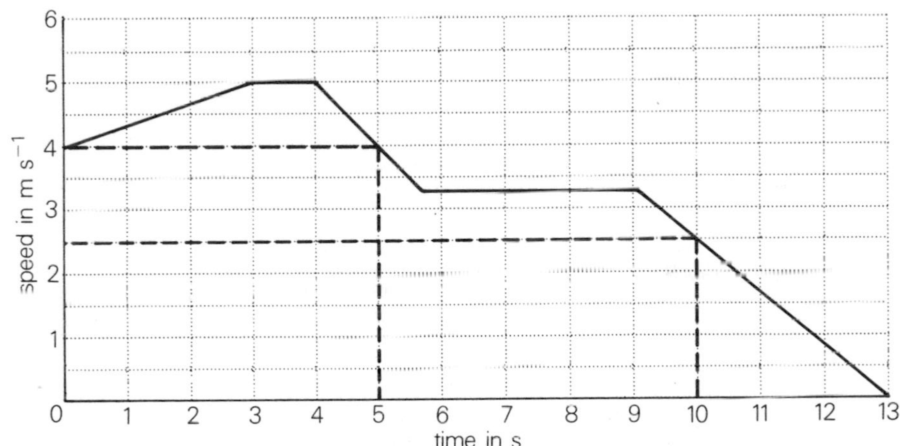

Figure 8.7

From the above graph

a) i) The speed is increasing from 0 s to 3 s.
ii) The speed is constant from 3 s to 4 s and also from 6 s to 9 s.
iii) The speed is decreasing from 4 s to 6 s and from 9 s to 13 s.

b) The speed of the sledge was $2{\cdot}5\ \text{m s}^{-1}$ after 10 s.

c) After 5 s the speed of the sledge was $4\ \text{m s}^{-1}$.

d) The speed of the sledge at the beginning of the time interval was $4\ \text{m s}^{-1}$.

8.4 Speed measurement using 'strobe' photographs

We have seen that stroboscopes can be used to allow us to view a scene as a series of pictures at fixed time intervals and that a camera used with a stroboscope can give a photograph recording the positions of an object at fixed time intervals, Figure 8.8. Since the images are produced at fixed time intervals and a faster moving object moves greater distances in equal time intervals, we can see from the photograph where the object is moving fastest. In Figure 8.8 the images of the tennis racket are furthest apart just before impact showing that this is where the racket head was moving fastest.

The stroboscope provides images at fixed time intervals. If the strobe frequency is f_s, the time that passes between the formation of successive images is given by

$$t = \frac{1}{f_s}$$

If we want to calculate the speed of an object from the information on the strobe photograph, we must have some method of finding out how far the object moved during these known time intervals.

This is easily done if we include a distance scale on the photograph (Figure 8.9).

Figure 8.8 Strobe photograph

Figure 8.9

Example 2

Figure 8.9 shows a strobe photograph taken with a stroboscope flashing ten times per second.

a) Estimate the average speed
- i) over the whole of the journey;
- ii) between the formation of the first two images;
- iii) between the formation of the last two images.

b) What is the greatest source of error in these estimations?

a) Strobe frequency = 10 Hz

time between formation of successive images $= \frac{1}{10}$ s
$= 0{\cdot}1$ s

i) For the whole journey:

Number of spaces between the images = 5
Number of time intervals = 5
time $= 5 \times 0{\cdot}1$
$= 0{\cdot}5$ s
Distance = 75 cm
$= 0{\cdot}75$ m

$$\bar{v} = \frac{s}{t}$$
$$= \frac{0{\cdot}75}{0{\cdot}5}$$
$$= 1{\cdot}5 \text{ m s}^{-1}$$

ii) Between the first two images:

$s = 0{\cdot}05$ m

$t = 0{\cdot}1$ s

$$\bar{v} = \frac{s}{t}$$

$$\bar{v} = \frac{0{\cdot}05}{0{\cdot}1} = 0{\cdot}5 \text{ m s}^{-1}$$

iii) Between the last two images:

$s = 0{\cdot}25$ m

$t = 0{\cdot}1$ s

$$\bar{v} = \frac{s}{t}$$

$$v = 2{\cdot}5 \text{ m s}^{-1}$$

b) The greatest source of error occurs in reading off the distances from the photograph.

8.5 Measurement of speed over a small time interval

If we are to measure a small time interval accurately, we need an accurate timer and a method of switching it on and off automatically, so that the reaction time of the operator is eliminated. Figure 8.10 shows an arrangement which automatically times how long it takes a card to pass between a light source and a photodetector. The photodetector is connected to the electric timer.

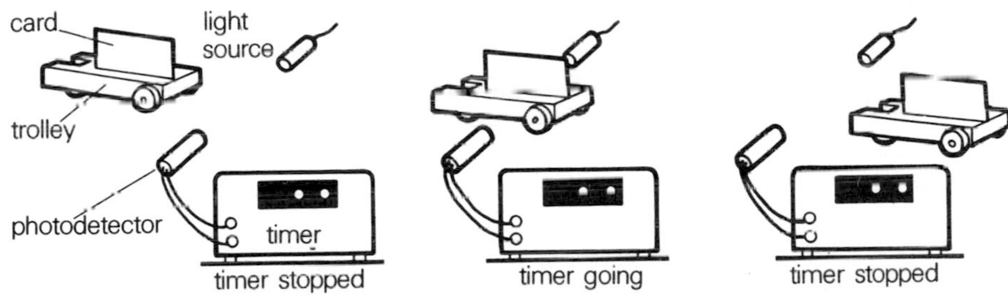

Figure 8.10

The timer only operates when the light beam is interrupted and the photodetector is not illuminated. When the light again shines on the photodetector, the timer stops.

With the arrangement shown in Figure 8.10, the time is measured only while the card interrupts the light beam falling on the photo detector. The reading obtained on the timer is that of the time taken for the card to pass between the light source and the photodetector.

If the length of the card is measured, the average speed of the trolley as it passes between the photodetector and the light source can be calculated from the equation

$$\bar{v} = \frac{s}{t} = \frac{\text{length of the card}}{\text{reading on the timer}}$$

Measurement of the speed of an air rifle pellet

To measure the speed of something moving as fast as an air rifle pellet, we shall either have to time it over a large distance or to measure a short time interval. It is not very useful to measure the average speed over a large distance because the speed will change considerably as the pellet moves through the air.

To measure the time that it takes to travel a relatively short distance (e.g. 1 m), we again require an accurate timer with a method of switching it on and off automatically. Figure 8.11 shows such an arrangement.

Figure 8.11 Speed of an air rifle pellet

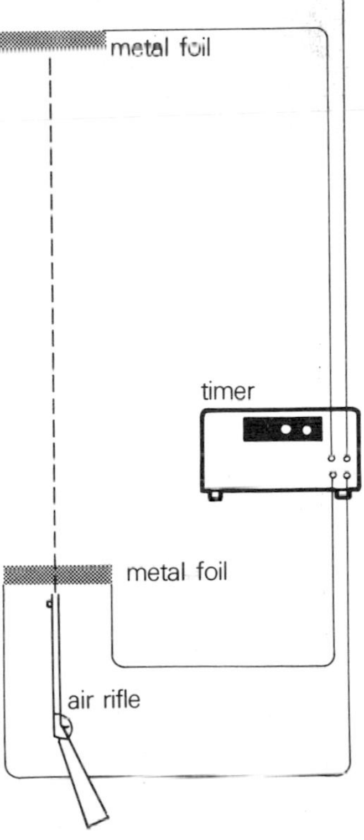

The distance between the two strips of foil is 1 m and they are lined up so that the pellet will break both strips when the air rifle is fired. The metal foil is thin and easily broken so that it will not greatly affect the speed of the pellet.

When the first strip is broken, one set of terminals on the timer is disconnected and the timer starts. When the second strip is broken, another set of terminals is disconnected and the timer stops.

So the reading on the timer gives the time taken for the pellet to travel the distance of 1 m between the two strips of metal foil, and the average speed for the pellet over this time can be calculated as follows.

$$\bar{v} = \frac{s}{t}$$

$s = 1$ m

t = timer reading

$$\bar{v} = \frac{1}{\text{timer reading}} \text{ m s}^{-1}$$

8.6 Area under a speed-time graph

Let us consider a strip of tape used to make up a ticker-tape chart, Figure 8.12.

We calculated our scales so that the vertical scale was a measure of the average speed and the horizontal scale was a measure of the time.

$$\text{speed} = \frac{\text{distance travelled}}{\text{time taken}}$$

Multiplying both sides of the equation by 'time' gives

speed × time = distance

If we multiply the vertical reading (speed) by the horizontal reading (time), we find the distance travelled in that time interval. Because the height multiplied by the width of the strip is equal to its area, we say that the area of the strip is equal to the distance travelled.

Caution: the true area is given by the height in units of distance multiplied by the width in units of distance. When we say that the area is equal to the distance travelled, we are referring to the height on the *speed scale* multiplied by the width on the *time scale*.

As we have seen, a speed-time graph can be made up from a number of strips and we can say that:

total distance travelled = area under a speed-time graph.

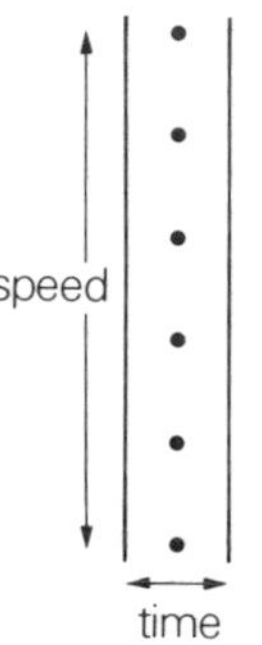

Figure 8.12

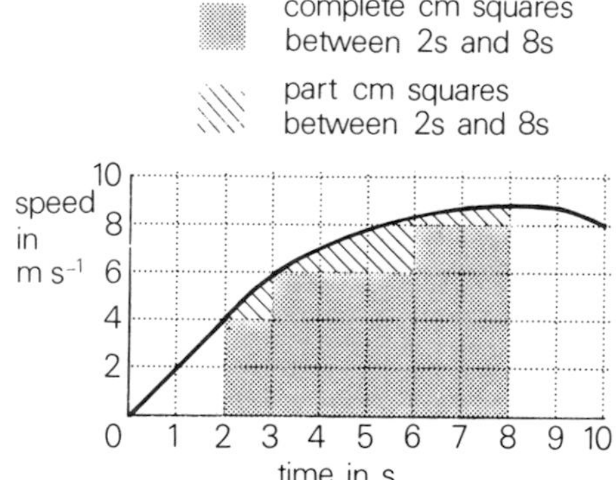

Figure 8.13 Speed-time graph of cyclist

Example 3

Figure 8·13 shows a speed-time graph for a cyclist. Calculate the distance he travelled between 2 and 8 seconds.

On the vertical scale, 1 cm represents 2 m s^{-1}.
On the horizontal scale, 1 cm represents 1 s.
1 cm^2 of graph paper represents 2 m distance travelled.
From the graph:
Area under the graph between 2 s and 8 s is made up of 19 complete cm squares + part cm squares equivalent to 3 cm squares.
Area under the graph between 2 s and 8 s = 22 cm^2
Since 1 cm^2 represents 2 m distance travelled,
distance travelled between 2 s and 8 s = 22 × 2
= 44

Distance travelled between 2 s and 8 s = 44 m.

8.7 Velocity

In the last chapter we saw that distance is a scalar quantity and that the corresponding vector quantity is displacement.

Speed is a scalar quantity. The corresponding vector quantity is called velocity.
Velocity is speed in a given direction.
The average velocity of a body is given by the equation

$$\text{average velocity (m s}^{-1}\text{)} = \frac{\text{total displacement (m)}}{\text{total time taken (s)}}$$

This can be written as $\bar{v} = \dfrac{s}{t}$

where $\bar{v}$ = average velocity
s = displacement
t = time

When the velocity is constant, the average velocity is equal to the actual velocity at any instant.

The direction of the velocity is the same as the direction of the displacement.

Example 4

An orienteer ran 6 km North and then 8 km East on level ground. The run took him 2 hours

a) What was his average speed?
b) What was his average velocity?

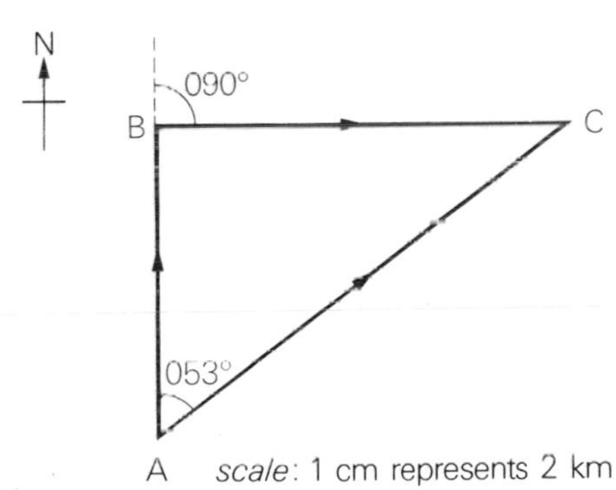

Using three figure bearings, North is (000°) and East is (090°). Draw a scale diagram to represent the run of the orienteer.

a) distance travelled s = AB + BC
= 6 + 8
= 14 km

$$\text{average speed } \bar{v} = \frac{\text{distance travelled}}{\text{time taken}} = \frac{14}{2} = 7{\cdot}0 \text{ km h}^{-1}$$

b) displacement = $\vec{AB} + \vec{BC} = \vec{AC}$
On vector diagram, length of $\vec{AC}$ = 5 cm
1 cm represents 2 km
⇒ AC represents 10 km
⇒ displacement = 10 km (053°)

$$\text{average velocity } \bar{v} = \frac{\text{total displacement}}{\text{time taken}} = \frac{10}{2}\,(053°) = 5{\cdot}0 \text{ km h}^{-1}\ (053°)$$

(a) Average speed = 7·0 km h^{-1}
(b) Average velocity = 5·0 km h^{-1} (053°).

Velocity is a vector and does have direction. However, velocities are often quoted without direction. A common example of this is in calculations involving ticker-tapes. This is because ticker-tapes are used to make measurements of movement along a straight line directly away from the ticker-timer and the direction is often assumed without being stated. To be strictly accurate, whenever a velocity is stated, it should be given as a speed in a certain direction. Often however, the speed is given and the direction is assumed.

Just as we can add two displacement vectors by head-to-tail addition, so we can add two velocity vectors by head-to-tail addition.

Example 5

A man can swim with a speed of 3 m s^{-1} through the water. If he swims upstream in a river flowing at 1 m s^{-1}, what is his velocity relative to the bank?

The velocity of the swimmer relative to the bank is equal to the sum of the velocity of the swimmer through the water and the velocity of the water.

On the vector diagram:
velocity of the swimmer is represented by $\vec{AB} + \vec{CD}$
$\vec{AB} + \vec{CD} = \vec{AD}$

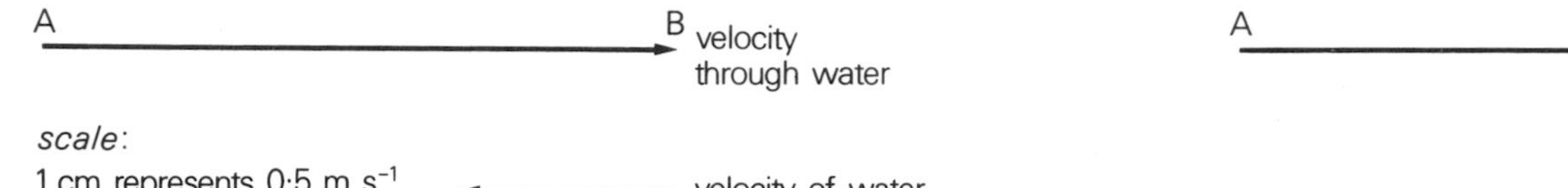

A B D C

Length of AD = 4 cm
$\Rightarrow$ AD represents $4 \times 0{\cdot}5$ m s^{-1}
$\Rightarrow$ AD represents 2 m s^{-1}
Velocity of swimmer relative to the bank = 2 m s^{-1} upstream.

Example 6

A man rows a boat across a river. The river is flowing East with a speed of 12 m s^{-1}. The man points the boat directly at the South bank and rows with a speed of 5 m s^{-1} through the water. What is his velocity relative to the bank?

scale: 1 cm represents 2 m s^{-1}

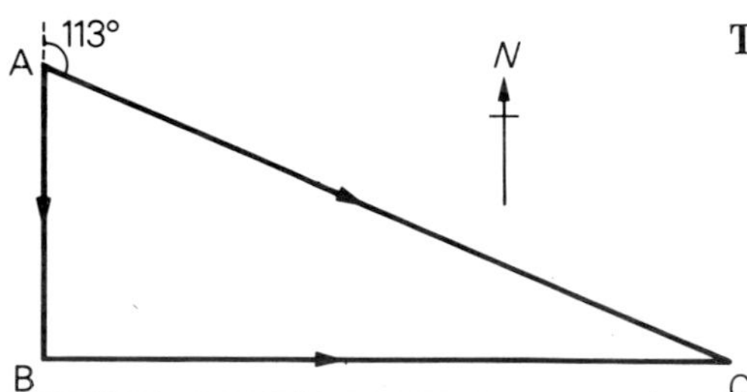

The man's velocity is the vector sum of the velocities 5 m s^{-1} South and 12 m s^{-1} East. Vector addition of these two velocities will give the man's velocity relative to the bank.

$\vec{AB}$ represents velocity 5 m s^{-1} (180°)
$\vec{BC}$ represents velocity 12 m s^{-1} (090°)
$\vec{AC} = \vec{AB} + \vec{BC}$
Length of AC = 6·5 cm
AC represents a velocity of 13 m s^{-1} (113°)
The man's velocity relative to the bank = 13 m s^{-1} (113°).

8.8 Acceleration

When we say that an object is accelerating, we mean that its speed in a given direction is increasing. Acceleration is the rate of change of speed, that is, the change of speed per second. If the speed of the object is decreasing, we say that it is decelerating.

Speed-time graphs present us with a picture of how the speed of an object is changing with time. Figure 8.14 shows three special cases of speed-time graphs, and the tables below the graphs show the readings of the speed of the object at intervals of one second.

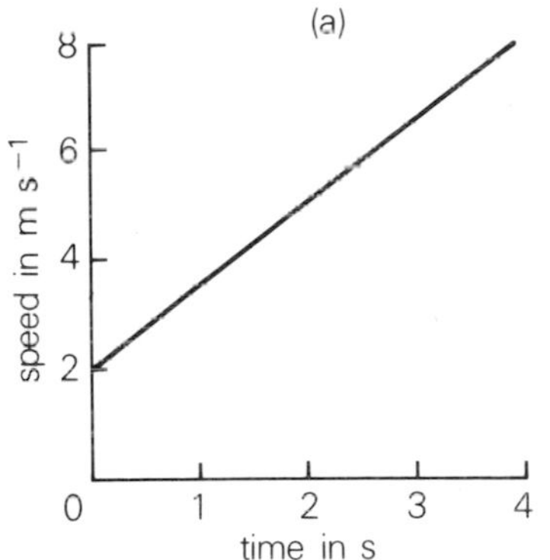

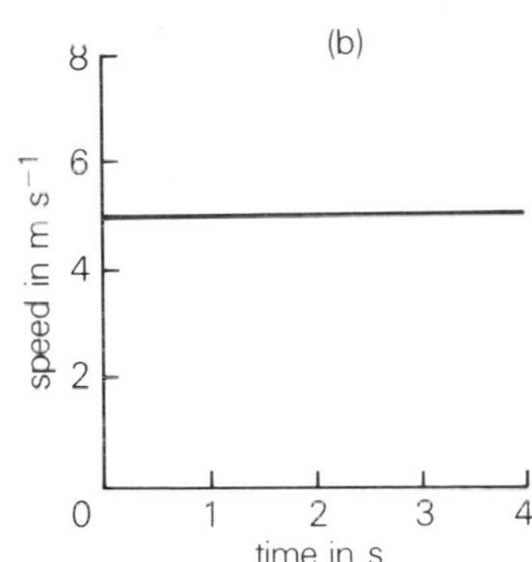

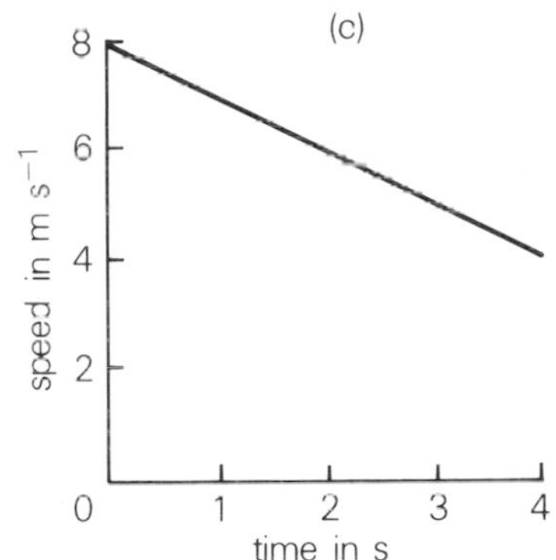

Figure 8.14 Speed-time graphs

time s	*speed* m s^{-1}	*time* s	*speed* m s^{-1}	*time* s	*speed* m s^{-1}
0	2·0	0	5·0	0	8·0
1	3·5	1	5·0	1	7·0
2	5·0	2	5·0	2	6·0
3	6·5	3	5·0	3	5·0
4	8·0	4	5·0	4	4·0

Graph (a). The readings show that the speed increases by 1·5 m s^{-1} every second. Because the change in speed for each time interval is the same, we say that the acceleration is constant (uniform). The speed is increasing and so the acceleration is positive.

Graph (b). The readings show that the speed is constant over the whole time interval. Because the change in speed is zero, we say that the acceleration is zero.

Graph (c). The readings show that the speed decreases by 1·0 m s^{-1} every second. Because the change in speed for each time interval is the same, we say that the acceleration is constant (uniform). The speed is decreasing and so the acceleration is negative and the object is decelerating.

Acceleration and vectors

The rate of change of speed is a measure of acceleration, but acceleration is a vector quantity and is therefore defined as the rate of change of velocity.

Acceleration is the rate of change of velocity

This means that, even though an object continues to travel at the same speed, it will be accelerating if its direction of travel changes. Because both velocity and acceleration are vector quantities, they have magnitude and direction. Where the direction is not stated, we may assume that the object is travelling in a straight line: in such cases, vector notation is not used in this book for simplicity. A negative velocity indicates that the object is travelling in the opposite direction along this line.

Uniform acceleration

Acceleration is the change in velocity per second. However, it is not often convenient to measure changes in velocity over time intervals of one second, and we shall use the more general equation

$$\text{acceleration} = \frac{\text{change in velocity}}{\text{time for the change}}$$

If the acceleration is **uniform**, we can write this equation as

$$a = \frac{v - u}{t} \quad \text{(uniform acceleration only)}$$

where a = acceleration, u = initial velocity, v = final velocity, t = time for velocity change from u to v

We see from the equation above that the units of acceleration are

$$\frac{\text{units for velocity}}{\text{units for time}} = \frac{\text{m s}^{-1}}{\text{s}} = \text{m s}^{-2}$$

Example 7

Figure 8.15 shows a speed-time graph for an object moving in a straight line.

Describe and calculate values for the acceleration for the parts of the journey marked (a), (b) and (c).

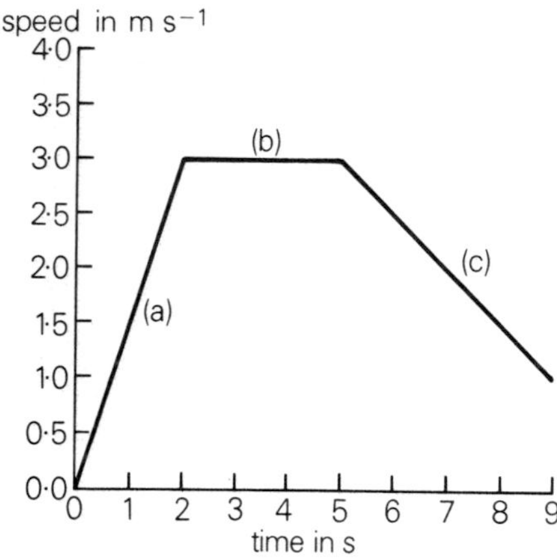

Figure 8.15

a) During the first two seconds the object accelerates uniformly from a velocity of 0 m s^{-1} to 3 m s^{-1}.
initial velocity $u = 0$
final velocity $v = 3$
time for velocity to change from u to v, $t = 2$

using $a = \dfrac{v - u}{t}$

$\Rightarrow \quad a = \dfrac{3 - 0}{2}$

$\Rightarrow \quad a = 1{\cdot}5$

From 0 s to 2 s the acceleration = 1·5 m s^{-2}

b) During the time interval from 2 s to 5 s the speed is constant
acceleration = 0

From 2 s to 5 s the acceleration = 0 m s^{-2}

c) During the time interval from 5 s to 9 s the object decelerates uniformly from 3 m s^{-1} to 1 m s^{-1}
initial velocity $u = 3$
final velocity $v = 1$
time for the velocity to change from u to v, $t = 4$

using $a = \dfrac{v - u}{t}$

$\Rightarrow \quad a = \dfrac{1 - 3}{4}$

$\Rightarrow \quad a = -\dfrac{2}{4}$

$a = -0{\cdot}5$

From 5 s to 9 s the acceleration = −0·5 m s^{-2}.

As we have seen, an upward slope on a speed-time graph indicates a positive acceleration while a downward slope indicates a negative acceleration (deceleration). The steepness of the slope indicates the magnitude of the acceleration: a steeper slope indicates a greater acceleration.

8.9 Average velocity and uniform acceleration

Let t = time in s and v = velocity in m s^{-1}

Taking readings at intervals of 1 s:

v	2·0	2·5	3·0	3·5	4·0	4·5	5·0	5·5	6·0
t	0	1	2	3	4	5	6	7	8

Taking readings at intervals of 2 s:

v	2·0	3·0	4·0	5·0	6·0
t	0	2	4	6	8

Taking readings at intervals of 4 s:

v	2·0	4·0	6·0
t	0	4	8

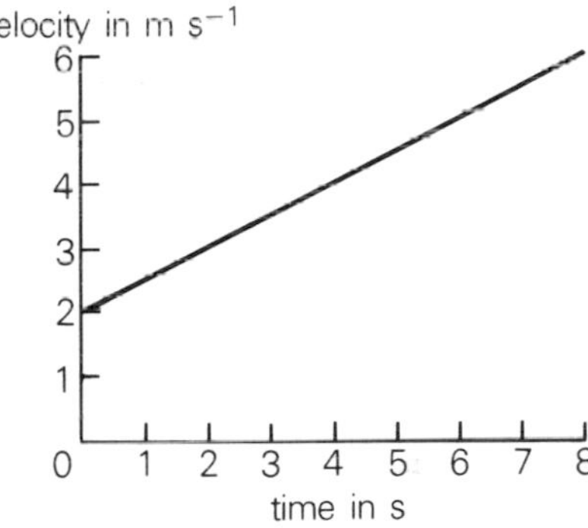

Figure 8.16

To find the average velocity of the motion represented on the graph in Figure 8.16, we can take a number of readings of the velocities at equal time intervals and average them. To find the average of a number of readings we use

$$\text{average of a number of quantities} = \frac{\text{the sum of all the quantities}}{\text{the number of quantities}}$$

Thus if we average the velocity readings taken over intervals of 1 s:

$$\text{average} = \frac{2{\cdot}0 + 2{\cdot}5 + 3{\cdot}0 + 3{\cdot}5 + 4{\cdot}0 + 4{\cdot}5 + 5{\cdot}0 + 5{\cdot}5 + 6{\cdot}0}{9} = \frac{36}{9} = 4$$

Averaging the readings of the velocities taken over intervals of 2 s:

$$\text{average} = \frac{2{\cdot}0 + 3{\cdot}0 + 4{\cdot}0 + 5{\cdot}0 + 6{\cdot}0}{5} = \frac{20}{5} = 4$$

Averaging the readings of the velocities taken over intervals of 4 s:

$$\text{average} = \frac{2{\cdot}0 + 4{\cdot}0 + 6{\cdot}0}{3} = \frac{12}{3} = 4$$

Finally, if we average the initial velocity u and the final velocity v obtained from the graph using average $= \frac{u + v}{2}$

$$u = 2; \quad v = 6;$$

$$\therefore \text{ average} = \frac{2 + 6}{2} = 4$$

The above calculations show that, for the motion with uniform acceleration represented by the graph in Figure 8.16, the average of velocity readings taken at equal time intervals is always the same, whatever the value of those time intervals. This average is equal to the average velocity of the object.

Thus for the motion illustrated by the graph, the average velocity is 4 m s^{-1}. This is equal to the velocity after 4 s which is halfway through the time interval considered.

For an object moving with uniform acceleration, the average velocity is equal to the velocity at the centre of the time interval over which the velocity is measured, and in terms of an equation:

$$\text{average velocity} = \frac{\text{initial velocity} + \text{final velocity}}{2} \quad \text{or } \bar{v} = \frac{u + v}{2}.$$

8.10 Acceleration down a slope

If the gradient of a slope is uniform, the acceleration of an object rolling down it is uniform. Figure 8.17 shows an experimental arrangement for obtaining the uniform acceleration of a trolley.

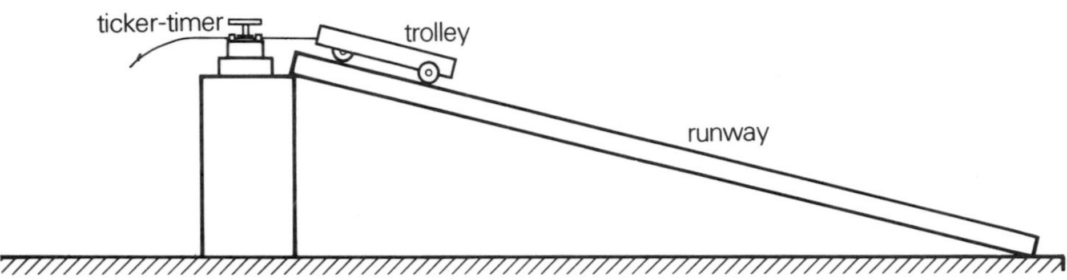

Figure 8.17 Acceleration down a slope

Figure 8.18 Ticker-tape of constant acceleration

start

Figure 8.18 shows a ticker-tape produced by such an arrangement. The increasing space between the dots shows that the speed of the trolley is increasing.

8.11 Calculation of uniform acceleration from ticker-tape

We have already seen on page 72 how a ticker-tape can be used to produce a speed-time graph from which we can calculate acceleration. It does, however, require a long piece of ticker-tape to produce a speed-time graph and it is often necessary or more convenient to calculate the acceleration directly from a shorter section of tape, such as that shown below.

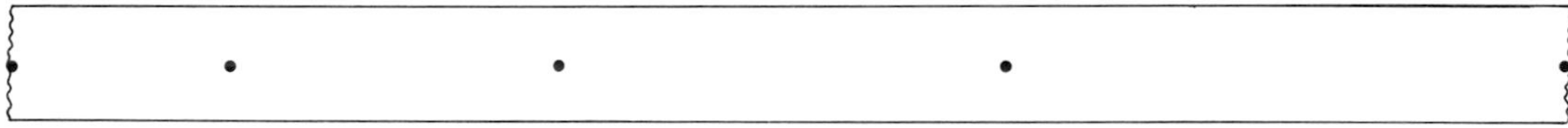

We can use the equation $a = \frac{v - u}{t}$

to calculate the acceleration, if we can find a velocity u at the start of the tape, a velocity v at the end of the tape and the time t that it takes for the velocity to change from u to v. The following detailed calculation of the acceleration for the tape shown above illustrates the method for calculating uniform acceleration from a ticker-tape:

Let u = the velocity at the middle of the first time interval
then u = the average velocity over the first time interval.
Let v = the velocity at the middle of the last time interval
then v = the average velocity over the last time interval.

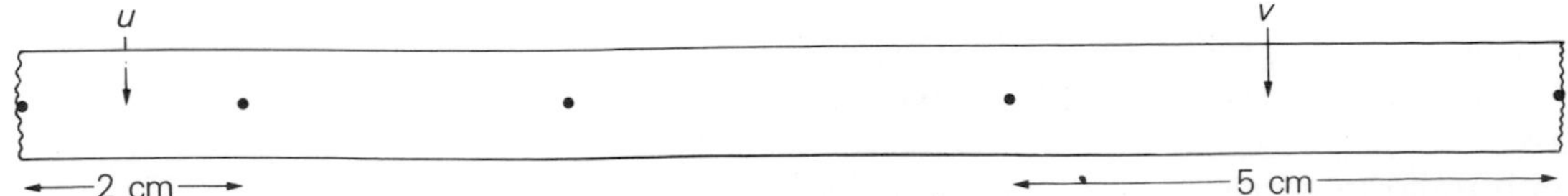

$$\text{velocity} = \frac{\text{total displacement}}{\text{time taken}}$$

$$u = \text{average velocity over the first time interval}$$

$$= \frac{\text{spacing between the first two dots}}{\text{one time interval}}$$

spacing between the first two dots = 2·0 cm = 0·02 m
Since the ticker-timer operates at mains frequency (50 Hz),
1 time interval = 0·02 s

$$\text{and } u = \frac{0{\cdot}02}{0{\cdot}02} = 1 \text{ m s}^{-1}$$

$$v = \text{average velocity over the last time interval}$$

$$= \frac{\text{spacing between the last two dots}}{\text{one time interval}}$$

spacing between the last two dots = 5·0 cm = 0·05 m
1 time interval = 0·02 s

$$\text{and } v = \frac{0{\cdot}05}{0{\cdot}02} = 2{\cdot}5 \text{ m s}^{-1}$$

t = time taken for the velocity to change from u to v
= time from the middle of the first time interval to the middle of the last time interval.

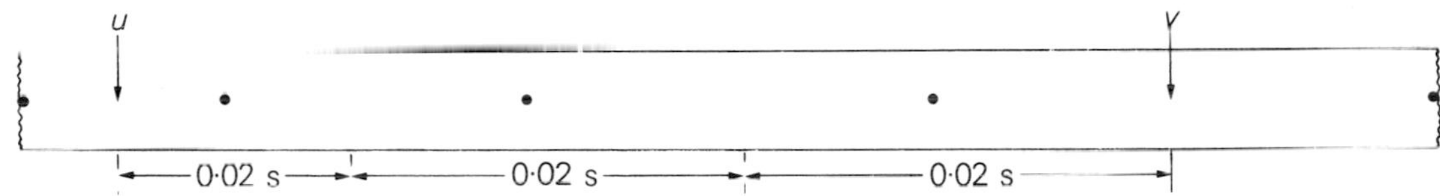

$$t = 3 \text{ time intervals} = 3 \times 0{\cdot}02 = 0{\cdot}06$$

Having calculated values for u, v and t, we can substitute into the equation $a = \frac{v - u}{t}$

$$a = \frac{2{\cdot}5 - 1{\cdot}0}{0{\cdot}06} = \frac{1{\cdot}5}{0{\cdot}06} = 25$$

Acceleration shown by the tape = 25 m s^{-2}.

Example 8

Figure 8.19 shows a tape produced by a ticker-timer operating at mains frequency (50 Hz).

Figure 8.19

Calculate the acceleration shown by the tape.

spacing between the first two dots = 3·0 cm = 0·03 m;
spacing between the last two dots = 7·0 cm = 0·07 m.

$$u = \frac{0{\cdot}03}{0{\cdot}02} = 1{\cdot}5 \text{ m s}^{-1}$$

$$\text{and} \quad v = \frac{0{\cdot}07}{0{\cdot}02} = 3{\cdot}5 \text{ m s}^{-1}$$

t = time from the middle of the first time interval to the middle of the last time interval
$= 2 \times 0{\cdot}02 = 0{\cdot}04$ s

$$a = \frac{v - u}{t} = \frac{3{\cdot}5 - 1{\cdot}5}{0{\cdot}04} = \frac{2{\cdot}0}{0{\cdot}04} = 50$$

Acceleration = 50 m s^{-2}.

8.12 Calculation of uniform acceleration from a strobe photo

The calculation is similar to that for a ticker-tape. With the strobe photograph, the basic time interval equals the period of the stroboscope (the time between pictures), and the displacements are read off a scale on the photograph.

Example 9

Figure 8.20 shows a strobe photograph of a ball rolling down a slope. The ball was illuminated by a stroboscope flashing with a frequency of 10 Hz. Calculate the acceleration of the rolling ball.

Figure 8.20

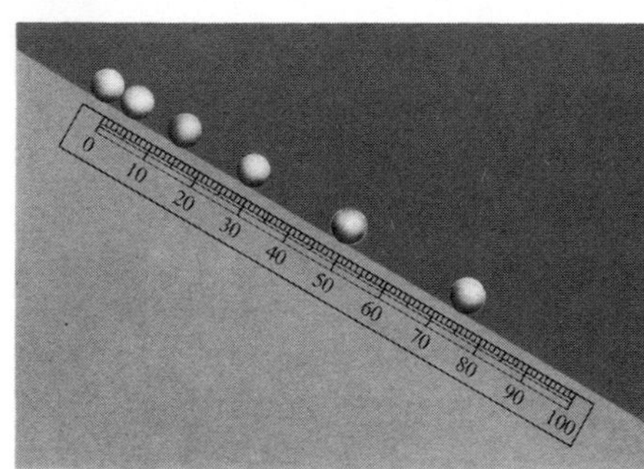

Frequency of the stroboscope = 10 Hz
period of the stroboscope $= \frac{1}{10}$
$= 0{\cdot}1$ s

From the photograph:—
displacement during the first time interval = 5 cm
= 0·05 m
displacement during the last time interval = 25 cm
= 0·25 m

$$u = \frac{\text{displacement during the first time interval}}{\text{one time interval between flashes}}$$

$$= \frac{0{\cdot}05}{0{\cdot}10} = 0{\cdot}5 \text{ m s}^{-1}$$

$$v = \frac{\text{displacement during the last time interval}}{\text{one time interval between flashes}}$$

$$= \frac{0{\cdot}25}{0{\cdot}10} = 2{\cdot}5 \text{ m s}^{-1}$$

t = time from the middle of the first time interval to the middle of the last time interval
= 4 time intervals
$= 4 \times 0{\cdot}10$
$= 0{\cdot}40$

$$a = \frac{v - u}{t}$$

$$= \frac{2{\cdot}5 - 0{\cdot}5}{0{\cdot}4} = \frac{2{\cdot}0}{0{\cdot}4} = 5{\cdot}0$$

Acceleration = 5·0 m s^{-2}

8.13 Acceleration of a falling object

Figure 8.21 shows how the apparatus to measure the acceleration of a falling ball is set up. As the ball falls, it pulls the tape through the ticker-timer. The acceleration of the ball is calculated from the tape either directly or by constructing a ticker-tape chart and hence a speed-time graph.

Figure 8.22 shows a speed-time graph made from a tape produced in this manner.

If we consider the time interval from 0 to 0·16 s on the graph,
initial velocity $u = 0{\cdot}8 \text{ m s}^{-1}$
final velocity $v = 2{\cdot}2 \text{ m s}^{-1}$
$t = 0{\cdot}16 \text{ s}$

$$a = \frac{v - u}{t} = \frac{2{\cdot}2 - 0{\cdot}8}{0{\cdot}16} = \frac{1{\cdot}4}{0{\cdot}16} = 8{\cdot}75 \text{ m s}^{-2}$$

The above experiment was carried out to find the acceleration of the object falling freely. However, the length of tape and its passage through the ticker-timer would probably have slowed the movement of the object.

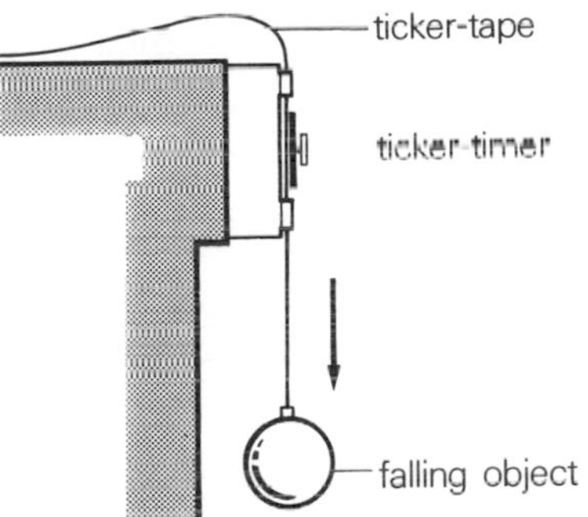

Figure 8.21

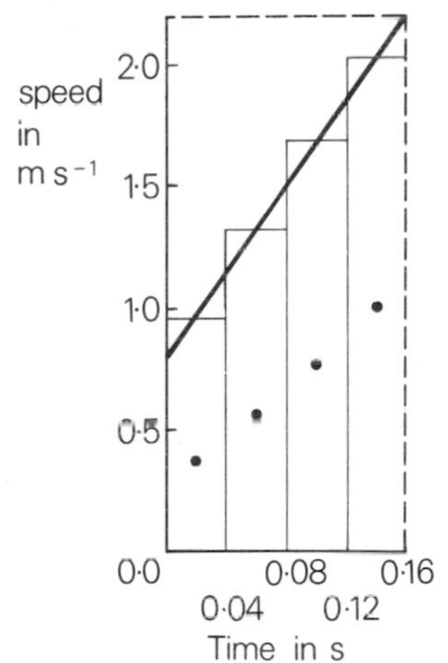

Figure 8.22

8.14 Acceleration of objects in free fall

An object's acceleration is reduced when it pulls a tape through a timer. We know that some falling objects accelerate rapidly towards Earth, while others (such as parachutes) float gently through the air. When an object, such as a feather or a parachute, is allowed to fall in a space from which most of the air has been removed, the results are surprising.

Figure 8.23 shows a photograph of a piece of metal and a feather falling in a glass tube from which most of the air has been pumped out. The metal and the feather are allowed to start falling at the same instant, and it can be seen from the photograph, that they fall together. In fact they both fall with the same acceleration. This experiment has been traditionally performed with a guinea and a feather and this is why it is called the 'Guinea and Feather Experiment'. From the experiment we can conclude that, in the absence of air resistance, all objects fall with the **same** acceleration. This has only been demonstrated for two objects falling together at the same place. In the next chapter we shall consider what factors do affect the acceleration due to gravity.

Figure 8.23 Guinea and Feather Experiment

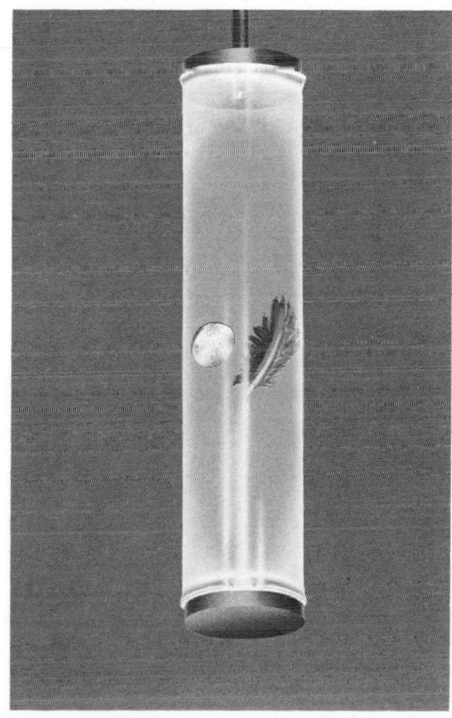

Summary

Average speed $= \dfrac{\text{total distance travelled}}{\text{time taken}}$; $\bar{v} = \dfrac{s}{t}$.

The SI unit for speed is the metre per second ($m\ s^{-1}$).

The speed at any instant can be found by taking the average speed during a very small time interval about that instant.

On a speed-time graph the area under the curve equals the distance travelled.

Speed is a scalar quantity; velocity is a vector quantity.

Average velocity $= \dfrac{\text{total displacement}}{\text{time taken}}$; $\bar{\mathbf{v}} = \dfrac{\mathbf{s}}{\mathbf{t}}$.

Acceleration is the rate of change of velocity.

Average acceleration $= \dfrac{\text{change in velocity}}{\text{time for the change}}$; $a = \dfrac{v - u}{t}$.

The SI unit for acceleration is the metre per second squared ($m\ s^{-2}$).

For uniform acceleration, $\bar{v} = \dfrac{u + v}{2}$

In the absence of air resistance, all objects at the same place fall with the same acceleration due to gravity.

Problems

1 A boy walks 400 m in 200 s. What is his average speed?

2 How long would it take a car travelling at 30 $m\ s^{-1}$ to travel 900 m?

3 How far will a jet aircraft travel in 2 minutes if it flies at 400 $m\ s^{-1}$?

4 Each of the following tapes was produced by a ticker-timer operating at 50 Hz. Assuming the tapes are drawn to scale, calculate the average speed in each case.

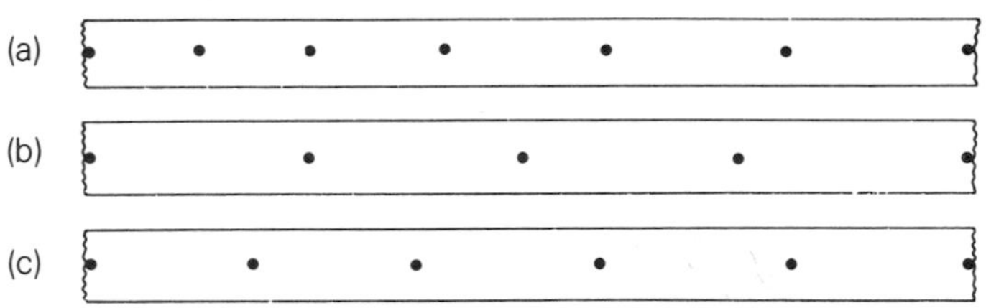

5 The following tapes are all drawn to the same scale, and were produced by ticker-timers operating at the same frequency.

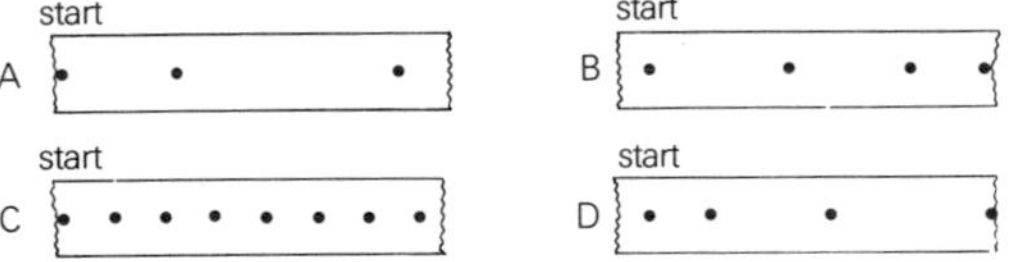

Which tape or tapes represent
- **5·1** the greatest average speed;
- **5·2** the lowest average speed;
- **5·3** constant speed;
- **5·4** decreasing speed;
- **5·5** increasing speed?

6 A ship pointing North, is travelling through the water at a speed of 24 $m\ s^{-1}$. If there is a cross current of velocity 10 $m\ s^{-1}$ (090°), what is the actual velocity of the ship?

7 A parachutist is falling with a constant vertical velocity of 16 $m\ s^{-1}$, but is being blown horizontally at 12 $m\ s^{-1}$. What is his actual velocity?
If he jumps from a height of 500 m, directly above where he intends to land, how far away from this point does he actually land?

8 The diagram shows the speed-time graph of part of a journey by car.

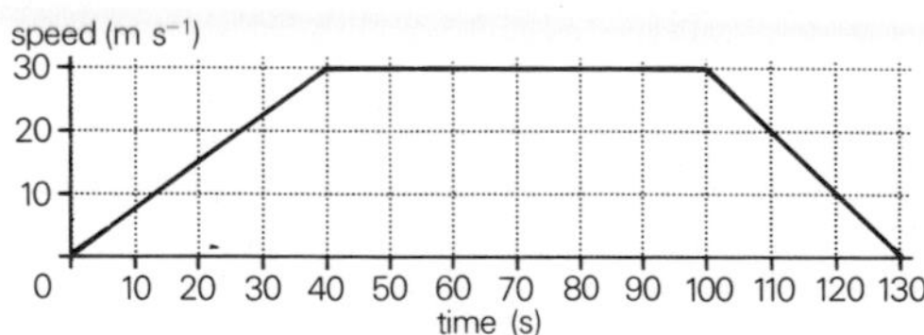

8·1 What was the speed of the car
a) at the start of this part of the car's journey;

b) after 20 s;
c) after 40 s;
d) after 75 s;
e) after 100 s;
f) after 120 s;
g) after 130 s?

8·2 During which time intervals was the car
a) increasing speed;
b) decreasing speed;
c) moving with constant speed?

8·3 Assuming that the car was moving in a straight line, calculate the average acceleration of the car
a) over the first 40 s;
b) over the next 60 s;
c) over the last 30 s;
d) over the whole 130 s.

8·4 What was the average speed
a) over the first 40 s;
b) over the next 60 s;
c) Over the last 30 s;
d) over the whole 130 s?

9 The following tapes are produced by a ticker-timer operating at a frequency of 50 Hz and all represent motion with uniform acceleration.

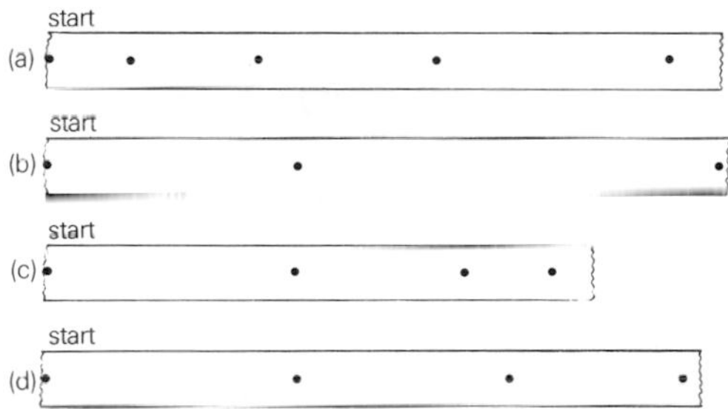

If the tapes are drawn to scale, calculate the acceleration in each case.

10 A track was tilted a few degrees from the horizontal. A trolley was released from rest at point A and allowed to run freely down the track until it passed a photo switch at B. AB was 1·5 m.

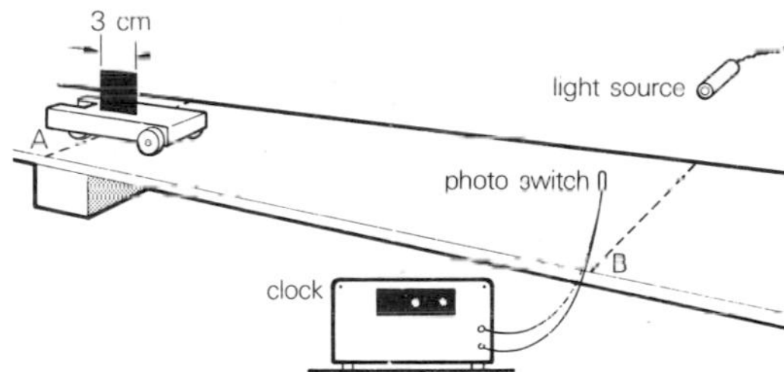

The time from A to B was measured on a stopwatch as 5·0 s. A 3·0 cm card on the trolley interrupted the light beam falling on the photo switch so that a clock connected to it registered 0·05 s as the time taken for the card to pass. The clock was accurate to one millisecond.

10·1 Find
a) the acceleration;
b) the average velocity between A and B.

10·2 Give two reasons why stopwatches could not be used for both the time measurements. *SCEEB*

11 In the apparatus shown in the diagram, the electric timing device prints 50 dots every second on the tape. When the ball is allowed to fall, a record of its motion is produced on the tape.

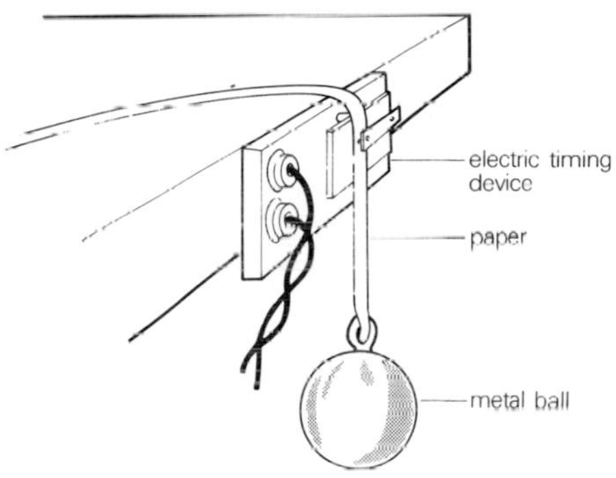

By cutting the tape at each alternate dot, the following tape-chart (actual size) is obtained.

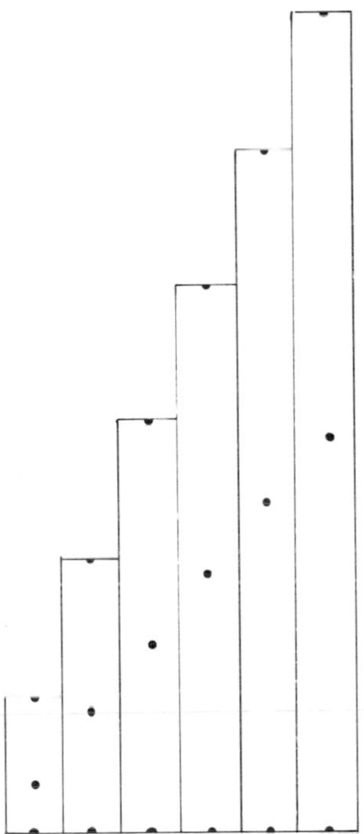

11·1 Calculate the average acceleration.

11·2 Compare your answer with the accepted value for the acceleration due to gravity and suggest an explanation for the difference. *SCEEB*

9 Acceleration, force and mass

9.1 Introduction

We have defined acceleration and the units in which we measure it. We shall now investigate what causes an object to accelerate and what determines the magnitude of the acceleration.

In 1686 Newton published his *Principia Mathematica* which contained his three Laws of Motion. His statement of the First Law was

> 'Every body persists in its state of rest or of uniform motion in a straight line unless it is compelled to change that state by forces impressed on it.'

We may simplify this statement by saying that, unless any unbalanced forces acts on it, a stationary object will remain at rest or a moving object will continue to move at a constant speed in a straight line.

Acceleration is the rate of change of velocity, and velocity is the speed in a certain direction. Thus, if we can change the speed or direction of movement of an object, we are causing it to change its velocity, i.e. to accelerate. We can change the speed or direction of movement of an object by pushing it or by pulling it. When we push or pull, we are exerting a **force**. A force can change the velocity of an object.

Acceleration is produced by a force, and we can measure acceleration. If we are to investigate the relationship between a force and the acceleration that it produces, we must have a standard for the measurement of force. For the following experiments we use the force exerted by a stretched elastic cord as our standard.

If we take an elastic cord and pull (exert a force) on it, it stretches. If we pull harder (exert a greater force) on it, it stretches further. We assume that, if the cord is always stretched by the same amount, the force is always the same.

The standard unit of force that we shall use for this experiment is the force needed to stretch an elastic cord to the end of the trolley, Figure 9.1.

If you have ever used a chest expander, you will know that it requires a greater force to stretch the chest expander with several springs than it does with one spring, Figure 9.2.

Similarly, it takes a greater force to stretch two elastic cords by a certain length than to stretch one by that length. If we always use identical cords and stretch them by the same amount, we can assume that 1, 2, and 3 cords produce 1, 2, and 3 units of force respectively.

To measure the change in velocity produced by a force, we need to create conditions under which an object moves with constant velocity when we do not exert such a force on it. Our experiments are carried out using a trolley, and we set up a runway on which the trolley moves with constant velocity when no force is exerted on it.

Figure 9.1 One unit of force

Figure 9.2

9.2 Production of constant velocity

If a trolley is set moving on a flat surface it will slow down and stop, Figure 9.3.
On the other hand, if we place a trolley on a steep slope, it will accelerate, Figure 9.4
We can adjust the slope somewhere between the horizontal and the steep slope so that a trolley set in motion will continue to move with constant velocity, Figure 9.5. We can check that its velocity is constant by studying a ticker-tape produced by a trolley set in motion down the slope. The dots should be equally spaced, Figure 9.6.

A slope arranged so that the trolley moves with constant velocity (zero acceleration) when no additional force is exerted on it is called a **friction-compensated slope**. We shall discuss the term friction later in the chapter.

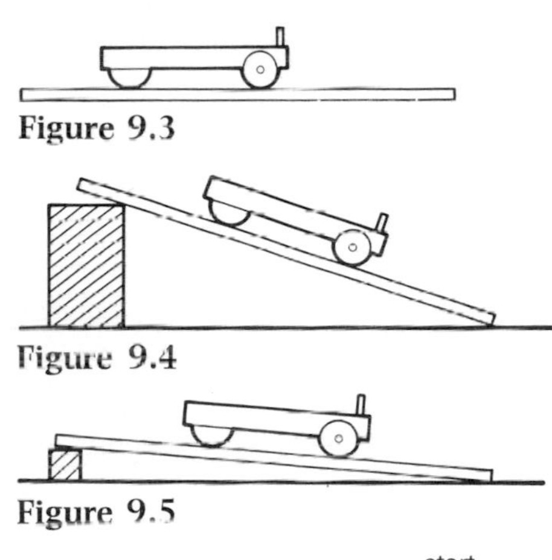

Figure 9.3

Figure 9.4

Figure 9.5

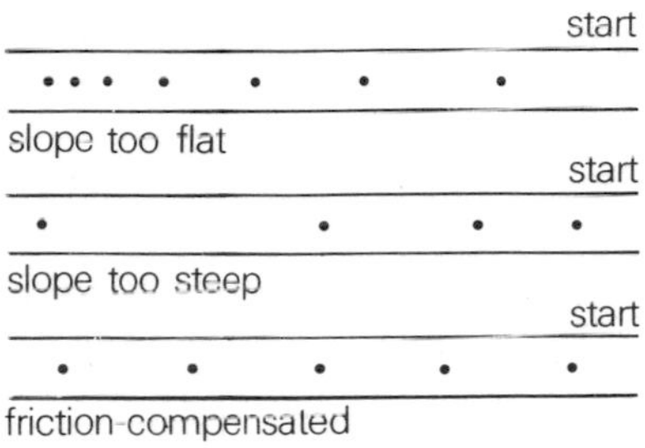

Figure 9.6

9.3 Acceleration and force

Using one elastic cord, we exert a force of one unit on the trolley. The elastic cord is fixed at one end of the trolley and held stretched the length of the trolley while it runs all the way down the friction-compensated slope. As it runs down the slope it pulls a ticker-tape, Figure 9.7

We repeat this experiment using two cords and then three cords. We use the three ticker-tapes to construct speed-time graphs and get the results shown in Figure 9.8.

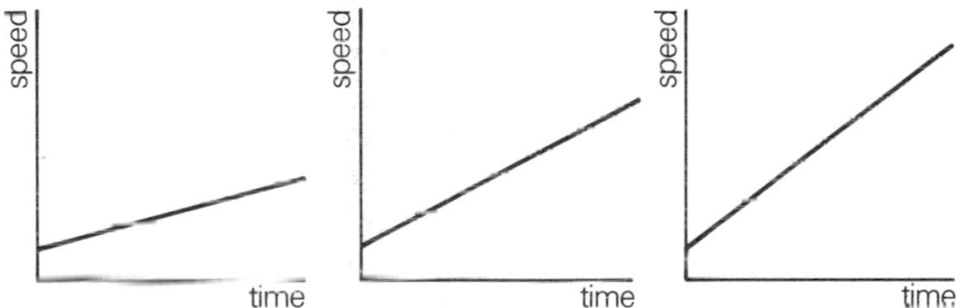

Figure 9.8 Speed-time graphs for various forces

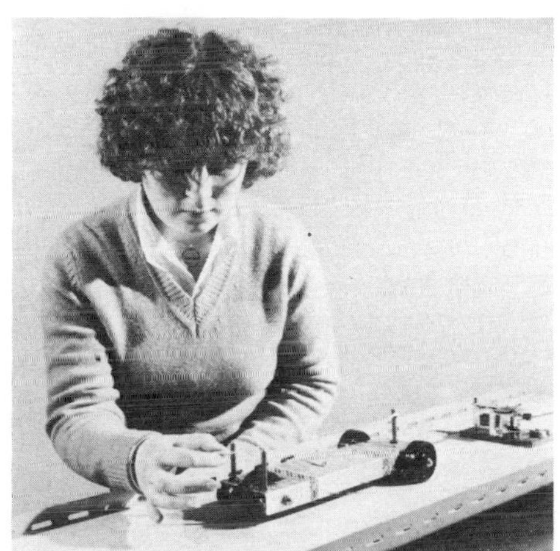

Figure 9.7

We can see that the acceleration increases as the force increases because the slope of the graph get steeper.

From these graphs, calculation of the acceleration produced the results shown in Table 1.

Force F	*Acceleration a* m s^{-2}	$\frac{F}{a}$
1 unit	0·20	5
2 units	0·40	5
3 units	0·60	5

Table 1

We can see from these results that

a) $\frac{F}{a}$ = constant

b) double the force produces double the acceleration;

c) three times the force produces three times the acceleration.

This is summarized by saying that the acceleration **varies directly as** the force; in other words, the acceleration is directly proportional to the force.

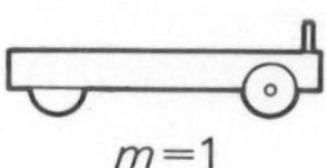

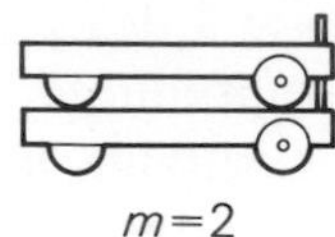

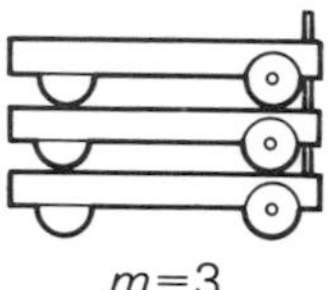

Figure 9.9

9.4 Acceleration and mass

We have investigated the relationship between force and acceleration by applying different forces to a trolley and measuring the acceleration produced. We shall now investigate how the acceleration is affected by the object on which the force is acting.

We would not expect to produce the same acceleration when we push a bus as when we push a car. The bus is harder to push because there is a lot more of it. The amount of matter in an object is its **mass**. We have yet to consider the measurement of mass and the unit for mass, but, for our immediate investigation of the effect of mass on acceleration, we may assume that each identical trolley has the same mass; we shall call this a mass of one unit.

Using the same experimental set-up as in section 9.3, we apply a constant force of two units (two elastic cords) to 1, 2, and 3 trolleys (Figure 9.9) and measure the acceleration produced. The force must be kept constant so that we know that any change in acceleration is produced by a change in mass.

Thus we are keeping the force constant and seeing how the acceleration is related to the mass.

The speed-time graphs produced from the ticker-tapes obtained in this experiment are shown in Figure 9.10.

We can see that the acceleration decreases as the mass increases, because the slope of the graph gets less.

From these graphs we may calculate the acceleration in each case as shown in Table 2.

Figure 9.10 Speed-time graphs for various masses

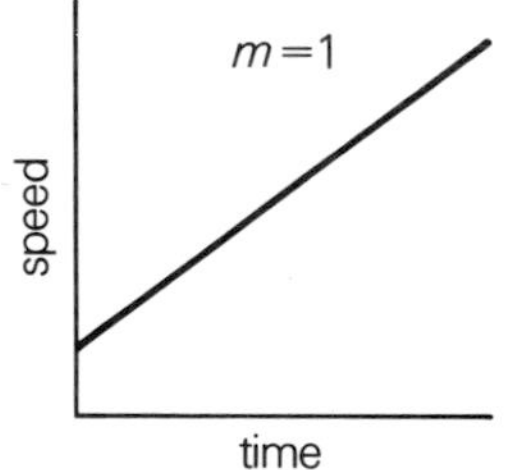

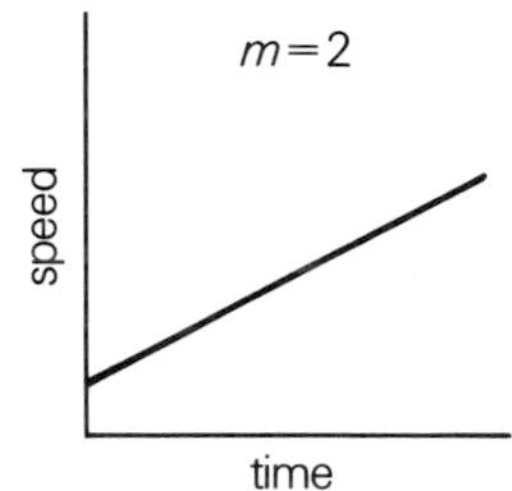

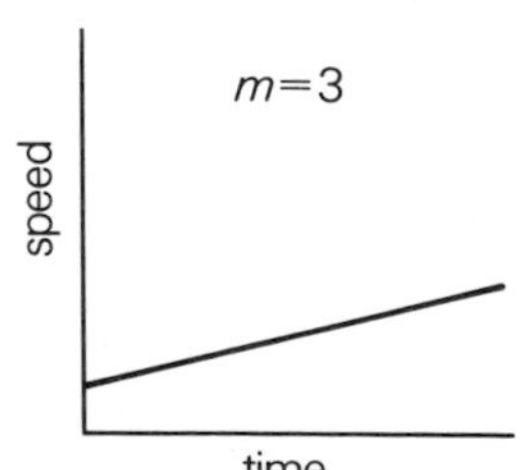

Mass m	*Acceleration a* m s^{-2}	*m × a*
1 unit	0·40	0·4
2 units	0·20	0·4
3 units	0·13	0·4

Table 2

We can see from these results that

a) $m \times a$ = constant

b) double the mass has half the acceleration;

c) three times the mass has one third the acceleration.

This is summarized by saying that the acceleration **varies inversely as** the mass; in other words, the acceleration is inversely proportional to the mass.

From the two experiments that we have just described, we have found that:

a) $\frac{F}{a}$ = constant when the **mass** is kept constant;

b) $m \times a$ = constant when the **force** is kept constant.

In order to find one equation relating all three quantities, force, mass and acceleration, it is useful to combine the results of both experiments.

In the first experiment, the constant **mass** was a mass of one unit and, in the second experiment the constant **force** was a force of two units. The results of both experiments are combined in Table 3.

Force F	*Mass m*	*Acceleration a* m s^{-2}	$\frac{F}{m \times a}$
1 unit	1 unit	0·20	5
2 units	1 unit	0·40	5
3 units	1 unit	0·60	5
2 units	1 unit	0·40	5
2 units	2 units	0·20	5
2 units	3 units	0·13	5

Table 3

These results show that $\frac{F}{m \times a}$ = constant

We may multiply both sides of the equation by $m \times a$ to give the equation in another form: F = **constant** $\times m \times a$

These experiments have been carried out with measurements of mass and force in units that we chose for convenience. The value of the constant would have been different if we had chosen different units.

9.5 The unit for mass

We chose the mass of one trolley as being a convenient unit of mass for our experiment. But this is hardly a practical and generally accepted unit for mass. The SI unit for mass is the kilogram (kg). The standard kilogram is the mass of a platinum-iridium cylinder kept at Sèvres in France.

9.6 The unit for force

If we consider the relationship

F = constant $\times m \times a$

that we found from our experiments, we now have units for two of the quantities, mass and acceleration, and we have said that the value of the constant depends on the units in which we measure the three quantities. We use this information to define the unit for force.

We define the **newton** (N) as the force that gives a mass of 1 kilogram an acceleration of 1 m s^{-2}.

Thus when $m = 1$ kg
and $a = 1\ \text{m s}^{-2}$
then $F = 1$ N

Substituting these values in our equation

$$F = \text{constant} \times m \times a$$

we have

$$1 = \text{constant} \times 1 \times 1$$
$$1 = \text{constant}$$

Thus, when we use the above units, the value of the constant is equal to one, and our equation $F = \text{constant} \times m \times a$

becomes

$$F = 1 \times m \times a$$
$$F = m \times a$$

This important equation expresses **Newton's Second Law**, and it can be applied to the motion of all objects.

We have so far considered the case in which the acceleration is produced by a single force. Acceleration is a vector quantity, and the direction of the acceleration produced is the same as the direction of the force which produces it. Thus force has direction and is a **vector** quantity.

9.7 The measurement of mass

If we have a standard mass, we must have some method of comparing other masses with it, and this means that we must have a method of measuring mass.

Earlier in our discussions we realized that it was easier to accelerate a car than a bus because the car had a smaller mass. One way of thinking of the mass of an object is as its resistance to change in motion. This resistance to change in motion is called **inertia**. Measurement of the inertia of an object is a means of measuring its mass.

One experimental arrangement in which a mass is continually being accelerated and decelerated is that shown in Figure 9.11. The trolley or trolleys are moved to one side and then released to oscillate from side to side. The time for ten oscillations is measured for various trolley masses. Some typical results are shown in Table 4.

This shows that the period of oscillation increases with the number of trolleys (mass) that are being pulled from side to side.

Figure 9·12 shows an inertial balance which uses the same principle for comparing masses. The balance is supplied with four identical

Number of trolleys	*Time for 10 oscillations* (s)
1	8
2	15
3	23

Table 4

Figure 9.12 Inertial balance

Figure 9.11 Oscillating trolley

cylinders. The unit of mass is the mass of one cylinder. If we measure the time for ten oscillations of the balance when it is loaded with 1, 2, 3 and 4 cylinders, we can plot the results on a graph, Figure 9.13.

Having determined how the time for ten oscillations varies with the mass in kilograms, we have calibrated the instrument so that we can use it to measure mass. If we wish to measure the mass of an object, we can place it on the empty balance and measure the time for ten oscillations. Using the graph in Figure 9.13 we can read off what mass takes that time to complete ten oscillations and hence we find the mass of the object. For example, a mass of 1·05 kg will take 5·5 s for 10 oscillations.

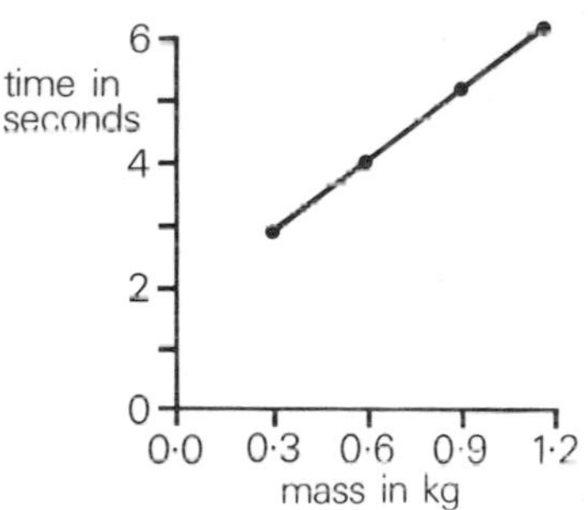

Figure 9.13

9.8 Weight is a force

We saw at the end of chapter 8 that, in the absence of air resistance, all objects fall with the same acceleration. The magnitude of this acceleration is produced by the force (pull) due to gravity acting on an object. The force of gravity on an object is called the weight of the object.

From Newton's Second Law:

force = mass × acceleration

⇒ force due to gravity = mass × (acceleration due to gravity)

⇒ weight = mass × (acceleration due to gravity)

This is written $W = m \times g$

where W = weight; m = mass; g = acceleration due to gravity.

Multiplying both sides of the equation by $\frac{1}{m}$:

$$\frac{W}{m} = g$$

$$\Rightarrow \quad \frac{\text{unit for weight}}{\text{unit for mass}} = \text{unit for acceleration}$$

$$\Rightarrow \quad \frac{\text{N}}{\text{kg}} = \text{unit for acceleration}$$

$$\Rightarrow \quad \text{N kg}^{-1} = \text{unit for acceleration}$$

We already know that the unit for acceleration is m s^{-2}, and we can see that another SI unit for acceleration is N kg^{-1}. These two units are exactly equivalent to each other.

$$1 \text{ N kg}^{-1} \equiv 1 \text{ m s}^{-2}$$

9.9 Gravitational field strength

The mass of an object is an unchanging quantity. An inertial balance with a certain mass would have the same period on the Moon as on Earth. Thus, a lump of lead labelled 'one kilogram' on Earth should also be labelled '1 kilogram' on the surface of the Moon – or anywhere else in the Universe.

When a mass is acted on by gravity, it experiences the force which we call weight. For example, the weight of a 1 kg mass on Earth is about

10 N. On the Moon the same mass would weigh only about 1·7 N. So the weight of an object depends on the strength of the gravitational field that acts on it.

The gravitational field strength g tells us how many newtons that a kilogram weighs at a certain place. It is defined by the equation

$$\text{gravitational field strength} = \frac{\text{weight (in newtons)}}{\text{mass (in kilograms)}}$$

$$\Rightarrow \qquad g = \frac{W}{m}$$

Because the unit of weight is the newton and the unit of mass is the kilogram, the unit of gravitational field strength is the newton per kilogram (N kg^{-1}). The gravitational field strength on Earth is about 10 N kg^{-1}.

9.10 Weight and mass

Unfortunately the everyday use of the word 'weight' is not the same as the precise definition used by the scientist. People speak of the weight of goods expressed in kilograms, yet the unit for weight (which is a force) is the newton.

The unit for weight is the newton
The unit for mass is the kilogram

Mass is a measure of the quantity of matter or of its inertia. It is a property of the object and is therefore everywhere constant. The weight of an object is a property of the object which also depends on the gravitational field strength. In places of different field strength, an object will have a different weight, but the mass of the object will not vary.

Example 1

An object has a weight of 50 N on Earth. If the acceleration due to Earth's gravity is 10 m s^{-2};
the acceleration due to Jupiter's gravity is 26 m s^{-2};
the acceleration due to Mars' gravity is 3·6 m s^{-2};
a) what is the mass of the object on Earth?
b) what would be the mass of the object on Jupiter?
c) what would be the wight of the object on Jupiter?
d) what would be the mass of the object on Mars?
e) what would be the weight of the object on Mars?

a) On Earth, $g = 10$; $W = 50$.

$$W = m \times g$$
$$50 = m \times 10$$

Multiply both sides of the equation by $\frac{1}{10}$

$$5 = m$$

mass of the object = 5 kg

b) Because the mass of the object is the same everywhere,
mass of the object on Jupiter = 5 kg

c) On Jupiter, $m = 5$; $g = 26$.

$$W = m \times g$$
$$W = 5 \times 26 = 130$$

weight of the object on Jupiter = 130 N

d) Because the mass of the object is the same everywhere,
mass of the object on Mars = 5 kg

e) On Mars, $m = 5$; $g = 3{\cdot}6$.

$$W = m \times g$$
$$W = 5 \times 3{\cdot}6 = 18{\cdot}0$$

weight of the object on Mars = 18·0 N

9.11 Opposing forces

All objects on the surface of the Earth have weight, but they are not all accelerating downwards. Why are you not accelerating downwards at the moment? No doubt you are being held up by a chair or the ground. In fact, the chair is exerting an upward force on you, and this upward force is balancing out the downward pull of gravity on you, Figure 9.14. When two forces acting on an object are equal and opposite, they balance each other out, and we say that the unbalanced force is zero. If two unequal forces act on an object in opposite directions, the unbalanced force is equal to the difference between the forces, Figure 9.15. Force is a **vector** quantity; if we choose one direction as being the positive direction, forces in the opposite direction are negative.
When we investigated the effect of force on acceleration, we considered the effect of a single force. In fact, Newton's Second Law applies to the unbalanced force acting on an object.
Newton's Second Law: $F = m \times a$
where F = unbalanced force, m = mass, a = acceleration.

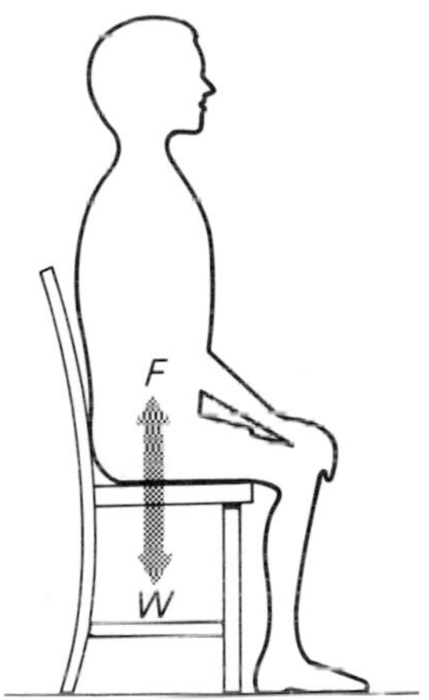

Figure 9.14

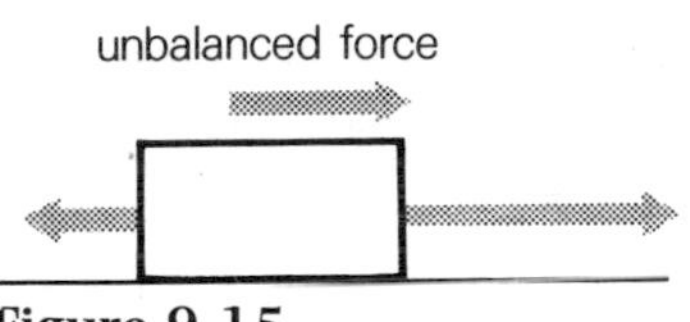

Figure 9.15

9.12 Galileo's experiments

Galileo was interested in the acceleration of falling objects. As part of his experiments he investigated the movement of a ball rolling down an incline.

He noted that, when a ball was released on a slope such as that shown in Figure 9.16a, it continued to roll until it nearly regained its original height (limit of the journey). This was the case whatever slopes were involved. Galileo reasoned that if the track levelled out, the ball would not reach the limit of its journey because it would never regain its original height, Figure 9.16d.

Galileo's reasoning was correct if the unbalanced force on the ball on the level track is zero. The weight of the ball is balanced by the force of the track on the ball. While Galileo's reasoning suggests that the ball would continue to roll forever on a level track, experience tells us that it would eventually stop. We shall now consider what causes it to stop.

Figure 9.16

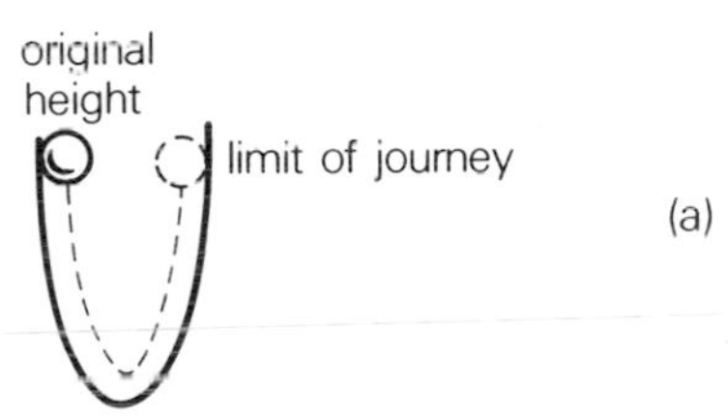

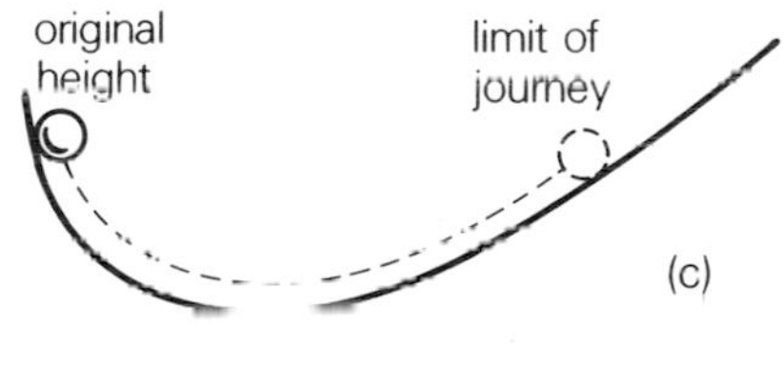

9.13 Friction

An object moving with constant speed has zero acceleration.
Because $F = m \times a$, we know that when a is zero, F also is zero.

Thus in the absence of an unbalanced force, an object has a constant velocity. If it is at rest, it will remain at rest; if it is moving, it will continue to move with the same velocity. This is a statement of Newton's First Law.

A cyclist has to continue pedalling to keep a bicycle moving with constant velocity along a flat road, so there must be a resisting force that acts against his pedalling. Any moving object tends to slow down unless a force is exerted to keep it moving. Since objects tend to slow down (decelerate), there must be a force tending to slow them down. This force which opposes motion is the force of **friction**.

A force is needed to drag an object over the ground. The force needed is less if the object is pulled over ice. This is because the frictional force between the ice and the object is less than that between the object and the ground. Whenever two surfaces rub against each other, there are frictional forces which try to stop the two surfaces moving past each other.

Friction is a force which **opposes** motion.

A sledge is a useful form of transport over ice or snow, because ice and snow are slippery: the frictional force on objects moving over them is small. When we wish to pull or push a vehicle over rougher surfaces, it is easier if we put it on wheels. We are not dragging the vehicle over the surface of the ground and do not have the same frictional force to overcome. This does not mean that frictional forces are eliminated, because the wheels turning in their bearings will experience a frictional force opposing their rotation.

On page 91 we determined the relationship between force, mass and acceleration and we used a friction-compensated slope. The slope was adjusted so that the effect of gravity tending to accelerate the trolley down the slope just balanced the effect of the frictional forces tending to decelerate the trolley.

In these experiments we were able to compensate for friction, but in others we try to reduce friction as much as possible. Figure 9.17 shows a linear air-track in which air is blown out through the holes in the air-track and the vehicle rides on a cushion of air. The friction between the vehicle and the air is much less than that between two solid surfaces, and the vehicle experiences only very small frictional forces. We can take a strobe photograph of a vehicle which has been set in motion on the linear air-track. The equal spaces between the images show that the vehicle is moving with constant velocity. The frictional forces are insufficient to decelerate the vehicle noticeably.

The hovercraft (Figure 9.18) is a vehicle in which the effects of friction are reduced in a similar way. The vehicle travels on a cushion of air which separates it from the surface of sea or land along its route.

Lubrication with oil provides a similar cushion between two moving surfaces. Generally when two solid surfaces rub on each other, the frictional force is relatively high. This force can be reduced if the two solid surfaces are separated by oil or grease which forms a thin layer between the two surfaces, so reducing friction.

Figure 9.17 Linear air track

Figure 9.18

9.14 Terminal velocity

While the frictional forces between a solid and a liquid or a solid and a gas are less than those between two solids, they still exist. We have already seen in section 8.14 on page 85 how the frictional force (air resistance) on the feather affects its motion through the air. As any cyclist knows, the faster you move the more you feel the effects of friction and the harder you have to pedal. Frictional forces of all kinds increase with speed.

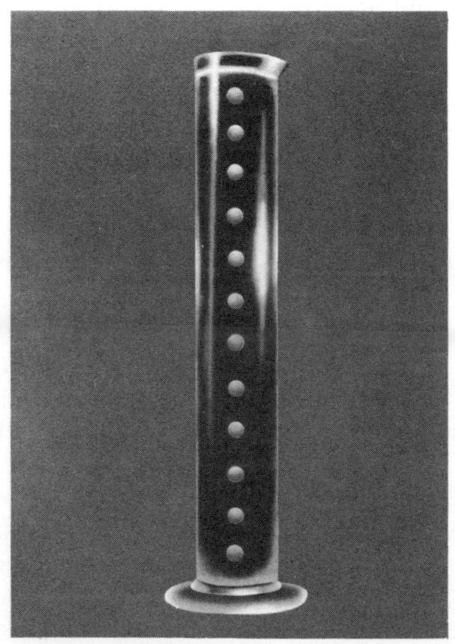

Figure 9.19

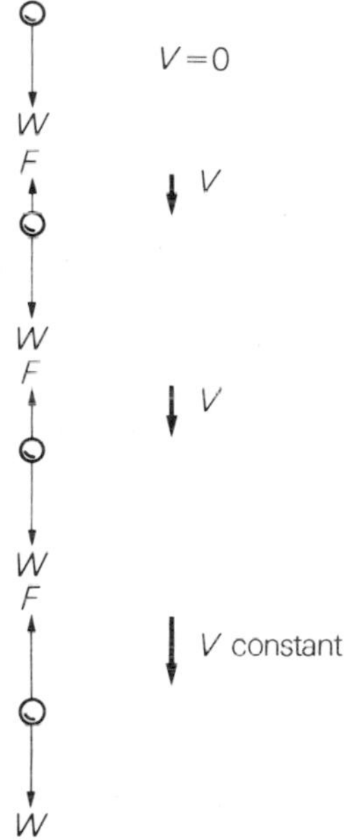

Figure 9.20 Achieving terminal velocity

Figure 9.19 shows a strobe photo of a ball-bearing falling in oil. The images of the ball-bearing are equally spaced, showing that it was moving with constant velocity (zero acceleration). This constant velocity is called the **terminal** velocity.

If the acceleration is zero, the unbalanced force acting on it is zero. The ball-bearing has the pull of gravity (its weight) acting downwards on it and the frictional force acting up on it. If these two opposing forces are equal and cancel each other out, the unbalanced force is zero and the velocity is the terminal velocity (constant).

Figure 9.21 shows how an object released from rest accelerates until it reaches a certain velocity and then continues to fall with constant velocity. The arrows in the diagram represent vectors.

At the instant of release the object is not moving and there is no friction. The weight of the object accelerates it downwards.

As its velocity increases the frictional force increases.

So long as the weight is greater than the frictional force, there is an unbalanced downward force and the object accelerates downwards. As the object accelerates, the frictional force increases.

When the frictional force becomes equal in magnitude to the weight of the object, the unbalanced force equals zero, the acceleration is zero, and the object falls with constant velocity. Because the velocity is constant, the frictional force is constant and the object continues to fall with constant velocity with an unbalanced force of zero.

The constant velocity is the **terminal velocity**.

Raindrops soon reach their terminal velocity because of air resistance. If there were no friction acting on a raindrop, it would fall with an acceleration of approximately 10 m s^{-2}. After falling from a height of 2 000 m, its speed would be as high as 200 m s^{-1}.

A parachutist falls with constant velocity because the upward frictional force due to air resistance balances his weight. The large surface area of a parachute increases the frictional force and the terminal velocity is reduced.

9.15 Addition of forces

Forces are vector quantities and are therefore added as vectors. The combined effect of two or more forces is called the **resultant** of the forces and is their vector sum.

Example 2

Two boys are pulling a sledge at a constant speed over flat ice. They are both pulling at an angle of 30° to the path of the sledge. If they both exert a force of 100 N, what is the frictional force acting on the sledge?

Scale: 1 cm represents 40 N

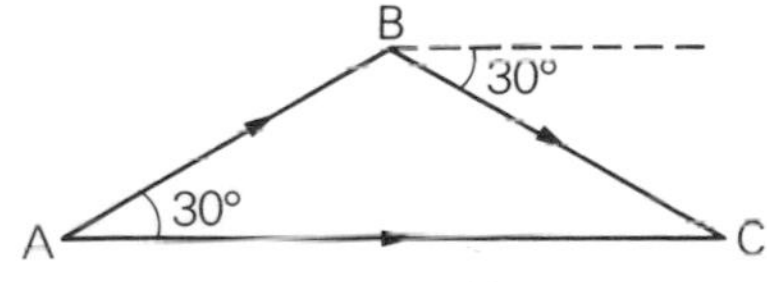

The resultant of the two forces is found by vector addition.

$\vec{AB} + \vec{BC} = \vec{AC}$

Length of AC = 4·35 cm

resultant force = 4·35 × 40 = 174

The resultant force exerted by the two boys is 174 N.

Since the sledge moves with constant speed, the acceleration is zero and the unbalanced force is zero.

The frictional force is equal and opposite to the resultant of the forces exerted by the two boys.

The frictional force is **174 N** in the direction **opposite** to the path of the sledge.

We shall only use vector notation for forces when we are dealing with forces not acting along the same straight line.

9.16 Resolution of forces

Two forces can be replaced by a single force which is the vector sum of those two forces. In the same way, a single force can be replaced by two forces. These two forces are called **components**. The splitting up of a force into two component forces is called **resolution**. The vector sum of the two components is equal to the single force that they replace.

A force has **no** effect in a direction **perpendicular** to itself. To find the effect of a force in a certain direction, we resolve it into two components: one in that direction and one perpendicular to that direction.

The general case

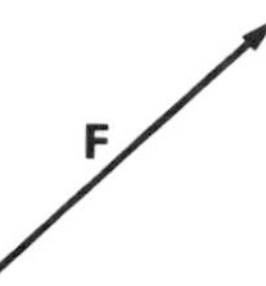

To find the effect of a force at an angle θ to itself, we find the component in that direction.

If the force is $\mathbf{F}$, the component at an angle θ is $\mathbf{F}_1$ and the other component at right angles is $\mathbf{F}_2$,

then $\mathbf{F}_1 + \mathbf{F}_2 = \mathbf{F}$

$\mathbf{F}_1$ and $\mathbf{F}_2$ are perpendicular to each other.
$\mathbf{F}_1$ is at an angle θ to $\mathbf{F}$.

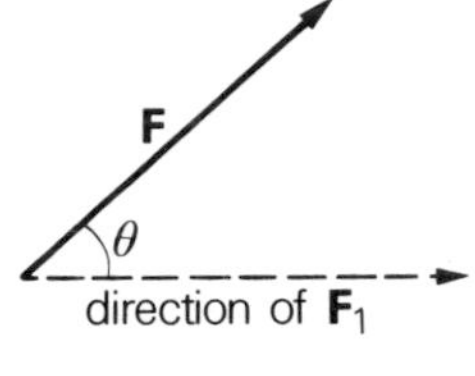

In a right-angled triangle, cosine of an angle $= \dfrac{\text{adjacent}}{\text{hypotenuse}}$

$$\Rightarrow \cos\theta = \frac{F_1}{F}$$

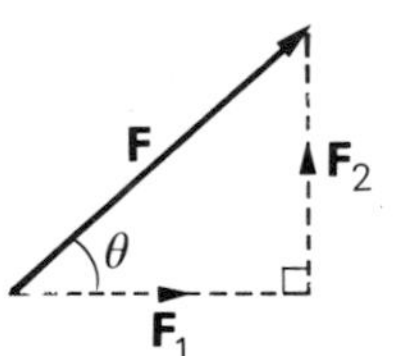

$\mathbf{F}_1 + \mathbf{F}_2 = \mathbf{F}$
$\mathbf{F}_2$ is perpendicular to $\mathbf{F}_1$

Multiply both sides by F: $F\cos\theta = F_1$

$\mathbf{F}_1$ is the component of $\mathbf{F}$ which makes an angle θ with $\mathbf{F}$.

From this general case we see that the effect of a force $\mathbf{F}$ in a direction at an angle θ to $\mathbf{F}$ is given by $F\cos\theta$.

Example 3

If an object of mass 4 kg is on a slope at 60° to the vertical, what is the component of its weight acting down the slope?

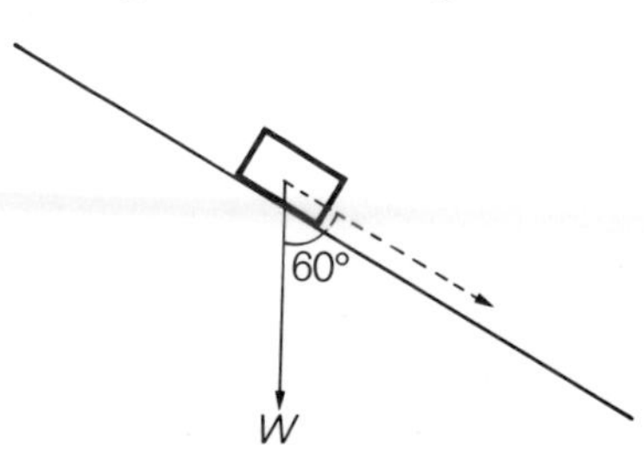

Mass m of the object = 4 kg
$W = m \times g = 4 \times 10 = 40$
Weight = 40 N
Weight acts vertically downwards.
Component of the weight at 60° to the vertical $= W\cos 60°$
$= 40 \times \cos 60°$
$= 40 \times 0{\cdot}5$
$= 20$

Component of the weight acting down the slope is 20 N.

Summary

Newton's First Law states that, if no unbalanced force acts on an object, its velocity does not change. If it is stationary, it remains at rest. If it is moving, it continues to move with constant speed in a straight line.

A friction-compensated slope is one along which a trolley will continue to move with constant velocity when no other force is applied.

Acceleration varies directly as the unbalanced force when the mass remains constant.

Acceleration varies inversely as the mass when the unbalanced force remains constant.

Newton's Second Law is expressed by the equation

$F = m \times a$

where F = unbalanced force acting on the object
m = mass of the object
a = acceleration of the object.

Mass is a property of an object alone.

Weight is the force of gravity of an object and is equal to the product of the mass of the object and the gravitational field strength.

$W = m \times g$

The SI unit for mass is the kilogram (kg). The SI unit for force is the newton (N).

Mass is the measure of the quantity of matter or the inertia of an object.

Force is a vector quantity. The combination of forces can be found by vector addition.

Friction is a force which opposes motion.

Terminal velocity is the constant velocity reached by a falling object when the frictional force is equal to the weight of the object.

The resolution of a force is the splitting up of the force into components. The component of a force F at an angle θ to itself is F cos θ.

Problems

1 A girl was setting up a friction-compensated slope. When testing different slopes by allowing a trolley to run down the slope, she obtained the following tapes:

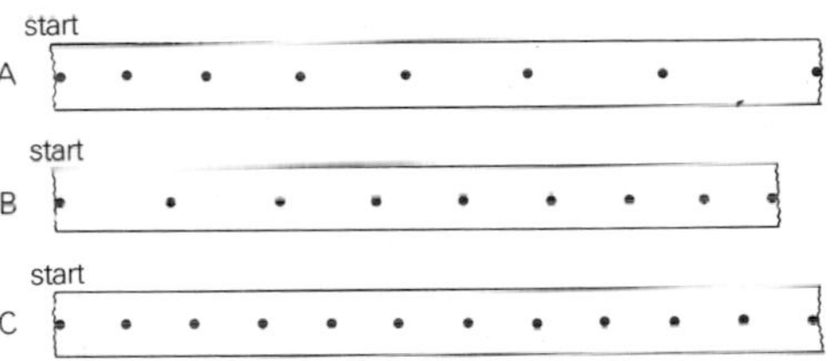

1·1 Which tape indicated that the slope was properly friction compensated?
1·2 Which tape indicated that the slope was too steep?
1·3 Which tape indicated that the slope was not steep enough?

2 What average net force is required to accelerate a car of mass 1 200 kg from rest to 20 m s^{-1} in 10 s?
If an average braking force of 4 800 N is applied when the car is travelling at 20 m s^{-1}, how long will it take to stop the car?

3 Earth's gravitational field strength = 10 N kg^{-1}.
Jupiter's gravitational field strength = 26 N kg^{-1}.
Mars' gravitational field strength = 3·6 N kg^{-1}.
What would be the weight of the following objects on each of the planets?
3·1 An apple of mass 0·1 kg.
3·2 A man of mass 75 kg.

4 **4·1** What force (size and direction) is required to lift a mass of 5 kg vertically upwards near the surface of the earth at a constant speed of 4 m s^{-1}?
4·2 What is the resultant of two forces of 6 N and 8 N acting on an object at right angles to each other as shown? (size and direction are required).

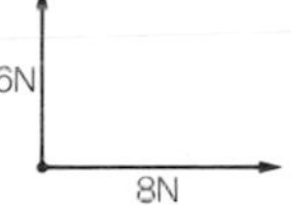

SCEEB

5 A track is tilted so that a trolley, when given a small initial push, runs down the track at constant speed. The trolley is now pulled along the track by an elastic band kept at constant tension. The motion of the trolley is recorded using a vibrator and paper tape.

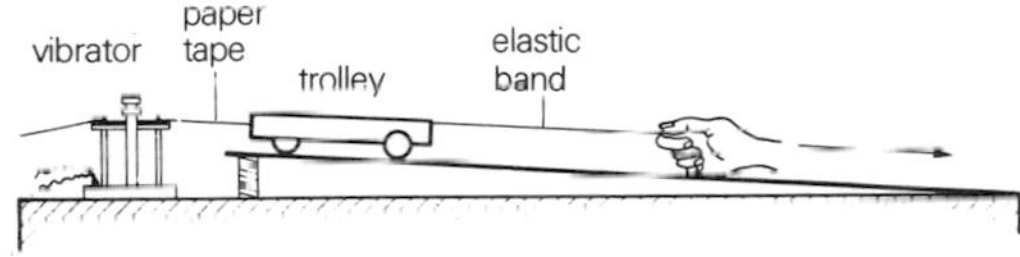

The tape is divided into sections as shown in the sketch, each section corresponding to a time interval of 0·1 second. A middle portion of the tape is then analysed.

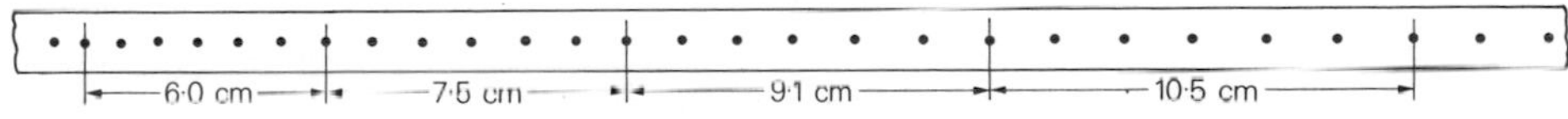

5·1 Copy the following analysis table into your answer book and fill in the blank spaces.

Distance moved in successive 0·1 s intervals	6·0 cm	7·5 cm	9·1 cm	10·5 cm
Time interval	0·1 s	0·1 s	0·1 s	0·1 s
Average speed in each 0·1 s interval	60 cm s^{-1}	75 cm s^{-1}		
Increase in speed between successive 0·1 s intervals	–	15 cm s^{-1}		
	–	150 cm s^{-1}		
Average acceleration	–			

5·2 How many vibrations per second did the vibrator make?
5·3 Suggest **two** changes you could make in carrying out the experiment each of which would double the acceleration recorded.
5.4 Predict what the corresponding figures in the first row of the above table might be if the track were set horizontally.
SCEEB

6 This speed-time graph represents the motion of a stone dropped from a low bridge.

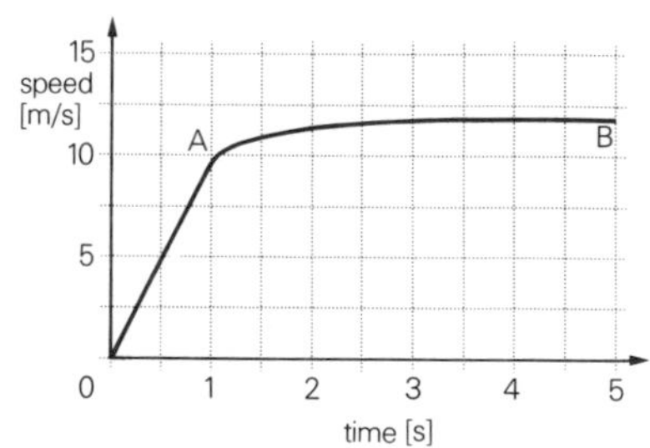

6·1 Suggest what happened to the stone at A.
6·2 Describe both the speed and the acceleration during
a) the first second (OA),
b) the next four seconds (AB).
6·3 Name the forces acting on the stone
a) during OA,
b) during AB.
SCEEB

7 A girl on a sledge slides down a slope. The total mass of the girl and the sledge is 100 kg. The record of their journey from A to D is indicated on a combined stopwatch-speedometer attached to the sledge. The readings of this instrument at positions A, B, C and D are shown.
7·1 Which stage of the journey takes the shortest time: AB, BC or CD?

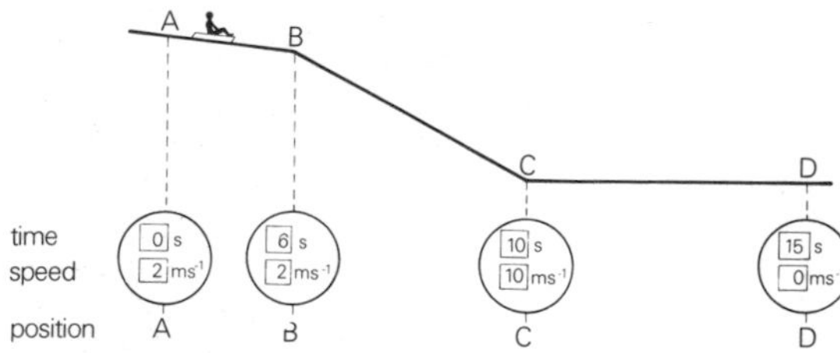

7·2 Describe the motion of the sledge during each of these stages.
7·3 Calculate the average retarding force acting on the sledge during stage CD.
7·4 If the girl lies flat along the sledge after passing A, suggest possible new readings on the stopwatch-speedometer as the sledge passes B.
SCEEB

8 A steel ball is allowed to fall in front of a vertical screen on which is marked a series of horizontal lines 10 cm apart. A stroboscopic photograph is taken of the motion of the ball. The frequency of flashing of the stroboscope is 10 Hz. The resulting photograph is shown.
8·1
a) What is the time interval between successive ball positions?
b) Calculate the average speed of the ball between positions A and B; B and C; C and D; and D and E.
c) Show that the ball falls with uniform acceleration.
8·2 An air-filled balloon falls vertically in place of the steel ball.
a) State two differences in the motion of the balloon compared with the motion of the steel ball.
b) Explain the reasons for these differences.
SCEEB

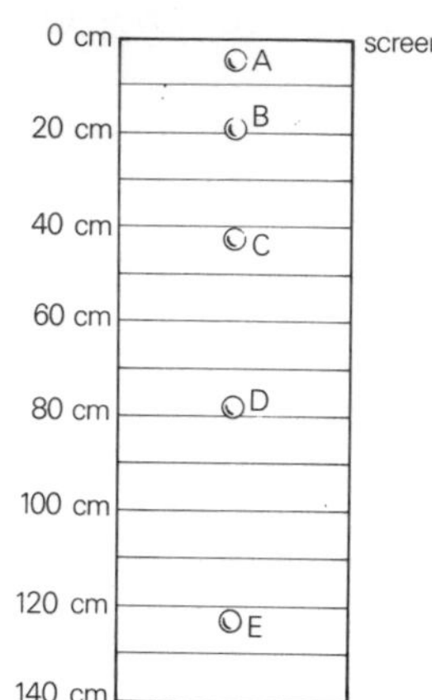

9 A trolley is at rest on a runway. A constant force, acting parallel to the runway, is applied to the trolley and the resulting acceleration of the trolley along the runway is measured. This procedure is repeated in turn with one, two and three identical trolleys stacked on the top of the first trolley. The graph shows how a pupil recorded the results of the experiment.

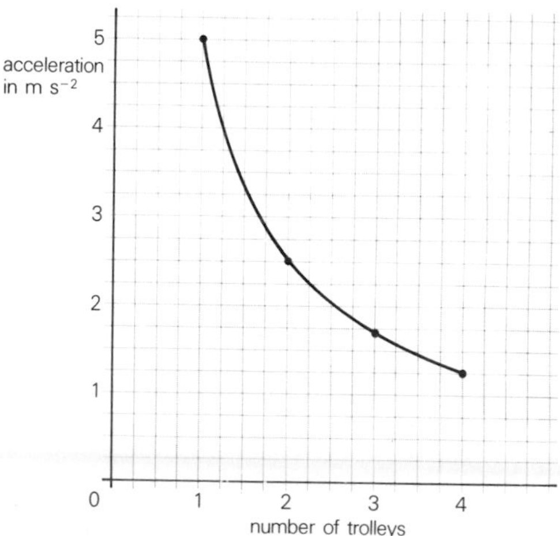

9·1 Use the graph
a) to show that the accelerating force remains constant throughout the experiment,
b) to calculate the magnitude of this force if the mass of each trolley is 0·8 kg.
9·2 Care was taken to ensure that the apparatus was compensated for friction throughout the experiment. Describe **fully** how you would do this.
9·3 Describe a method of applying a constant accelerating force to the trolleys in this experiment.
SCEEB

10 Gravity and projectiles

10.1 Falling objects

We saw at the end of Chapter 8 that all objects fall with the same acceleration unless there is air resistance or some other liquid or gas acting on the surface of the falling object.

A few hundred years ago, most people thought that a heavy object fell at a faster rate than a light object. However in the seventeenth century an Italian scientist, Galileo, is said to have carried out experiments which showed that in fact a heavy object falls at almost the same rate as a light object.

You can repeat this experiment very easily by simultaneously dropping a marble and a somewhat heavier lump of lead on to a hard floor. You will hear only one click as they both reach the floor at the same time. This shows that, although the objects have different masses, they both fall at about the same rate: otherwise two separate clicks would have been heard.

Figure 10.1 is a stroboscopic photograph showing a light and a heavy ball accelerating downwards due to gravity. It is fairly obvious that both balls are falling at about the same rate.

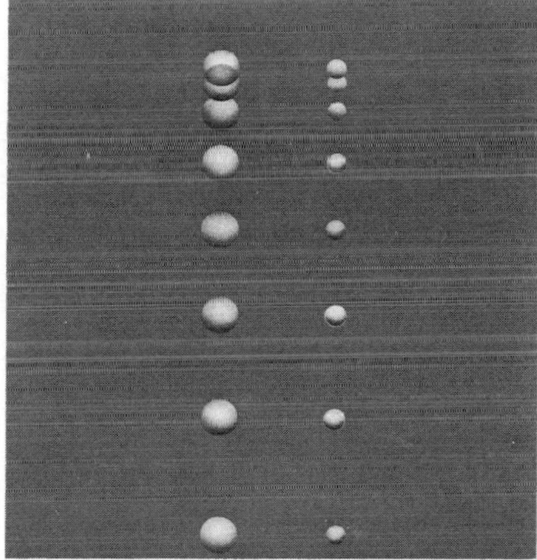

Figure 10.1

10.2 Projectile motion

We will now consider what happens when a falling object has a horizontal velocity. A falling object with a horizontal velocity is called a **projectile** and it moves in a curved path called a **trajectory**.

You can do a simple experiment to find whether the horizontal velocity has any effect on the vertical acceleration due to gravity, using two identical marbles. If one is thrown horizontally and the other is simultaneously dropped (Figure 10.2), only one click is heard, showing that they both fall at the same rate. This means that the horizontal velocity of a projectile has no effect on its vertical acceleration.

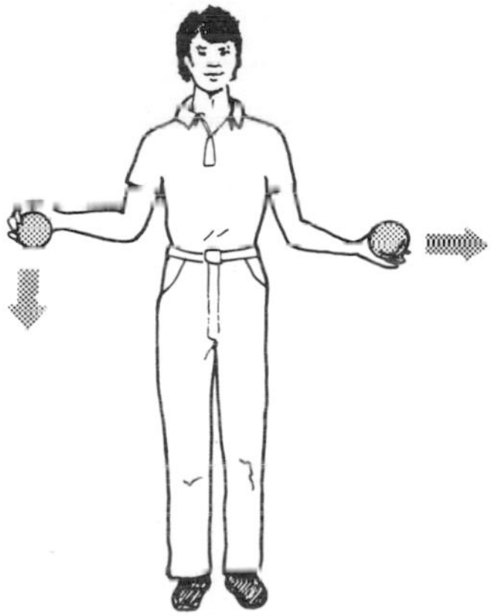

Figure 10.2

Figure 10.3 shows a stroboscopic photograph of a ball which has been projected with a horizontal velocity. Vertical lines have been superimposed on the consecutive positions of the ball.

It is clear that the vertical lines are equally spaced. This means that the horizontal displacement of the ball increases by the same amount in each time interval between the flashes of the stroboscope, i.e. the horizontal motion is at a **constant** velocity. If we neglect air resistance, there are no horizontal forces acting on the ball. We could have predicted that the horizontal velocity is constant from Newton's First Law of Motion (page 88).

Figure 10.3 Stroboscopic photograph of a projectile

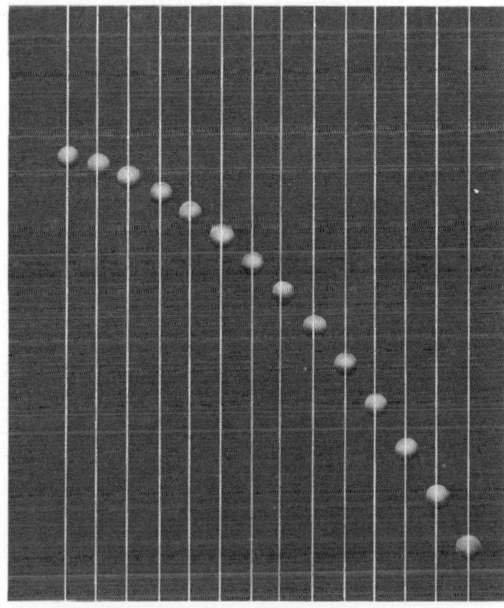

Figure 10.4 shows an apparatus which projects one ball horizontally and instantaneously allows the other ball to fall vertically.

A stroboscopic photograph obtained using this apparatus is shown in Figure 10.5.

Horizontal lines have been superimposed on the consecutive positions of the balls. It is clear that:

a) as the spacing of the lines increases with time, the two balls are accelerating vertically downwards.

b) since both balls have the same downward acceleration, the horizontal velocity does not affect this acceleration.

We conclude that the vertical motion is entirely independent of the horizontal motion.

When we are solving problems on projectile motion, we will therefore consider the horizontal motion independently of the vertical motion.

Figure 10.4 Horizontal projection apparatus

Figure 10.5

Example 1

Figure 10.6 below represents a stroboscopic photograph of a projectile.

If the stroboscope flashes every 0·1 s, find:

a) the horizontal velocity,

b) the vertical acceleration.

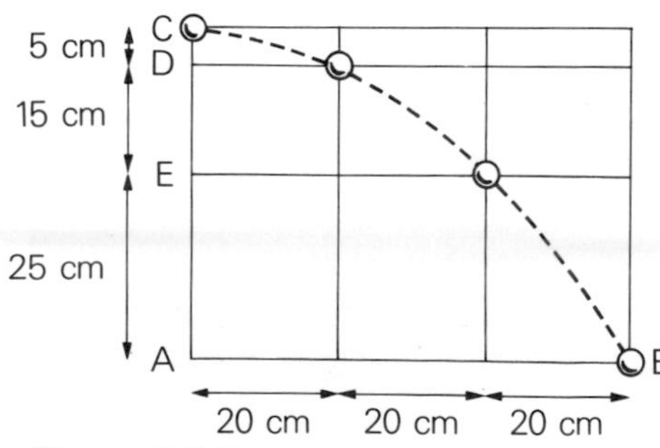

Figure 10.6

a) The horizontal velocity is constant.

$$\text{Horizontal velocity} = \frac{\text{horizontal displacement (AB)}}{\text{total time for the displacement}}$$

$$= \frac{\text{horizontal displacement (AB)}}{\text{three time intervals between flashes}}$$

$$= \frac{0{\cdot}6}{0{\cdot}3} = 2$$

The horizontal velocity is 2 m s^{-1}

b) Average vertical velocity, v_1, during the first 0·1 s, is given by,

$$\text{average vertical velocity, } (v_1) = \frac{\text{vertical displacement (CD)}}{\text{total time for this displacement}}$$

$$\Rightarrow \quad v_1 = \frac{\text{vertical displacement (CD)}}{\text{one time interval between flashes}}$$

$$\Rightarrow \quad v_1 = \frac{0{\cdot}05}{0{\cdot}10} = 0{\cdot}5 \text{ m s}^{-1}$$

Average vertical velocity, v_2, during the second 0·1 s, similarly, is given by,

$$\text{average vertical velocity, } (v_2) = \frac{\text{vertical displacement (DE)}}{\text{one time interval between flashes}}$$

$$\Rightarrow \quad v_2 = \frac{0{\cdot}15}{0{\cdot}10} = 1{\cdot}5 \text{ m s}^{-1}$$

Therefore the acceleration a is given by,

$$a = \frac{\text{change in velocity } (v_2 - v_1)}{\text{time for the change}}$$

$$= \frac{\text{change in velocity } (v_2 - v_1)}{\text{one time interval between flashes}}$$

$$= \frac{1{\cdot}5 - 0{\cdot}5}{0{\cdot}1} = \frac{1}{0{\cdot}1} = 10$$

The vertical acceleration is 10 m s^{-2}.

Example 2

A bomb dropped from an aeroplane flying horizontally at 100 m s^{-1} reaches the ground after 3 s, Figure 10.7.

If $g = 10$ m s^{-2}, calculate

a) the horizontal velocity of the bomb just before it hits the ground.

b) the vertical velocity of the bomb just before it hits the ground,

c) the velocity with which the bomb strikes the ground,

d) the horizontal displacement of the bomb (AB),

e) the altitude of the plane (AP), assuming that air resistance is negligible.

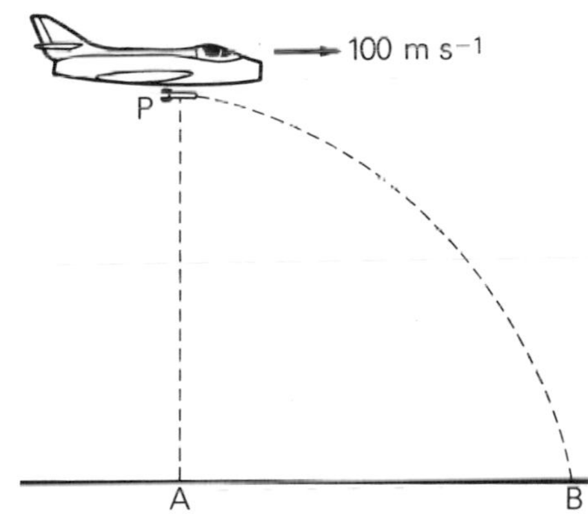

Figure 10.7

a) Because the horizontal velocity of the aeroplane is 100 m s^{-1}, the bomb also has a horizontal velocity of 100 m s^{-1} just as it is released. Ignoring air resistance, the horizontal velocity remains constant. Hence the horizontal velocity of the bomb just before it hits the ground is **100 m s^{-1}.**

b) The acceleration a is constant and is given by

$$a = \frac{v - u}{t},$$

where v = final vertical velocity and u = initial vertical velocity,

$$\Rightarrow 10 = \frac{v - 0}{3}$$

$$\Rightarrow 30 = v$$

The vertical velocity of the bomb on hitting the ground is **30 m s^{-1}.**

c) The resultant velocity of the bomb just before it hits the ground is the vector sum of its horizontal and vertical velocities. To calculate this resultant velocity we draw a vector diagram:

Scale: 1 cm represents 10 m s^{-1}

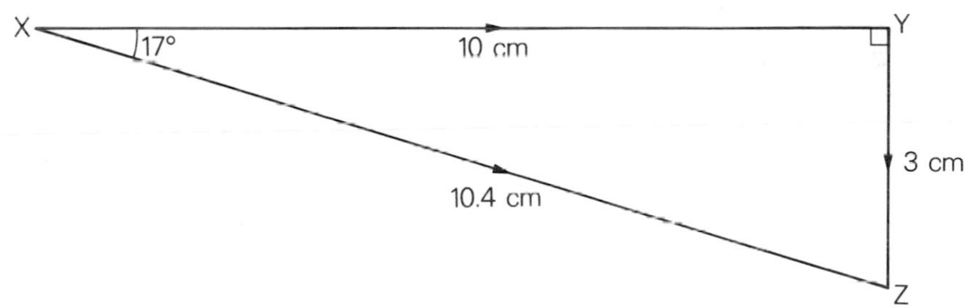

$\vec{XY}$ represents the horizontal velocity

$\vec{YZ}$ represents the vertical velocity

The resultant velocity is represented by $\vec{XZ}$, where

$\vec{XZ} = \vec{XY} + \vec{YZ}$

By measurement from the diagram, $|XZ| = 10{\cdot}4$ cm and $\angle YXZ = 17°$. Therefore the velocity with which the bomb strikes the ground is **104 m s^{-1}** at 17° below the horizontal.

d) Horizontal velocity $= \dfrac{\text{horizontal displacement } \vec{AB}}{\text{time of fall}}$

$$\Rightarrow \quad 100 = \frac{\vec{AB}}{3}$$

$$\Rightarrow \quad 300 = \vec{AB}$$

The horizontal displacement of the bomb is 300 m

e) For uniform acceleration,

$$\text{average velocity} = \frac{(\text{initial vertical velocity}) + (\text{final vertical velocity})}{2}$$

$$\bar{v} = \frac{0 + 30}{2} = 15$$

Vertical displacement $(\vec{AP}) = \bar{v} \times \text{time}$

$$|AP| = 15 \times 3 = 45$$

The altitude of the plane is 45 m.

10.3 Pulsed water-drop experiment

The apparatus shown in Figure 10.8 can be used to display the trajectory of a series of water droplets with interesting results.

A reservoir of water is used to obtain a constant flow of water from the jet. The flow rate can be adjusted by a clip. The rubber tubing connecting the reservoir to the jet is compressed by the arm of a ticker timer 50 times every second. This tends to separate the water stream into 50 drops every second. When the experiment is performed in a dark room and the water is illuminated with a stroboscope flashing at 50 Hz the drops are seen to be stationary, Figure 10.9.

If the frequency of flashing is increased or decreased slightly, the drops appear either to move back to, or away from the jet! Vertical lines have been superimposed on the black background of Figure 10.9. From this you can see that the horizontal velocity of a water drop remains constant.

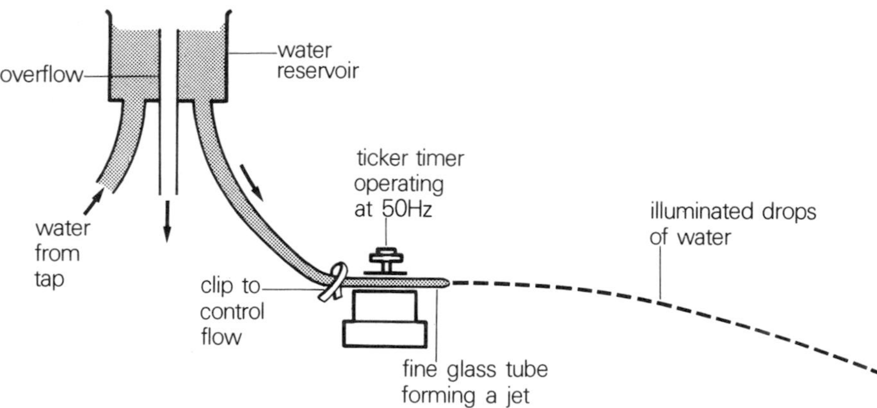

Figure 10.8 Pulsed water-drop experiment

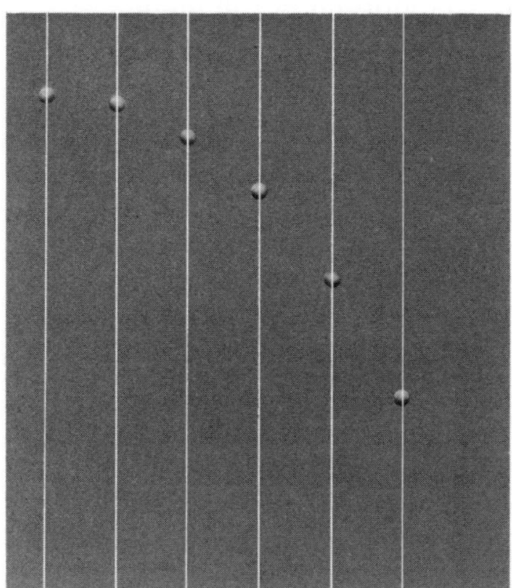

Figure 10.9 Stroboscopic photograph of pulsed water drops

10.4 Monkey-and-hunter experiment

Imagine a hunter intends to shoot a monkey hanging from the branch of a tree, Figure 10.10.

The hunter aims straight at the monkey. However, at the instant he fires, the monkey lets go of the branch. Let us consider whether the monkey is hit. The bullet can be thought of as a projectile and both the bullet and the monkey accelerate downwards at the same rate due to gravity. Because the bullet has a horizontal velocity, the horizontal separation of the monkey and bullet decreases with time. Since the monkey and bullet are falling at the same rate, they are always at the same height. The monkey will therefore be hit.

Figure 10.10 A monkey and hunter

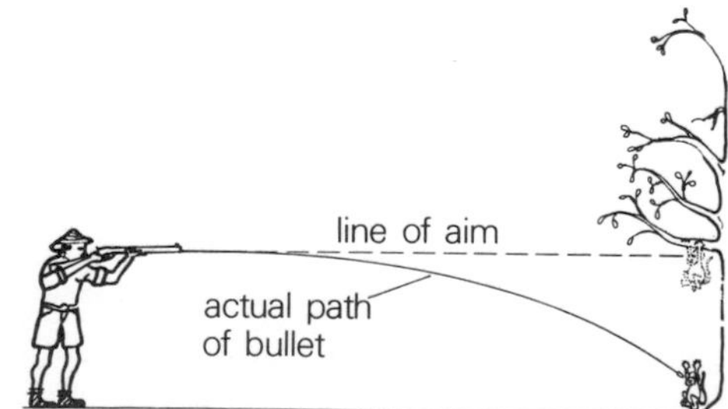

The above situation is obviously a 'tall story'. However, an experiment called the 'monkey-and-hunter experiment' can be set up to show that the collision really does happen, Figure 10.11.

One ball is held stationary by an electromagnet. A second ball, moving along a horizontal ramp, breaks a metal foil strip at the instant it leaves the ramp. This breaks the circuit and cuts off the power to the electromagnet, so allowing the first ball to fall. The result is that both balls start to fall at the same instant.

Figure 10.12 is a stroboscopic photograph obtained from this experiment. It can be seen that both balls are falling at the same rate due to gravity. Since the ball on the left has a horizontal velocity, they collide. This experiment confirms that the horizontal motion of a projectile is independent of its vertical motion.

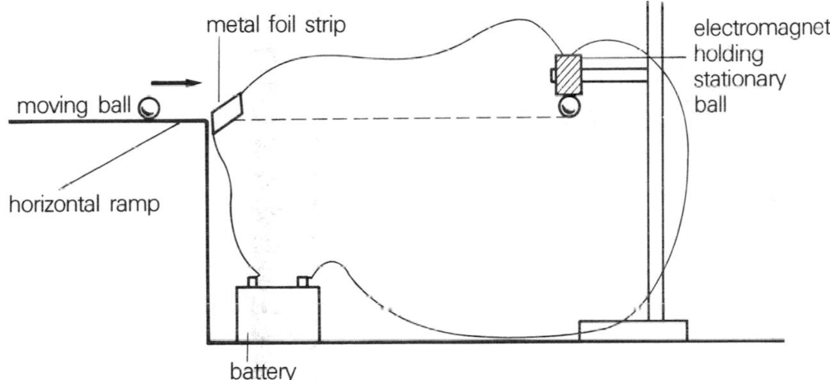

Figure 10.11 Monkey-and-hunter experiment

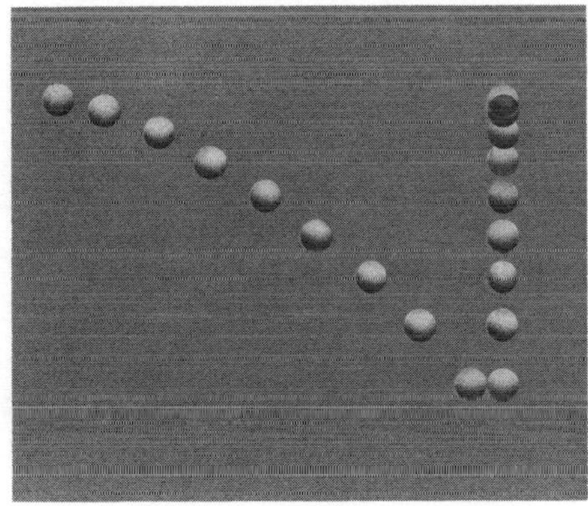

Figure 10.12 Collision of falling bodies

10.5 Newton's thought experiment

This 'experiment' was originally thought of by Sir Isaac Newton. He tried to imagine what path a cannonball would take if fired horizontally from the top of a mountain, Figure 10.13.

A cannonball fired from R might follow the curved path RS. If the cannonball was fired horizontally at a higher speed, it might follow the path RT or RU, each time going further round the Earth before hitting it.

The next step was ingenious. Newton reasoned that if a cannonball could be fired fast enough, it would go right round the Earth and eventually arrive back at the starting point R. At an even faster speed the cannonball could orbit the Earth, so becoming a satellite. In practice this would not happen because air resistance would drastically slow the cannonball from the very high speeds required.

In all the cases we have considered, whether the path taken is RS, RT, RU or even RV, the cannonball is a projectile. It is accelerating all the time under the acceleration due to gravity g which is directed towards the centre of the Earth.

Finally, if the projectile is fired faster still, it could follow the path RW. In other words, this experiment implies the possibility of an object being able to escape from Earth altogether.

We now know that a satellite can be placed in orbit round the Earth by a rocket. The satellite remains in orbit outside the atmosphere where friction is very small. It can be thought of as a projectile: it must be accelerating towards the Earth because its direction is changing all the time.

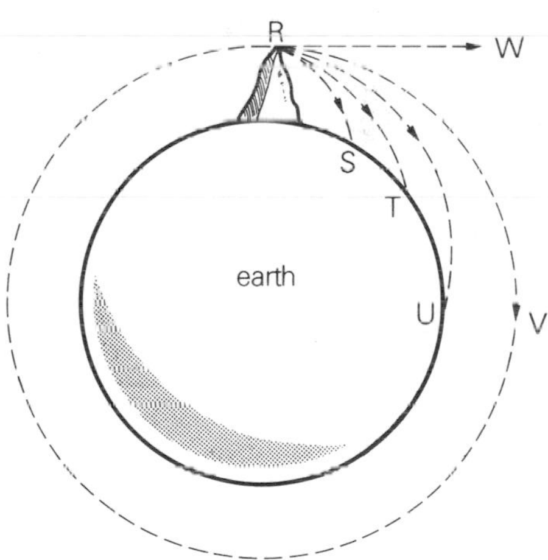

Figure 10.13 Newton's 'thought experiment'

10.6 Weightlessness

You have probably seen film on television of astronauts floating round a space-ship. We say that in this situation they are weightless, Figure 10.14.

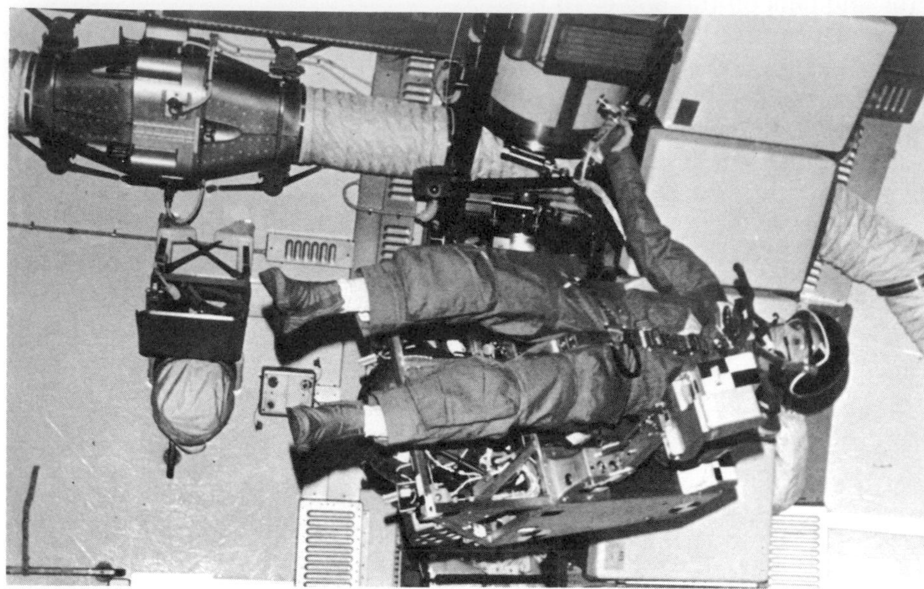

Figure 10.14 Weightless astronaut

Figure 10.15 Weightlessness

a)

b)

Imagine trying to pour milk on your cornflakes in this space-ship. As everything is weightless, the milk would not fall into the bowl. In fact it would probably go all over the ship. To overcome this problem an astronaut squeezes all his food from a transparent plastic bag directly into his mouth.

How is weightlessness in a space-ship explained? Weight is a force which we experience because it tends to pull us down towards the ground. We exert a force on the ground when we are standing on it and so we feel the effect of it supporting us.

Consider an astronaut in a space-ship orbiting the Earth one thousand kilometres above its surface. At this distance from Earth the gravitational field strength is still strong (about $7{\cdot}4\ \mathrm{N\ kg^{-1}}$). This means that there is a force tending to accelerate the astronaut down to the floor of the space-ship. However, at the same time, there is a force on the space-ship accelerating it downwards, so the floor tends to fall away from the astronaut.

You can think of the astronaut and the space-ship as two projectiles in a continual state of free-fall, both accelerating with the same acceleration due to gravity directed towards the centre of the Earth. This state is called weightlessness.

The effect of weightlessness can be shown in the laboratory using a weighing scales and a 1 kg mass mounted in a box as shown in Figure 10.15 a.

When the box is dropped, a photograph shows that the scales read zero as the box falls, Figure 10.15 b. The 1 kg mass is 'weightless' with respect to the box, although it still has a weight of about 10 N because it is near the Earth's surface.

True weightlessness could only be experienced by a space-ship in a region of outer space where the gravitational field strength is zero.

Summary

In the absence of air resistance, an object's acceleration due to gravity is independent of its mass.

A falling body with a horizontal velocity is called a projectile and its path is called its trajectory.

The motion of a projectile consists of
a constant velocity in the horizontal direction;
and the uniform acceleration due to gravity in the vertical direction.
The two motions are independent.

Newton's thought experiment first suggested the possibility of a satellite orbiting the Earth.

An object is said to be 'weightless' when it is falling freely.

Problems

1 The diagram below represents a stroboscopic photograph of a freely falling ball. The stroboscope is flashing at 6 Hz. From the diagram calculate the acceleration of the ball in m s^{-2}. Give a reason for your answer differing from the accepted value for the acceleration due to gravity.

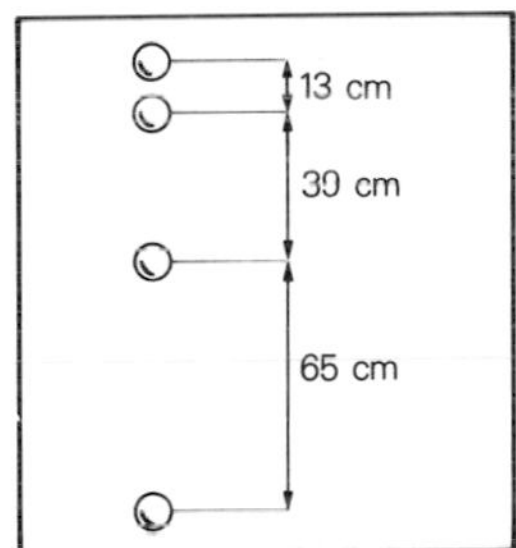

2 The diagram below represents a stroboscopic photograph of a projectile, which falls off a horizontal table at point A. If the stroboscope flashes every 0·1 seconds find:

2·1 the horizontal velocity at D,
2·2 the vertical acceleration,
2·3 the vertical velocity at D.

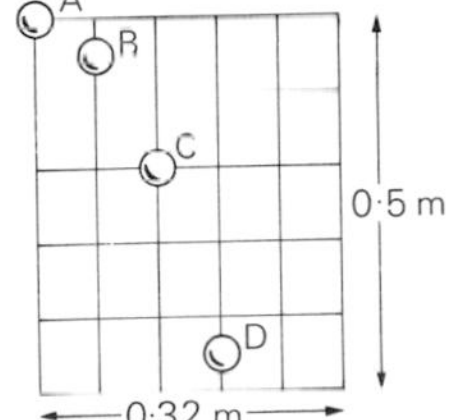

3 A package is jettisoned from a helicopter flying at 30 m s^{-1} horizontally and reaches the ground after 4 s. Taking $g = 10$ m s^{-2}, calculate the following:

3·1 the horizontal velocity of the package just before it touches the ground,
3·2 its vertical velocity at that time,
3·3 the altitude (AP) of the helicopter,
3·4 the horizontal displacement (AB),
3·5 the average vertical velocity of the package,
3·6 the resultant displacement of the package from the instant it is dropped until it hits the horizontal ground.

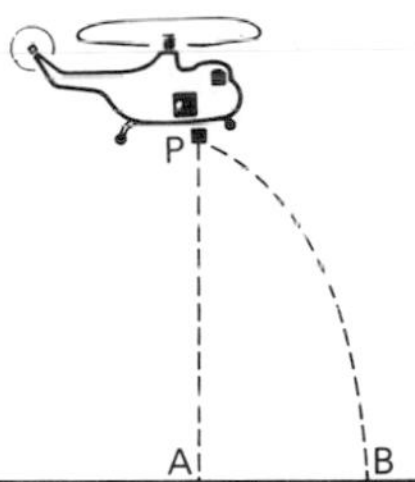

4 A ball rolls down a ramp and is projected horizontally off the end of a table. It falls a vertical height of 0·45 m and hits the ground 2·1 m away at point G.

4·1 How long does it take to fall 0·45 m?
4·2 What is its horizontal velocity as it leaves the ramp?

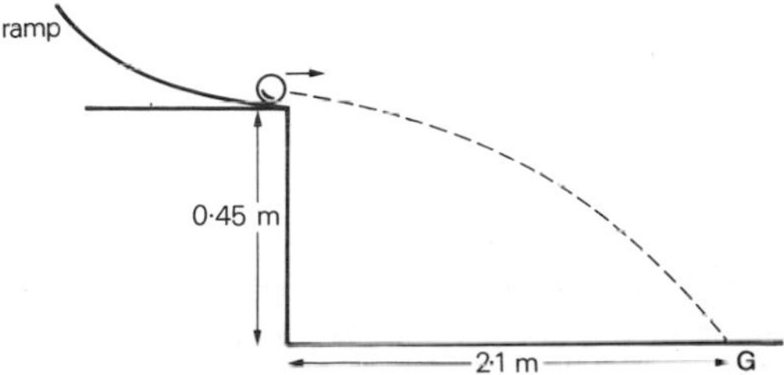

11 Momentum

11.1 Introduction

In this chapter we are concerned with objects which collide. If we try to imagine the collision of a vehicle with a wall, the amount of damage caused will help us to decide what are the important quantities in collisions. Let us consider the following vehicles:

a) a bus travelling at 30 m s^{-1}

b) a car travelling at 30 m s^{-1}

c) a bus travelling at 20 m s^{-1}

If we compare (a) and (b), they are both travelling at the same velocity but since (a) has the greater mass it will cause more damage when it collides.

If we compare (a) and (c), they both have the same mass but since (a) has the greater velocity it will cause most damage.

Mass and velocity, then, are two quantities which we measure in collisions experiments. We can find out about colliding systems by taking measurements before and after the collisions have taken place.

11.2 Inelastic collisions

When two objects stick together after a collision it is described as an **inelastic** collision. An example would be the collision of a lump of putty with the floor.

We can use trolleys and ticker-timers to investigate collision situations. We assume that all the trolleys are of equal mass.

When referring to collision systems and recording results in tables we will use the following abbreviations.

A subscript $_A$ refers to trolley A and $_B$ to trolley B.

m is the mass of the trolley

u is the velocity of the trolley before the collision

v is the velocity of the trolley after the collision

e.g. m_A = mass of trolley A

v_A = velocity of trolley A after the collision

One method of producing an inelastic collision between trolleys is shown in Figure 11.1.

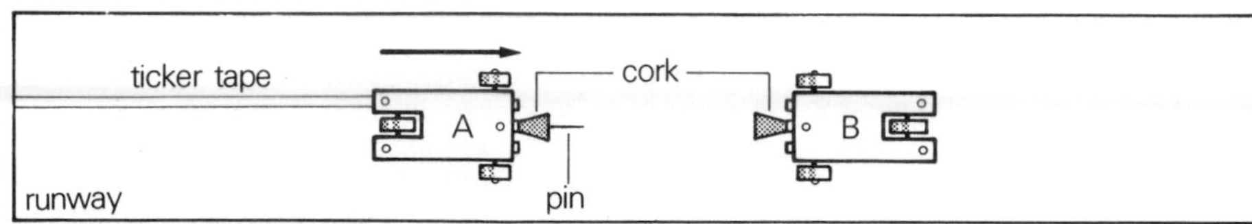

Figure 11.1 Inelastic collision

Figure 11.2 Diagrammatic inelastic collision

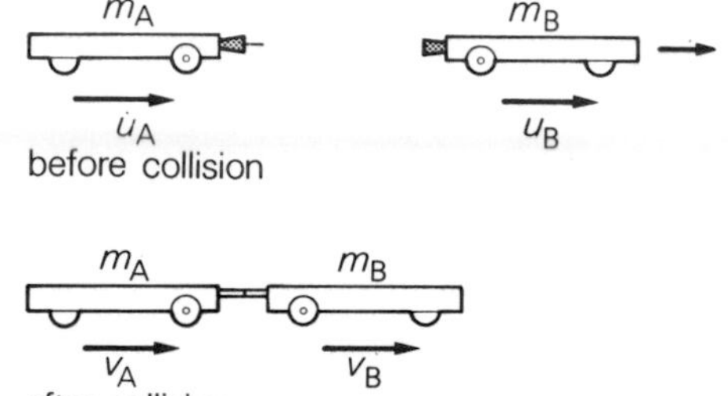

If we fit a cork to one trolley and a cork with a pin to the other, then on colliding the trolleys join together in an inelastic collision. The experiment is carried out on a friction compensated runway.

Trolley B is stationary and trolley A is started with a push so that it moves down the runway at a constant velocity. It collides inelastically with trolley B, Figure 11.2.

Since we deal with collisions in a straight line, the velocity has only two possible directions, left and right. When we record numerical values in our tables and equations we will indicate a velocity value from left to right as positive and one from right to left as negative.

A typical set of tapes for inelastic collisions is shown in Figure 11.3. Point X indicates the instant of collision.

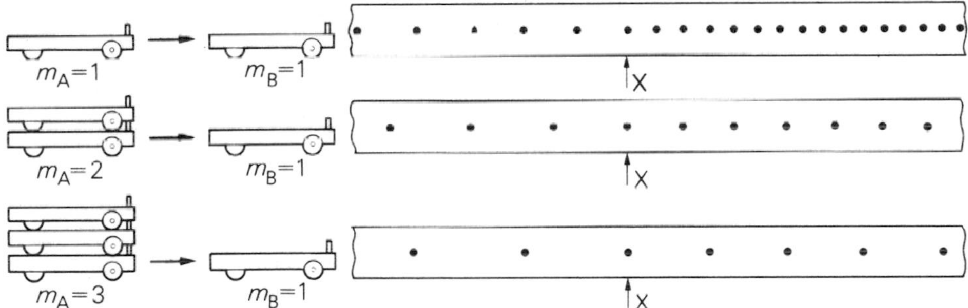

Figure 11.3 Inelastic collision tapes

The results obtained from a set of tapes are shown in Table 1.

		Before collision			*After collision*		
m_A	m_B	u_A	u_B	$m_Au_A + m_Bu_B$	v_A	v_B	$m_Av_A + m_Bv_B$
1	1	10	0	10 + 0 = 10	5	5	5 + 5 = 10
2	1	15	0	30 + 0 = 30	10	10	20 + 10 = 30
3	1	8	0	24 + 0 = 24	6	6	18 + 6 = 24

Table 1 Inelastic collision results

If we calculate the sum $(m_Au_A + m_Bu_B)$ to describe the system before the collision, and $(m_Av_A + m_Bv_B)$ for the system after the collision, then we can see that these two totals are the same before and after the collision.

We will call the quantity mv the **momentum** of an object. The conclusion from this experiment is that, if the trolleys collide inelastically, the total momentum before the collision is equal to the total momentum after the collision. We say that momentum is **conserved** in an inelastic collision. The conservation equation is $m_Au_A + m_Bu_B = m_Av_A + m_Bv_B$.

11.3 Elastic collisions

In an elastic collision the objects do not stay together after impact. An experimental arrangement for investigating elastic collisions between trolleys is shown in Figure 11.4.

Trolley A has an aluminium 'nose' attached to it and trolley B has a rubber band stretched between its upright supports. The nose and rubber band are arranged so that they cause the trolleys to bounce apart, and so the collision is **elastic**. Each trolley has a tape attached to a tape bracket so that the tapes can be pulled without getting in the way of the trolleys. One timer can be used to mark both tapes.

Figure 11.4 Elastic collision

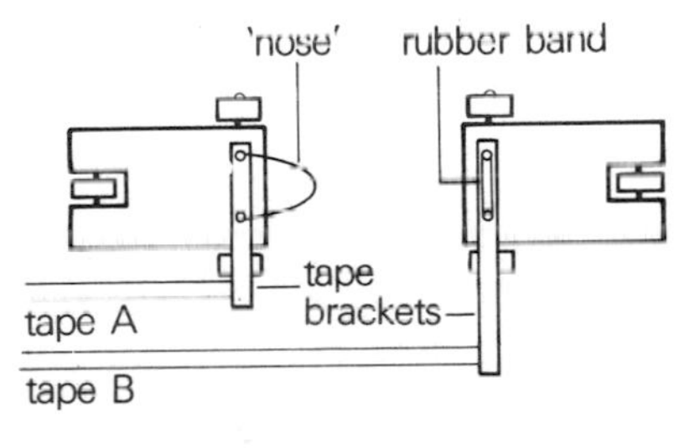

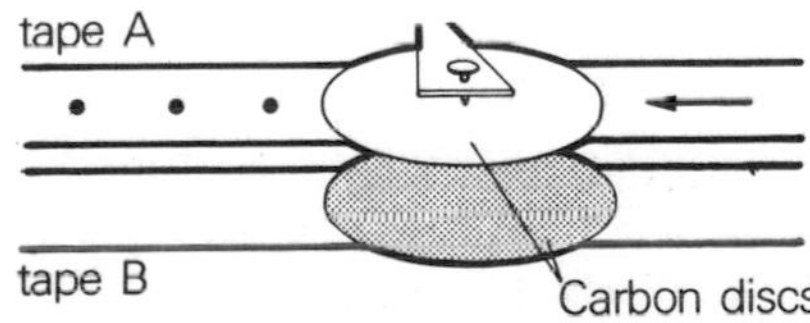

Typical pairs of tapes for elastic collisions between A and B are shown in Figure 11.5.
The results obtained from a set of pairs of tapes are shown in Table 2.

		Before collision				After collision	
m_A	m_B	u_A	u_B	$m_Au_A + m_Bu_B$	v_A	v_B	$m_Av_A + m_Bv_B$
1	1	5	0	$5 + 0 = 5$	0	5	$0 + 5 = 5$
2	1	6	0	$12 + 0 = 12$	2	8	$4 + 8 = 12$
3	1	4	0	$12 + 0 = 12$	2	6	$6 + 6 = 12$

Table 2 Elastic collision results

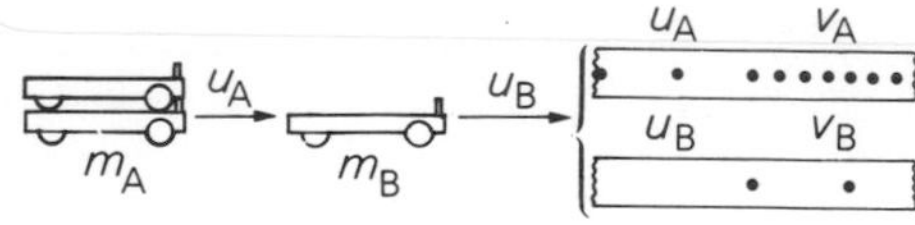

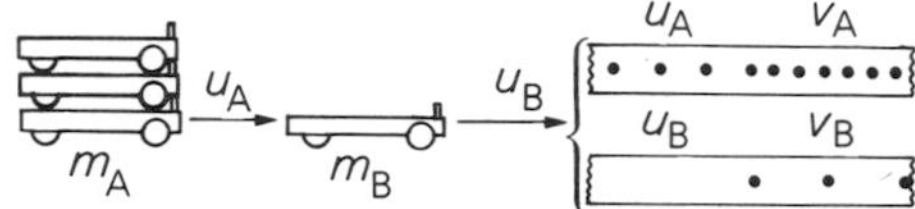

Figure 11.5 Elastic collision tapes

The results show us that the total momentum after the collision is the same as the total momentum before the collision:

$m_Au_A + m_Bu_B = m_Av_A + m_Bv_B$

Momentum is **conserved** in elastic collisions as well as in inelastic collisions.

11.4 Explosions

If we use a stationary trolley which has a spring-loaded plunger placed in contact with another trolley, then we can allow the trolleys to explode apart by triggering off the plunger release mechanism, Figure 11.6.
Some typical tape pairs are shown below, Figure 11.7.

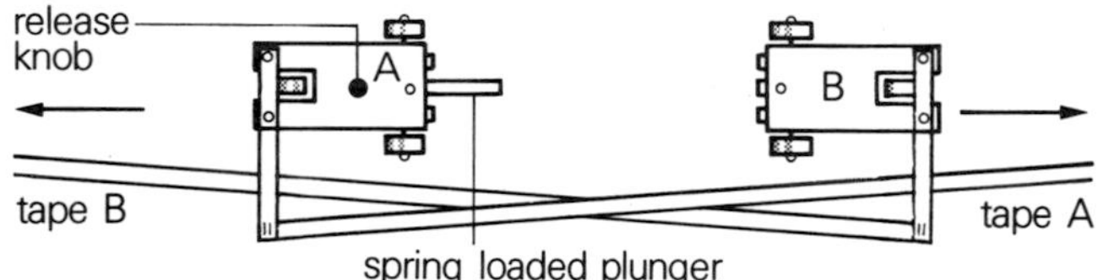

Figure 11.6 Explosive trolley arrangement

Figure 11.7 Explosive trolley tapes

m_A=1 tape A
m_B=1 tape B
m_A=1 tape A
m_B=2 tape B
m_A=1 tape A
m_B tape B

Table 3 shows results obtained from a set of tapes. Note that all the values of v_A are **negative** since they are to the left.

		Before				After	
m_A	m_B	u_A	u_B	$m_Au_A + m_Bu_B$	v_A	v_B	$m_Av_A + m_Bv_B$
1	1	0	0	$0 + 0 = 0$	−13	13	$-13 + 13 = 0$
1	2	0	0	$0 + 0 = 0$	−20	10	$-20 + 20 = 0$
1	3	0	0	$0 + 0 = 0$	−18	6	$-18 + 18 = 0$

Table 3 Explosion results

We can see that since both trolleys were stationary before the explosion, the total momentum before the explosion was zero. When the direction of v_A is taken into account then the total momentum after the explosion is zero too.

$$m_A u_A + m_B u_B = m_A v_A + m_B v_B$$

We can conclude that momentum is conserved in an explosion.

We may summarize our results by saying that it is reasonable to conclude that **momentum is conserved** for elastic and inelastic collisions and explosions. This is called the Law of Conservation of Momentum.

We have defined the momentum of an object as the product mass times velocity $m \times v$ and so we can determine the units of momentum.

The unit of mass is the kilogram and the unit of velocity is the metre per second. Therefore the unit of momentum is the kg m s^{-1}.

Momentum is a vector quantity. In the above explosion experiment the magnitude of the final momentum of both trolleys is the same but in opposite directions. Vector addition gives a total momentum of zero.

11.5 Worked examples

Example 1

a) A car has a mass of 180 kg and travels at 10 m s^{-1}. How much momentum has it?

b) A vehicle has a mass of 200 kg and momentum 380 kg m s^{-1}. At what velocity is it travelling?

a) mass $m = 180$ kg velocity $v = 10$ m s^{-1}

$$\text{Momentum} = m \times v = 180 \times 10 = 1\,800$$

The car has momentum of 1 800 kg m s^{-1}.

b) mass $m = 200$ kg $m\,v = 380$ kg m s^{-1}

$$\text{Momentum} = m \times v$$

$\Rightarrow$ $380 = 200 \times v$

$\Rightarrow$ $\frac{380}{200} = v$

$\Rightarrow$ $1{\cdot}9 = v$

The vehicle has a velocity of 1·9 m s^{-1}.

Example 2

A trolley of mass 3 kg travelling at a velocity of 4 m s^{-1} collides inelastically with a trolley of mass 1 kg which is at rest.

At what velocity do the trolleys move after the collision?

before collision
m_A u_A m_B u_B

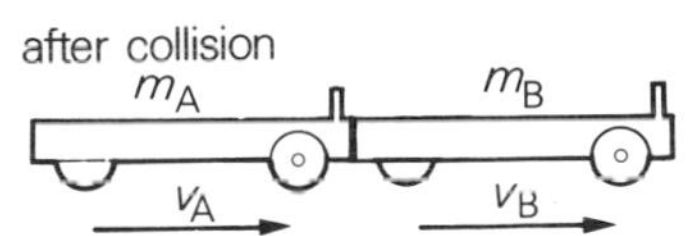

Total momentum before collision $= m_A u_A + m_B u_B$

Total momentum after collision $= m_A v_A + m_B v_B$

Momentum is conserved in an inelastic collision

Therefore $m_A u_A + m_B u_B = m_A v_A + m_B v_B$, the conservation equation.

$m_A = 3$ kg $u_A = 4$ m s^{-1} $m_B = 1$ kg $u_B = 0$ $v_A = v_B$

Substituting these values in the conservation equation above

$(3 \times 4) + (1 \times 0) = (3 \times v_A) + (1 \times v_B)$

$\Rightarrow$ $12 = (3 \times v_A) + (1 \times v_A)$ since $v_A = V_B$

$\Rightarrow$ $12 = 4v_A$

$\Rightarrow$ $v_A = 3$

The trolleys both travel at 3 m s^{-1} to the right after colliding.

Example 3

A trolley of mass 2 kg travelling from left to right at 4 m s^{-1} collides elastically with another trolley of mass 4 kg which is travelling from right to left at 1 m s^{-1}.
If the 2 kg trolley bounces back at 2 m s^{-1} after the collision, what is the velocity of the 4 kg trolley after the collision?

Momentum is conserved in an elastic collision.
Therefore total momentum before collision = total momentum after collision $m_A u_A + m_B u_B = m_A v_A + m_B v_B$
The value of a velocity to the left is negative.
$m_A = 2$ kg $\quad u_A = 4$ m s^{-1} $\quad v_A = -2$ m s^{-1} $\quad m_B = 4$ kg $\quad u_B = -1$ m s^{-1}
Substituting these values into the conservation equation
$(2 \times 4) + (4 \times -1) = (2 \times -2) + (4 \times v_B)$
$\Rightarrow \quad 4 = -4 + 4v_B$
$\Rightarrow \quad 8 = 4v_B$
$\Rightarrow \quad \frac{8}{4} = v_B$
$\Rightarrow \quad v_B = 2$
The 4 kg trolley travels at 2 m s^{-1} to the right after the collision.

Example 4

A 2 kg explosive trolley is placed next to a 5 kg trolley. When they are exploded apart, the 5 kg trolley travels to the right at 2·6 m s^{-1}. What is the velocity of the 2 kg trolley after the explosion?

Momentum is conserved.
$\Rightarrow m_A u_A + m_B u_B = m_A v_A + m_B v_B$
$m_A = 2$ kg $\quad m_B = 5$ kg $\quad u_A = u_B = 0 \quad v_B = 2{\cdot}6$ m s^{-1}
Substituting in the conservation equation:
$2 \times 0 + 5 \times 0 = 2 \times v_A + 5 \times 2{\cdot}6$
$0 = 2v_A + 5 \times 2{\cdot}6$
$\Rightarrow \quad -2v_A = 5 \times 2{\cdot}6$
$\Rightarrow \quad v_A = \frac{5 \times 2{\cdot}6}{-2}$
$\Rightarrow \quad v_A = -6{\cdot}5$
The 2 kg trolley travels to the left at 6·5 m s^{-1} after the collision.

Example 5

A 1 kg trolley is travelling along at a constant velocity of 0·6 m s^{-1} when an object of mass 0·5 kg is dropped on to it. What is the trolley's velocity then?

Momentum is conserved.
$\Rightarrow m_A u_A + m_B u_B = m_A v_A + m_B v_B$
$m_A = 1$ kg $\quad u_A = 0{\cdot}6$ m s^{-1}
$m_B = 0{\cdot}5$ kg $\quad u_B = 0$
$v_A = v_B$ because the 0·5 kg object stays on the trolley.
The addition of this mass is equivalent to an inelastic collision.
Substituting in the conservation equation:
$1 \times 0{\cdot}6 + 0{\cdot}5 \times 0 = 1 \times v_A + 0{\cdot}5 \times v_B$
$\Rightarrow \quad 0{\cdot}6 = 1{\cdot}5 \times v_A \quad$ since $v_A = v_B$
$\Rightarrow \quad \frac{0{\cdot}6}{1{\cdot}5} = v_A$
$\Rightarrow \quad 0{\cdot}4 = v_A$
The trolley travels at 0·4 m s^{-1} to the right.

11.6 Linear air-track collisions

The linear air-track is an alternative friction-free track on which we can investigate collisions in a straight line. The motion can be analysed photoelectrically or with strobe photography, Figure 11.8.

Figure 11.9 shows an inelastic collision between an air-track vehicle of mass 1 unit with a stationary vehicle of mass 2 units.
The left-hand vehicle approaches at a constant velocity since the straw image positions are equally spaced before the collision. After the collision both vehicles travel to the right with the same constant velocity.
We find the average velocity u_A from the equation

$$\text{average velocity} = \frac{\text{total displacement}}{\text{time taken}}$$

$$\Rightarrow \qquad u_A = \frac{0{\cdot}24}{0{\cdot}40} = 0{\cdot}6 \text{ m s}^{-1}$$

Similarly $$v_A = \frac{0{\cdot}24}{1{\cdot}20} = 0{\cdot}2 \text{ m s}^{-1}$$

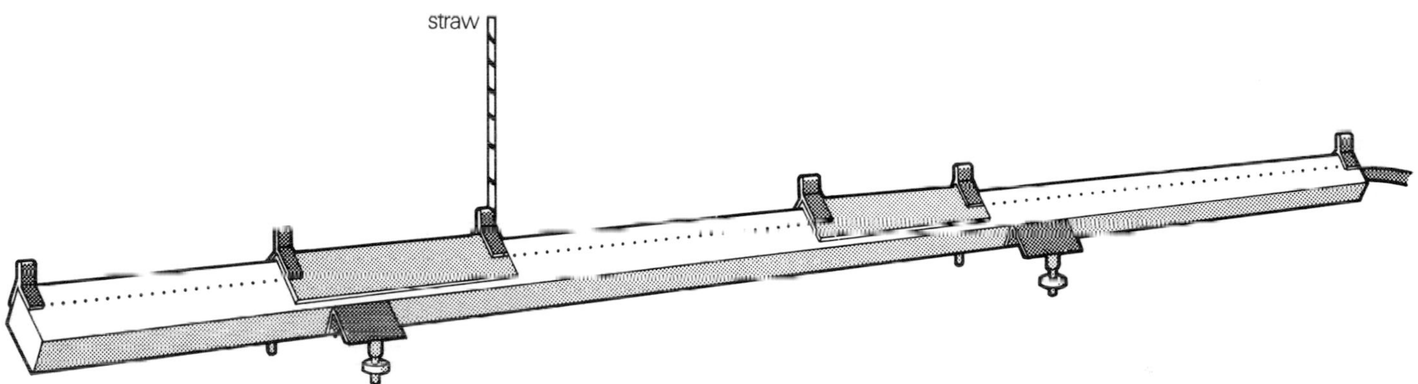

Figure 11.8 Linear air-track collisions

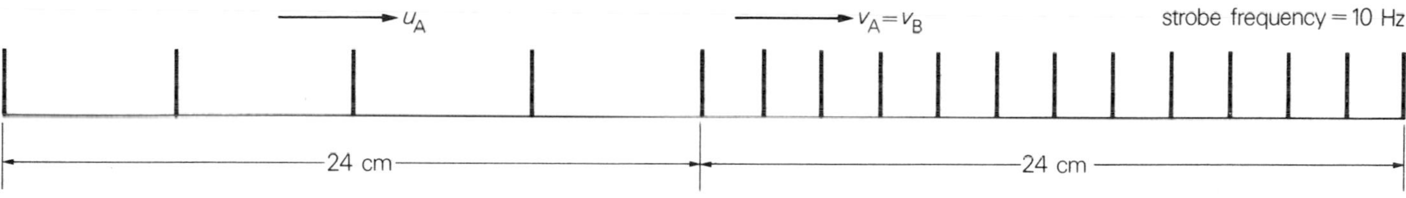

Figure 11.9 Linear air-track vehicle collision (inelastic)

Table 4 shows the total momentum before and after the collision.

		Before collision			*After collision*		
m_A	m_B	u_A	u_B	$m_Au_A + m_Bu_B$	v_A	v_B	$m_Av_A + m_Bv_B$
1	2	0·6	0	0·6 + 0 = 0·6	0·2	0·2	0·2 + 0·4 = 0·6

Table 4

This set of results confirms that the total momentum before and after the collision is conserved.

The velocity of an air-rifle pellet

On page 75 in chapter 8 we referred to the measurement of the velocity of an air-rifle pellet. We can now calculate this velocity indirectly if the air-rifle pellet is allowed to 'collide' with a linear air-track vehicle and we consider the momentum of the system, Figure 11.10.

The pellet is fired at a lump of plasticine on the vehicle to give an inelastic collision. The velocity of the vehicle after the collision is measured by a 10 cm card attached to it passing between a photo-detector and light source. The time can be registered on a digital timer or on an electric stopclock.

By using the equation for the conservation of momentum we can calculate the velocity of the air-rifle pellet.

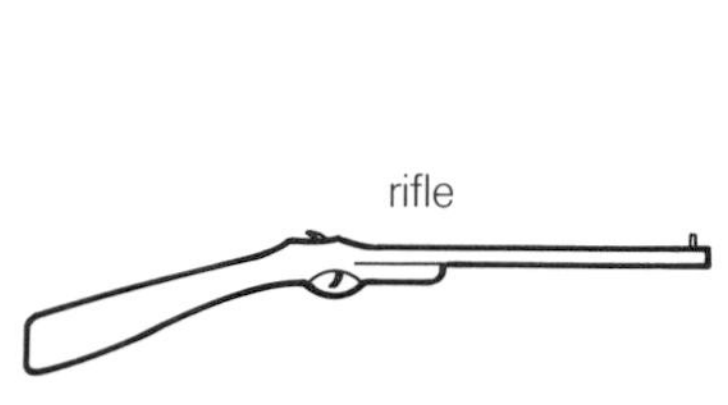

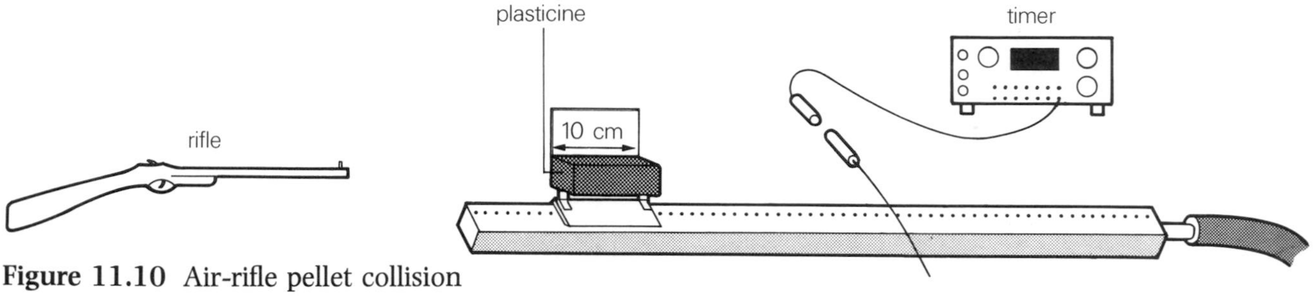

Figure 11.10 Air-rifle pellet collision

Example 6

Use the results given below for the collision of an air-rifle pellet with plasticine on a linear air-track vehicle to find the velocity of the pellet.
mass of one pellet = 0·0005 kg
mass of vehicle + plasticine + card = 0·240 kg
time for 10 cm card to pass through light beam = 0·375 s

Momentum is conserved,
so total momentum before collision = total momentum after collision

$$\Rightarrow \quad m_A u_A + m_B u_B = m_A v_A + m_B v_B$$

where mass of pellet $m_A = 0{\cdot}0005$ kg
mass of vehicle + plasticine + card $m_B = 0{\cdot}240$ kg
$u_A = ?\ \text{m s}^{-1} \quad u_B = 0\ \text{ms}^{-1}$

$$v_A = v_B = \frac{\text{length of card}}{\text{time to pass through light beam}} = \frac{0{\cdot}1}{0{\cdot}375}\ \text{m s}^{-1}$$

$$\Rightarrow 0{\cdot}0005 \times u_A + 0{\cdot}240 \times 0 = 0{\cdot}0005 \times \frac{0{\cdot}1}{0{\cdot}375} + 0{\cdot}240 \times \frac{0{\cdot}1}{0{\cdot}375}$$

$$\Rightarrow \quad 0{\cdot}0005\ u_A = 0{\cdot}2405 \times \frac{0{\cdot}1}{0{\cdot}375}$$

$$\Rightarrow \quad u_A = \frac{0{\cdot}2405 \times 0{\cdot}1}{0{\cdot}0005 \times 0{\cdot}375} \approx 128$$

The velocity of the pellet ≈ 128 m s^{-1} to the right.

11.7 Paired forces and Newton's Third Law

Imagine yourself wearing roller skates and standing facing a wall, Figure 11.11. If you exert a force on the wall, what happens? The wall exerts a force in the opposite direction on you and you travel backwards.

A sprinter at the start of a race wants to go forwards and so he pushes backwards on his starting blocks, Figure 11.12.
The sprinter exerts a force on his blocks and his blocks exert a force on him in the opposite direction.

Figure 11.11

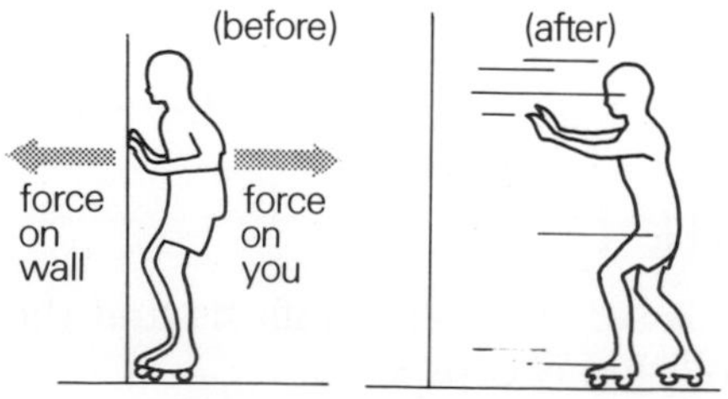

(before)
force on wall
force on you
(after)

When a rifle is fired the gun exerts a force on the bullet (Figure 11.13), but the bullet exerts a force in the opposite direction on the gun.

All these examples have one thing in common; when there is a force exerted by A on B, there is a force exerted by B on A in the opposite direction. **Forces occur in pairs.**

Let us think again about the explosive trolley collision on page 110. It might have seemed that only one of the trolleys should have moved, but in fact they both did in different directions. We now see that both the left-hand trolley exerted a force on the right-hand trolley in one direction and the right-hand trolley exerted a force on the left-hand trolley in the opposite direction. The trolleys travelled in opposite directions with the same velocity. It seems therefore, that the pair of forces involved in this collision are equal in value and opposite in direction. This principle is summarized by **Newton's Third Law**:

'To every action force there is an equal and opposite reaction force'.

Note that the action force acts on one object and the reaction force acts on the other object, but it does not matter which force we label as the action force and which as the reaction force: the one does not exist without the other.

We can illustrate this with a trolley which fires a billet on releasing a plunger, Figure 11.14. When a billet is fired, the trolley exerts a force on the billet and the billet exerts a force on the trolley. A check on the final velocity of billet and trolley shows that momentum is conserved.

A heavy steel billet gives the trolley a higher velocity than a lighter wooden billet. If the plunger is released with no billet loaded on the trolley, there is no reaction force on the trolley and it does not move.

Newton's Third Law describes the way in which objects 'interact', and is another way of expressing the conservation of momentum.

Another illustration is provided by the water rocket apparatus, Figure 11.15. The air in the water rocket exerts a force downwards on the water (action). There is an equal and opposite force (reaction) exerted by the water on the air inside the rocket. This causes the rocket to move upwards.

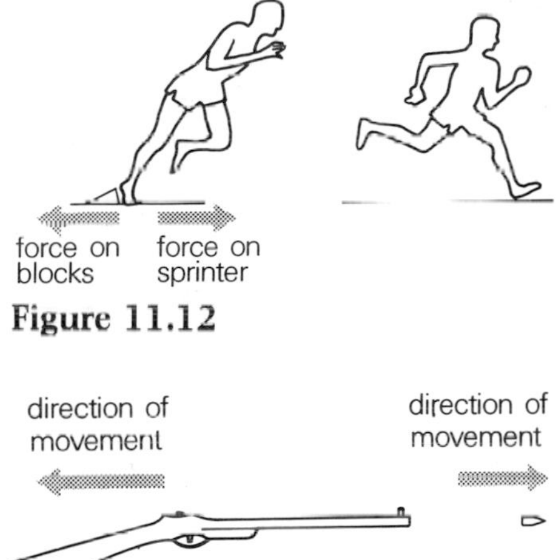

Figure 11.12

Figure 11.13

Figure 11.14 Firing billets

Figure 11.15 Water rocket

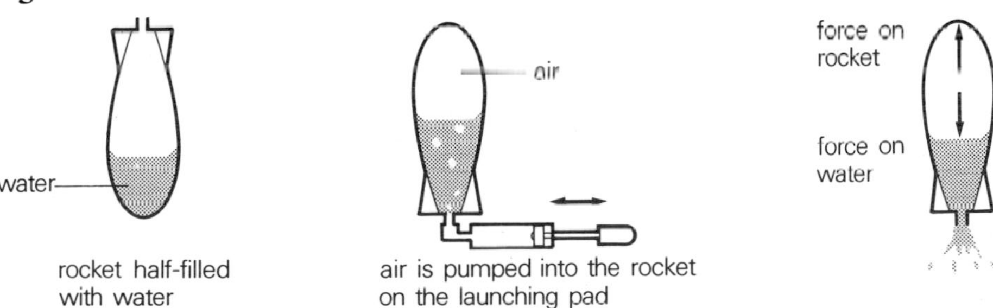

Summary

Momentum is a vector quantity.

Momentum is the product of mass times velocity. The SI unit of momentum is kg m s^{-1}.

In any collision or explosion
total momentum before = total momentum after.
This is the Law of Conservation of Momentum.

Newton's Third Law states that
To every action force there is an equal and opposite reaction force.
Thus when two objects interact, they exert forces on each other which are of equal magnitude and act in opposite directions.

Problems

1 Three railway wagons are stationary some distance apart on a friction-compensated railway line. All the wagons are of the same mass and a fourth wagon approaches at 24 m s^{-1}. If a series of three inelastic collisions result, what will be the final velocity of the four wagons coupled together?

2 A man of mass 80 kg stands on ice and throws an object of mass 2 kg horizontally at a velocity of 10 m s^{-1}. At what velocity does he recoil?

3 A target of mass 4 kg hangs from a tree by a long string. An arrow of mass 100 g is fired with a velocity of 100 m s^{-1} and embeds itself in the target. At what velocity does the target begin to move after the impact?

4 A bus of mass 4000 kg travelling at 20 m s^{-1} collides inelastically with a coach which is approaching from the opposite direction at 32 m s^{-1}. If the coach has a mass of 2 500 kg, at what velocity do the bus and coach travel after the collision?

5 Two vehicles are fired from opposite ends of a horizontal frictionless air-track using catapults X and Y constructed from single identical elastic bands. Vehicle A has a mass of 0·3 kg and vehicle B has a mass of 0·1 kg.

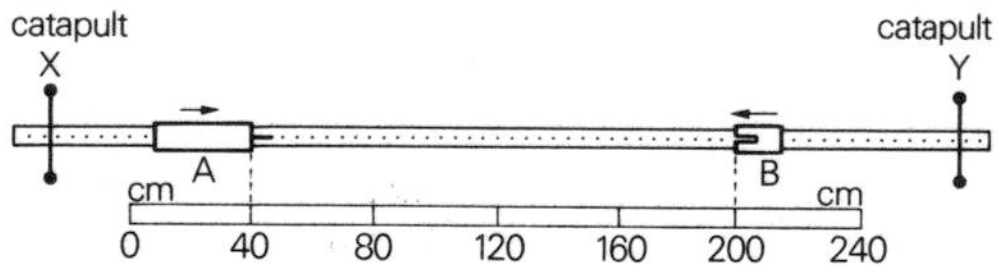

Figure A

Two seconds after passing the positions shown in Figure A, the vehicles collide and stop opposite the 80 cm mark as shown in Figure B.

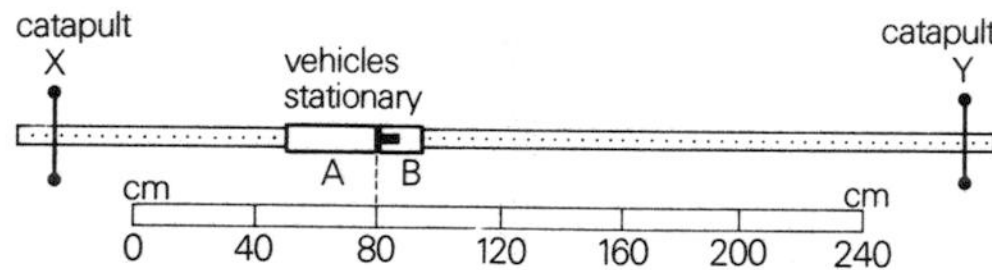

Figure B

5·1 Copy and complete the following table for the motion of the vehicles before the collision.

Quantity	*Unit*	*Vehicle* A	*Vehicle* B
mass	kg	0·3	0·1
speed	m s^{-1}		
momentum	kg m s^{-1}		

5·2 What is the total momentum
a) before the collision and
b) after the collision?

SCEEB

6 A bullet is fired horizontally at a target fixed to a trolley which is resting on a horizontal surface. The bullet embeds itself in the target.

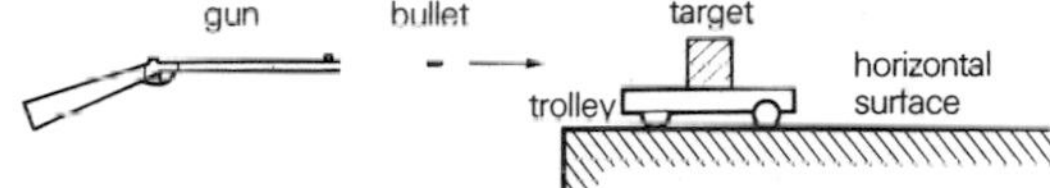

6·1 Describe the motion of the trolley and target from the time the bullet strikes the front surface of the target until several seconds later. (Neglect the effect of air resistance and friction in the wheels.)
6·2 List all the measurements you would make to estimate the speed of the bullet using this apparatus.
6·3 Why does the gun recoil when the bullet is fired?
6·4 How would the final speed of the trolley and target compare with that in **6·1** if the bullet had
a) passed through the target and emerged from the other side with reduced speed,
c) bounced backwards off the target instead of penetrating it? *SCEEB*

7 This is a simplified diagram of a rocket of mass 12×10^3 kg with small fine adjustment rocket motors A, B, C, D, each exerting a thrust of 600 N when fired. The diagram also shows a probe which has to dock (i.e. join together) with the rocket.

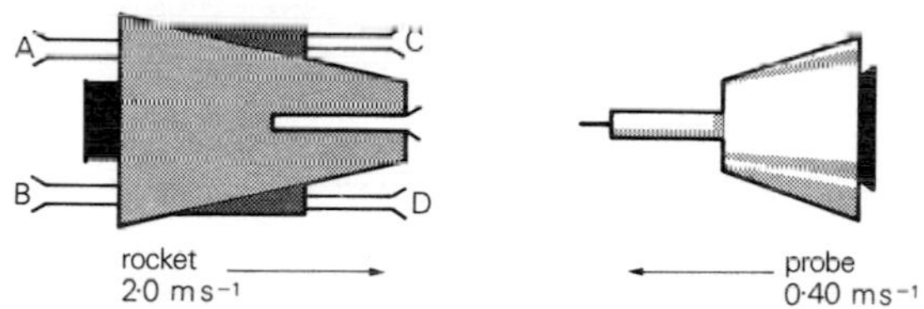

7·1 When the probe comes into view of the rocket all the motors are switched off and the velocities are 2·0 m s^{-1} and 0·40 m s^{-1} in the directions shown.
In order to dock, it is decided to slow the rocket down to 0·20 m s^{-1} by firing motors C and D. How long must this firing last?
7·2 After the motors have been fired the rocket and probe approach slowly and dock. The combination of rocket and probe is then found to be at rest. What must the mass of the probe have been for this to happen?
7·3 Describe what firings of the motors would be required to shift the combination of rocket and probe several metres to the right and leave it again at rest? *SCEEB*

12 Energy, work and power

12.1 Energy

Introduction

Earlier in your science course you will have heard of various forms of energy. In this chapter we will be concerned with three forms of energy:
kinetic energy,
gravitational potential energy,
and strain energy.
However, we will also be mentioning chemical energy, heat energy and sound energy. We will start by describing the first three energy forms.

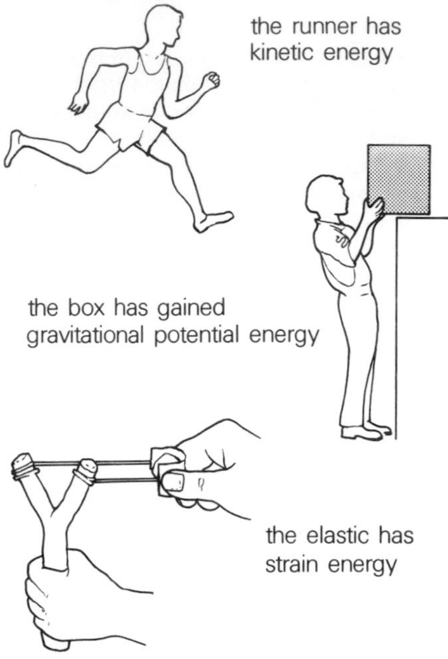

Figure 12.1 Forms of energy

Kinetic energy (energy of movement)

Any object which is moving has kinetic energy. When we run, some of our chemical (food) energy is changing into kinetic energy.

Gravitational potential energy (stored energy)

If you lift a heavy box on to a shelf, some of the chemical energy from your food is used up. However this amount of energy has not disappeared but has been stored in the box as gravitational potential energy. You could convince yourself that the box has stored energy by pushing it off the shelf. The box will fall and the gravitational potential energy will change into kinetic energy.

Strain energy

When you pull back the elastic of a catapult some of the chemical energy from your food is changed into strain energy in the elastic. Strain energy is another form of potential energy. If the elastic is released, the strain energy changes into kinetic energy. Strain energy is therefore the energy possessed by stretched elastic or by springs.

Figure 12.1 shows examples of objects possessing these energy forms.

Energy changes

You may carry out some experiments involving energy changes as indicated in Figure 12.2.

In Figure 12.2 (a) a trolley is given a push. Chemical energy from food is changed into kinetic energy.

In Figure 12.2 (b) a stretched catapult is used to launch a trolley. Strain energy in the catapult is changed into the kinetic energy of the trolley.

In Figure 12.2 (c) a trolley rolls from the top of a steep slope. Gravitational potential energy is changed into kinetic energy.

In Figure 12.2 (d) two trolleys are joined by a stretched spring. They move towards each other and strain energy is changed into kinetic energy. When they collide, however, the trolleys become stationary. The kinetic energy is therefore now zero. However, this amount of energy has not disappeared: it has been changed from kinetic energy mainly into heat (and also a little sound energy).

Figure 12.2 Energy changes

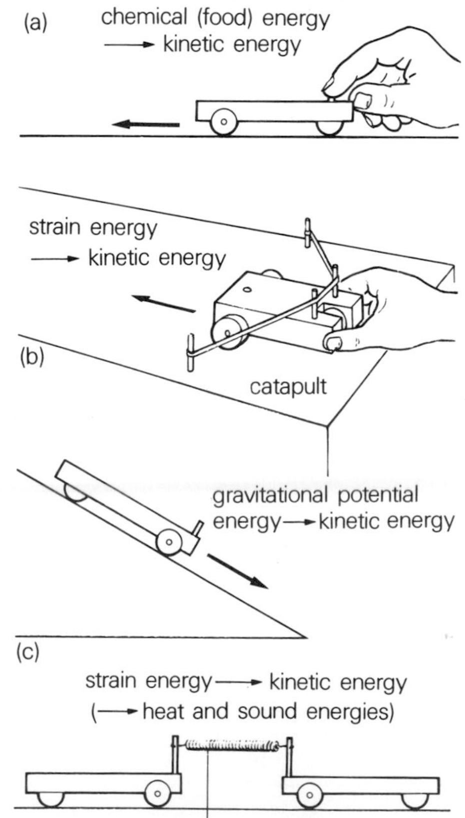

12.2 Work

We are all very familiar with the word 'work'. It is used to describe many different kinds of activities. For example, you may think that doing physics problems or lifting a heavy box are both hard work! However, in physics, we have to be more specific about what we mean by work.

We say that **work is done** only when **energy is transferred.** It follows that the second of our two examples is closer to what we mean by work in physics. You have to exert a force to lift the box and energy is transferred from you to the box. The amount of work you have to do is greater either if the box is lifted higher (greater displacement), or if the box is heavier (greater force required). In fact, work can only be done when a force moves an object through a certain distance. The amount of work required is given by

work = force × displacement in the direction of the force

i.e. work = $\mathbf{F} \times \mathbf{s}$

where $\mathbf{F}$ = applied force
$\mathbf{s}$ = displacement in direction of force

You will remember that both force and displacement are vector quantities. However, work or energy do not have any particular direction: they are therefore scalar quantities.

Note that our definition of work tells us that no work is done unless a force actually moves an object. For example, although you may become tired standing holding a heavy shopping bag at a bus stop, you are not doing any work on the shopping bag.

Units

Since the unit for force is the newton N and the unit for displacement is the metre m, the unit for work is the newton-metre N m. This unit is called the **joule**, written as J. Hence, when a force of 1 newton is used to move an object a distance of 1 metre, 1 joule of work is done,

i.e. 1 joule = 1 newton × 1 metre
or 1 J = 1 N m

Since work is a measure of the amount of energy transferred, we use the same unit, the joule, to measure both energy and work. Table 1 shows some representative energy values in megajoules (1 MJ = 1 000 000 joules).

Energy involved	*Typical value*
Energy output in 1 second of a power station	2 000 MJ
Domestic energy used by a British family in one day	150 MJ
Daily chemical (food) energy for a schoolboy	13 MJ
One day's heavy manual work	3 MJ
Kinetic energy of a car on a motorway	0·5 MJ
Chemical (food) energy of a sugar lump	0·06 MJ

Table 1

Modern man is very dependent on energy for factories and transport and to keep his house warm, among many other things. There is world-wide concern about the rate at which sources of energy such as coal, oil and gas are being used up.

Example 1

A horse applies a constant force of 100 N on a cart over a distance of 80 m.
Calculate the amount of work done by the horse.

Work = force × displacement in the direction of the force
= 100 × 80
= 8 000

Work done is 8 000 J

Figure 12.3

12.3 Machines

Many machines change energy from one form to another. For example, a bicycle dynamo changes some of the wheels' kinetic energy into electrical energy which is changed into heat and light in the bicycle lamps. However, some machines do not change the form of energy, but simply allow us to do work or to transfer energy more easily. One common example of this type of machine is the lever.

Lever

We can use a lever to exert a much larger force than we can actually apply. Figure 12.4 shows an experimental arrangement to investigate how a lever works.

By applying a force called the **effort**, an object (called the **load**) can be raised. The effort required to raise the load is measured on a spring balance (or newton balance) calibrated in newtons. The usefulness of the lever in Figure 12.4 is that a large load of weight 10 N can be raised by applying a much smaller effort force of 1 N: the force exerted on the load is ten times the effort.

The lever is therefore often referred to as a **force multiplier**. However, the lever does not multiply the energy as the following example shows.

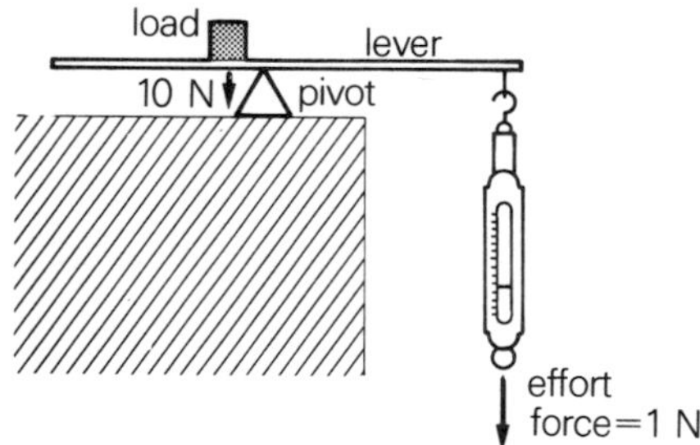

Figure 12.4 Lever

Example 2

The effort required to raise the 10 N load in Figure 12.4 is 1 N. The load is raised by 0·1 cm when the effort moves 1 cm.

a) Calculate the energy required to raise the load. We call this the useful output energy.

b) Calculate the work done by the effort to raise the load by this distance. We call this the input energy.

c) Comment on the answers to parts (a) and (b).

a) Work = **F** × **s**

work to raise load = weight of load × displacement of load
= $10 \times 0{\cdot}1 \times 10^{-2}$

⇒ work to raise load = 0·01 J

b) Work done by effort = effort × displacement of effort
= 1 × 0·01

⇒ work done by effort = 0·01 J

c) The work done by the effort (input energy) is equal to the work required to raise the load (useful output energy). This machine therefore transfers all the input energy to useful output energy, i.e. there are no energy 'losses'.

In example 2, note that, although the load is lifted by applying a much smaller effort, the effort has to move a much larger distance than the load to compensate. In this sense the lever does not give 'something for nothing', Figure 12.5.

In fact this applies to all machines. The St. Gotthard Pass in Switzerland (Figure 12.6) changes a steep hill into a much more gentle one so that car engines do not have to exert such a large force to reach the top. However, it is obvious from the photograph that the car must travel a much greater distance to reach the top.
Figure 12.7 shows some everyday applications of the lever.

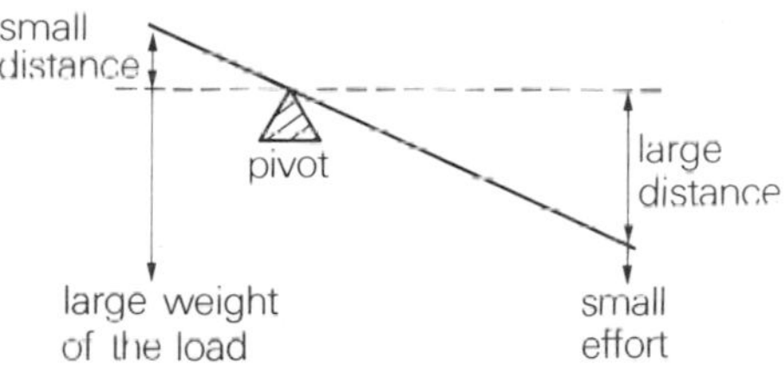

Figure 12.5 The lever in action

Figure 12.6 St. Gotthard Pass

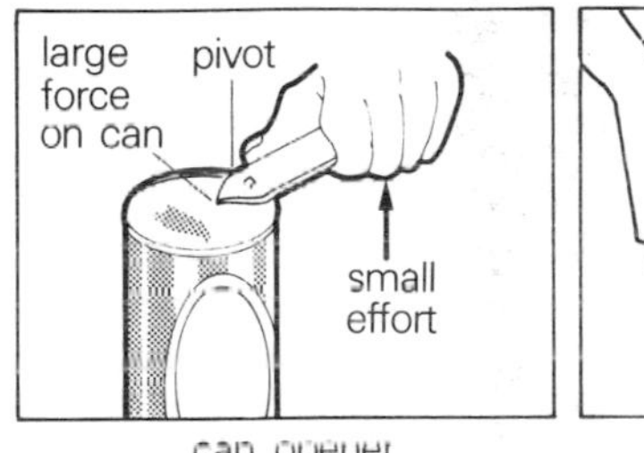

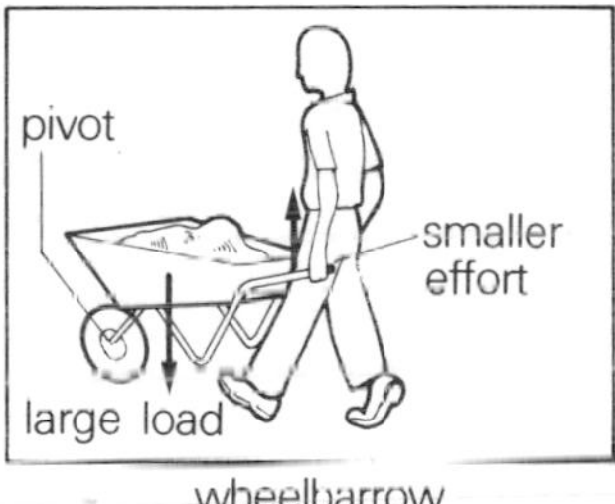

can opener moving a boulder wheelbarrow

Figure 12.7 Applications of the lever

Figure 12.8

Pulleys

A pulley or a system of pulleys is another commonly used machine. For example, a set of pulleys called a block and tackle is used in garages for lifting an engine out of a car. Pulleys are also used in a crane for lifting cargo on to a ship, Figure 12.8.

Figure 12.9 shows an experimental arrangement where a block and tackle is being used to raise a load. Using this arrangement a heavy load of 90 N can be lifted by a much smaller effort. However to compensate for this, the effort has to move a great distance to raise the load by a small amount. In the example shown, the effort has to move three times as far as the load.

Figure 12.9 Using a system of pulleys

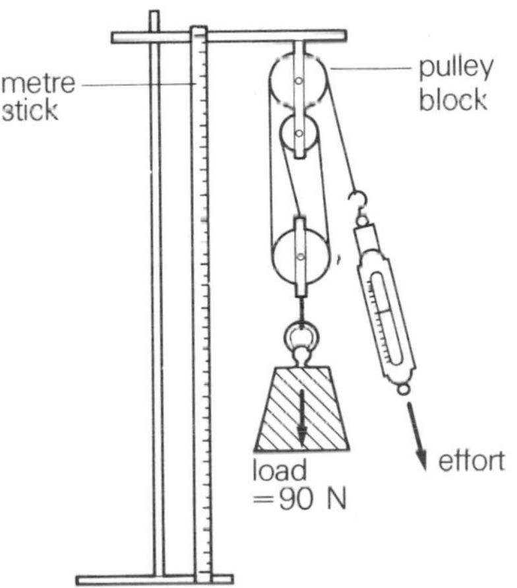

Example 3

In Figure 12.9, a 90 N load is lifted 2 cm by applying an effort, measured by the newton balance. If the effort moves a distance of 6 cm, calculate its value.

Useful output energy = load × upward displacement of load
= 90 × 0·02
⇒ useful output energy = 1·8 J
Input energy = effort × displacement of newton balance
⇒ input energy = F × 0·06 (where F = effort)

Assume that all the input energy is transferred to become useful output energy.

Input energy = useful output energy
⇒ F × 0·06 = 1·8 (divide both sides by 0·06)

$$\Rightarrow \quad F = \frac{1{\cdot}80}{0{\cdot}06}$$

⇒ F = 30 **Effort required is 30 N.**

Efficiency

In practice, the effort required for the machine in example 3 is found to be larger than 30 N. This means that the input energy actually needed is greater than the useful output energy. Not all the input energy is transferred by the pulleys to become useful output energy.

This is true for all machines. The useful energy we get from a machine is never as great as the energy we put in. Some of the input energy is always changed into other energy forms such as heat (and possibly some sound energy). Even in the lever a tiny amount of the input energy is changed into heat energy at the pivot.

However, it is important to realize that no energy is actually destroyed, so that the total amount of energy does stay the same. This means that the total energy is always conserved.

We use the term **efficiency** as a way of expressing how well a machine converts input energy into useful output energy. It is usually expressed as the percentage of the input energy that finally appears as useful output energy:

$$\text{efficiency (\%)} = \frac{\text{useful output energy}}{\text{input energy}} \times \frac{100}{1}$$

Example 4

Suppose that 100 J of energy (not necessarily in the same energy form), is fed into three machines,
a) a motor car,
b) a block and tackle, and
c) a lever.

The useful output energies are as follows: (a) 10 J, (b) 50 J, and (c) approximately 100 J. Calculate the percentage efficiency of each machine.

a) efficiency (%) $= \frac{\text{useful output energy}}{\text{input energy}} \times \frac{100}{1}$

⇒ efficiency of car $= \frac{10}{100} \times \frac{100}{1} = 10\%$

b) efficiency (%) $= \frac{\text{useful output energy}}{\text{input energy}} \times \frac{100}{1}$

⇒ efficiency of block and tackle $= \frac{50}{100} \times \frac{100}{1} = 50\%$

c) efficiency (%) $= \frac{\text{useful output energy}}{\text{input energy}} \times \frac{100}{1}$

⇒ efficiency of lever $= \frac{100}{100} \times \frac{100}{1} = 100\%$

The efficiency of all real machines is less than 100%. However, unless otherwise stated in a problem, you can only assume that the efficiency of the machine is 100% and the useful output energy is equal to the input energy.

12.4 Kinetic energy

You now know that kinetic energy is the energy that an object has because it is moving. In this section we will investigate what exactly we mean by kinetic energy. Figure 12.10 shows the apparatus that we use.

A linear air-track vehicle is launched by means of an elastic thread. The time for which a card attached to the vehicle breaks a light beam is measured on an electronic timer. Hence the speed of the vehicle along the track can be calculated using the equation,

$$\text{speed of vehicle} = \frac{\text{length of card}}{\text{time for which beam is broken}} = \frac{\text{length of card}}{\text{reading on timer}}$$

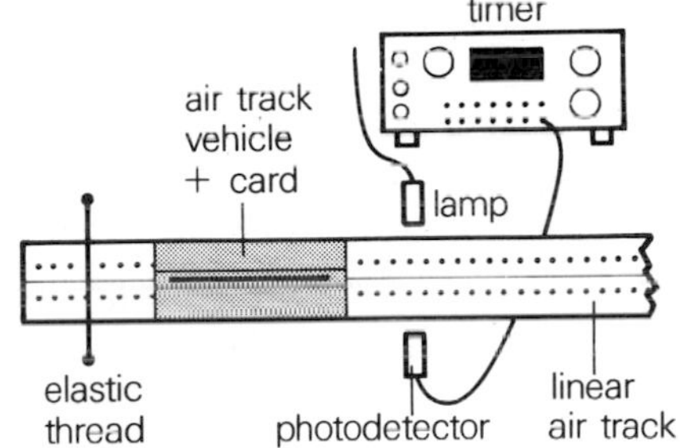

Figure 12.10

We will say that it takes 1 unit of work to extend an elastic string by a distance s. This energy is now stored in the elastic as strain energy. If this elastic is used to launch an air-track vehicle, then the strain energy is transferred to the vehicle as 1 unit of kinetic energy, Figure 12.11. The resulting speed of the vehicle can then be measured. The experiment is then repeated with 2, 3 and 4 identical elastic strings extended by the same distance s which we will assume gives the vehicle 2, 3 and 4 units of kinetic energy respectively. Table 2 shows some typical results.

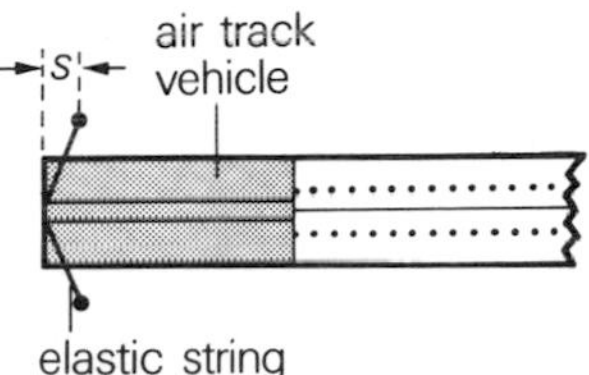

Figure 12.11 Method of launching vehicle

kinetic energy of vehicle	1	2	3	4
speed m s^{-1}	0·20	0·28	0·35	0·40
$(\text{speed})^2$	0·04	0·08	0·12	0·16

Table 2

A graph of kinetic energy against speed is shown in Figure 12.12.

The results lie on a smooth curve. This suggests that a fairly simple relationship exists between kinetic energy and speed. In fact, a graph of kinetic energy against $(\text{speed})^2$ produces a graph which is a straight line through the origin, Figure 12.13.

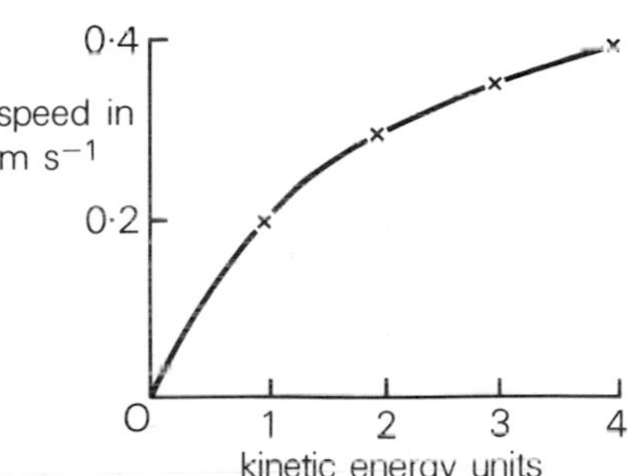

Figure 12.12 Kinetic energy against speed

This indicates that kinetic energy varies directly as the speed squared, i.e. kinetic energy $\propto (\text{speed})^2$ for a given mass.

In a second experiment, the speed v of the vehicle launched by one elastic stretched a distance s is again measured. When the mass m of the vehicle is doubled to $2m$, it is found that two elastics stretched by a distance s give practically the same speed v to the vehicle. This means that if the mass is doubled, the kinetic energy must be doubled to produce the same speed. Tests with vehicles of mass $3m$ and $4m$ lead to the following conclusion. The kinetic energy required to produce the same speed varies directly as the mass of the vehicle,

i.e. kinetic energy $\propto$ mass for a given speed.

The exact relationship between kinetic energy, mass and speed can be derived as follows.

Figure 12.13 Kinetic energy against $(\text{speed})^2$

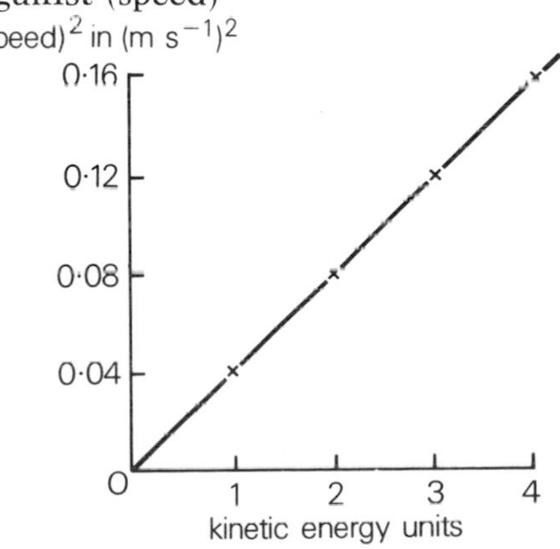

Consider an object of mass m at rest on a frictionless surface. When a constant unbalanced force F is applied over a distance s (Figure 12.14), the object will accelerate with uniform acceleration a to a final velocity v. The amount of work done is given by $F \times s$ and if all the work is transformed into kinetic energy, the object gains an amount of kinetic energy equal to $F \times s$,

i.e. kinetic energy gained = $F \times s$.

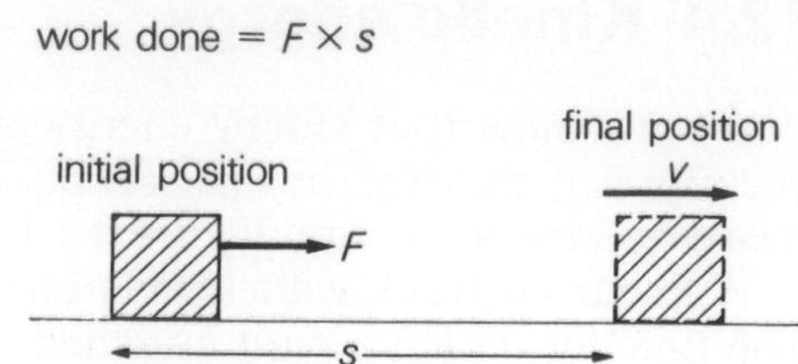

Figure 12.14

The average velocity $\bar{v}$ for uniform acceleration is given by,

$\bar{v} = \frac{u+v}{2}$ where u = initial velocity, v = final velocity.

$\Rightarrow \bar{v} = \frac{v}{2}$ since the initial velocity u is zero.

The average velocity $\bar{v}$ is also given by the equation,

$$\bar{v} = \frac{\text{total displacement}}{\text{time taken}}$$

$$\Rightarrow \bar{v} = \frac{s}{t}$$

It follows that the average velocity $\bar{v}$ is

$$\bar{v} = \frac{s}{t} = \frac{v}{2}$$

$\Rightarrow \quad \frac{s}{t} = \frac{v}{2}$ multiply both sides by t.

$$\Rightarrow \quad s = \frac{v}{2} \times t$$

Also the acceleration a is given by

$$a = \frac{v-u}{t}$$

$\Rightarrow \quad a = \frac{v}{t}$ since $u = 0$.

The kinetic energy gained $= F \times s$

$= F \times \frac{v}{2} \times t$

$= ma \times \frac{v}{2} \times t$ since $F = ma$

$= m\frac{v}{t} \times \frac{v}{2} \times t$ since $a = \frac{v}{t}$

$= mv \times \frac{v}{2}$

$\Rightarrow$ kinetic energy $= \frac{1}{2} mv^2$

This equation explains our experimental results as it shows that kinetic energy varies directly as both mass and (speed)^2. The symbol for kinetic energy is E_k.

$$E_k = \frac{1}{2} mv^2$$

Units

The units of kinetic energy must be the units of mass × (units of speed)^2.

$\Rightarrow$ units of kinetic energy $= \text{kg} \times (\text{m s}^{-1})^2$

$= \text{kg} \times \text{m}^2\ \text{s}^{-2}$

We can show that the unit of work (the **joule**) can be expressed in an identical way.

$$\begin{aligned} 1\ \text{J} &= 1\ \text{N} \times 1\ \text{m} \\ &= 1\ \text{kg m s}^{-2} \times 1\ \text{m} \\ &= 1\ \text{kg m}^2\ \text{s}^{-2} \end{aligned}$$

The units of work and of kinetic energy are the same because they can be expressed in exactly the same way. We conclude that the unit of kinetic energy is the joule, (J).

Example 5

A car, mass 1 000 kg, is moving at a speed of 20 m s^{-1}.
a) Calculate its kinetic energy at that speed.
b) Explain what happens to this kinetic energy when the brakes are applied and the car stops.

a)

$$\begin{aligned} E_k &= \frac{1}{2} mv^2 \\ &= \frac{1}{2} \times 1\,000 \times (20)^2 \\ &= 500 \times 400 \\ &= 200\,000 \\ &= 2 \times 10^5 \end{aligned}$$

Kinetic energy of car is 2 × 10^5 J.

b) The kinetic energy is not destroyed, but is changed into other forms of energy in the brakes of the car, e.g. heat energy, sound energy.

Example 6

A constant unbalanced force F of 50 N accelerates a packing case of mass 10 kg along a factory floor. If the case is accelerated from 10 m s^{-1} to 20 m s^{-1}, find the displacement s of the case.

Assuming that all the work done increases the kinetic energy of the packing case,

Work done = Increase in E_k of the case

$$\Rightarrow F \times s = \frac{1}{2} m\ v^2 - \frac{1}{2} m\ u^2$$

(where m = mass of case, v = final speed of case, and u = initial speed of case)

$$\begin{aligned} \Rightarrow 50 \times s &= \frac{1}{2}\ 10 \times (20)^2 - \frac{1}{2}\ 10 \times (10)^2 \\ &= 5 \times 400 - 5 \times 100 \\ &= 5(400 - 100) \\ &= 1\,500 \\ \Rightarrow \quad s &= 30 \end{aligned}$$

The displacement of the case is 30 m.

Example 7

A rocket of mass 10 kg has a kinetic energy of 12 500 J. Calculate the speed of the rocket.

$$\begin{aligned} E_k &= \frac{1}{2} mv^2 \\ \Rightarrow 12\,500 &= \frac{1}{2} \times 10 \times v^2 \\ \Rightarrow 12\,500 &= 5 \times v^2 \\ \Rightarrow 2\,500 &= v^2 \\ \Rightarrow 50 &= v \end{aligned}$$

Speed of the rocket is 50 m s^{-1}.

Collisions

In Chapter 11 we found that momentum is conserved in all types of collision. We shall now re-examine collisions in terms of energy. An inelastic collision was defined rather vaguely as one where the objects stuck together. We can now redefine an **inelastic** collision as one where the total kinetic energy is reduced by the collision. However, the difference in energy is not destroyed but is changed by the impact into other energy forms (mainly heat and a little sound). When no kinetic energy is lost in the collision, the collision is said to be **perfectly elastic**. These facts are summarized below.

	Momentum conserved?	*Kinetic energy conserved?*
Perfectly elastic collision	yes	yes
Inelastic collision	yes	no

Example 8

Trolley A of mass 2 kg moving with a speed of 3 m s^{-1} collides with trolley B of mass 1 kg which is stationary. After the collision, trolley A has a speed of 1 m s^{-1}, and trolley B moves off at 4 m s^{-1} as shown in the diagram.

Show that the collision is perfectly elastic.

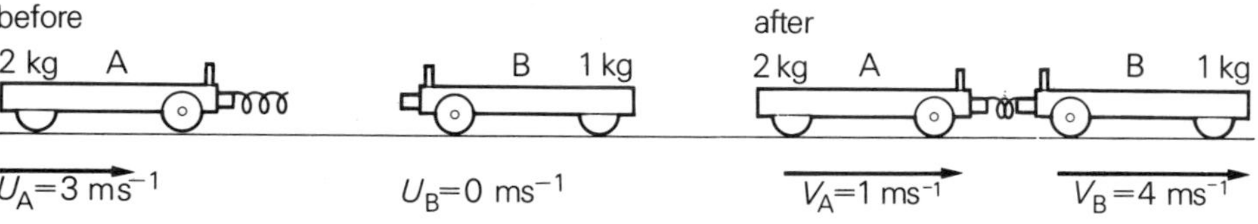

We shall use the same notation for collisions as in Chapter 11.

$$\text{Total kinetic energy before the collision} = \frac{1}{2} m_A u_A^2 = \frac{1}{2} \times 2 \times (3)^2 = \frac{1}{2} \times 2 \times 9 = 9 \text{ J}$$

$$\text{Total kinetic energy after the collision} = \frac{1}{2} m_A v_A^2 + \frac{1}{2} m_B v_B^2 = \frac{1}{2} \times 2 \times (1)^2 + \frac{1}{2} \times 1 \times (4)^2 = 1 + 8 = 9 \text{ J}$$

The total kinetic energy is unchanged by the collision. Therefore the collision is perfectly elastic.

Example 9

Trolley A of mass 1 kg moving at 1 m s^{-1} collides with an identical stationary trolley B. The trolleys stick together and move off with a common speed of 0·5 m s^{-1} after the collision.
a) Find the total kinetic energy before the collision.
b) Find the total kinetic energy after the collision.
c) State whether the collision is perfectly elastic or inelastic, giving your reasons.

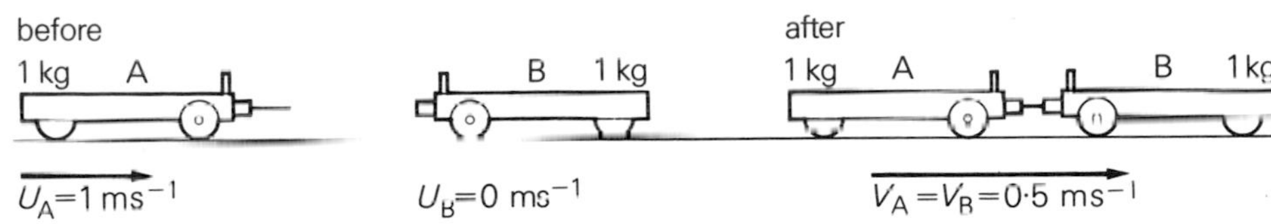

a) Total kinetic energy before the collision $= \frac{1}{2} m_A {u_A}^2$

$= \frac{1}{2} \times 1 \times (1)^2$

$= 0{\cdot}5$

Total kinetic energy before the collision is 0·5 J

b) Total kinetic energy after the collision $= \frac{1}{2} m_A {v_A}^2 + \frac{1}{2} m_B {v_B}^2$

$= \frac{1}{2} \times 1 \times (0{\cdot}5)^2 + \frac{1}{2} \times 1 \times (0{\cdot}5)^2$

$= (0{\cdot}5)^2$

$= 0{\cdot}25$

Total kinetic energy after the collision is 0·25 J

c) The collision is not perfectly elastic because the total kinetic energy is reduced by the collision.

12.5 Gravitational potential energy

Another form of energy you know about is gravitational potential energy (stored energy). This is the energy possessed by an object because of its position. For example, when a box is lifted up from the floor and put on a high shelf, work has to be done. This amount of work is stored as the gravitational potential energy of the box with respect to the floor. We now establish the relationship between the gravitational potential energy, the mass and the height raised.

Consider a box of mass m which is to be raised a vertical height h, Figure 12.15.

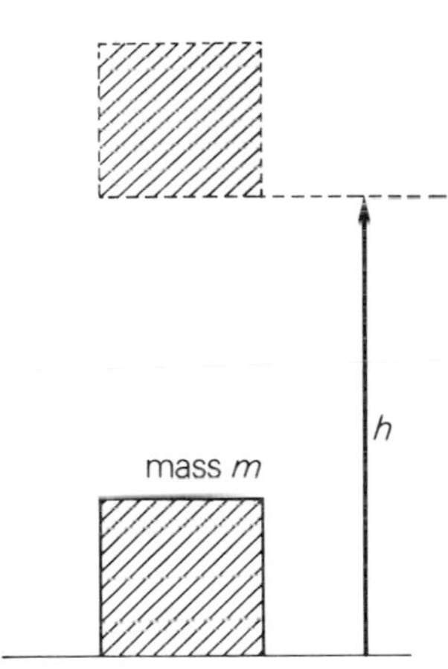

Figure 12.15

The weight of the box W is given by

$W = mg$

where 'g' is the gravitational field strength in N kg^{-1}.
In order to raise the box, an upward force must be applied to balance out the weight acting downwards.
The minimum upward force F required to lift the box is equal and opposite to the weight W

$\Rightarrow F = mg$

The work done in raising the box a distance h is given by

work required = force × upward displacement
$= mg \times h$

All the work done is stored as gravitational potential energy in the box once it has been lifted,

gravitational potential energy of box = mgh
The symbol for gravitational potential energy is E_p.

$E_p = mgh$

Units

The units of gravitational potential energy are the units of force × the units of distance, i.e. the units of gravitational potential energy are the same as the units of work. Hence the unit of gravitational potential energy is the joule, (J).

Example 10

A ball of mass 0·5 kg is thrown vertically upwards, and it reaches a height of 45 m above the ground.

Calculate the gravitational potential energy of the ball at its highest point.

$E_p = mgh = 0{\cdot}5 \times 10 \times 45 = 225$
assuming $g = 10\ \mathrm{N\ kg^{-1}}$

Gravitational potential energy of the ball is 225 J

12.6 Energy conversion

In this section we consider some further situations involving conversion of potential energy into kinetic energy.

1. Bouncing ball

If a ball is dropped vertically downwards, its gravitational potential energy changes into kinetic energy. Assuming that no energy is lost due to air resistance, its total energy (i.e. $E_p + E_k$) remains constant as it falls. It therefore follows that all the potential energy of the ball before it is dropped becomes its kinetic energy just as it touches the ground. If the ball collides elastically with the ground (i.e. no energy is changed into heat or sound), it will bounce up with the same kinetic energy that it had on hitting the ground. It will therefore rise to its original height. Figure 12.16 shows the downward motion, impact and upward motion separately.

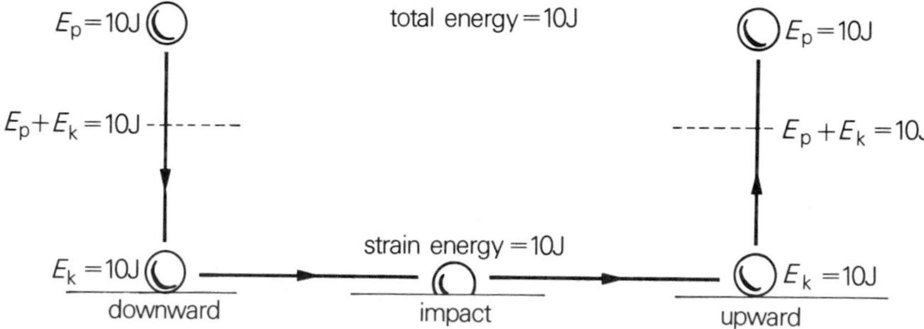

Figure 12.16 Energy of a bouncing ball

During the collision, the kinetic energy is transformed into strain energy in the ball.

Example 11

A ball of mass 2 kg is dropped from a height of 20 m.
a) Calculate the gravitational potential energy of the ball before it is dropped.
b) Calculate the speed of the ball on hitting the ground.
c) If the ball bounces to a height of 5 m, with what speed does it leave the ground?
d) Explain why the ball does not reach its original height after the bounce.
Take $g = 10 \text{ N kg}^{-1}$

a) $E_p = mgh = 2 \times 10 \times 20 = 400$

Gravitational potential energy is 400 J

b) The kinetic energy of the ball on hitting the ground is equal to the ball's original potential energy (neglecting air resistance).

E_k of ball on hitting the ground $= \frac{1}{2}\, m\, v^2$

(where the ball of mass 'm' hits the ground with speed v)

$\Rightarrow \quad \frac{1}{2} mv^2 = 400$

$\Rightarrow \frac{1}{2} \times 2 \times v^2 = 400$

$\Rightarrow \quad v^2 = 400$

$\Rightarrow \quad v = 20$

Ball hits the ground with a speed of 20 m s^{-1}

c) The kinetic energy of the ball on leaving the ground is equal to the gravitational potential energy of the ball on rising to 5 m (neglecting air resistance).

E_p of ball 5 m above the ground $= mgh$

$= 2 \times 10 \times 5 = 100$

$\Rightarrow E_k$ of ball on leaving the ground $= 100$ J

$\Rightarrow \quad \frac{1}{2} mv^2 = 100$

$\Rightarrow \quad \frac{1}{2} \times 2 \times v^2 = 100$

$\Rightarrow \quad v^2 = 100$

$\Rightarrow \quad v = 10$

Ball leaves the ground with a speed of 10 m s^{-1}

d) The ball does not reach its original height after the bounce because kinetic energy is changed into other energy forms during the bounce (mainly heat and a little sound).

2. Potential energy and kinetic energy

Figure 12.17 shows a method of directly changing the gravitational potential energy of a falling mass into the kinetic energy of moving trolleys.

The mass m is allowed to fall from rest through a vertical distance h. The falling mass loses gravitational potential energy, mgh which becomes the kinetic energy of the moving trolley system. Once the falling mass hits the floor, the trolleys continue down the friction-compensated runway at a constant velocity which can be measured from the ticker-tape attached to them. The kinetic energy of the trolleys can then be calculated and its value compared with the gravitational potential energy lost by the falling mass.

Figure 12.17 Potential energy → kinetic energy

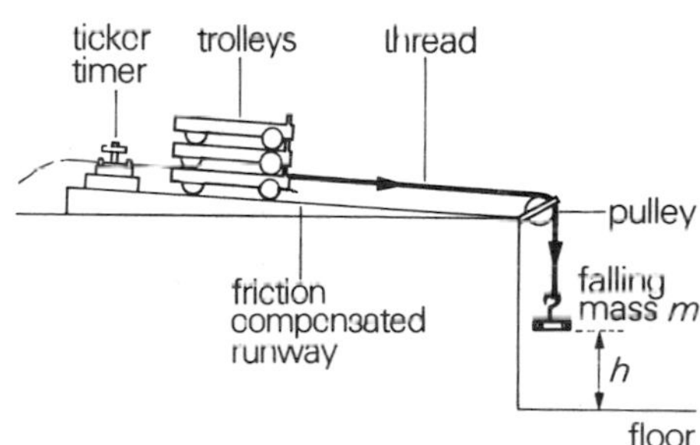

Example 12

In Figure 12.17 a falling mass of 0·10 kg accelerates a set of trolleys of mass 2·25 kg by falling through a distance of 0·20 m. By analysing the ticker-tape the trolleys are found to move at 0·40 m s^{-1} after the mass hits the floor.

a) Calculate the gravitational potential energy lost by the falling mass.

b) Calculate the gain in kinetic energy of the trolleys.

c) Account for any difference between your answers to (a) and (b).

a) $E_p = mgh$

For the falling mass, $E_p = 0{\cdot}1 \times 10 \times 0{\cdot}20 = 0{\cdot}20$

Gravitational potential energy lost by falling mass is 0·20 J

b) $E_k = \frac{1}{2} m_T v^2$

For the set of trolleys, $E_k = \frac{1}{2} \times 2{\cdot}25 \times (0{\cdot}40)^2 = 0{\cdot}18$

Kinetic energy gained by trolleys is 0·18 J

c) The kinetic energy gained by the trolleys is slightly less than the potential energy lost by the falling mass. The difference is due to, (i) errors in measuring the speed from the ticker-tape, and (ii) the fact that the falling mass also gains kinetic energy which is lost mainly as heat when it hits the floor.

3. Energy changes in a simple pendulum

We will now reconsider the simple pendulum in terms of energy. If the pendulum bob is displaced to one side so that it has risen by vertical height h and then released, it is seen to swing to the same vertical height on the other side, Figure 12.18.

This is because the total energy E of the pendulum bob (i.e. $E_p + E_k$) remains constant. At its highest point the pendulum bob is stationary and all its energy is gravitational potential energy. As it falls, this energy changes into kinetic energy and at its lowest point **all** its energy is kinetic energy, Figure 12.19. Hence the bob has its maximum speed at its lowest point.

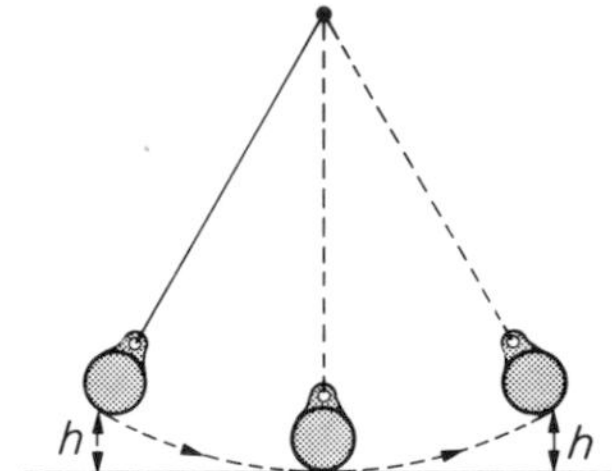

Figure 12.18 Simple pendulum

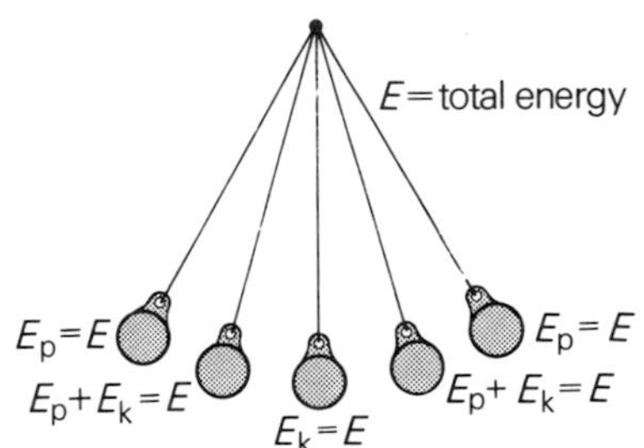

Figure 12.19 Energy changes in a pendulum

Example 13

A pendulum bob of mass 1·0 kg is moved sideways until it has risen by a vertical height of 0·45 m. Calculate the speed of the bob at its lowest point. (Take g = 10 N kg^{-1}.)

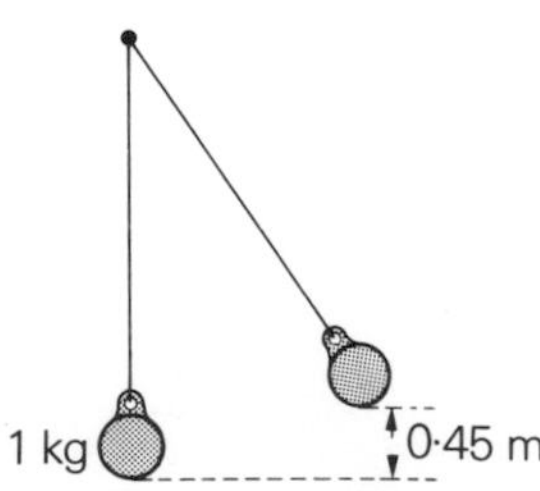

E_p of bob at highest point $= mgh$

$= 1{\cdot}0 \times 10 \times 0{\cdot}45$

$= 4{\cdot}5$ J

$\Rightarrow$ total energy at highest point = 4·5 J

$\Rightarrow$ total energy at lowest point = 4·5 J

$\Rightarrow$ E_k at lowest point = 4·5 J

$$\Rightarrow \quad \frac{1}{2} m v^2 = 4{\cdot}5$$

$$\Rightarrow \quad \frac{1}{2} \times 1{\cdot}0 \times v^2 = 4{\cdot}5$$

$$\Rightarrow \quad v^2 = 9{\cdot}0$$

$$\Rightarrow \quad v = 3{\cdot}0$$

Speed of the bob at its lowest point is 3·0 m s^{-1}

An experiment can be set up as shown in Figure 12.20 to confirm that it is the **energy** of the pendulum bob that determines the height it will reach.
The pendulum bob is released from a certain height and a pin interrupts the swing when the bob is at its lowest point. However, although the length of the pendulum has effectively been changed, the bob is seen to rise to the same height from which it was originally released. This experiment is known as 'Galileo's pin and pendulum experiment'.

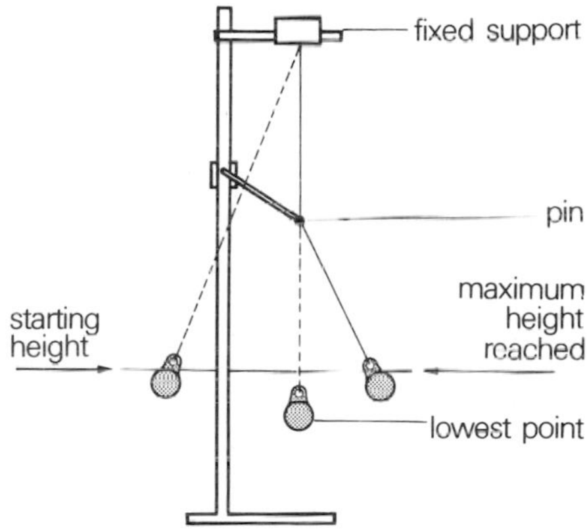

Figure 12.20 Pin-and-pendulum experiment

4. Ball on a curtain rail

The following situation is a case where there is a continual interconversion from E_p to E_k to E_p, and so on. Consider a ball on a curtain rail, Figure 12.21.

When it is released from a vertical height h, the ball rolls down the rail and is seen to reach nearly the same height on the right-hand side of the rail. This is because the total energy of the ball, $E_p + E_k$, remains nearly constant. There are small energy losses due to friction. Its potential energy at its highest point on the right-hand side of the rail is therefore equal to its original potential energy and so it must reach the same vertical height h.

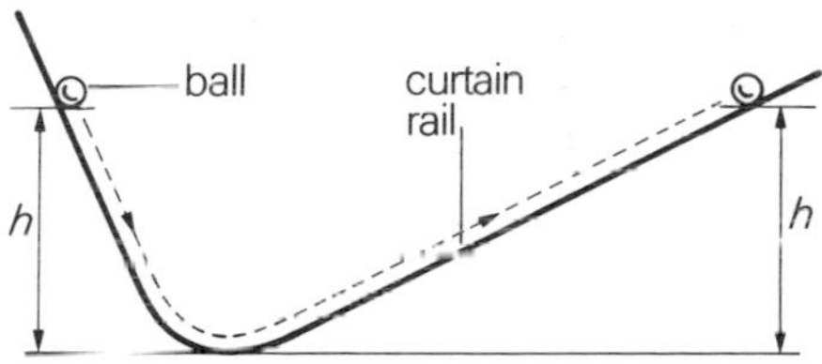

Figure 12.21

12.7 Power

The photograph below shows two externally similar motor cars, Figure 12.22. However the rally car on the right can accelerate to 70 miles per hour much more rapidly than the standard car on the left. In doing so it changes a given amount of chemical energy (obtained from the petrol) into kinetic energy in a shorter time. Although it is perhaps less economical, the rally car is much more **powerful** than the standard car.

Figure 12.22

In general, power is defined as the rate at which a machine can do work,

i.e. $$\text{power} = \frac{\text{work done}}{\text{time taken}}$$

Power does not have any particular direction associated with it and is therefore a scalar quantity.

Units

As the unit for work is the joule (J) and the unit for time is the second (s), the SI unit for power is the joule per second ($J\ s^{-1}$). This unit is called the **watt**, written as W, after James Watt the inventor of the steam engine in the eighteenth century.

i.e. 1 watt = 1 joule per second
or $1\ W = 1\ J\ s^{-1}$

Another common unit for measuring power is the horsepower (hp). This unit was originated by James Watt so that he could express the power of the engines he sold in terms of how many horses they would replace. Actually one horse-power is equal to 746 watts.

Table 3 shows some typical powers in watts.

Object	*Typical power*
Flea at take-off	0·1 mW
Human sleeping	70 W
Human walking	300 W
Colour television	300 W
Small car	40 kW
Large power station	2 000 MW

Table 3

Measuring power

If you run up a long flight of stairs you have to do work to raise your body the vertical height of the stairs. In doing so, your gravitational potential energy is increased. You can find your power by getting a partner to time how long you take to do this amount of work.

Example 14

A girl who weighs 500 N runs up a flight of stairs in a time of 20 s. If each step is 0·2 m high and there are 60 steps in the flight, calculate the girl's power.

Vertical height of one step = 0·2 m
Number of steps = 60
⇒ Total vertical height of stairs = 60 × 0·2 = 12 m

Weight of girl = 500 N

Work done by girl = force exerted × upward displacement
= 500 × 12
⇒ Work done by girl = 6 000 J

Time taken to climb stairs = 20 s

$$\text{Power} = \frac{\text{work done}}{\text{time taken}} = \frac{6\,000}{20} = 300$$

Power of the girl climbing the stairs = 300 W

Summary

Kinetic energy E_k is possessed by any moving object.
Gravitational potential energy E_p is stored when objects are lifted upwards.
Strain energy is potential energy stored in a stretched elastic or spring.

Work = force × displacement in direction of force
$W = F \times s$
The joule is the SI unit of work; 1 joule (J) is 1 newton-metre.

A lever is a force multiplier.

$$\text{Efficiency (\%)} = \frac{\text{useful output energy}}{\text{input energy}} \times \frac{100}{1}.$$
$E_k = \frac{1}{2}mv^2$; where m = mass, v = speed.

In a perfectly elastic collision, both momentum and kinetic energy are conserved.
$E_p = mgh$; where m = mass, h = vertical height, and g = gravitational field strength.

$$\text{Power} = \frac{\text{work done}}{\text{time taken}}$$
The watt is the SI unit of power; 1 watt (W) is 1 joule per second.

Problems

Take $g = 10 \text{ m s}^{-2}$

1 A girl drops a china vase which falls and breaks into pieces. What energy changes are involved?

2 A boy uses a catapult to fire a stone which accidentally smashes a greenhouse window. List the possible energy changes.

3 A labourer on a building site exerts a force of 300 N on a wheelbarrow over a distance of 20 m. Calculate the work done by the labourer.

4 A car pulls a trailer along a level road 1 km long. If the amount of work done on the trailer by the car is 4×10^5 J, what force does the car exert on the trailer?

5 A man raises a load of 2 000 N by exerting an effort force of 50 N on one end of a lever. Assuming 100% efficiency, what distance does the effort move to raise the load by 0·1 m?

6 A system of pulleys is used to raise a load of 100 N by applying an effort of 50 N. The effort moves by 0·80 m when the load is raised by 0·20 m. Calculate the efficiency of the machine in lifting the load.

7 A car of mass 800 kg is travelling at a speed of 20 m s^{-1}. Find its kinetic energy.

8 A rocket of mass 3 000 kg has a kinetic energy of 6×10^5 J. At what speed is it travelling?

9 A golf ball of mass 0·1 kg rises vertically to 40 m above the ground. Calculate its gravitational potential energy at that height.

10 A packing case has a gravitational potential energy of 400 J when it is 2 m above the ground. What is the mass of the case?

11 A stone of mass 0·1 kg is thrown vertically upwards. It arrives back at its starting point 4 s later. Calculate:
- **11·1** the initial speed.
- **11·2** the initial kinetic energy.
- **11·3** the potential energy at its greatest height.
- **11·4** the maximum height of the stone above the starting point.
- **11·5** the speed as it arrives back at the starting point.

12 A steel ball of mass 0·4 kg falls to the ground from a height of 1 m and rebounds to a height of 0·8 m. Calculate:
- **12·1** the initial gravitational potential energy.
- **12·2** the kinetic energy just as the ball reaches the ground.
- **12·3** the kinetic energy just after it leaves the ground after bouncing.
- **12·4** the speed with which the ball hits the ground.
- **12·5** the amount of energy changed into heat and sound energies as a result of the bounce.

13 A ball of mass m is released on a curtain rail at a height of 0·20 m vertically above the bottom of the rail.

13·1 Calculate the maximum speed of the ball.

13·2 Describe a possible method for measuring this speed.

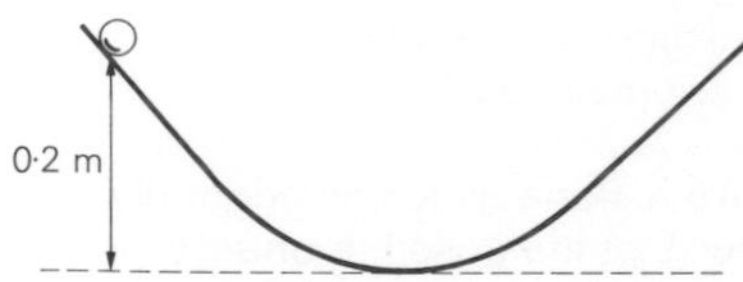

14 A constant unbalanced force changes the speed of a 10 kg sledge from 10 m s^{-1} to 30 m s^{-1} in a distance of 40 m. Calculate the size of the unbalanced force acting on the sledge.

15 A man mows a lawn in 10 minutes by exerting an average force of 100 N over a distance of 100 m. Calculate his average power.

16 An electric motor has an output power of 4 kW. How long would the motor take to raise a load of 1 000 N through a distance of 50 m?

17 A dam is situated at a height of 550 m above sea level and supplies water to a hydro-electric generating station which is at a height of 50 m above sea level. 2 000 kg of water pass through the turbines per second.

17·1 How much potential energy is converted into other forms of energy each second in this system?

17·2 What would be the maximum electrical power output of the station if the whole system was 80% efficient?

17·3 In what form of energy would most of the 'wasted' energy leave the hydro-electric station? *SCEEB*

18 The diagram shows the positions of a vehicle after equal successive intervals of time of 1/6 second. The vehicle was moving in the direction of the arrow. It collided with a stationary vehicle of mass 3 kg and the vehicles stuck together at the time of impact.

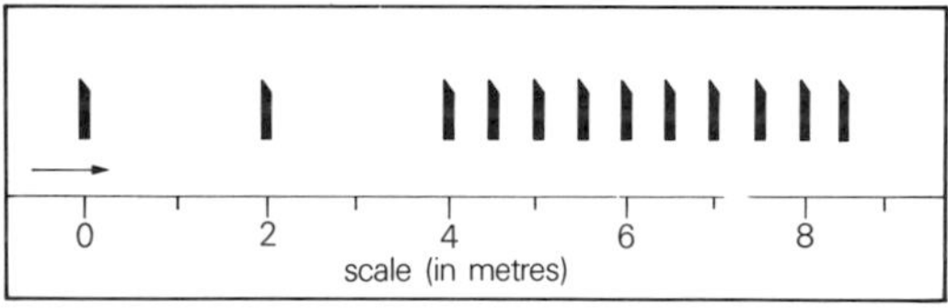

18·1 What was

a) the speed of the moving trolley before the collision?

b) the speed of both trolleys after the impact?

c) the mass of the trolley which collided with the stationary trolley?

d) the total kinetic energy of the vehicles before the collision?

e) the total kinetic energy of the trolleys after the collision?

18·2 Account for the loss in kinetic energy at the collision.

18·3 After the collision the vehicles encountered an uphill slope. What would be the maximum vertical height to which the vehicles might rise?

State any assumption that you have made. *SCEEB*

19 In a laboratory experiment, a moving trolley collided with, and stuck to a stationary one. The following results were obtained. (The units are omitted.)

before				*after*			
Mass	Speed	I	II	Mass	Speed	III	IV
1	12·0			2	6·1		
1	9·8			3	3·3		
2	10·2			4	5·0		
2	11·5			3	7·6		

19·1 Copy this table and complete columns I and III to test for momentum conservation. Comment on these results.

19·2 Complete columns II and IV to test for kinetic energy conservation. Is kinetic energy conserved here?

19·3 In a head-on collision between two cars,

a) is total momentum conserved?

b) is the kinetic energy of the cars conserved?

If the answer in either case is 'No' explain the change. *SCEEB*

20 An 800 kg car was equipped with a meter which accurately measured the quantity of petrol used. It travelled at constant speed along a horizontal road for 4·0 km and used 0·30 kg of petrol. It then travelled at the same speed up a hill for 4·0 km, where the total vertical climb was 300 m and used considerably more petrol for this part of the journey.

20·1 Why was extra petrol required?

20·2 Each kilogram of petrol could produce $1{\cdot}2 \times 10^7$ J of useful energy when it burned. What was the least extra quantity of petrol which would have been required?

After this hill the road levelled out again and the driver accelerated uniformly for 1·0 km.

20·3 Sketch a graph showing how the force exerted by the engine to drive the car forward varied throughout the whole 9·0 km journey. Values are required on the distance axis but not on the force axis. *SCEEB*

21 A large packing case of mass 500 kg rests on the ground. A fork-lift truck raises it 1·5 m, transports it at a steady speed of 2·0 m s^{-1} and deposits it on the loading platform of a lorry.

21·1

a) What is the minimum upward force exerted by the fork-lift?

b) How much potential energy is gained by the packing case?

21·2

a) Calculate the kinetic energy of the packing case while it is being transported at the steady speed.

b) What happens to this kinetic energy when the fork-lift truck stops?

21·3 If the fork-lift uses energy at the rate of 25 kW and the lifting operation takes 3·0 seconds, calculate the apparent efficiency of this operation. *SCEEB*

22 Figure 1 shows a pendulum in its rest position A. The pendulum bob has a mass of 0·3 kg. The bob is pulled to one side as shown in Figure 2 and held at position B which is 0·8 m above the rest position.

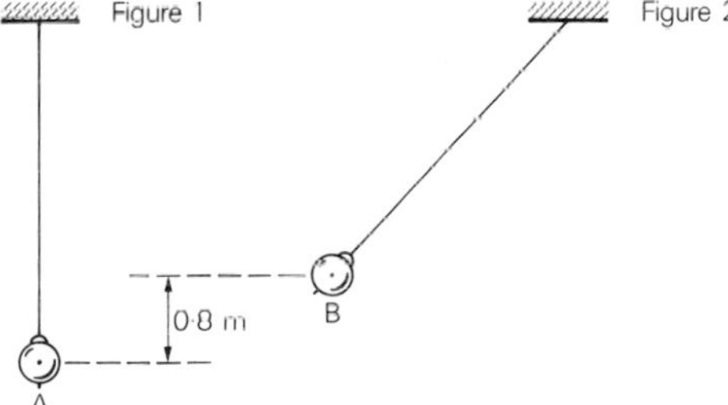

The bob is released from position B and swings to and fro until it comes to rest.

22·1 Find the gain in potential energy of the bob when it is moved from position A to position B.
22·2 When does the bob have greatest kinetic energy?
22·3 Estimate the maximum speed of the bob.
22·4 Describe the energy changes which take place from the time the bob is released until it eventually comes to rest. *SCEEB*

23 A loaded mine car of total mass 5 tonnes (5×10^3 kg) is being hauled up an incline to the surface when the cable breaks and the car runs back down the incline.

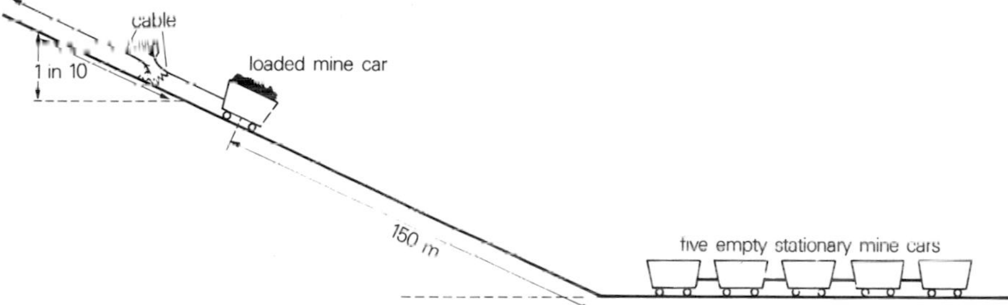

At the bottom of the incline the car collides with a row of five stationary empty cars, each of mass 1 tonne. The cars automatically couple together and move off with a speed of 4 m s^{-1}.

23·1 Use the law of conservation of momentum to calculate the speed of the loaded car just before impact.
23·2 Calculate the kinetic energy of the loaded car just before impact.
23·3 For every 10 m it is hauled up the incline, the car rises a vertical height of 1 m. If the cable broke when the car was 150 m up the incline, what was its potential energy at that point?
23·4 Account for any difference between the answers to **23·2** and **23·3**. *SCEEB*

13 Temperature

13.1 Temperature scales

We use terms such as hot, cold and warm to describe a quantity which we call temperature. **Temperature** is a measure of the hotness or coldness of an object. However, a warm room may seem hot to someone who has come inside from a cold street, but cold to someone else who has just stepped out of a hot bath. We need a more reliable means of describing temperature.

Until the eighteenth century, thermometers were uncalibrated, i.e. they had no standard temperature scales marked on them. It occurred to scientists about this time that pure ice always melted at a particular temperature and pure water always boiled at another definite temperature. This meant that these two temperatures could be used as **fixed points** on a temperature scale.

It was the Swedish scientist Celsius who first defined a standard temperature scale which used the melting point of ice and the boiling point of water as two fixed points. The melting point of pure ice was taken to be zero and the boiling point of pure water to be one hundred with the range of temperature between these points divided into one hundred equal intervals called **degrees**.

This scale used to be called the centigrade scale, literally meaning 'one hundred divisions'. However, the scale has been extended to include temperatures lower than zero and higher than one hundred degrees and is now known after its originator as the **Celsius temperature scale**.

Each interval on the Celsius scale is known as a **degree Celsius** which we abbreviate to °C. Figure 13.1 shows a part of the Celsius scale. Note that the boiling point of water is 100 °C, the melting point of ice is 0 °C, and that 20 degrees colder than 0 °C is written as −20 °C.

Situation	*Celsius temperature*
oxygen becomes a liquid	−183 °C
lowest recorded air temperature (Antarctica)	−88 °C (−126 °F)
carbon dioxide solidifies forming 'dry ice'	−78 °C
bitterly cold day	−10 °C (14 °F)
melting ice	0 °C (32 °F)
warm classroom	18 °C (64 °F)
hot summer day	27 °C (81 °F)
human body temperature	37 °C (98 °F)
highest recorded air temperature (Mexico)	58 °C (136 °F)
boiling water	100 °C (212 °F)
bunsen burner flame	900 °C
centre of the Earth	6 000 °C
centre of the Sun	14 000 000 °C

Table 1 A range of temperatures

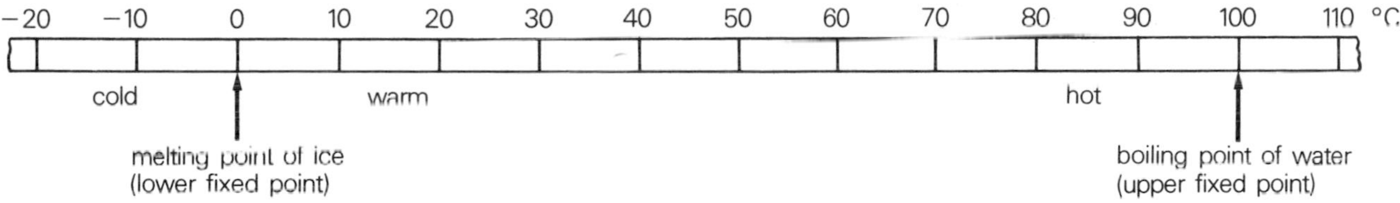

Figure 13.1 The Celsius temperature scale

Some typical temperatures on the Celsius scale are shown in Table 1. For comparison some temperatures are also shown in degrees Fahrenheit (°F).

13.2 Measuring temperature

1. Mercury-in-glass thermometer

The thermometer which we are most familiar with is the mercury-in-glass thermometer, Figure 13.2.

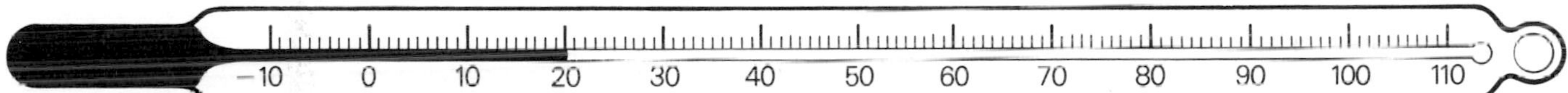

Figure 13.2 Mercury-in-glass thermometer

When mercury becomes hotter it expands and so in the mercury thermometer the mercury column rises up the narrow capillary tube. A disadvantage of the mercury thermometer is its limited range. It cannot record temperatures below the freezing point of mercury (−39 °C) or above the boiling point of mercury (357 °C).

Alcohol with a freezing point of −112 °C and a boiling point of 78 °C is sometimes used instead of mercury in this type of thermometer. The alcohol is usually coloured to make it easier to see. The alcohol-in-glass thermometer can record lower temperatures than the mercury thermometer. However, the mercury thermometer should give you a clue to other methods of measuring temperature. The mercury thermometer uses the fact that mercury expands as its temperature rises. In fact, any property which changes with temperature can form the basis of a thermometer. Provided the thermometer is first calibrated, measurement of how the property has changed indicates the temperature.

2. Rotary thermometer

The rotary thermometer consists of a coiled bi-metallic strip. A bi-metallic strip is made by strongly attaching together two strips of different metals, which expand by different amounts when heated. Figure 13.3 shows a bi-metallic strip made of copper and steel. Copper expands more than steel for the same temperature rise.
Therefore when the strip is heated the only way that the copper can become longer than the steel is for the strip to bend so that the copper is on the outside of the bend. This effect is magnified by using a coil of bi-metallic strip and the amount of bending can be indicated by a

Figure 13.3 Bi-metallic strip

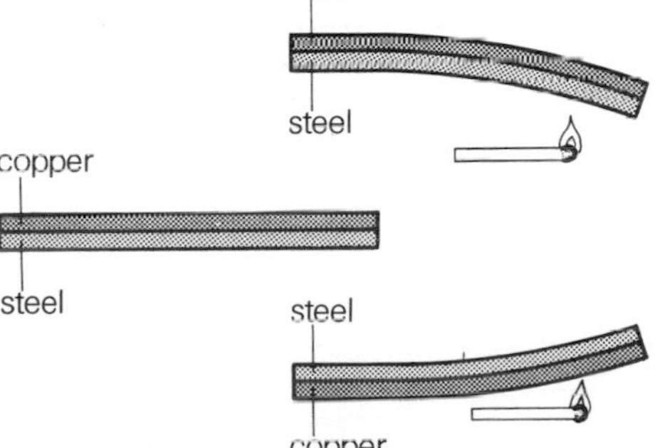

pointer, Figure 13.4. One end of the coil is fixed so that when the temperature rises the coil unrolls and the pointer moves round the scale. This instrument is called a **rotary thermometer** and is cheap and robust.

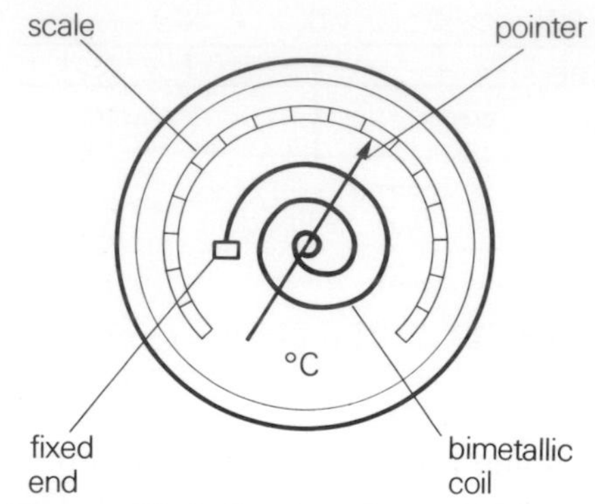

Figure 13.4 Rotary thermometer

3. Thermocouple thermometer

The thermocouple consists basically of two wires of different metals, e.g. copper and constantan, joined together at the ends to form two junctions. If the junctions are at different temperatures, a voltage is produced between the junctions which causes a small current to flow along the wires. A microammeter can be included in the loop to measure this current, Figure 13.5.

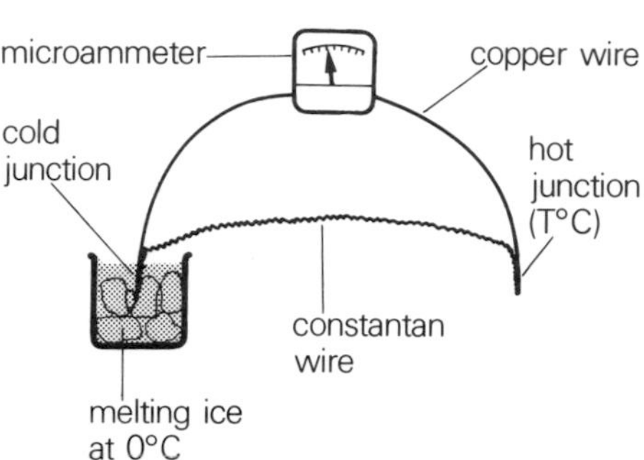

Figure 13.5 Thermocouple thermometer

Calibration
The 'cold' junction can be kept in melting ice at 0 °C. To calibrate this thermometer the 'hot' junction can be immersed in boiling water at 100 °C, and the current reading on the microammeter noted. This gives the current for a temperature difference of 100 °C. The current is found to vary directly with temperature difference. For example, a temperature difference of 50 °C produces half the current reading obtained with a temperature difference of 100 °C. This means that the thermocouple can be calibrated by inscribing equal intervals on to the meter, between the 0 °C and 100 °C positions.

Example

One junction of the thermocouple shown is kept at 0 °C. When the 'hot' junction is held at 100 °C, the current reading on the microammeter is 50×10^{-6} A.

If the microammeter indicates a current of 40×10^{-6} A when the 'hot' junction is held in hot water, find the temperature of the hot water.

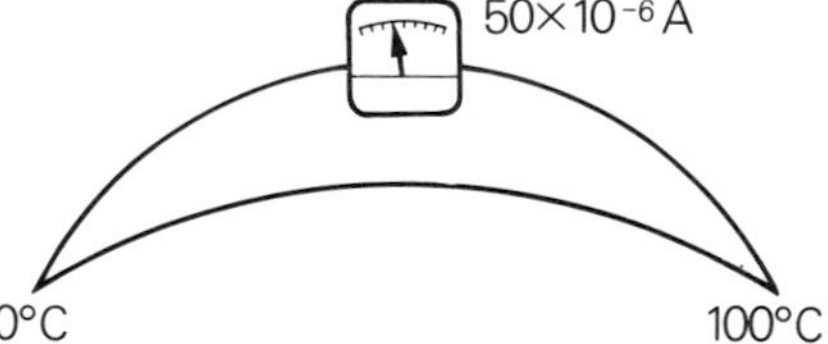

Current of 50×10^{-6} A indicates a temperature difference of 100 °C
$\Rightarrow$ current of 40×10^{-6} A indicates a temperature difference of 80 °C

Temperature of hot water is 80 °C

Applications
The temperature range over which a thermocouple can operate is from −200 °C to 1 700 °C. It can therefore be used to measure temperatures of ovens and furnaces. The size and direction of the current depends on the two metals used.

A **thermopile** is a group of thermocouples connected together and is more sensitive than a single thermocouple. It is therefore more commonly used than a single thermocouple.

4. Resistance thermometer

The resistance thermometer uses the fact that the electrical resistance of a wire increases as the temperature rises. This type of thermometer consists of a coil of wire connected to a power supply. An ammeter is included in the circuit to measure the current flowing through the coil. Figure 13.6 shows a model of a resistance thermometer.

When the coil is put near the flame of a Bunsen burner, the coil becomes hot and its electrical resistance increases. This means that less current now flows through the coil and this is indicated by a fall in the current reading on the ammeter. Thus a smaller current indicates a higher temperature and from this the thermometer can be calibrated against a standardized thermometer.

Resistance thermometers usually include special circuits which make them more sensitive. This type of thermometer is often used to measure high temperatures in furnaces and ovens and in this case the coil is often made of platinum. The temperature range over which this thermometer can operate is then from −200 °C to 1 200 °C.

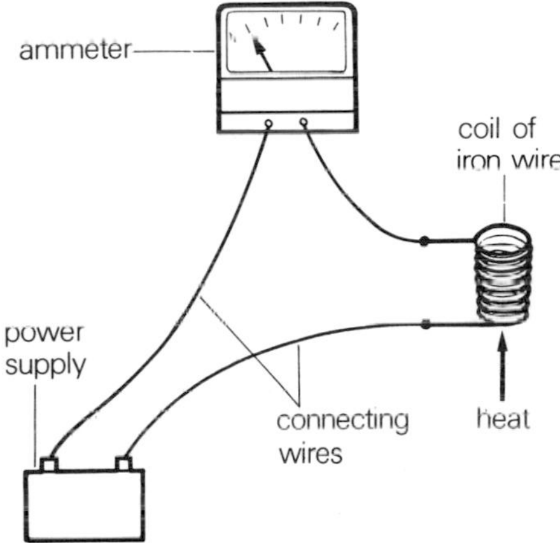

Figure 13.6 Model of a resistance thermometer

5. Thermistor thermometer

The thermistor thermometer is similar to the resistance thermometer in that it depends on a change in electrical resistance with temperature. The difference is that the coil of wire is replaced by a special component called a thermistor, Figure 13.7.

The electrical resistance of a thermistor **decreases** as the temperature rises, i.e. it works in the **opposite** way to a resistance thermometer. When the thermistor becomes hot, its electrical resistance falls. Therefore more current flows through it and the current reading on the ammeter increases. Thus a larger current indicates a higher temperature and from this the thermometer can be calibrated against a standardized thermometer.

Figure 13.7 Model of a thermistor thermometer

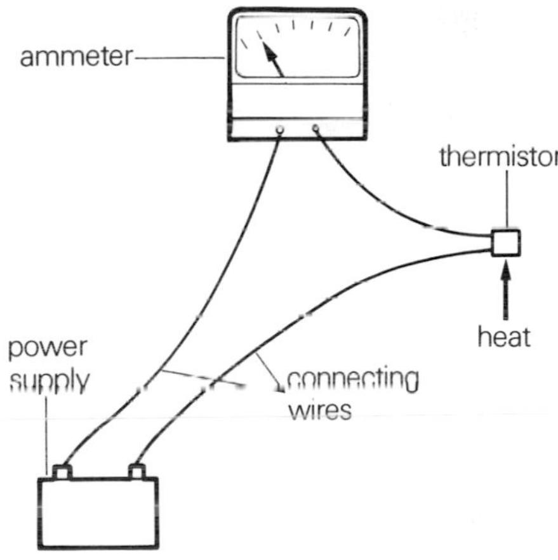

Summary

Temperature is a measure of the hotness or coldness of an object.

On the Celsius scale, the fixed points are the melting point of pure ice at 0 °C and the boiling point of pure water at 100 °C.

Temperature can be measured by the following thermometers.

The mercury-in-glass thermometer based on the fact that mercury expands when heated.

The rotary thermometer based on the fact that the two different metals in a bi-metallic strip expand by different amounts when heated.

The thermocouple thermometer which consists of two wires of different materials joined at the ends to form two junctions; a temperature difference between the junctions causes a current to flow round the loop.

The resistance thermometer is based on the fact that the electrical resistance of a wire increases as it is heated. This results in a current flowing through the wire decreasing as the temperatures rises. The thermistor thermometer is based on the fact that the electrical resistance of a thermistor decreases as it is heated; this results in a current flowing through the wire increasing as the temperature rises.

Problems

1 Liquid oxygen boils at −183 °C, while liquid nitrogen boils at −196 °C. Which of the two liquids has the higher boiling point?

2 What is the temperature difference between the inside of an igloo at 10 °C and outside the igloo where the temperature is −15 °C?

3 Describe two types of thermometer that you have met, explaining in each case how it works.

4 A thermocouple thermometer was calibrated by immersing the cold junction in melting ice at 0 °C, and keeping the hot junction at various temperatures.
The table below shows how the current flowing round the thermocouple varied with the temperature of the hot junction.

temperature of hot junction (°C)	0	25	50	75	100
current flowing in the thermocouple (mA)	0	0·05	0·10	0·15	0·20

Draw a graph of the temperature of the hot junction against current. Hence estimate the temperature of hot oil if the current produced is 0·18 mA when the hot junction is immersed in the oil.

5 A supply of pure melting ice and of water boiling steadily at 99 °C are provided. By using these, together with suitable lengths of iron and copper wire and a sensitive meter, show how it would be possible to determine the temperature of a beaker of warm oil, no other kind of thermometer being available. *SCEEB*

14 Heat Energy

14.1 Introduction

Figure 14.1 James Joule

In previous chapters we have accepted heat as a form of energy but until the start of the nineteenth century, scientists were not at all agreed on what heat actually was. Although the whole story is fascinating and could easily form a book by itself, we will discuss only briefly the history of the idea of heat.

In the eighteenth century, heat was regarded as a substance and hot objects were thought to have more of it. However, in 1798 Count Rumford of Bavaria published a paper which disagreed with this idea. He came to his conclusions during the time he was in charge of producing a cannon for the Bavarian army. The cannon was made by boring a hole in a solid metal cylinder; as a result, large amounts of heat were produced by friction. Count Rumford repeated the process with a blunt borer and found that heat was produced even though the borer was not actually cutting the metal. He decided that heat could not be a substance as he could go on obtaining heat endlessly, so long as the borer was turning. He concluded that heat was being produced purely by the **motion** of the borer.

By the middle of the nineteenth century several scientists had begun to view this process as an **energy change** from work (mechanical energy) to heat. Perhaps the first person to realize this was Carnot, a French engineer. Eventually it was realized that **all** forms of energy were equivalent, and that when an amount of energy disappeared the process was always accompanied by the appearance of the **same** amount of energy in another form (or forms). This leads to the conclusion that the total amount of energy remains constant, which is called the **Principle of Conservation of Energy**.

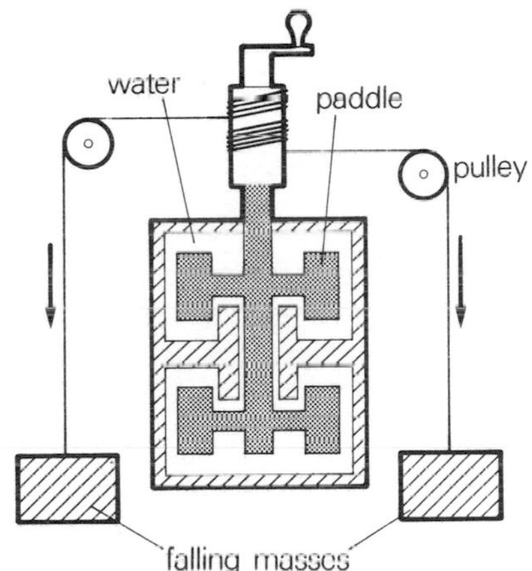

Figure 14.2 Joule's original apparatus

At about the same time, James Joule, a wealthy English amateur scientist, carried out extremely careful experiments to measure very precisely the quantity of heat produced by a certain amount of work (mechanical energy), Figure 14.1.
The apparatus he used for his most famous experiments is shown in Figure 14.2.
A metal paddle was rotated by falling masses and this churned water around in a can. The amount of work done was calculated by multiplying the weight of the falling masses by the distance they fell. The heat generated was calculated from careful measurements of the mass of water and the temperature rise. Joule found that exactly the same quantity of heat was always produced by a certain amount of work. He concluded from his experiments that heat was simply another form of energy. Later in this chapter we will describe an experiment which is similar to Joule's experiment and which you can perform in the laboratory.

14.2 Heat is energy

Heat must **not** be confused with temperature. They are entirely different quantities as the following discussion shows.

Suppose a large beaker and a small beaker are both filled with cold water. The water in the beakers is then heated using two identical bunsen burners, Figure 14.3.
As the two burners are identical, they both supply the same amount of heat energy to the water in two minutes. However, after two minutes the thermometer in the small beaker indicates a **much** larger temperature rise than the thermometer in the large beaker. It is therefore clear that heat and temperature are two entirely different quantities.

Heat, like all forms of energy, is measured in joules (written J), while we will measure temperature (or the degree of hotness) in degrees Celsius (°C).

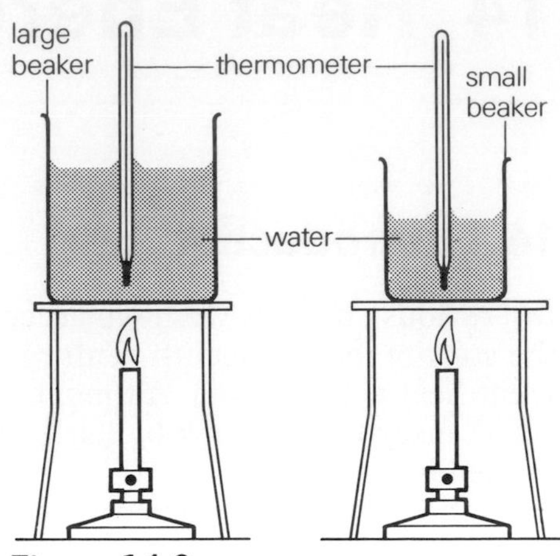

Figure 14.3

'Joule-type' heat-energy experiment

Using the apparatus shown in Figure 14.4 we can supply heat to a copper drum in two different ways.

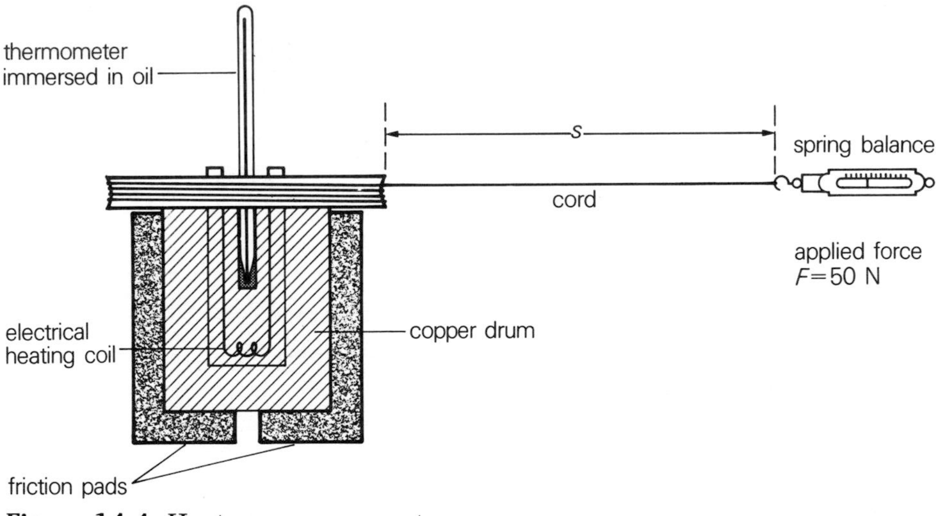

Figure 14.4 Heat-energy apparatus

a) When a spring balance is pulled, the cord unwinds and the copper drum rotates against the stationary friction pads. Work (mechanical energy) has to be done by the applied force because of the frictional resistance.
b) There is an electrical heating coil inside the drum. When the heater is connected to a power supply, electrical energy can be supplied to the heater.

If we can show that the same amount of heat is gained by the drum regardless of whether the heat is transformed from mechanical energy or from electrical energy, then we can surely accept heat as a definite form of energy.

We shall measure the temperature rise in both cases using a thermometer which fits into the drum. Oil on the thermometer bulb ensures good thermal contact with the drum. If the same temperature rise is produced from mechanical energy as from electrical energy, we shall assume that the same amount of heat has been transferred.

From mechanical energy to heat

Using mechanical energy, the drum is rotated by applying a force F which unwinds the cord over a distance s. The amount of work done can be calculated using the equation, work = $F \times s$. The cord can be rewound by hand and the experiment repeated to obtain a more accurate estimate of the temperature rise produced by a certain amount of work.

When an applied force of 50 N unwinds 4 m of cord and the experiment is carried out six times, we find that the total temperature rise produced is 10 °C. We can calculate the total amount of work done as follows:

Applied force	$F = 50$ N
Distance cord is unwound each time	$s = 4$ m
Number of times cord is unwound	$n = 6$
Total distance moved	$n \times s = 6 \times 4 = 24$ m
Total work done	$F \times n \times s = 50 \times 24 = 1\,200$ J

From electrical energy to heat

Using electrical energy, the amount of electrical energy supplied to the heater can be measured on a joulemeter.

The initial joulemeter reading is noted and the heater is switched on. When 1 200 J of electrical energy has been transferred to the heater, the apparatus is switched off. Again the maximum temperature rise produced is found to be about 10 °C.

We conclude that the same amount of heat has been transferred in both cases. Because heat has been shown to be equivalent to two different forms of energy, it follows that heat itself must be a form of energy.

14.3 Measuring heat

When heat is supplied to cold water in a kettle, its temperature rises. The greater the amount of heat supplied, the larger is the temperature rise. If there is only a small amount of water in the kettle, its temperature rises faster than a full kettle of water. From this example we see that the amount of heat, the temperature change and the mass of the water must be related in some way. In order to investigate how these quantities are related, we will heat up water in a container. We must choose a container which cuts down heat losses to the surroundings to a very small amount. The following experiment explains our choice.

The same amount of water at 60 °C is added to four different containers. The temperature of the water in each container is measured after twenty minutes and the results are as shown in Figure 14.5.

The temperature of the water in the polystyrene cup with the lid has only fallen by a small amount compared with the others. It is therefore clear that to cut down heat losses in our experiments to a minimum, we should use a polystyrene cup with a lid. The complete apparatus is shown in Figure 14.6.

The power supply is connected through a joulemeter which measures the amount of electrical energy supplied to the heater. As the electrical energy is changed into heat energy by the heater, we can take the joulemeter readings as a measurement of the heat energy supplied to the water. To find a relationship between two of the quantities under investigation we must keep the third quantity constant.

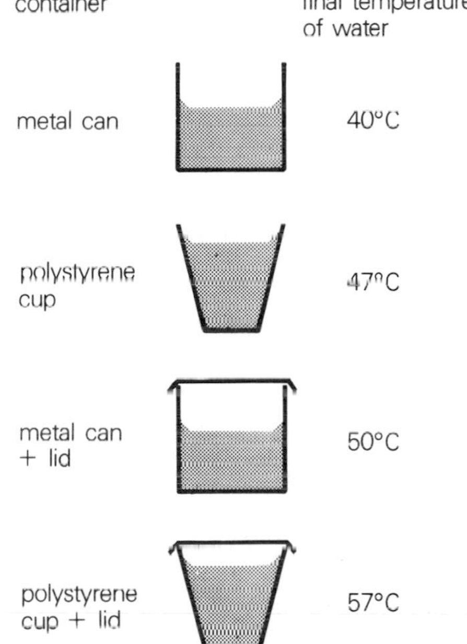

Figure 14.5

Figure 14.6 Investigating heat

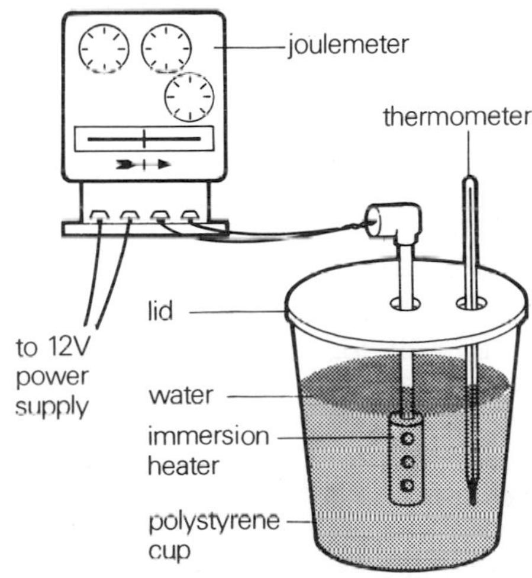

Heat and temperature change

We will look first at how heat and temperature change are related (keeping the mass of water constant). Using the apparatus shown in Figure 14.6, the initial joulemeter reading is noted and the apparatus is switched on. The water is stirred to ensure that it is at an even temperature. Joulemeter readings are taken by momentarily switching off the apparatus every time the temperature of the water rises by 3 °C.

Table 1 shows results obtained by heating a mass of 0·2 kg of water, initially at 12 °C.

Total heat supplied from the start of the experiment (J)	0	2 520	5 050	7 580	10 100	12 640
Temperature (°C)	12	15	18	21	24	27
Change in temperature from the initial temperature (°C)	0	3	6	9	12	15

Table 1

A graph of heat supplied against change in temperature is a straight line through the origin, Figure 14.7. This means that the heat supplied is directly proportional to (varies directly as) the change in temperature.

i.e. $E_h \propto \Delta T$, for a constant mass,
(where E_h = heat supplied, ΔT = change in temperature)

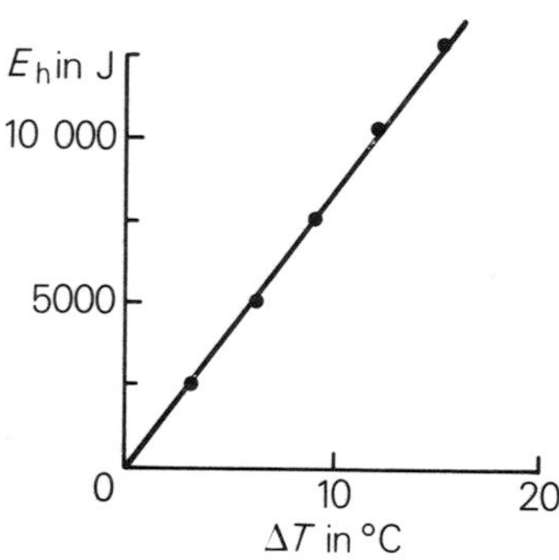

Figure 14.7 Heat supplied against temperature

Heat and mass

We will now look at how the amount of heat and the mass of the water are related. In this experiment we must keep the temperature change constant throughout.

The amount of heat required to raise the temperature of different masses of water by 10 °C is typically as shown in Table 2.

Total heat supplied (J)	6 305	8 410	10 515	12 625
Mass of water (kg)	0·15	0·20	0·25	0·30

Table 2

A graph of heat supplied (to produce a temperature change of 10 °C) against mass is a straight line through the origin, Figure 14.8. This means that the heat energy supplied is directly proportional to (varies directly as) the mass of water.

i.e. $E_h \propto m$, for the same temperature change,
(where E_h = heat supplied, m = mass of water)

Figure 14.8 Heat supplied against mass

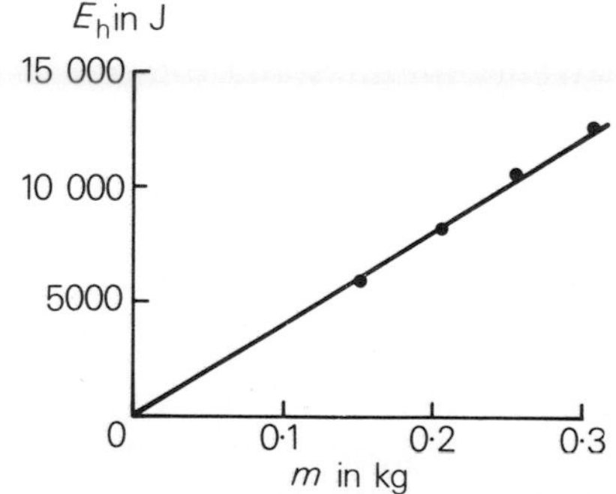

14.4 Specific heat capacity

Table 3 shows some results of the two experiments combined in one table, where E_h = heat supplied, m = mass of water, and ΔT = change in temperature.

E_h	m	ΔT	$m \times \Delta T$	$\frac{E_h}{m \times \Delta T}$
2 520	0·2	3	0·6	4 200
5 050	0·2	6	1·2	4 208
7 580	0·2	9	1·8	4 211
6 305	0·15	10	1·5	4 203
10 515	0·25	10	2·5	4 206

Table 3

From Table 3 it is clear that the quantity in the last column is constant in all our experiments, allowing for experimental error.

i.e. $\frac{E_h}{m \times \Delta T}$ = constant

We have found that the value of this constant for water is about 4 200. When the same experiments are repeated for different substances, the value of $\frac{E_h}{m \times \Delta T}$ is found to be a constant for each substance. It appears that the value of the constant is a property of the substance.

This constant is called the **specific heat capacity** of the substance, written as c. Think of the word 'specific' as meaning 'per kilogram.'

Hence $\frac{E_h}{m \times \Delta T} = c$ (Multiply both sides by $m \times \Delta T$)

$\Rightarrow \quad E_h = c \times m \times \Delta T$

Units

As the unit of heat E_h is the joule, the unit of mass m is the kilogram, and the unit of temperature change ΔT is the degree Celsius, the unit of c is given by,

$$\frac{\text{J}}{\text{kg} \times {}^\circ\text{C}} = \text{J kg}^{-1}\ {}^\circ\text{C}^{-1}$$

This rather unwieldy looking unit is called the joule per kilogram per degree celsius.

Note that when m = 1 kg and ΔT = 1 °C, $E_h = c \times 1 \times 1$, i.e. E_h and c are numerically equal. Specific heat capacity is therefore the amount of heat required to change the temperature of 1 kg of a substance by 1 °C.

The specific heat capacity of water is approximately 4 200 J kg^{-1} °C^{-1} as in Table 4 where approximate values of the specific heat capacities of some common substances are shown.

Substance	*Specific heat capacity* (J kg^{-1} °C^{-1})
water	4 200
methylated spirit	2 300
paraffin	2 200
aluminium	900
iron	480
copper	385
brass	370
lead	130

Table 4

Example 1

6 900 J of heat is supplied to 1 kg of methylated spirit in a polystyrene cup. Calculate the rise in temperature produced. Take the specific heat capacity of methylated spirit to be 2 300 J kg^{-1} °C^{-1}.

$$E_h = c \times m \times \Delta T$$
$$\Rightarrow 6\,900 = 2\,300 \times 1 \times \Delta T$$
$$\Rightarrow \frac{6\,900}{2\,300} = \Delta T \qquad \text{dividing both sides by 2 300}$$
$$3 = \Delta T$$

Rise in temperature of the methylated spirit is 3 °C.

Example 2

When $1{\cdot}9 \times 10^4$ J of heat is supplied to 2 kg of paraffin at 12·0 °C in a saucepan the temperature increases to 16·0 °C.

a) Calculate the specific heat capacity of paraffin.

b) Explain why the result in part (a) is different from the value quoted in Table 4.

a) $\Delta T = 16{\cdot}0 - 12{\cdot}0 = 4{\cdot}0$ °C

$$E_h = c \times m \times \Delta T$$
$$\Rightarrow 1{\cdot}9 \times 10^4 = c \times 2{\cdot}0 \times 4{\cdot}0$$
$$\Rightarrow 0{\cdot}24 \times 10^4 = c$$
$$\Rightarrow 2{\cdot}4 \times 10^3 = c$$

Specific heat capacity of paraffin is $2{\cdot}4 \times 10^3$ J kg^{-1} °C^{-1}.

b) The value of the specific heat capacity of paraffin found in (a) is higher than the expected value of 2 200. This is because some of the energy supplied in the experiment is 'lost' to the saucepan and to the surroundings. If the energy loss could be calculated and **subtracted** from $1{\cdot}9 \times 10^4$, the energy actually used to heat up the paraffin by 4 °C would be found. This would lead to a **lower** answer for the specific heat capacity.

From Table 4 you should realize that water has a high specific heat capacity compared with other common substances. This means that when a certain amount of heat is supplied to 1 kg of water, the temperature rise is much smaller than that produced in 1 kg of another common substance. Because water can take in a lot of heat with only a small temperature rise, it is ideal for the cooling systems of cars although it can corrode the engine.

Finally, although specific heat capacity is different for different substances, it does not matter what form of energy is used to heat up the substance. For example when Joule performed his experiments on heat, it would have made no difference to his results if he had used another form of input energy rather than mechanical energy.

Power rating of heater

Now that we can measure heat using the formula $E_h = c \times m \times \Delta T$, we can find the power rating of an immersion heater if we also measure the time for which it is switched on. The equation for power which we used in Chapter 12 was,

$$\text{power} = \frac{\text{work done}}{\text{time taken}}$$

Although this equation is for mechanical power, replacing work done (mechanical energy) by electrical energy allows us to calculate the electrical power of the heater.

If we use a heater to heat up some water in a polystyrene cup, the electrical energy supplied to the heater is changed to the heat given to the water. Hence we can rewrite the power equation as,

$$\text{power rating of heater} = \frac{\text{heat supplied}}{\text{time taken}}$$

Example 3

An immersion heater is used to heat 0·5 kg of water in a polystyrene pot. If the temperature rise after 600 s is 14 °C, calculate the power rating of the heater. Take the specific heat capacity of water to be 4 200 J kg^{-1} °C^{-1}.

$$E_h = c \times m \times \Delta T$$
$$= 4\,200 \times 0{\cdot}5 \times 14$$
$$= 4\,200 \times 7$$
$$= 29\,400$$
$$\text{Power rating of heater} = \frac{\text{heat supplied}}{\text{time taken}}$$
$$\frac{29\,400}{600}$$
$$= 49$$

Power rating of heater is 49 W.

Specific heat capacity of metals

Now that we know the power rating of our heaters, we can use them to determine specific heat capacities without the need for a joulemeter. The specific capacity of a metal can be measured experimentally using a cylindrical block of the metal with holes drilled for a heater and thermometer, Figure 14.9.

Oil on the thermometer bulb ensures good thermal contact with the block.

The power rating P of the heater is given by

$$P = \frac{E_h}{t} \quad \text{where } P = \text{power rating of heater}$$
$$E_h = \text{heat supplied}$$
$$t = \text{time taken}$$

Multiplying both sides by t gives,

$$P \times t = E_h$$

i.e. heat supplied = power rating × time

$$\Rightarrow \quad c \times m \times \Delta T = P \times t$$

Typical results obtained from this type of experiment using an aluminium block are shown below:

Figure 14.9 Specific heat capacity of a metal

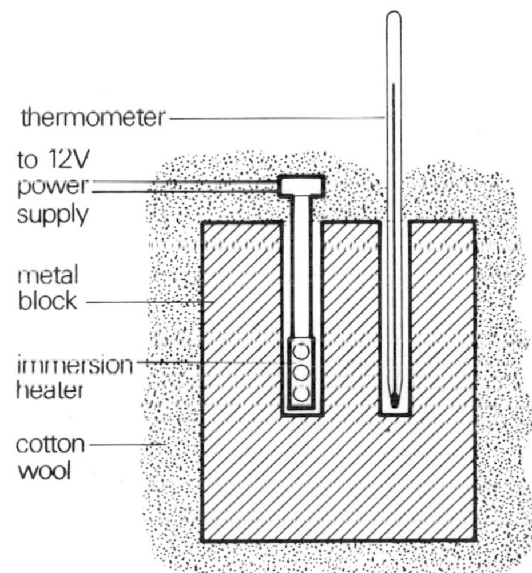

Mass of aluminium block $m = 1$ kg
Initial temperature of block $= 15$ °C
Highest final temperature reached $= 35$ °C
Change in temperature of block $\Delta T = 20$ °C
Power rating of heater $P = 50$ W
Time for which heater is switched on $t = 360$ s

Using $c \times m \times \Delta T = P \times t$

$\Rightarrow \quad c \times 1 \times 20 = 50 \times 360$

$\Rightarrow \quad 20\,c = 50 \times 360$

$\Rightarrow \quad c = 900$

Specific heat capacity of aluminium is 900 J kg^{-1} °C^{-1}.

14.5 Mixtures

If you want to cool down a cup of black coffee, a simple way is to add some cold water. The final temperature of the mixture is somewhere between the original temperature of the coffee and the temperature of the cold water. In this section we will investigate how these three temperatures are related.

Suppose we have two polystyrene cups, one containing 0·2 kg of hot water at 70 °C and the other containing 0·3 kg of cold water at 20 °C. When the contents of both are added to a large polystyrene cup and the resulting 'mixture' stirred, the final temperature is found to be 39·8 °C, Figure 14.10.

To find the relation between these temperatures we will work out the heat lost by the hot water and the heat gained by the cold water.

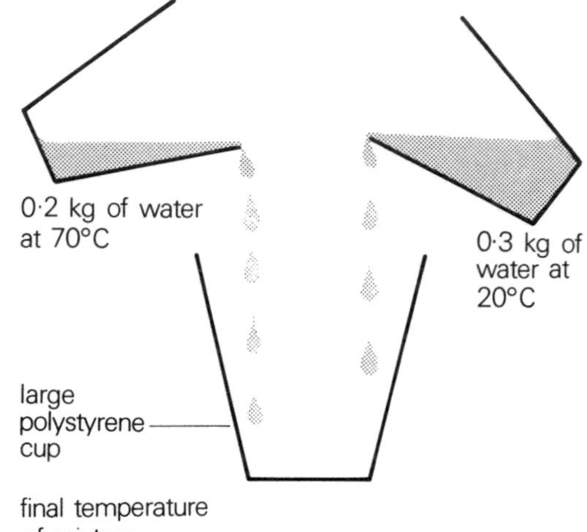

Figure 14.10 Mixing hot and cold water

Change in temperature of the hot water,

$\Delta T_{hot} = (70 - 39{\cdot}8) = 30{\cdot}2$ °C

Heat lost by the hot water,

$E_h = c \times m_{hot} \times \Delta T_{hot}$ (m_{hot} = mass of hot water)

$= 4\,200 \times 0{\cdot}2 \times 30{\cdot}2$

$\approx 2{\cdot}5 \times 10^4$ J

Change in temperature of the cold water,

$\Delta T_{cold} = (39{\cdot}8 - 20) = 19{\cdot}8$ °C

Heat gained by the cold water,

$E_h = c \times m_{cold} \times \Delta T_{cold}$ (m_{cold} = mass of cold water)

$= 4\,200 \times 0{\cdot}3 \times 19{\cdot}8$

$\approx 2{\cdot}5 \times 10^4$ J

We have shown that when hot and cold water are added, allowing for experimental errors,

heat lost by the hot water = heat gained by the cold water

The main error in this experiment is that some heat is inevitably lost from the hot water to the surroundings, during the mixing process. Note that again it appears that the total amount of energy is conserved which agrees with the Principle of Conservation of Energy.

Example 4

60 kg of hot water at 82 °C is run into a bath and it is found to be too hot. If 300 kg of cold water at 10 °C is added to lower the temperature, calculate the final temperature of the bath water.

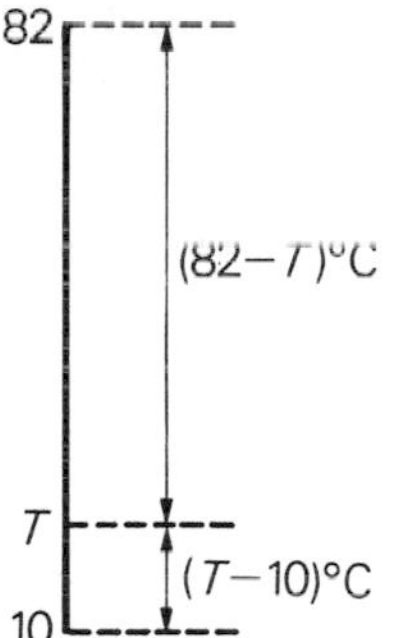

Let the final temperature of the mixture be T °C

Change in temperature of the hot water, $\Delta T_{hot} = (82 - T)$ °C

Change in temperature of the cold water, $\Delta T_{cold} = (T - 10)$ °C

Heat lost by the hot water $= c \times m_{hot} \times \Delta T_{hot}$
$= 4\,200 \times 60 \times (82 - T)$

Heat gained by the cold water $= c \times m_{cold} \times \Delta T_{cold}$
$= 4\,200 \times 300 \times (T - 10)$

Assuming no heat is lost to the surroundings,

heat lost by hot water = heat gained by cold water

$\Rightarrow 4\,200 \times 60 \times (82 - T) = 4\,200 \times 300 \times (T - 10)$

$\Rightarrow \quad 60(82 - T) = 300(T - 10)$

$\Rightarrow \quad 82 - T = 5(T - 10)$

$\Rightarrow \quad 82 - T = 5\,T - 50$ Add 50 to both sides

$\Rightarrow \quad 132 - T = 5\,T$ Add T to both sides

$\Rightarrow \quad 132 = 6\,T$

$\Rightarrow \quad 22 = T$

Final temperature of the bath is 22 °C.

14.6 Energy changes

As we can now measure heat, we can do calculations involving energy changes which are concerned with heat. There are obviously many examples of this but we will only consider one situation.

Suppose a lump of lead is dropped on to a hard surface without rebounding. The kinetic energy of the lead disappears on impact. Most of this energy will change into heat (some sound is also produced) and the total amount of energy will be conserved. The following example involves a calculation based on this energy change.

Example 5

A lump of lead of mass 1 kg is dropped onto a hard surface without rebounding and its temperature rises by 2 °C. Calculate the speed with which the lead hit the surface.

Assuming that all the kinetic energy of the lead is changed on impact to heat in the lead,

kinetic energy lost by lead = heat produced in lead

$\Rightarrow \quad \frac{1}{2}\,m \times v^2 = c \times m \times \Delta T$

$\Rightarrow \quad \frac{1}{2} \times 1 \times v^2 = 130 \times 1 \times 2$

$\Rightarrow \quad \frac{1}{2}\,v^2 = 260$

$\Rightarrow \quad v^2 = 520$

$\Rightarrow \quad v = 23$

Speed of the lead on hitting the surface is 23 m s^{-1}.

14.7 Changes of state

Introduction

When cold water is heated its temperature rises. However if a heater continues to supply heat to the water, the temperature only rises until the water starts to boil. Once the water is at its boiling point of 100 °C, the energy provided by the heater is used to change the water into steam. Using a thermometer it is easy to show that the temperature of the steam is also 100 °C. The energy supplied by the heater is therefore needed to change the water into steam without any change in the temperature. Energy is required to change the state from liquid into gas without change in temperature. When the process is reversed and steam at 100 °C changes back into water at 100 °C, it can be shown that heat is given out without change in temperature. This heat is therefore latent (or hidden) in the steam. For example, steam tends to produce more severe burns than water at the same temperature because this latent heat is transferred to the skin. An expresso coffee machine uses a steam jet to heat up the water. The water is heated by the release of the latent heat as the steam changes to water.

When ice at its melting point of 0 °C is heated, it changes into water also at 0 °C. Energy is therefore required to change the state from solid to liquid without change in temperature. Again, when the process is reversed and water at its freezing point of 0 °C changes into ice at 0 °C, it can be shown that this latent heat is given out without change in temperature.

Cooling curve of naphthalene

When solid naphthalene in a test-tube is heated by a Bunsen burner, it melts and becomes liquid. Using the apparatus shown in Figure 14.11, the temperature of the naphthalene can be recorded at 60 s time intervals until it has cooled down to room temperature.

Table 5 shows some of the results obtained in this experiment. A graph of temperature against time is shown in Figure 14.12.

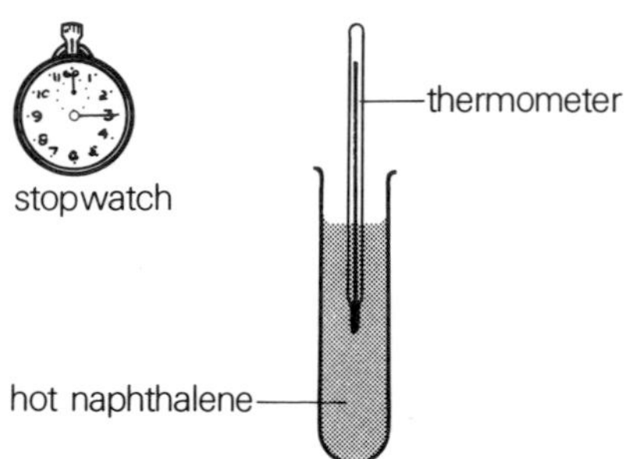

Figure 14.11 Naphthalene cooling

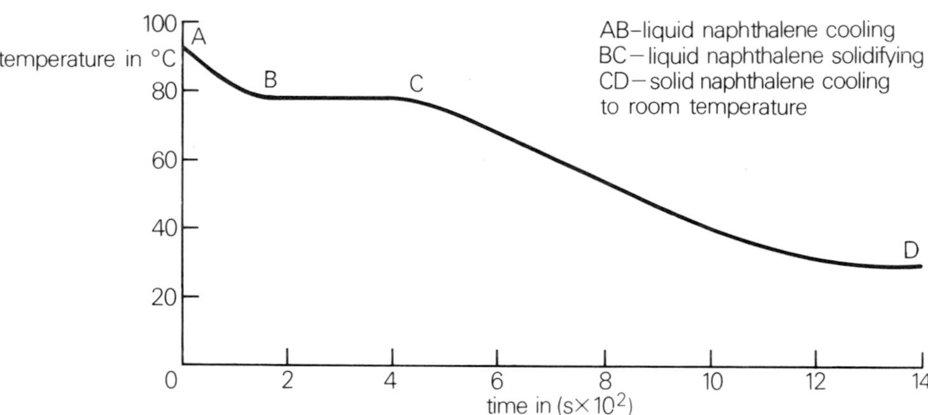

Figure 14.12 Cooling curve of naphthalene

Table 5

Time (s)	*Temperature* (°C)
0	94
60	86
120	82
180	79
240	79
300	79
360	79
420	79
480	76
540	72
600	69

When the naphthalene is hotter than its surroundings, it gives out heat. However during part BC of the graph the temperature of the naphthalene remains constant even though it is giving out heat. Thus while the naphthalene is changing state from a liquid to a solid, heat is being given out without any temperature change.

The heat being given out by the naphthalene during its change of state is its **latent heat.**

14.8 Specific latent heat

With twice the mass of naphthalene, twice the amount of latent heat would be given out during the change of state. It is useful to define the amount of latent heat given out by one kilogram of a substance. This quantity is called **specific latent heat**, written as l. Think of the word 'specific' as meaning per kilogram. The specific latent heat of a substance is the energy involved in changing the state of 1 kg of the substance without any temperature change.

Hence $l = \frac{E_h}{m}$, where E_h = latent heat, m = mass

$\Rightarrow m \times l = E_h$ multiplying both sides by m

i.e. $E_h = m \times l$

Units

As the unit of latent heat is the joule and the unit of mass is the kilogram, the unit of l is the joule per kilogram, written J kg^{-1}.

When we are talking about the change of state from solid to liquid at the melting point, we refer to the **specific latent heat of fusion** l_f, and when referring to the change of state from liquid to gas at the boiling point we use the term **specific latent heat of vaporization**, l_v.

Example 6

A mass of 5·0 kg of ammonia at its boiling point is vaporized when 6·5 MJ of heat is supplied to it. Calculate the specific latent heat of vaporization of ammonia.

$E_h = 6{\cdot}5 \text{ MJ} = 6{\cdot}5 \times 10^6$ J

$E_h = m \times l_v$

$\Rightarrow 6{\cdot}5 \times 10^6 = 5{\cdot}0 \times l_v$

$\Rightarrow 1{\cdot}3 \times 10^6 = l_v$

The specific latent heat of vaporization of ammonia is $1{\cdot}3 \times 10^6$ J kg^{-1}.

Specific latent heat of fusion of ice

The specific latent heat of fusion of ice l_f is the number of joules of energy required to change 1 kg of ice to water without change in temperature.

Figure 14.13 shows the apparatus which we can use to find the value of l_f.

A 'control' apparatus is needed because some of the ice in a filter funnel would melt **without** being heated by a heater. The control beaker therefore allows us to measure the amount of ice melted due only to the heat supplied by the surroundings. When the drip rate becomes steady the beakers are put in position, the experimental heater is switched on for a measured time, and then switched off. When the drip rate again becomes steady the beakers are removed and the masses of water in both are measured. The mass of water actually melted by the heater is found from the difference in the two masses. The following results were obtained from this experiment:

Mass of water in experimental beaker = 0·039 kg
Mass of water in control beaker = 0·010 kg
Mass of water melted by the heater m = 0·029 kg
Power rating of heater P = 50·0 W
Time for which heater was switched on t = 200 s
Energy supplied by the heater $P \times t = 50{\cdot}0 \times 200$ J

Figure 14.13 Specific latent heat of fusion of ice

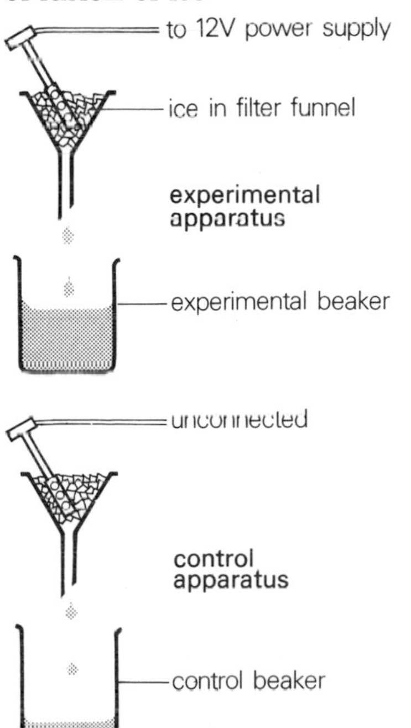

Using $E_h = m \times l_f$
$\Rightarrow 50 \times 200 = 0{\cdot}029 \times l_f$
$\Rightarrow 10\,000 = 0{\cdot}029 \times l_f$
$\Rightarrow \dfrac{10\,000}{0{\cdot}029} = l_f$
$\Rightarrow 3{\cdot}45 \times 10^5 = l_f$
Specific latent heat of fusion of ice is $3{\cdot}45 \times 10^5$ J kg^{-1}

This value is only approximate because of experimental inaccuracies. The accepted value for the specific latent heat of fusion of ice is $3{\cdot}34 \times 10^5$ J kg^{-1}.

Specific latent heat of vaporization of water

The specific latent heat of vaporization of water l_v is the number of joules of energy required to change 1 kg of water to steam without change in temperature.

Figure 14.14 shows the apparatus which we can use to find a value for l_v.

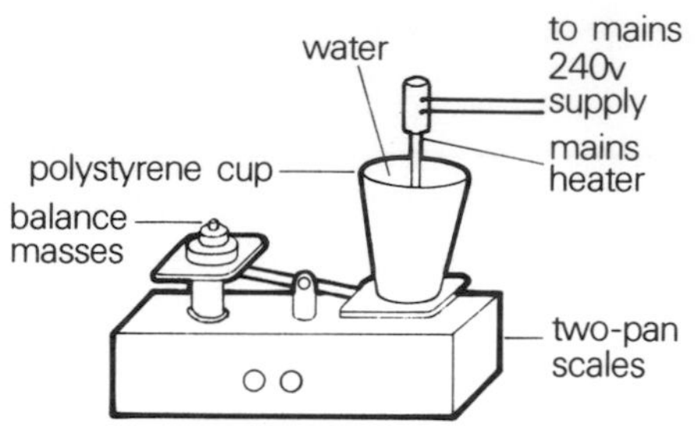

Figure 14.14 Specific latent heat of vaporization of water

A mains heater with a high power rating is used in this experiment because more heat is required to change 1 kg of water into steam than it takes to change 1 kg of ice into water. Balance masses are put on one pan of a two-pan scales and water is poured into a large polystyrene cup on the other pan until the mass of the water is just greater than that of the balance masses.

The rest of the experiment is explained using diagrams to show the position of the scales at various times. Figure 14.15.

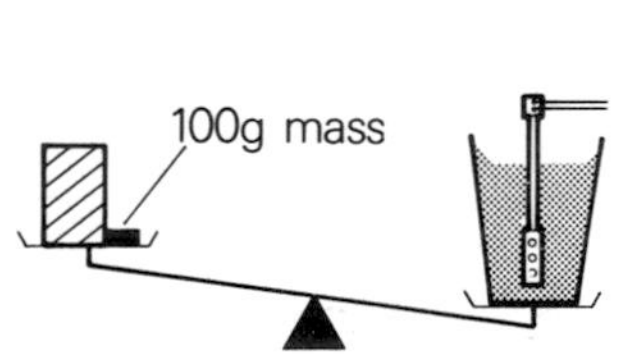

The heater is switched on and eventually the water starts to boil

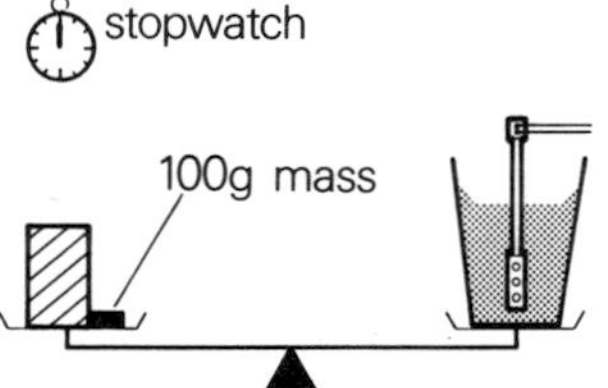

When some water has boiled away, the scales balance. The stopwatch is now started

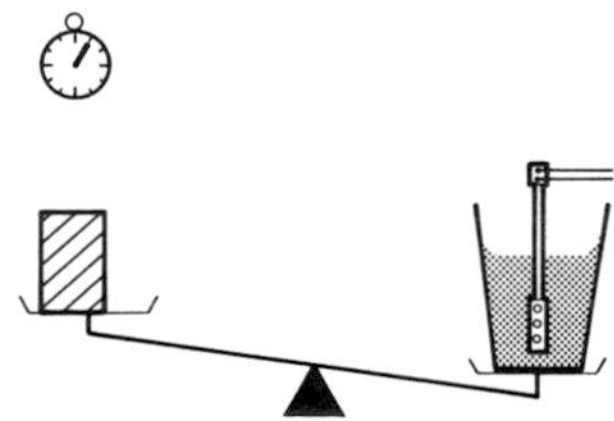

The 100g mass is removed

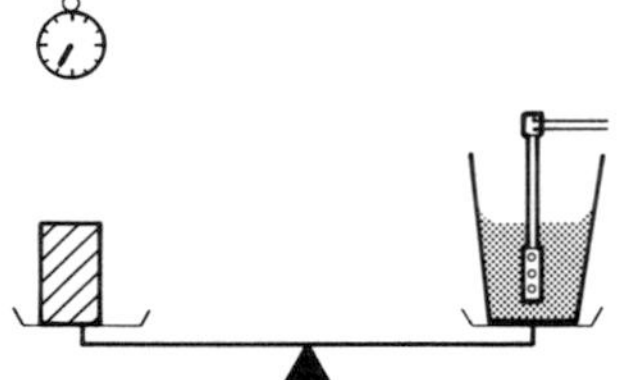

When a further 100g of water has boiled off, the scales will again balance. The watch is now stopped

Figure 14.15

The energy required to boil off 100 g of water can be calculated and so the specific latent heat of vaporization of water l_v can be found. The following results were obtained from this experiment:

Mass of water boiled off $m = 0{\cdot}1$ kg
Power rating of heater, $P = 600$ W
Time taken to boil off 0·1 kg of water $t = 380$ s
Energy supplied by heater $P \times t = 600 \times 380$

Using $E_h = m \times l_v$
$\Rightarrow 600 \times 380 = 0{\cdot}1 \times l_v$
$\Rightarrow 228\,000 = 0{\cdot}1\ l_v$
$\Rightarrow 2\,280\,000 = l_v$
$\Rightarrow 2{\cdot}28 \times 10^6 = l_v$
Specific latent heat of vaporization of water is $2{\cdot}28 \times 10^6$ J kg^{-1}.

This value is only approximate because of experimental inaccuracies. Two inaccuracies in this experiment are:
a) some steam condenses on the top part of the heater and drips back into the cup.
b) because the boiling is vigorous, some drops of water manage to jump out of the cup.
The accepted value for the specific latent heat of vaporization of water is $2{\cdot}26 \times 10^6$ J kg^{-1}.

Example 7

Calculate the amount of heat which must be supplied to change a 2 kg block of ice at 0 °C into water at 40 °C.

The total amount of heat required is the sum of the latent heat required to change the ice at 0 °C to water at 0 °C + the heat required to change the temperature of the water from 0 °C to 40 °C.

i.e. total heat required $= (m \times l_f) + (c \times m \times \Delta T)$
$= (2 \times 3{\cdot}34 \times 10^5) + (4200 \times 2 \times 40)$
$= (6{\cdot}68 \times 10^5) + (336\,000)$
$= (6{\cdot}68 \times 10^5) + (3{\cdot}36 \times 10^5)$
$\approx 10{\cdot}0 \times 10^5$

Total heat required is $10{\cdot}0 \times 10^5$ J.

Example 8

A 1 kg lump of solid wax is heated uniformly by a 100 W heating element until after it has all melted. The graph below shows how the temperature of the wax varies with time.

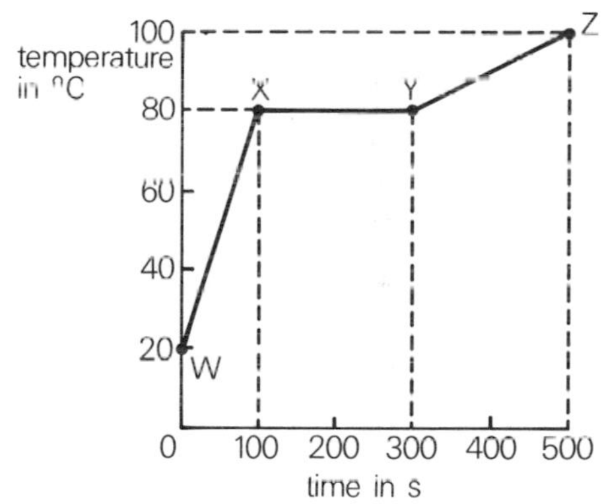

a) Explain what is happening in the regions WX, XY and YZ.
b) Calculate the specific heat capacity of the solid.
c) Calculate the specific latent heat of fusion of the solid.

a) WX – the solid is heating up.
XY – the solid is melting at 80 °C and changing to liquid without change in temperature.
YZ – the liquid is heating up.

b) The solid heats up from 20 °C to 80 °C in a time of 100 s. The energy supplied by the heater in that time is given by power × time, and this energy is changed to heat which raises the temperature of the solid.
Hence,

$P \times t = c \times m \times \Delta T$
$\Rightarrow 100 \times 100 = c \times 1 \times 60$
$\Rightarrow 10\,000 = 60\,c$
$\Rightarrow 167 \approx c$

The specific heat capacity of the solid is 167 J kg^{-1} °C^{-1}.

c) The solid melts during the time interval 100 s to 300 s. The energy supplied by the heater in that time is changed into the heat to melt the solid.
Hence,

$P \times t = m \times l_f$
$\Rightarrow 100 \times 200 = 1 \times l_f$
$\Rightarrow 20\,000 = l_f$

The specific latent heat of fusion of the solid is 20 000 J kg^{-1}.

Summary

The Principle of Conservation of Energy states that the total amount of energy remains constant.

Heat is a form of energy. It is measured in joules (J).

The specific heat capacity of a substance is the amount of heat required to change the temperature of 1 kg of the substance by 1 °C.

$E_h = c \times m \times \Delta T$

where E_h = heat
m = mass
ΔT = temperature change
c = specific heat capacity

The unit of specific heat capacity is the J kg^{-1} °C^{-1}.

The energy supplied by a heater is given by

$E_h = P \times t$

where E_h = energy supplied
P = power rating
t = time taken

When hot and cold water are mixed,
heat lost by hot water = heat gained by cold water

The specific latent heat of a substance is the amount of energy involved in changing the state of 1 kg of the substance without any temperature change,

$E_h = m \times l$

where E_h = latent heat
m = mass
l = specific latent heat.

The unit of specific latent heat is the J kg^{-1}.

The word fusion refers to melting.

The word vaporization refers to boiling or evaporating.

Problems

1 A paddle wheel is used to churn 1·0 kg of water in an insulated can. A total load of 200 N turns the paddle wheel by falling through a distance of 1·0 m. The experiment is repeated 90 times and a total temperature rise of 4 °C is produced. Calculate the specific heat capacity of water.

2 Calculate the amount of heat required to raise the temperature of 0·2 kg of lead by 60 °C. Specific heat capacity of lead is 130 J kg^{-1} °C^{-1}.

3 If 74 000 J of heat is given out when a brass ball of mass 1 kg cools from 300 °C to 100 °C, calculate the specific heat capacity of brass.

4 2 200 joules of heat raises the temperature of paraffin in a plastic beaker from 25 °C to 35 °C. Calculate the mass of the paraffin. Specific heat capacity of paraffin is 2 200 J kg^{-1} °C^{-1}.

5 When 1·0 kg of paraffin wax is heated a graph of temperature against time is as shown.

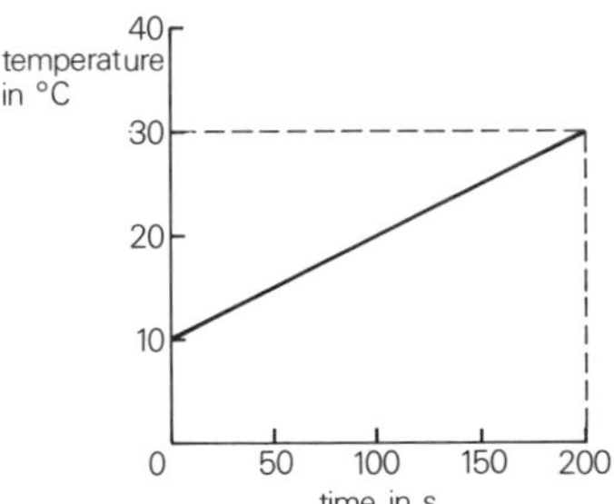

If the power of the heater is 300 W, calculate,

5·1 the amount of energy supplied by the heater in 200 s.
5·2 the specific heat capacity of paraffin wax.
5·3 the time taken by the heater to heat the paraffin wax from 30 °C to its boiling point of 55 °C.

6 An aluminium can containing 4·0 kg of water is heated by a bunsen burner for 600 s. The temperature rise of the water is 10 °C.

6·1 Calculate a value for the power rating of the burner.
6·2 Explain whether the true power rating is higher or lower than this value. Specific heat capacity of water is 4 200 J kg^{-1} °C^{-1}.

7 A car of mass 1 000 kg is brought to rest from a speed of 20 m s^{-1}. Assuming that all its kinetic energy is changed into heat in the disc brakes, find the temperature rise produced if each of the four brakes has a mass of 5·0 kg.
Specific heat capacity of the iron of which the brakes are made is 480 J kg^{-1} °C^{-1}.

8 0·40 kg of water at 85 °C is poured into a plastic bucket containing 0·60 kg of water at 20 °C. What is the temperature of the mixture?

9 3·0 kg of cold water at 10 °C is added to 1·0 kg of hot water and the final temperature of the warm mixture is 30 °C. Calculate the original temperature of the hot water.

10 In a central-heating system, steam at 100 °C enters a radiator, and water at 100 °C leaves the radiator. Explain whether this process can warm a room.

11 Calculate the amount of heat required to melt 0·10 kg of ice at 0 °C.
Specific latent heat of fusion of ice is $3{\cdot}34 \times 10^5$ J kg^{-1}.

12 Calculate the specific latent heat of fusion of naphthalene given that $2{\cdot}98 \times 10^5$ J of heat are given out when 2·0 kg of naphthalene at its melting point changes into solid.

13 Calculate the time taken for a 500 W heater to melt 1·0 kg of ice at 0 °C.

14 A 200 W heater is used to melt ice at 0 °C in a filter funnel. After 300 s the mass of water collected is 0·188 kg. If 0·018 kg of the ice melted purely due to heat taken in from the surroundings (and not from the heater), calculate the specific latent heat of fusion of ice.

15 Calculate the amount of heat required to change 0·50 kg of liquid ethanol to vapour at its boiling point.
Specific latent heat of vaporization of ethanol is $8{\cdot}5 \times 10^5$ J kg^{-1}.

16 A 2·5 kW kettle is left switched on for 100 s after the water has started to boil. What mass of water is boiled off in that time? Specific latent heat of vaporization of water is $2{\cdot}26 \times 10^6$ J kg^{-1}.

17 A bullet of mass 0·20 kg travelling at 400 m s^{-1} embeds itself in a large block of ice which is at 0 °C. Assuming that the bullet's kinetic energy is totally changed into heat, calculate the mass of ice which will melt.
Specific latent heat of fusion of ice is $3{\cdot}34 \times 10^5$ J kg^{-1}.

18 Calculate the amount of heat required to change 2 kg of ice at 0 °C into water at 10 °C.
Specific heat capacity of water is 4 200 J kg^{-1} °C^{-1}.
Specific latent heat of fusion of ice is $3{\cdot}34 \times 10^5$ J kg^{-1}.

19 An immersion heater is immersed in a large beaker full of water at an initial temperature of 10 °C. The heater supplies heat at a constant rate. The heater is switched on and the temperature of the water is taken every 30 seconds, the water being stirred each time before reading the thermometer. The results are:

Time (s)	0	30	60	90	120	150	180
Temperature (°C)	10	27	42	56	69	79	87

19·1 Draw (on the graph paper provided) a graph of temperature (y-axis) against time (x-axis). From this graph estimate how long the water takes to reach its boiling point from the time when the heater was switched on. (Show on your graph how you made your estimate.)
19·2 Explain why the graph is **not** a straight line.
19·3 Which part of the graph would you use in order to estimate as accurately as possible the rate at which heat is given out by the heater? Explain your choice. *SCEEB*

20 An immersion heater is placed in a vacuum flask containing a quantity of liquid and switched on.

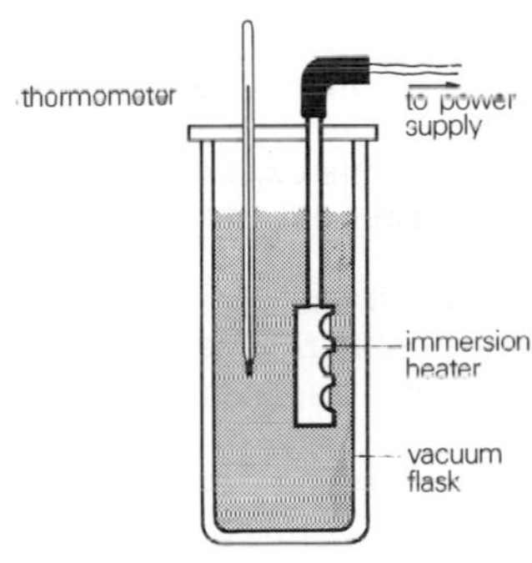

The temperature of the liquid is recorded every 50 seconds and the following table of results is obtained.

Time (s)	0	50	100	150	200	250
Temperature (°C)	20	51	82	114	130	130

20·1 Plot a graph of these results.
20·2 Estimate from the graph as accurately as you can
a) the boiling point of the liquid;
b) the time for the liquid to reach its boiling point.
20·3 Explain how the information on the graph may be used to calculate the specific heat capacity of the liquid, listing any extra data needed to do this. *SCEEB*

21 **21·1** An experiment was performed to test the heat insulation of various materials. Hot water was put into each of the cans A, B and C (shown below) and their fall in temperature recorded. The cans were identical apart from the insulating material between the can and the jacket.

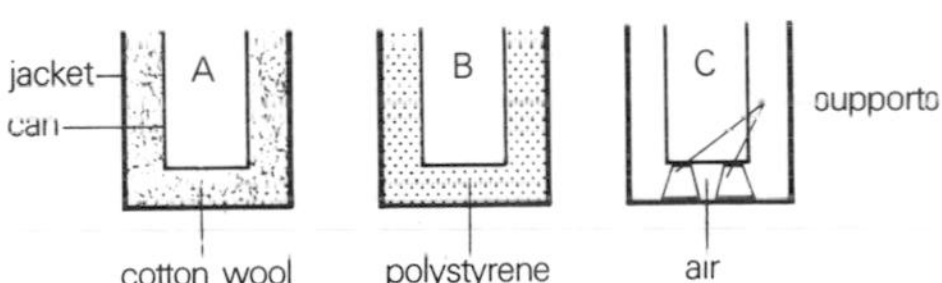

a) In order to make a fair comparison between these three arrangements, certain factors should be kept the same in each case. State **two** of these factors.
b) Suggest a modification to the apparatus which would greatly reduce the heat losses.
21·2
a) An immersion heater placed in 0·2 kg of water contained in a well insulated can raised the temperature of the water from 20 °C to 35 °C.
How much heat was supplied to the water by the immersion heater?
The specific heat capacity (specific heat) of water is 4200 J kg^{-1} °C^{-1}.
b) If the immersion heater was rated at 50 watts, for how long was it switched on?

21·3
a) Describe an experiment that you might perform to find the power output of a bunsen flame in watts. Your description should include a diagram of your apparatus, the measurements you would make and the method for calculating your results.
b) What is the main source of error in this experiment?
c) State how this error would affect your answer. *SCEEB*

22 A text book contained a graph showing the temperature plotted against time for 100 g of paraffin heated by a 50 W immersion heater (graph A). A pupil tried this experiment and obtained graph B.

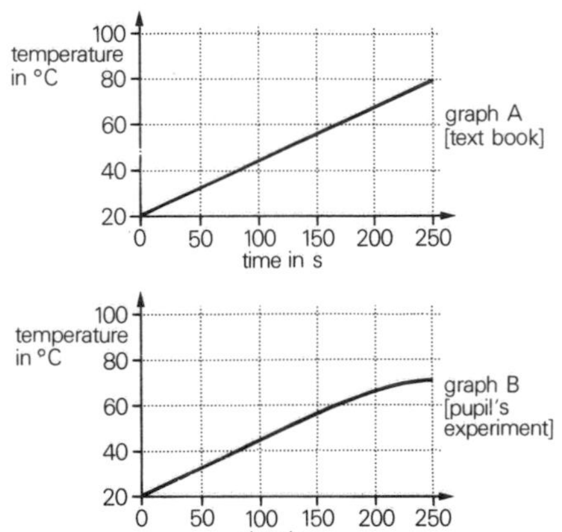

22·1 Using the text book graph A calculate the specific heat capacity of paraffin.
22·2 How might the pupil test to see if his heater was supplying 50 joules each second?
22·3 Suggest **one** reason which could account for graph B bending over. (An incorrectly rated heater is **not** a reason.) Explain how the experiment could be modified so that the graph looked more like graph A. *SCEEB*

23 In an energy conversion experiment, an electric motor rotates a copper cylinder at a steady speed against two friction pads. The cylinder is initially at room temperature.

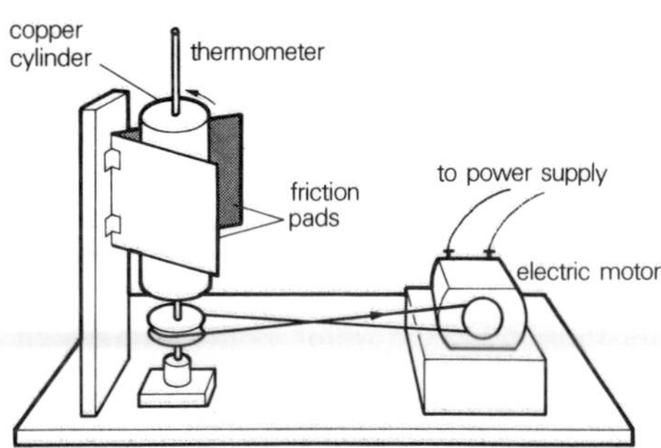

The following experimental results are obtained during a five minute test run.
Power input to motor = 1·2 W
Mass of copper cylinder = 0·3 kg
Rise in temperature of copper cylinder = 2·5 °C

23·1 How much energy is supplied to the motor during the test run?
23·2 How much heat is generated in the cylinder during the test run?
23·3 Explain why the answer to **23·2** is less than the answer to **23·1**.
23·4 Discuss whether a ten minute test run under the same conditions would produce a temperature rise of 5 °C in the copper cylinder. *SCEEB*

24 A metal cylinder of mass 0·3 kg is rotated between two friction pads by pulling a cord with a steady force of 60 N. While rotating, the cylinder rubs against the friction pads and, as a result, the temperature of the cylinder rises. When the cord is pulled a distance of 4 m, a rise in temperature of 2 °C is produced in the cylinder.

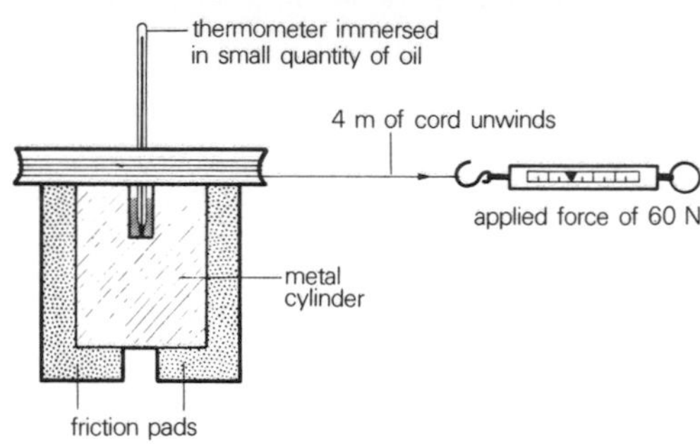

24·1 Assuming no heat losses, calculate the specific heat capacity of the metal.
24·2 It is a good experimental procedure to have a small quantity of oil surrounding the thermometer bulb. Give a reason for this.
24·3
a) Suggest two ways in which a higher rise in temperature would be produced in the metal cylinder in this experiment.
b) Discuss whether a higher rise in temperature would lead to a more accurate value for the specific heat capacity of the metal. *SCEEB*

25 The graph shown represents the curve for a substance cooling from a high temperature down to room temperature.

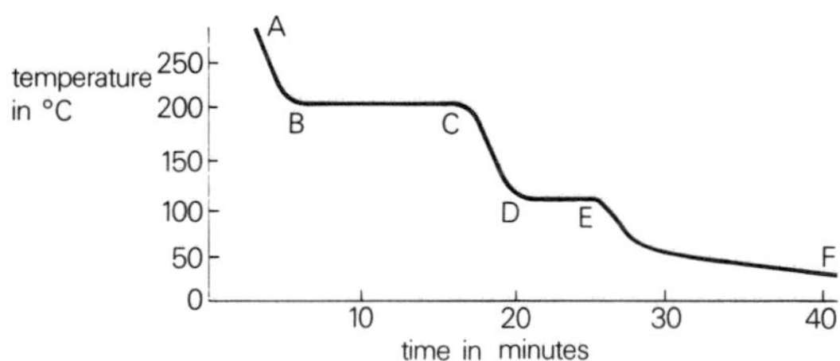

25·1 What is the boiling point of the substance?
25·2 What is happening at region DE?
25·3 Why is region BC longer than region DE? *SCEEB*

26 **26·1** A test tube containing 0·10 kg of a powder is heated for several minutes by a bunsen burner which is supplying heat at a constant rate of 50 joules per second. The temperature of the contents is noted at equal time intervals and a graph of temperature against time drawn.

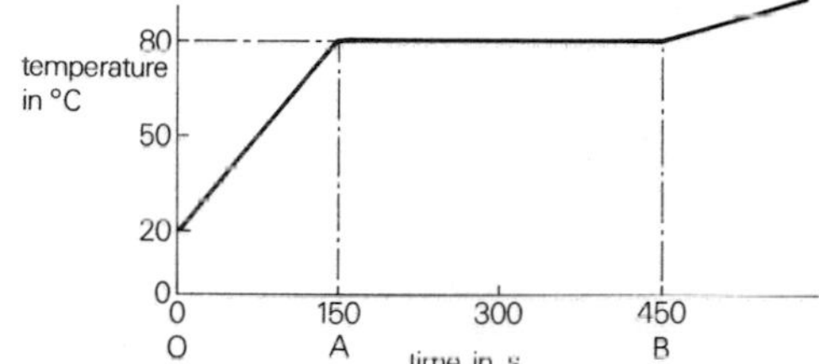

a) Express the power output of the bunsen burner in watts.
b) What is happening to the substance during the interval OA?
c) What is happening to the substance during the interval AB?
d) What is the significance of the time AB being greater than the time OA?
e) Calculate the specific heat capacity of the powder, assuming that all the heat supplied by the bunsen is transferred to the powder.

26·2 A characteristic of certain cooking pots is that when they are removed from the source of heat, the contents may continue to boil for a short time. Suggest an explanation for this effect. *SCEEB*

15 Gas laws and kinetic theory

15.1 Introduction

In this chapter we investigate how gases behave and we will try to explain their behaviour in terms of particles. We consider the three different quantities pressure, temperature and volume, and how they are related. We begin by discussing each quantity in turn.

Pressure

Like the word 'work', pressure has many meanings in everyday use. For example, we talk of the pressure of exams, or pressure being exerted by the Government. In physics however, pressure is defined in terms of two other quantities, force and area. The following example illustrates the effects of different pressures.

When you stand in soft snow you sink into it. This is because your weight is acting over a small area, the area of your feet. However, if you wear snow shoes you find that you can walk over the snow without sinking in, Figure 15.1.

Your weight has not changed but has simply been spread over a much larger area by the snow shoes.

Without snow shoes a force is exerted on a small area and so the force on unit area is large. In this case we say that the **pressure** is large and so you sink into the snow. When wearing snow shoes, the same force is exerted over a much larger area and so the force on unit area is much smaller. We say that the pressure is smaller and so this time you do not sink. To be more exact, we define pressure as follows,

$$\text{pressure} = \frac{\text{force acting at right angles to an area}}{\text{area}}$$

i.e. $p = \frac{F}{A}$

where p = pressure, F = force, A = area

Figure 15.1 Soldier using snow shoes

Units

As the SI unit of force is the newton (N) and the SI unit of area is the square metre (m^2) it follows that the unit of pressure is the newton per square metre, written as $N\ m^{-2}$. This unit has been named the pascal (Pa) after Blaise Pascal, a Frenchman who did many experiments involving gas and liquid pressures.

$1\ Pa = 1\ N\ m^{-2}$

However the pascal is a very small unit and so we will often use the kilopascal (kPa) where 1 000 Pa = 1 kPa.

Example 1

Calculate the pressure produced by a force of 50 N acting down on a metal sheet of area 0·01 m^2.

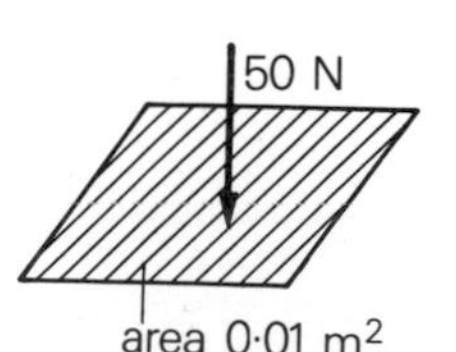

$p = \frac{F}{A}$

$p = \frac{50}{0{\cdot}01}$

$p = 5\,000$

The pressure produced is 5 000 Pa.

Gas in a balloon exerts a pressure on the walls of the balloon which keeps the balloon inflated. Gas pressure is produced because the gas particles are moving and bombarding the container walls from all directions. Because the pressure acts in all directions and is not associated with one particular direction, pressure is a scalar quantity.

Temperature

We have already dealt with the property of temperature in Chapter 13. It is defined as the hotness or coldness of an object. The unit of temperature is the °C.

Volume

The volume of an object is the amount of space it occupies. The SI unit for volume is the cubic metre written as m^3. The cubic centimetre written as cm^3 is commonly used for measuring smaller volumes, ($1\ cm^3 = 10^{-6}\ m^3$).

15.2 The Bourdon pressure gauge

In the experiments discussed in this chapter we shall use a Bourdon pressure gauge to measure gas pressures. Figure 15.2 shows front and rear views of the pressure gauge.

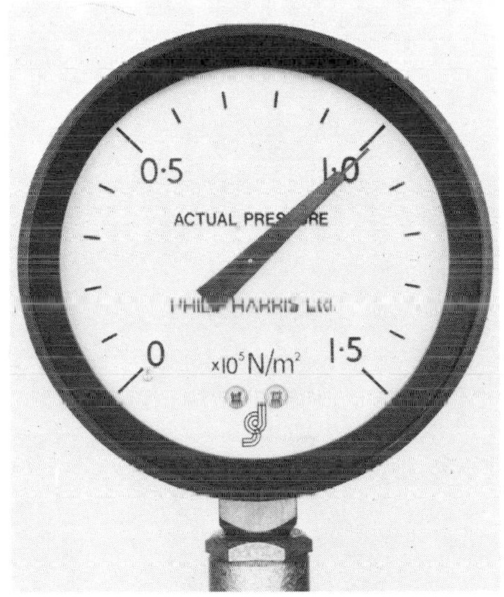

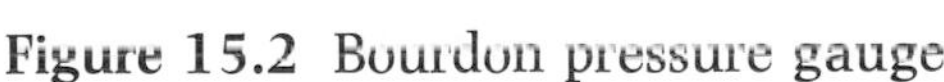
Figure 15.2 Bourdon pressure gauge

Figure 15.3

This pressure gauge works on the same principle as the familiar party-tooter often found in Christmas crackers, Figure 15.3. The harder you blow into the party-tooter, the more the paper tube uncurls. The Bourdon gauge consists of a hollow curved metal tube. An increase in pressure in the tube causes it to uncurl and a system of cogwheels makes a pointer move round a scale. A decrease in pressure results in the pointer moving in the opposite direction. In some Bourdon gauges, although the gauge is not connected to anything, the pointer is not at zero. This is because the air around us exerts a pressure which we call **atmospheric pressure**. The value of atmospheric pressure changes from day to day but is approximately 1×10^5 Pa or 100 kPa. The gauge in Figure 15.2 is indicating the atmospheric pressure.

The Bourdon gauge has many uses in industry, for example on gas cylinders, pumps and boilers. Figure 15.4 shows a type of Bourdon gauge which can be seen on many car dashboards. It is used to measure the oil pressure in the engine.

Figure 15.4 Oil pressure gauge

Checking the calibration

Using the apparatus shown in Figure 15.5, the pressure of a gas can be changed by applying a force on a piston in a syringe.

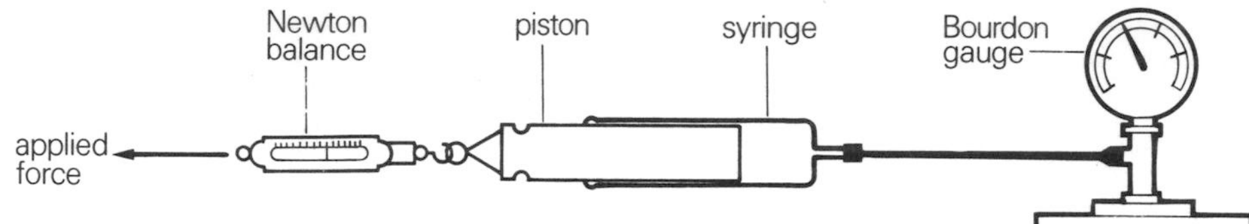

Figure 15.5 Calibration of bourdon gauge

If we measure the force and the area of the piston, we can calculate the change in pressure from the equation $p = \frac{F}{A}$. If the calibration of the gauge is correct, the change in pressure recorded by the gauge should equal our calculated pressure change.

Typical results from this experiment are shown below:

Original pressure reading on gauge	p_1 = 102 000 Pa
Final pressure reading on gauge	p_2 = 72 000 Pa
Change in pressure recorded by gauge	$(p_1 - p_2)$ = 30 000 Pa
Force applied to piston	F = 30 N
Radius of piston	r = 0·018 m
Area of circular piston	$(A = \pi r^2)$ = 0·001 m^2
Calculated pressure change	F/A = 30 000 Pa

These results show that the pressure change calculated from F/A agrees with that recorded by the gauge. We have therefore checked that the calibration of the Bourdon gauge is correct.

15.3 Pressure and temperature

We will now investigate how the pressure and temperature of a gas are related. To do this, the other variables must be kept constant. We therefore use a constant mass and volume of gas in a sealed flask. The pressure of air in a flask is measured by a Bourdon gauge while the temperature of the air is indicated by a thermometer as in Figure 15.6.

The flask is totally immersed initially in a beaker containing dry ice (solid carbon dioxide) at −78 °C and the pressure reading noted. The dry ice is then replaced by melting ice at 0 °C and the pressure reading is again noted. The water can then be heated using a Bunsen burner and the pressure readings noted at various water temperatures up to 100 °C. To ensure that the air in the flask is at the temperature indicated by the thermometer, the pressure readings are only taken after the temperature has remained constant for about a minute. There is, of course, some air in the pressure tubing which is not at the correct temperature. This is a source of error which can be reduced by using a short length of narrow tubing.

Figure 15.6 Variation of pressure with temperature

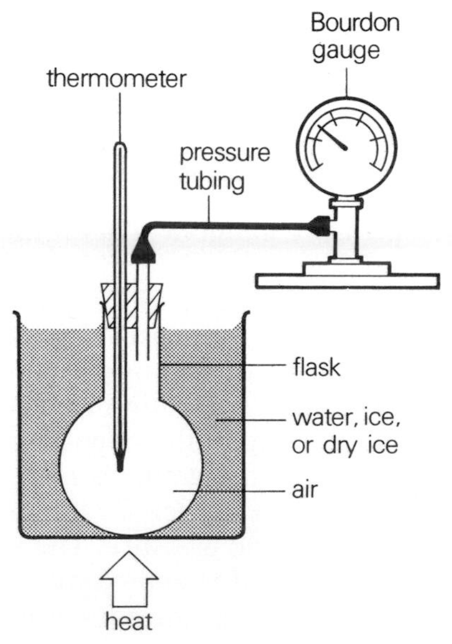

Typical results obtained from this experiment are shown in Table 1.

	dry ice	*ice/water*	*water*			
Temperature (°C)	−78	0	20	50	80	100
Pressure (kPa)	67	93	100	110	120	127

Table 1

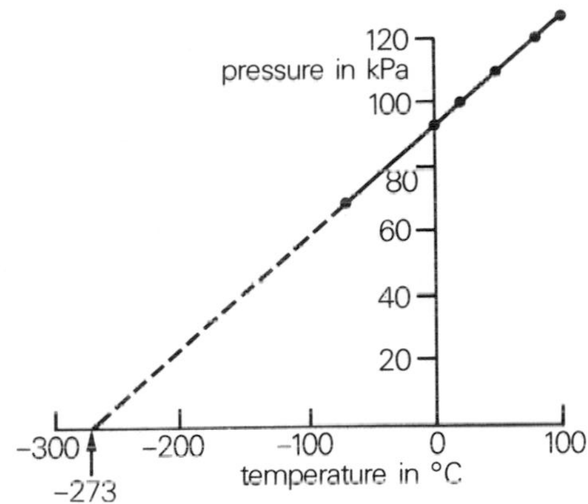

Figure 15.7 Pressure against temperature in °C

Figure 15.7 shows a graph of the air pressure plotted against the temperature of the air in °C.

It is clear that although the graph does not pass through the origin, the points do lie on a straight line. By extending the graph back, it is possible to find the temperature in °C at which the pressure of the gas would be zero if it carried on behaving in the same way. From the graph this temperature is about −273 °C. It is therefore possible to make this graph a straight line through the origin by shifting the origin of the temperature scale to −273 °C. At this temperature the gas pressure would be zero. According to our particle theory, this implies that the particles exert no pressure and so are no longer moving. This temperature is therefore the lowest possible temperature attainable and so is called **absolute zero** temperature. Temperatures measured on this scale are called **absolute** or **Kelvin temperatures**. The kelvin (K) is the same size as the degree Celsius (°C). A graph of gas pressure against the Kelvin temperature is as shown in Figure 15.8.

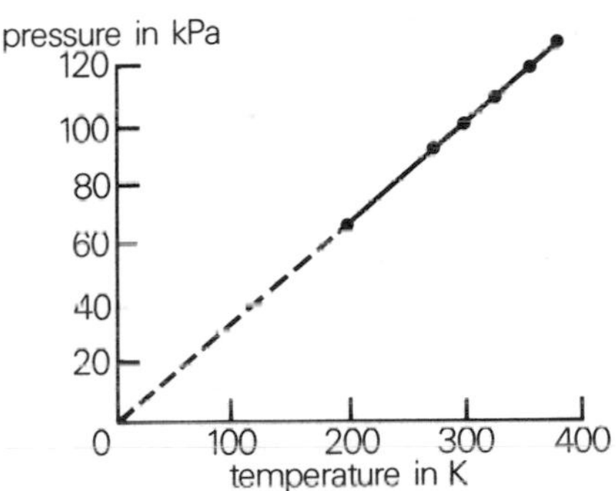

Figure 15.8 Pressure against temperature in kelvin

The Kelvin temperature scale

The Kelvin temperature scale is named after Lord Kelvin, a famous Scottish scientist who in the nineteenth century established the standard scale of temperature. As a kelvin (K) is exactly the same size as a degree Celsius (°C), and absolute zero is 0 K or −273 °C, it follows that we can change any temperature in degrees Celsius into kelvin simply by adding 273. The two scales are shown together in Figure 15.9.

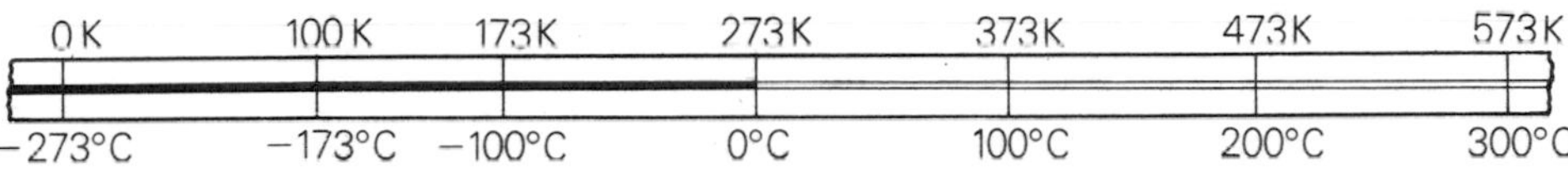

Figure 15.9 Kelvin and Celsius temperature scales

Example 2

a) Change the following Celsius temperatures into Kelvin temperatures, (i) 127 °C, (ii) 27 °C, (iii) 0 °C, (iv) −100 °C, (v) −173 °C.
b) Change the following Kelvin temperatures into Celsius temperatures, (i) 350 K, (ii) 273 K, (iii) 173 K, (iv) 150 K, (v) 27 K.

a) To change Celsius temperatures to Kelvin temperatures, we add 273. Hence the Kelvin temperatures are

(i) 127 + 273 = 400 K
(ii) 27 + 273 = 300 K
(iii) 0 + 273 = 273 K
(iv) −100 + 273 = 173 K
(v) −173 + 273 = 100 K

b) To change Kelvin temperatures to Celsius temperatures, we must subtract 273. Hence the Celsius temperatures are

(i) 350 − 273 = 77 °C
(ii) 273 − 273 = 0 °C
(iii) 173 − 273 = −100 °C
(iv) 150 − 273 = −123 °C
(v) 27 − 273 = −246 °C

To return to our experiment, Figure 15.8 shows that a graph of pressure against Kelvin temperature is a straight line through the origin. This means that the pressure of a fixed mass of gas at constant volume is directly proportional to (varies directly as) its Kelvin temperature. This is a statement of what is often called the **Pressure Law**,

i.e. $p \propto T$ (constant volume and mass)

where p = gas pressure, T = temperature in kelvin.

$\Rightarrow p = \text{constant} \times T$ (Divide both sides by T)

$\Rightarrow \frac{p}{T} = \text{constant}$ (constant volume and mass)

Although our experiment was carried out with air, the pressure law applies to all gases. Different graphs might be obtained, but when they are extended back they all meet the temperature axis at absolute zero (−273 °C). You must realize, however, that a real gas does not always behave in the manner indicated by the pressure law. Remember that at low temperatures all gases liquefy and so the gas pressure will disappear long before absolute zero can be reached. A gas which obeyed the pressure law all the way to absolute zero would be called an **ideal gas.** An ideal gas does not actually exist, but most gases behave like an ideal gas over a fairly wide range of temperatures and pressures.

As a gas is cooled to near absolute zero, it becomes more and more difficult to cool it further. However, at the Clarendon Laboratory in Oxford, temperatures within a few millionths of a degree above absolute zero have been produced.

According to the pressure law, if the values of pressure and temperature of a fixed mass of gas at constant volume are initially p_1 and T_1, and these values change to p_2 and T_2, then

$\frac{p_1}{T_1} = \frac{p_2}{T_2}$ (constant volume and mass)

Note however that the temperatures are measured in kelvin. It is a common mistake to forget to change Celsius temperatures into Kelvin temperatures before using the above formula.

Example 3

A sealed flask contains gas at a temperature of 27 °C and a pressure of 90 kPa. If the temperature rises to 127 °C what will be the new pressure?

Initial pressure $p_1 = 90 \times 10^3$ Pa; final pressure $= p_2$.

Initial temperature $T_1 = 27$ °C = 300 K

final temperature $T_2 = 127$ °C = 400 K

$$\frac{p_1}{T_1} = \frac{p_2}{T_2}$$

$$\Rightarrow \frac{90 \times 10^3}{300} = \frac{p_2}{400} \text{ (Multiply both sides by 400)}$$

$$\Rightarrow \frac{90 \times 10^3 \times 400}{300} = p_2$$

$$\Rightarrow 120 \times 10^3 = p_2$$

The new pressure will be 120 kPa.

15.4 Volume and temperature

In our second investigation we will find how the volume and temperature of a gas are related. In this experiment we use a constant mass of gas and keep the pressure constant throughout the experiment. Figure 15.10 shows the apparatus that we will use.

A column of air is trapped in a capillary tube by a mercury index. The top end of the tube is open to the air so the pressure of the air is always atmospheric pressure. The capillary tube is immersed in water so that the entire trapped column of air is under water. When the water is heated the air expands and pushes the mercury index up the tube. The pressure of the air in the tube therefore remains constant as it is always at atmospheric pressure. The temperature can be varied and readings of the length of the trapped air column (which is proportional to its volume), can be taken. The heater is removed and the water is stirred before taking readings, to ensure that the air column is at the temperature indicated by the thermometer.

Typical results obtained from this experiment are shown in Table 2.

Figure 15.10 Variation of volume with temperature

Temperature (°C)	0	20	40	60	80	100
Temperature (K)	273	293	313	333	353	373
Length of air column (cm) (proportional to the volume of air)	20·0	21·5	22·9	24·4	25·9	27·3

Table 2

Figure 15.11 shows a graph of the volume of air plotted against its temperature in °C.

This graph is similar to the graph in Figure 15.7. When the graph is extended back it also meets the temperature axis at −273 °C (0 K). This suggests that we draw a graph of volume against Kelvin temperature which, as Figure 15.12 shows, is a straight line passing through the origin when extended back.

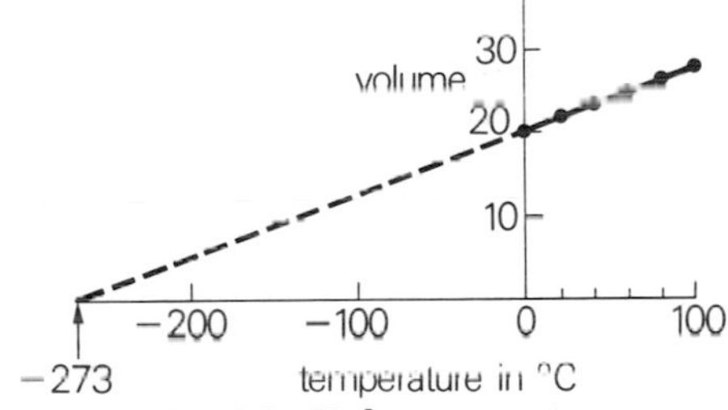

Figure 15.11 Volume against temperature in °C

This means that the volume of a fixed mass of gas at constant pressure is directly proportional to (varies directly as) its Kelvin temperature. This is a statement which is often called **Charles' Law.**

i.e. $V \propto T$ (constant pressure and mass)

(where V = volume of gas, T = (temperature in kelvin)

$\Rightarrow V = \text{constant} \times T$ (Divide both sides by T)

$\Rightarrow \frac{V}{T} = \text{constant}$ (constant pressure and mass)

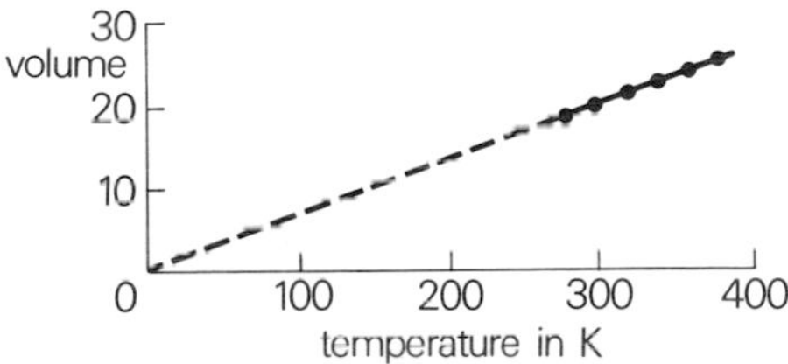

Figure 15.12 Volume against temperature in kelvin

Again, although our experiment was carried out with air, most gases obey Charles' law. Different gases would again give graphs which when extended back would meet the temperature axis at absolute zero.

When a real gas is cooled, it will liquefy and cease to obey Charles' law. However, most gases behave like an ideal gas and obey Charles' law over a wide range of temperatures.

According to Charles' law, if the values of volume and temperature of a fixed mass of gas at constant pressure are initially V_1 and T_1, and these values change to V_2 and T_2, then

$$\frac{V_1}{T_1} = \frac{V_2}{T_2} \text{ (constant pressure and mass)}$$

Note again that the temperatures are measured in **kelvin.**

Example 4

100 cm^3 of air is at a temperature of 0 °C. At what temperature will the volume be 125 cm^3 if the pressure of the air does not change?

Initial volume $V_1 = 100$ cm^3; final volume $V_2 = 125$ cm^3.
Initial temperature $T_1 = 0$ °C $= 273$ K; final temperature $= T_2$.

$$\frac{V_1}{T_1} = \frac{V_2}{T_2}$$

$$\Rightarrow \frac{100}{273} = \frac{125}{T_1} \text{ (Multiply both sides by } 273 \times T_1\text{)}$$

$$\Rightarrow 100\ T_1 = 125 \times 273$$

$$\Rightarrow T_1 = \frac{125 \times 273}{100}$$

$$\Rightarrow T_1 \approx 341$$

The volume will be 125 cm^3 at 341 K (68 °C).

15.5 Pressure and volume

In our final investigation we will find how the pressure and volume of a gas are related. In this experiment we use a constant mass of gas and keep the temperature constant. Figure 15.13 shows the apparatus that we will use.

A column of air is trapped above oil in a glass tube. When air is pumped into the oil reservoir, more oil is pushed into the glass tube which increases the pressure on the trapped air. A Bourdon gauge measures the pressure.

When the pressure of the trapped air is increased, its volume decreases. Using the valve, the pressure can gradually be reduced and readings taken of the volume of air. The temperature of the oil remains constant at room temperature throughout the experiment. However, the temperature of the air tends to change when its volume is changed so readings are only taken after the temperature of the air column has had time to become steady.

Typical results obtained from this experiment are shown in Table 3.

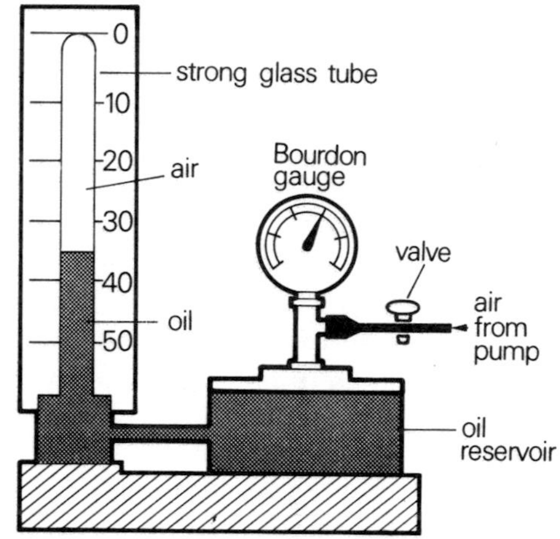

Figure 15.13 Variation of pressure with volume

Pressure of air p kPa	100	111	125	143	167	250
Volume of air V cm^3	50·0	45·0	40·0	35·0	30·0	20·0
$p \times V$	5 000	4 995	5 000	5 005	5 010	5 000
$\frac{1}{V}$	0·020	0·022	0·025	0·029	0·033	0·050

Table 3

Figure 15.14 Pressure against volume

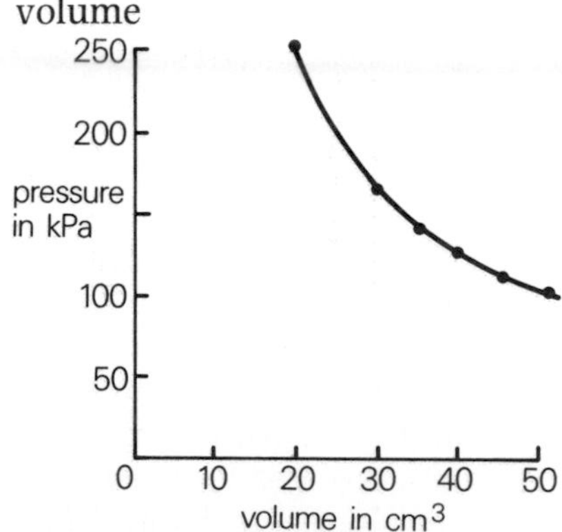

A graph of pressure against the volume of air is shown in Figure 15.14.

The graph is obviously not a straight line through the origin, but the results do lie on a smooth curve. The relationship between pressure and volume can be found by multiplying the two quantities together. The third row of Table 3 shows that $p \times V$ is approximately constant.

Figure 15.15 shows that a graph of pressure against 1/volume is a straight line through the origin.

This means that the pressure of a fixed mass of gas at constant temperature is directly proportional to (varies directly as) the reciprocal of its volume. This is a statement of **Boyle's Law**.

$$p \propto \frac{1}{V} \quad \text{(constant temperature and mass)}$$

(where p = pressure of gas, V = volume of gas)

$$\Rightarrow \quad p = \frac{\text{constant}}{V} \quad \text{(Multiply both sides by } V\text{)}$$

$$\Rightarrow p \times V = \text{constant} \quad \text{(constant temperature and mass)}$$

Although our experiment was carried out with air, most gases obey Boyle's law over a wide range of pressures. Different gases and different masses of the same gas would, however, give different values for the constant.

According to Boyle's law, if the values of pressure and volume of a fixed mass of gas at constant temperature are initially p_1 and V_1 and these values change to p_2 and V_2, then

$$p_1 \times V_1 = p_2 \times V_2 \quad \text{(constant temperature and mass)}$$

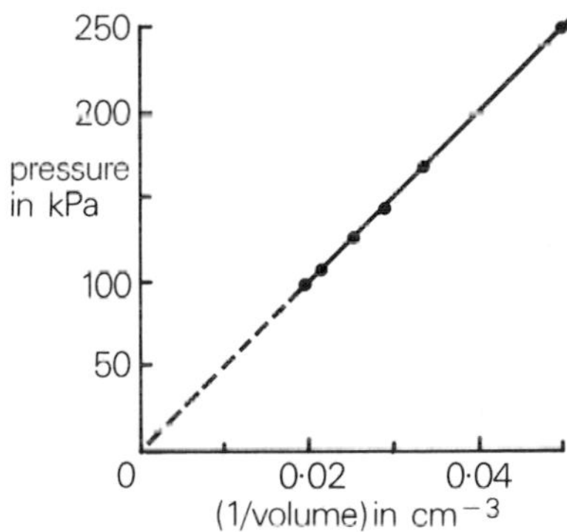

Figure 15.15 Pressure against 1/volume

Example 5

100 cm³ of gas at room temperature and a pressure of 120 kPa are trapped in a syringe by sealing the nozzle. The plunger is depressed until the volume is 20 cm³ and the temperature allowed to return to room temperature. What is the new pressure?

Initial pressure = 120×10^3 Pa; final pressure $= p_2$.
Initial volume = 100 cm³; final volume = 20 cm³.

$$p_1 \times V_1 = p_2 \times V_2$$

$$\Rightarrow 120 \times 10^3 \times 100 = p_2 \times 20 \text{ (Divide both sides by 20)}$$

$$\Rightarrow \quad 6 \times 10^3 \times 100 = p_2$$

$$\Rightarrow \quad 600 \times 10^3 = p_2$$

The new pressure is 600 kPa.

15.6 Combined Gas Equation

Having done three experiments in which either the volume, pressure or temperature of a fixed mass of gas was kept constant, we will now consider how these quantities are related when all three are allowed to vary. Suppose that we repeated the third experiment (p and V varying), at *different* temperatures. Typical results are shown in Table 4.

p	100	125	250	100	125	250
V	50	40	20	45·5	47	24·9
T (K)	300	300	300	273	353	373
$\frac{p \times V}{T}$	16·67	16·67	16·67	16·67	16·64	16·69

Table 4

It is clear from Table 4 that the quantity $p \times V/T$ is approximately constant,

i.e. $\dfrac{p \times V}{T}$ = constant (for a fixed mass of gas)

This equation is usually called the **Combined Gas Equation**, and is obeyed very well by most gases over a wide range of temperatures and pressures.

According to the Combined Gas Equation, if a fixed mass of gas has initial values of pressure, volume, and kelvin temperature given by p_1, V_1 and T_1, and these values change to p_2, V_2 and T_2, then

$$\frac{p_1 \times V_1}{T_1} = \frac{p_2 \times V_2}{T_2} \text{ (for a fixed mass of gas)}$$

Note that if the volume is held constant in the above equation, i.e. $V_1 = V_2$, then the equation simplifies to,

$$\frac{p_1}{T_1} = \frac{p_2}{T_2} \qquad \text{(constant volume and mass)}$$

which is a statement of the Pressure Law. Similarly, if the pressure is constant the equation reduces to Charles' Law, and if the temperature is constant it reduces to Boyle's Law. Thus the Combined Gas equation includes the three gas laws.

Example 6

A weather balloon contains 3 m^3 of helium at a pressure of 100 kPa and a temperature of 27 °C. If the pressure is doubled and the temperature changes to 127 °C, what will be the new volume of the balloon?

$p_1 = 100 \times 10^3$ Pa; $p_2 = 200 \times 10^3$ Pa; $T_1 = 27$ °C = 300 K; $T_2 = 127$ °C = 400 K; $V_1 = 3$ m^3; new volume = V_2

Using $\dfrac{p_1 \times V_1}{T_1} = \dfrac{p_2 \times V_2}{T_2}$

$$\Rightarrow \frac{100 \times 10^3 \times 3}{300} = \frac{200 \times 10^3 \times V_2}{400}$$

$$\Rightarrow \frac{3}{3} = \frac{2 \times V_2}{4} \text{ (cancelling and dividing both sides by } 10^3\text{)}$$

$$\Rightarrow 1 = \frac{V_2}{2}$$

$$\Rightarrow 2 = V_2$$

The new volume of the balloon is 2 m^3.

15.7 Kinetic Theory

The basic assumption of the Kinetic Theory is that all matter consists of particles which are in constant motion at any temperature above absolute zero. Air particles cannot be seen as they are extremely small. However, evidence for their motion was discovered in 1827 by Robert Brown.

Brownian motion

Robert Brown, a Scottish botanist, carried out an experiment which showed that when tiny pollen grains suspended in water are viewed under a microscope they are seen to be in constant motion. At first it was thought that this motion was a form of life but further experiments with non-living particles showed that the motion was always present.

In the laboratory we can look at the **Brownian motion** of smoke particles using the apparatus shown in Figure 15.16.

Light from the lamp is focused by the glass rod on the glass cell. A dropper filled with smoke from smouldering string is used to fill the glass cell and a glass cover will then seal the cell. When the microscope is correctly adjusted, the smoke which consists of tiny specks of ash can be seen reflecting the light from the lamp. The smoke particles are seen to move jerkily through short distances in many different directions. The first accurate explanation of this Brownian motion was given by Albert Einstein in 1905. According to the Kinetic Theory, the air particles surrounding a speck of ash are moving randomly and colliding with it. Although a speck of ash is much larger than the air particles, it is small enough to be affected by collisions with them. These randomly occurring collisions produce unbalanced forces which cause the speck of ash to move. The air particles are moving in all directions so that the collisions come from all directions. The motion of a speck of ash is therefore random, i.e. it moves, but not in any particular direction, Figure 15.17.

According to the Kinetic Theory, at a higher temperature the air particles have more kinetic energy and so are moving faster. This means that at higher temperatures the collisions are more violent and so the specks of ash are jostled around faster.

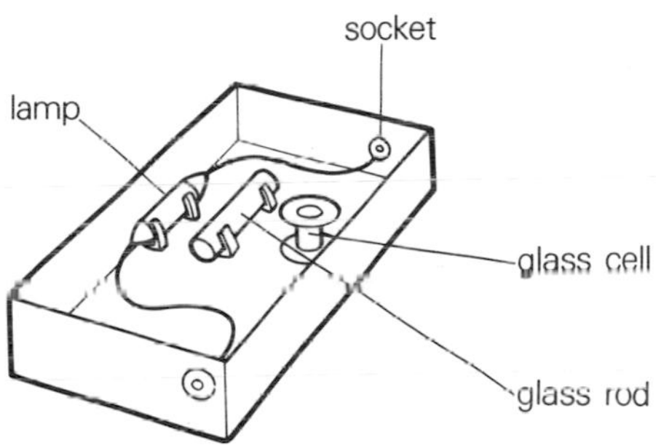

Figure 15.16 Brownian motion apparatus

Figure 15.17 Random motion of specks of ash

15.8 Explanations based on Kinetic Theory

Although the Kinetic Theory has its limitations, it is useful because it provides an explanation of many properties of solids, liquids, and gases.

Solids

The particles of a solid are close together and arranged in regular patterns. They are held together by strong attractive forces. Each particle vibrates to and fro about its average position. However, the vibrations are not vigorous enough to allow the particles to free themselves from the attraction of their neighbours. This means that the particles cannot change positions and so the solid keeps its shape.

Liquids

In a liquid the forces are not strong enough to hold the particles in position. This means that the well-ordered structure is destroyed and the particles can now change position and flow. However, the particles are still nearly as close together as in the solid state.

Gases

In a gas the particles are moving at high speed and colliding elastically with everything they meet. The particles are now much further apart and move much more freely. Between these elastic collisions the particles exert practically no forces on each other and so move with a uniform velocity. When the particles collide they do so violently and therefore do not stay together. The motion of the particles is **random** in that they move at different speeds and in different directions. Figure 15.18 may be useful to help you understand the arrangement of the particles in solids, liquids and gases, but remember that it does not show that the particles are moving.

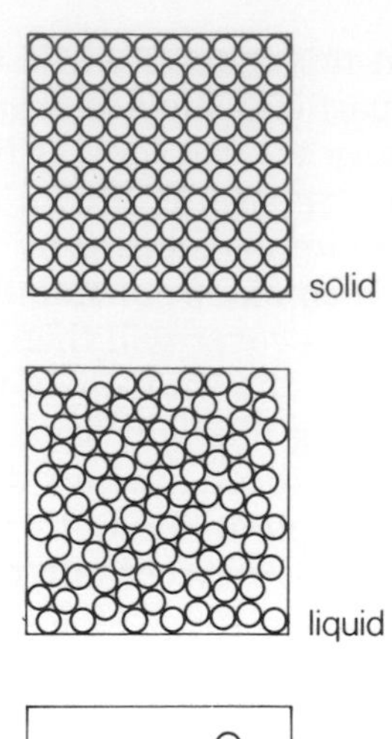

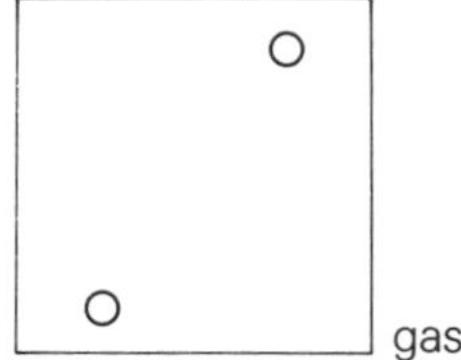

Figure 15.18 Particle spacing

Change in temperature

When energy is supplied to a substance (which is not about to change its state), the particles move faster. The average kinetic energy of the particles has been increased and we say the substance is hotter. In fact the average kinetic energy of the particles is directly proportional to (varies directly as) the Kelvin temperature,

i.e. $\overline{E}_k \propto T$

Changes of state

There is no change of temperature when a substance changes its state, so the average kinetic energy of the particles remains constant. However, work has to be done against the attractive forces between particles which hold the solid or liquid together. It requires energy to pull a particle away from the attraction of its neighbours. This means that although the kinetic energy of the particles has not changed, the potential energy has been increased. This energy which has to be supplied to change the state of a substance is its **latent heat**.

Internal energy

We can define the total **internal energy** of a solid, liquid or gas as the sum of its kinetic and potential energies. Figure 15.19 shows the changes in internal energy in a substance when heat is supplied to it.

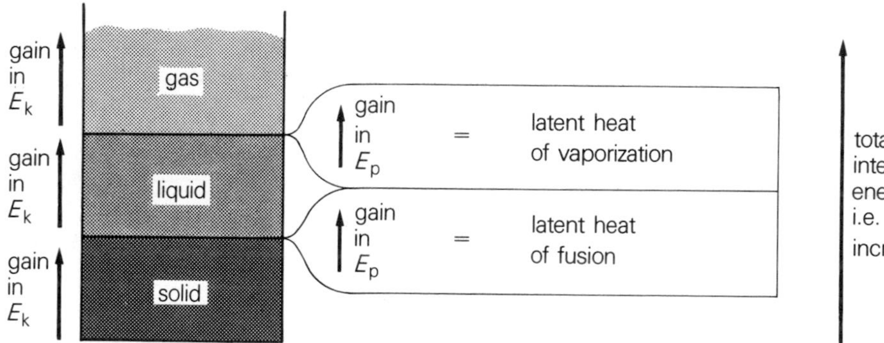

Figure 15.19 Internal energy changes

Figure 15.20

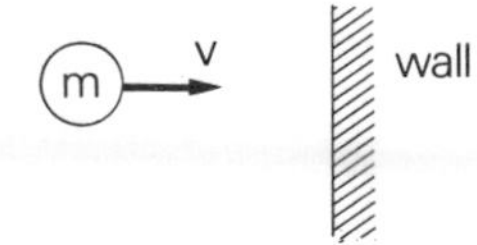

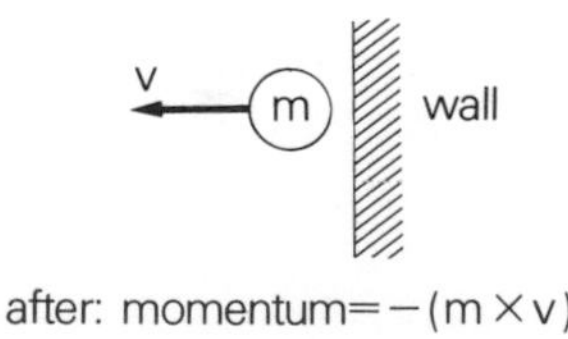

Pressure

When a gas particle collides elastically with a container wall, it rebounds with the same speed but in the opposite direction so that the change in momentum is $mv - (-mv) = 2\,mv$, Figure 15.20.

This change in momentum is caused by the force that the wall exerts on the particle. It follows from Newton's Third Law that the particle exerts an equal but opposite force on the wall. The force on unit area of the container's walls is the gas pressure.

The Gas Laws

Pressure Law: $p \propto T$ (at constant V)

Consider a fixed volume of gas at temperature T and pressure p. If the temperature of the gas increases, this means that the kinetic energy and therefore the speed of the particles must increase. As the volume is constant, this will lead to an increased gas pressure because the particles will be colliding more violently and more frequently with the container walls,

i.e. as T increases, p increases.

Charles' Law: $V \propto T$ (at constant p)

Consider a volume V of gas at temperature T. If the temperature of the gas increases, this means that the kinetic energy and therefore the speed of the gas particles must increase. If the volume were to remain the same, this would produce an increased gas pressure because the particles would be colliding more violently and more frequently with the container walls. Therefore if the pressure is to remain constant, the volume of the gas must increase to spread the increased force over a larger area,

i.e. as T increases, V increases.

Boyle's Law: $p \propto 1/V$ (at constant T)

Consider a volume V of gas at pressure p. If the volume is reduced without change in temperature, the particles will collide more frequently with the container walls. This will produce a larger force. Also the area of the container walls has been reduced and both the increased force and reduced area will lead to an increase in the gas pressure,

i.e. as V decreases, p increases.

Summary

$$p = \frac{F}{A}$$

where p = pressure, F = force, A = area.
The SI unit of pressure is the pascal (Pa); 1 Pa = 1 N m^{-2}.

A Bourdon gauge is an instrument for measuring gas or liquid pressure.

The unit of temperature on the Kelvin temperature scale is the kelvin (K). A kelvin is the same size as a degree Celsius. Absolute zero (0 K) on the Kelvin scale corresponds to −273 °C on the Celsius scale.

The three Gas Laws can be summarized as below, where p = gas pressure, V = volume, T = Kelvin temperature and m = mass of gas.

Pressure Law (constant V and m)

$$p \propto T$$

$$\Rightarrow \frac{p_1}{T_1} = \frac{p_2}{T_2}$$

Charles' Law (constant p and m)

$$V \propto T$$

$$\Rightarrow \frac{V_1}{T_1} = \frac{V_2}{T_2}$$

Boyle's Law (constant T and m)

$$p \propto \frac{1}{V}$$

$$\Rightarrow p_1 \times V_1 = p_2 \times V_2$$

The Combined Gas Equation (for constant m) relates all three of the other variables.

$$\frac{p \times V}{T} = \text{constant}$$

$$\Rightarrow \frac{p_1 \times V_1}{T_1} = \frac{p_2 \times V_2}{T_2}$$

The Brownian motion of specks of ash in air is due to collisions with the moving air particles.

According to the Kinetic Theory, all matter consists of particles which are in constant motion at any temperature above absolute zero. The average kinetic energy of the particles is directly proportional to (varies directly as) the Kelvin temperature. Gas pressure is produced by the particles colliding with the walls of their container.

Problems

1 Explain why the use of large tyres helps to prevent a tractor from sinking into soft ground.

2 If you want to rescue someone who has fallen through the ice on a pond, would it be safer to walk or crawl across the ice towards him? Explain.

3 An elephant exerts a force of 5 000 N by pressing his foot on the ground. If the area of his foot is 0·02 m^2, calculate the pressure exerted by his foot.

4 A tank contains 1 000 kg of water. If the base of the tank has an area of 20 m^2, calculate the pressure exerted by the water on the base.

5 The pressure of air in a car tyre is $2{\cdot}5 \times 10^5$ Pa at a temperature of 27 °C. After a motorway journey the pressure has risen to $3{\cdot}0 \times 10^5$ Pa. Assuming that the volume of the tyre has not changed,

5·1 calculate the resulting temperature of the air in the tyre.

5·2 explain the change in pressure in terms of the motion of the air particles in the tyre.

6 **6·1** Change the following Celsius temperatures into Kelvin temperatures,

a) −273 °C,

b) −150 °C,

c) 500 °C.

6·2 Change the following Kelvin temperatures into Celsius temperatures,

a) 0 K,

b) 272 K,

c) 500 K.

7 A fixed mass of air occupies a volume of 0·2 m^3 at a temperature of 0 °C. At what temperature will its volume be 0·4 m^3 if the pressure is unchanged?

8 A weather balloon contains 100 m^3 of helium when atmospheric pressure is 90 kPa. If atmospheric pressure changes to 100 kPa, calculate the new volume of helium at the same temperature.

9 A sealed syringe contains 50 cm^3 of gas at atmospheric pressure (10^5 Pa), and a temperature of 127 °C. When the piston is depressed the volume of the air is reduced to 20 cm^3 and this produces a temperature rise of 3 °C. Calculate the new pressure of the gas.

10 The word 'gas' comes from a Greek word meaning chaos. Explain why the word gas is therefore very suitable.

11 Explain why gases can be compressed (squeezed) much more easily than solids or liquids.

12 Explain why hydrogen molecules at 20 °C will be travelling on average faster than oxygen molecules at the same temperature.

13 In terms of particles explain

13·1 what happens when a solid melts,

13·2 why energy is required for a solid to melt.

14 Why does the air pressure in a bicycle tyre increase as more air is pumped in?

15 When air is pumped out of a metal can it collapses as indicated in the diagram.

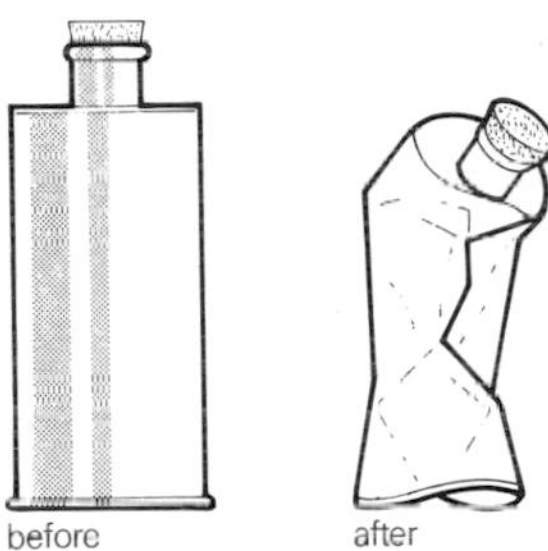

Explain this effect in terms of air pressure.

16 A boy used the following apparatus to test a hypothesis that 'the pressure of a gas is directly proportional to its temperature'.

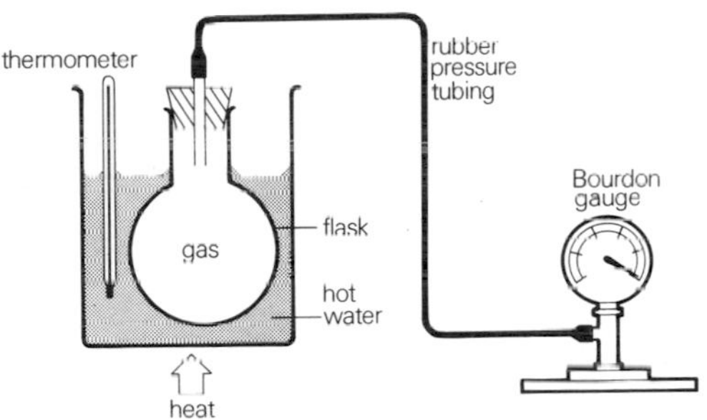

He obtained the following results:

Temperature °C	0	20	40	60	80	100
Pressure kN/m^2	90	96	103	110	117	123

16·1 Comment on the volume of the enclosed air during the experiment.
16·2 Draw a graph of the results and use the graph to show how a relationship between pressure and temperature of a gas may be deduced.
16·3 Rewrite the original hypothesis in a fully and more correct form.
16·4 A classmate criticized the experiment because some of the air was not at the temperature recorded on the thermometer. Suggest one way to reduce this error. *SCEEB*

17 Some gas at room temperature and a pressure of 100 kPa was trapped in a syringe by sealing the nozzle.
17·1 The plunger was depressed and the gas allowed to return to room temperature. The volume was found to have been halved. What was the new pressure?
17·2 In another experiment the pressure of the gas was kept constant. The temperature of the gas was 300 K (27 °C). To what temperature would the gas have to be heated to double its volume? *SCEEB*

18 A graduated glass tube containing air trapped below a mercury thread is immersed in a beaker of water. When the temperature of the water is 27 °C, the volume of the trapped air is 0·12 cm^3.

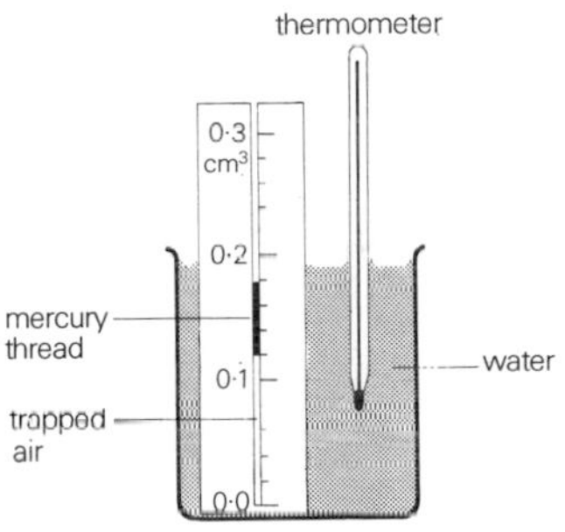

18·1 Draw the graph which shows how the volume in cm^3 of this trapped air changes with temperature in degrees Kelvin.
18·2 What volume does the trapped air occupy when its temperature is 77 °C?
18·3 Describe how this glass tube could now be used to find the temperature in degrees Celsius of a freezing mixture of ice and salt.
18·4 Explain why the accuracy of this 'thermometer' would be affected by changes in atmospheric pressure. *SCEEB*

19 **19·1** A syringe with a close fitting plunger is attached to an uncalibrated Bourdon gauge. The plunger can be pushed down in the syringe by placing weights on top of the plunger.

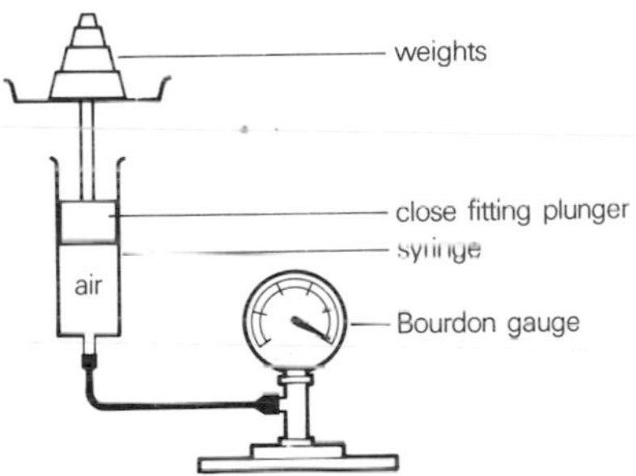

Describe how you would calibrate the Bourdon gauge using this arrangement.
19·2 An experiment is carried out to investigate how the pressure of a gas varies with temperature. The experimental arrangement is shown below.

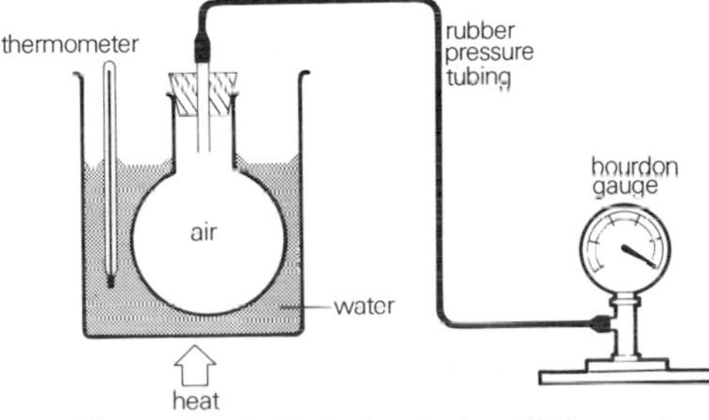

The water bath is heated until the water is boiling. As the water cools, readings of pressure are taken at a number of different temperatures.
a) Why is it desirable to allow the water to cool slowly?

b) Give an explanation in terms of the behaviour of the particles of the gas for the change in pressure as the gas cools.

c) Redraw the following axes and sketch the graph of the results you would expect from this experiment. *SCEEB*

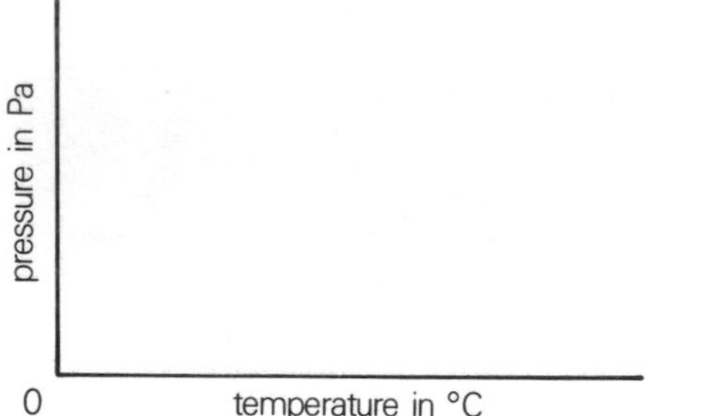

20 A quantity of gas is trapped in a cylinder by a close fitting piston. The cylinder also contains a heating coil.

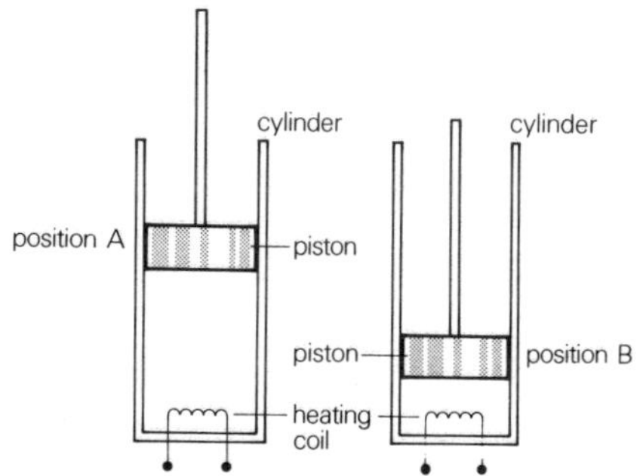

20·1 The piston is pushed down from position A to position B and held there.

a) State what change occurs in the pressure of the trapped gas.

b) Use a particle model of the gas to explain this change.

20·2 The heating coil is switched on and the piston is held in position B.

a) State what happens to the molecules of the gas during the time the heating coil is on.

b) State what change occurs in the pressure of the gas during this time.

c) Use a particle model of the gas to explain this pressure change.

20·3 The petrol-air mixture in the cylinder of a car engine is ignited when the piston is in the position shown.

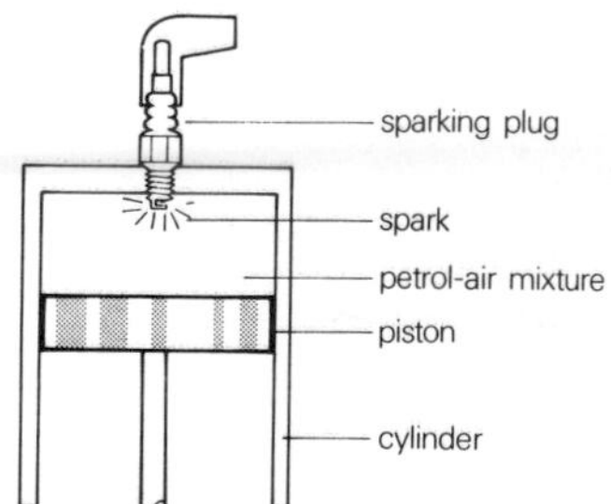

Explain in terms of the motion of the molecules of the gas in the cylinder why the piston moves downwards. *SCEEB*

21 'A particle model for matter is justified if it can describe the properties of matter in a reasonably simple manner.'
Justify your acceptance of the particle model by using it to explain

21·1 Brownian motion,
21·2 pressure of a gas,
21·3 melting,
21·4 latent heat of vaporization. *SCEEB*

22 **22·1** In an explanation of Brownian motion the expression 'random motion' is often used.

a) What is meant by 'random motion'?

b) Explain the change which would occur if the temperature of the contents of the vessel in which the Brownian motion was being observed were raised.

22·2 The graph shows the results of an examination of the pressure of a given mass of gas, kept at constant volume, as its temperature is varied.

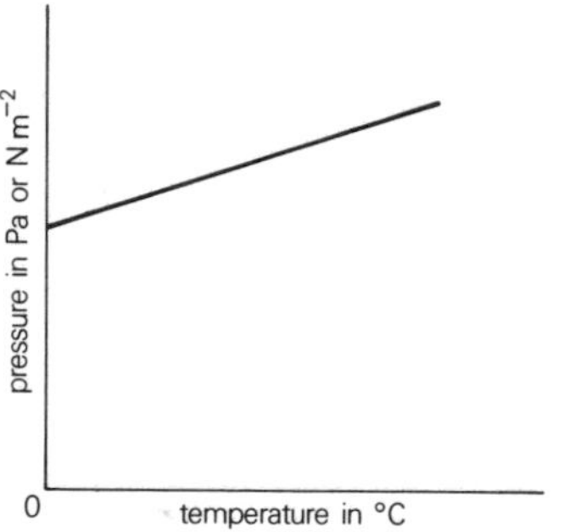

a) Explain how the graph can be used to establish a mathematical relationship between pressure and temperature.

b) Explain this relationship in terms of the motion of the particles of the gas. *SCEEB*

16 Electrostatics

16.1 Introduction

If you comb your hair and then hold the comb near to a few small pieces of tissue paper, they can be seen to jump up and down under the influence of some force. When you try to dust your gramophone records with a duster, this somehow makes them attract even more dust. We say that these objects have the property called **electric charge**. The study of the forces between stationary electrical charges is called **electrostatics.**

We can investigate further this electric charge using rods of different materials. In both the examples above, the objects which exhibited the charge had been rubbed with another material. When we rub a polythene rod with a cloth and then suspend it to move freely on a nylon thread, we find that it is repelled when another similarly rubbed polythene rod is brought near it. Both rods are polythene, so their charges are the same.

When we rub a nylon (or acetate) rod and hold it near to the charged polythene rod, we find that the rods are attracted, suggesting that there is a second type of charge. When we bring (charged) rubbed nylon rods near to each other – they repel.

No matter what other charged object we bring towards the charged polythene rod it either attracts or repels. We conclude that there are **two** types of charge which we call **positive** and **negative**. We say that polythene becomes charged negatively, and nylon (or acetate) becomes charged positively, Figure 16.1.

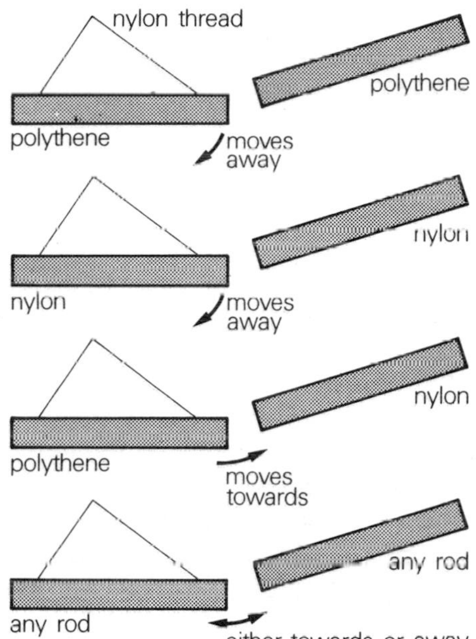

Figure 16.1 Behaviour of charged rods

Charges come from atoms which are the smallest particles of any element which can exist, Figure 16.2. An atom consists of a small positively charged **nucleus** with negatively charged **electrons** around it. The nucleus consists of **protons** and **neutrons**, each of similar mass, while the orbiting electrons have a very much smaller mass. The neutron is uncharged while the electron and proton have equal but opposite charges (negative and positive respectively). As the protons and neutrons are in the nucleus, it is the nucleus that carries all the positive charge and practically all the mass of the atom. However, as the electrons carry an equal amount of negative charge, the atom as a whole is described as electrically **neutral**. When we study electrostatics, it is the property of these particles (called the electric charge) which we investigate. The protons and neutrons are held together very tightly in the nucleus by a nuclear force. This force is so strong that protons are unable to move away from their atomic nucleus. The force holding electrons in the atom is much weaker.

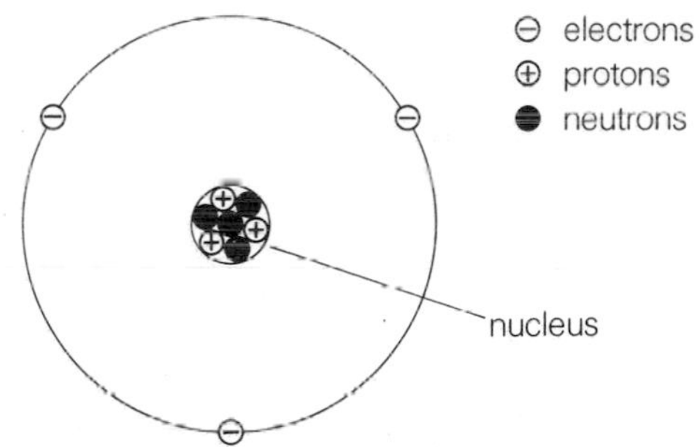

Figure 16.2 An atom

16.2 Charging by rubbing

We need energy to charge an object positively or negatively. We use this energy to remove charge from, or add charge to, an object. One simple method of moving the charges about is to rub one object against another. Let us take the example of the polythene rod which is rubbed with a cloth, Figure 16.3.

Figure 16.3 Charging by rubbing

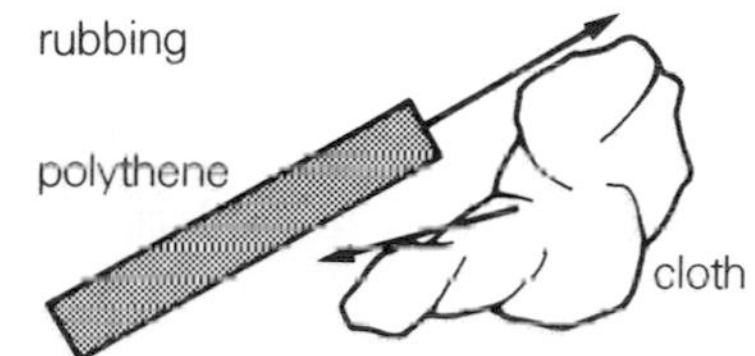

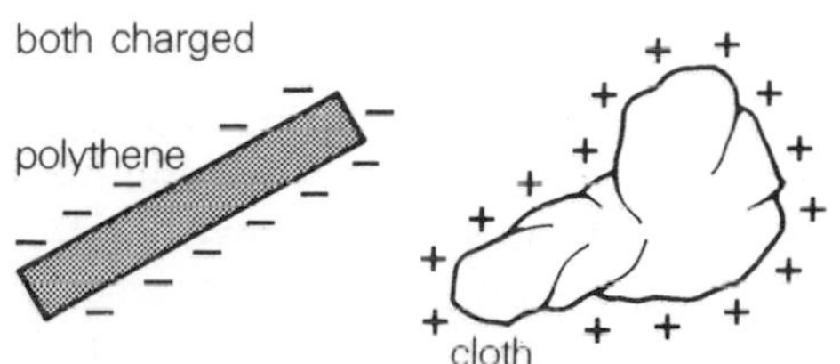

Both the rod and the cloth start off neutral, that is, they both have equal amounts of positive and negative charge. When we rub the two materials together we provide the energy to remove negative charges (electrons) from the cloth and transfer them to the polythene rod because it is easier to remove electrons from the cloth. The rod has then more negative charges than positive charges – the charge balance has been upset. We describe the polythene rod as being **negatively** charged. The cloth on the other hand, has lost negative charges and has therefore more positive charges than negative charges and we describe it as **positively** charged.

When a nylon (or acetate) rod is rubbed by a cloth, it is found that electrons are removed from the rod to the cloth because it is easier in this case to remove electrons from the rod, Figure 16.4. So the rod becomes positively charged and the cloth becomes negatively charged.

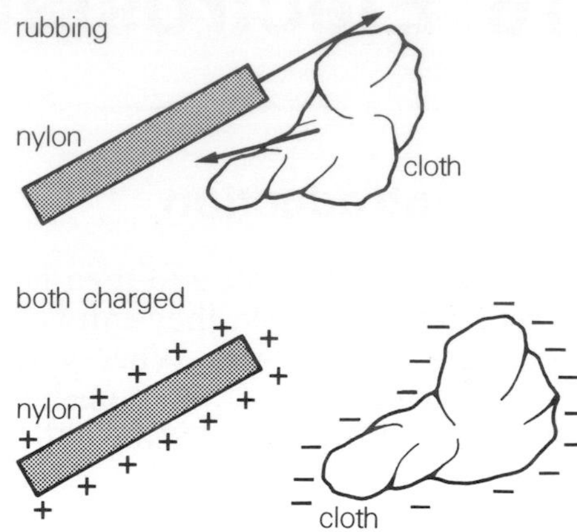

Figure 16.4 Charging by rubbing

16.3 Conductors and insulators

Materials can be divided into two electrical categories as follows:
a) Conductors are those materials through which charge can flow, e.g. metals, carbon and some liquids.
b) Insulators are those materials through which there is almost no flow of charge, e.g. plastics, ceramics, many non-metals.
The reason for the difference in the conduction properties of these materials is that relatively little energy is needed to 'free' electrons from their atoms and cause them to move through a conductor. Some of these electrons are so weakly attracted to the protons in the nuclei (as, for example, in metals) that they are called 'free electrons'. In fact at room temperature, a metal receives enough energy from its surroundings to free these electrons and so it is a conductor. An insulator on the other hand, has hardly any free electrons which can move through the material because its electrons are held more tightly by the atoms of the insulator.

While there is a difference in the conduction properties of these two categories of material we must not forget that they can both be charged by rubbing. In order to show the charging of a metal bar we would first have to give it an insulating handle so that the charge could not flow away.

In industry there has been a great increase in the use of very good insulators such as plastics, and there is a great deal more charge to contend with. Anything passing over guides, between rollers, down chutes or removed from a mould may end up charged. It is not so much that this itself is a problem. The difficulty is more that the charge tends to remain on the object because it is an insulator. When the build-up of charge is excessive, a spark can result which may ignite an explosive mixture. A dramatic effect of the flow of charge is a flash of lightning.

Figure 16.5 Gold-leaf electroscope

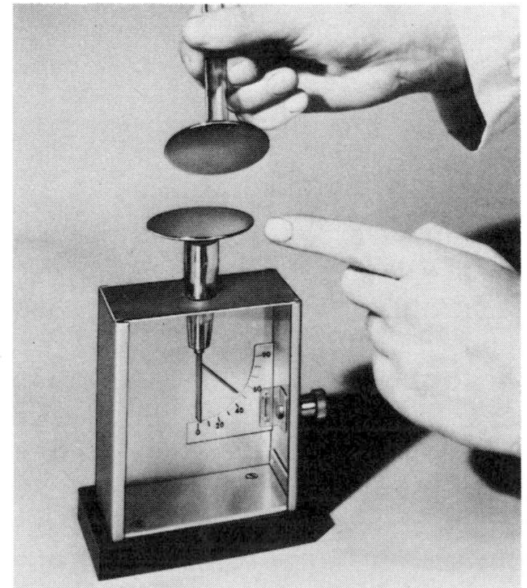

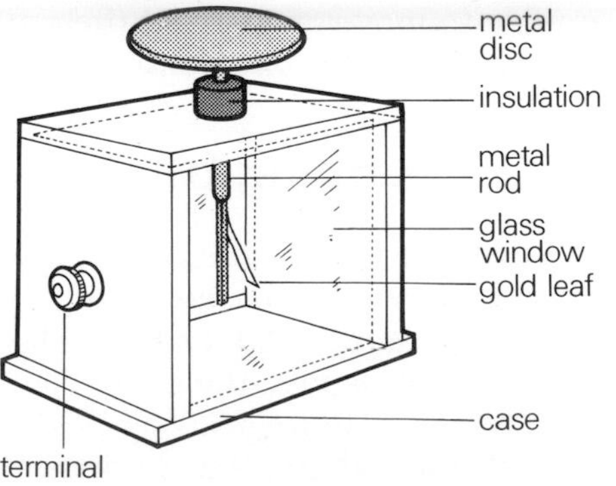

16.4 Gold-leaf electroscope

This instrument (Figure 16.5) consists of a metal disc and rod supported in a case by an insulating material. Attached to the lower end of the rod is a rectangle of thin metal foil (or leaf) which is hinged at its upper end and free to move at its lower end. The disc can be charged by the addition or removal of electrons. When charge is supplied, it is distributed over the disc, rod and leaf. Since charges which are the same repel each other, the rod repels the leaf and the lower end of the leaf rises, Figure 16.6.

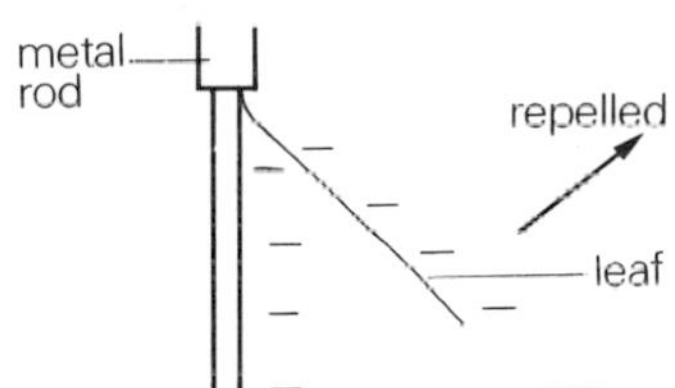

Figure 16.6 Electroscope leaf deflection

The gold-leaf electroscope can help us to determine which type of charge an object has. First we shall study how the instrument is affected by bringing a charged rod near to it.

If we take an uncharged electroscope and hold near its disc a negatively charged polythene rod, we find that there is a deflection due to the electrons in the disc being repelled to the lower end of the rod. When these electrons flow to the leaf and the rod then they repel each other and the leaf rises. When the polythene rod is removed again, these electrons are no longer repelled to the lower part of the electroscope and they return to their normal distribution. The electroscope was of course neutral during the whole of this experiment as no extra charge was supplied to it; its charge was only rearranged, Figure 16.7.

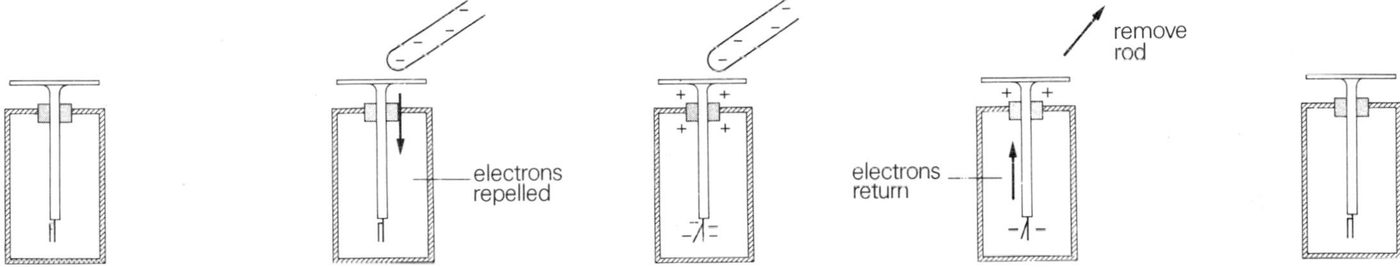

Figure 16.7 Negative charge near an electroscope

If we take an uncharged electroscope and instead hold near its disc a positively charged nylon rod, we find that there is also a deflection due to the electrons in the rod and leaf being attracted from the lower end towards the positive nylon. When these electrons flow up to the disc, the rod and leaf both have electrons missing and are therefore positively charged and repel each other. The gold leaf therefore rises. When the nylon rod is removed again, these electrons are no longer attracted to the upper part of the electroscope and they return to their normal distribution. This charge rearrangement process is shown in Figure 16.8.

We cannot tell what type of charge an object has merely by bringing it near to an electroscope, and must find another procedure which will provide us with this information.

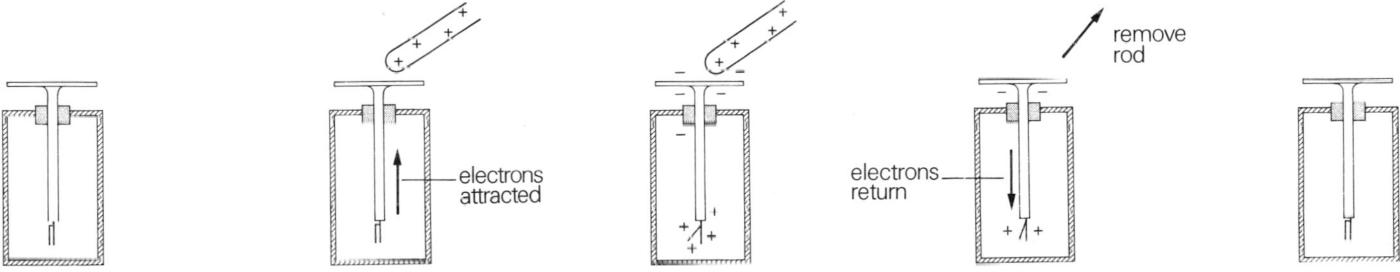

Figure 16.8 Positive charge near an electroscope

If we bring a charged rod near to a **charged** electroscope, we obtain a different result.

Let us assume that the electroscope is negatively charged and we bring a negatively charged object near to it. The electroscope has extra electrons distributed all over it; but when the negatively charged object is brought near, it repels electrons in the disc of the electroscope towards the leaf where they result in a greater deflection. If we bring the negatively charged object even closer, then we repel even more electrons to the lower end of the rod and we obtain an even greater deflection, Figure 16.9.

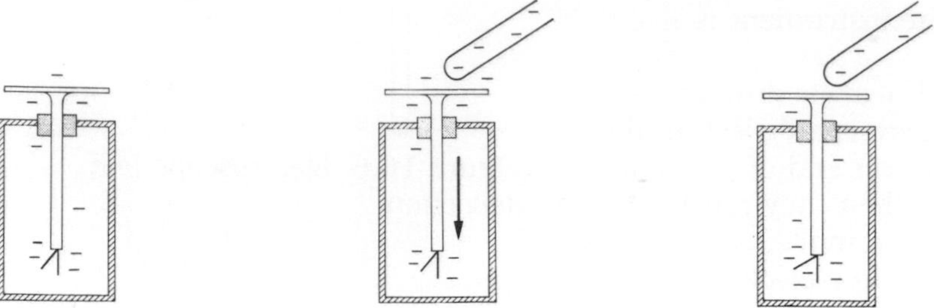
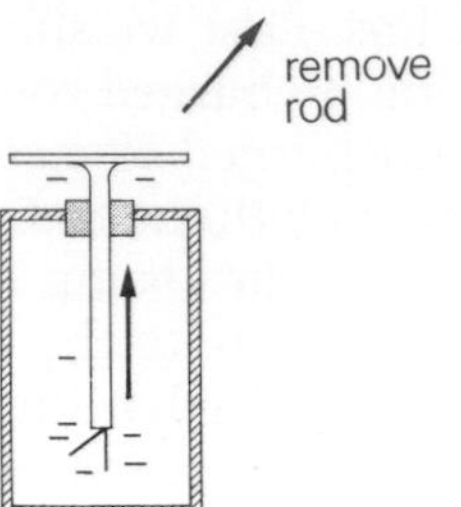

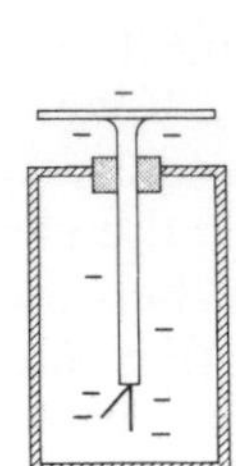

Figure 16.9 Negative charge near to a negative electroscope

If instead we bring a positively charged object near to the negatively charged electroscope, we find that the deflection due to the negative charge which is already there is reduced. This is because the positively charged object attracts electrons from the leaf end of the electroscope towards the disc, and it leaves fewer electrons at the lower end and hence less to cause a deflection, Figure 16.10.

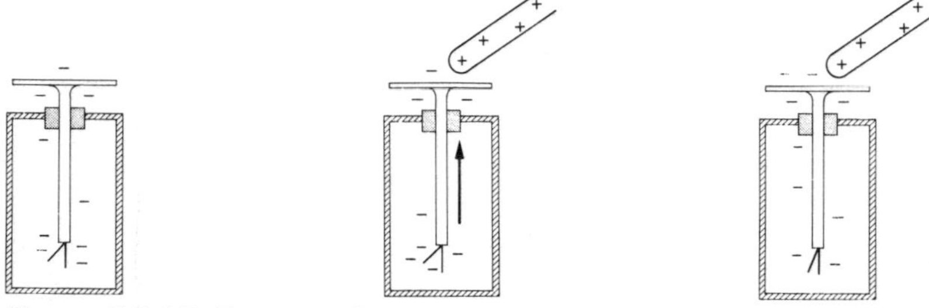
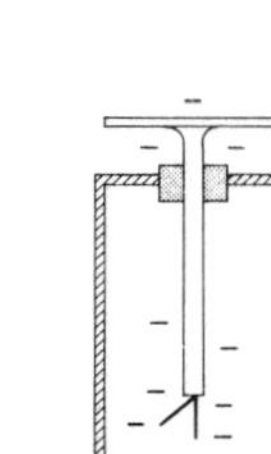

Figure 16.10 Positive charge near to a negative electroscope.

We find a similar behaviour when the electroscope is positively charged and we bring up a positive or a negative object.

When we bring an **uncharged** object near to a negatively charged electroscope or a positively charged electroscope, we find that there is a **reduction** in the deflection. This reduction in deflection results because the charge on the electroscope is either repelling or attracting electrons in the uncharged object. The charge is rearranged on the uncharged object and either attracts or repels electrons on the electroscope. This results in the deflection being reduced. Figure 16.11.

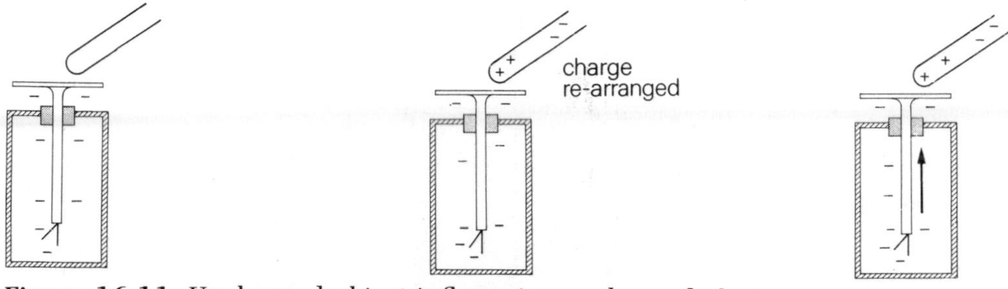

Figure 16.11 Uncharged object influencing a charged electroscope

We have already mentioned that uncharged objects, such as pieces of tissue paper, can be attracted by a charged object.

We can now summarize our method for determining the charge of an object. If we bring the object near a **negatively** charged electroscope

and we produce an **increased** deflection, then it is **negatively** charged. If we bring the object near a **positively** charged electroscope and we produce an **increased** deflection, then it is **positively** charged. If we bring an object near to a charged electroscope and there is a reduced deflection, we cannot tell whether the object is charged or not.

16.5 Conservation of charge

We can use the electroscope to investigate what actually happens to a polythene rod when it is rubbed with a cloth, Figure 16.12.

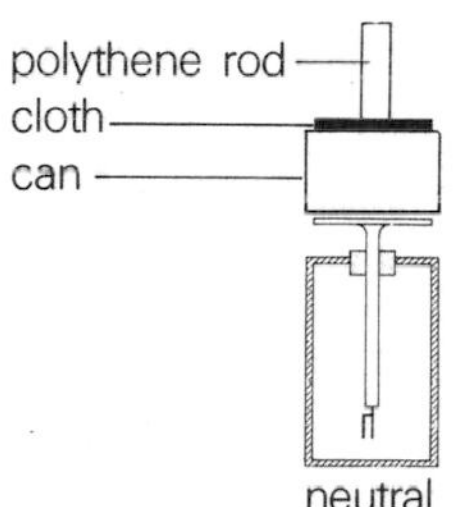

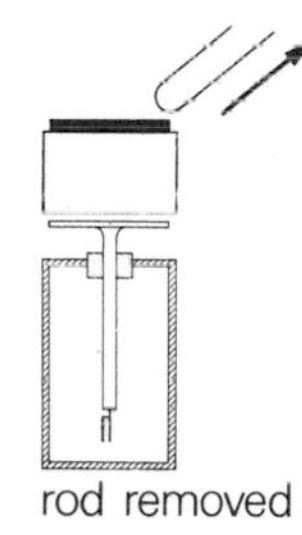

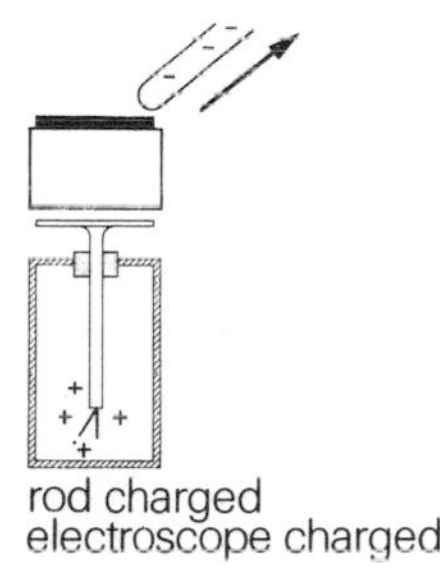

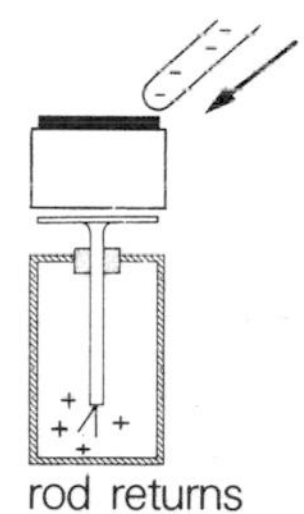

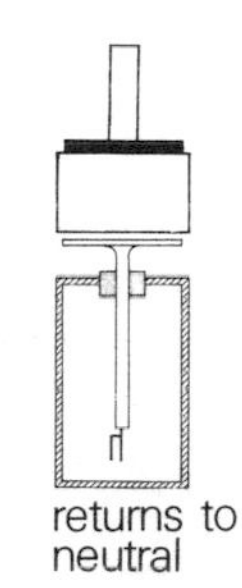

Figure 16.12 Charge separation

When we place a polythene rod with one end wrapped in a cloth in a metal can on top of an electroscope, there is no deflection because the system is neutral.

When we remove the rod from within the cloth, the electroscope shows a deflection. We can confirm that the electroscope is now positively charged by bringing a positively charged rod near it and observing the increased deflection. The cloth must therefore have lost electrons and the polythene has gained electrons.

When the polythene rod is returned to the cloth, the deflection falls to zero showing us that the rod has the same amount of negative charge as the electroscope system has of positive charge. The total charge is **conserved.**

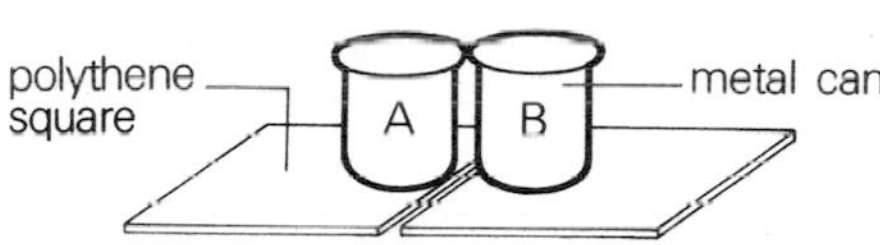

Figure 16.13 Metal cans on polythene squares

16.6 Charging by induction

If we wish to charge a metal object, we find this difficult to do because metals conduct charge. If we place the metal object on an insulating base, we can charge it. In fact, it is the conducting properties of metals which help us to charge them by **induction.**

We place two neutral metal cans in contact on two polythene squares, Figure 16.13.

When a negatively charged rod is brought close to can A, electrons in A are repelled to can B (the cans are metal and in contact so that charge can flow). While the negatively charged rod is still there, can B is moved to one side so that it is no longer in contact and the electrons from A cannot return. When the charged rod is removed, we are left with can B negatively charged and can A positively charged due to its loss of electrons, Figure 16.14.

The procedure of charging objects without actually touching them is called charging by induction. A variation of this procedure can be used to charge an electroscope.

Figure 16.14 Charging by induction

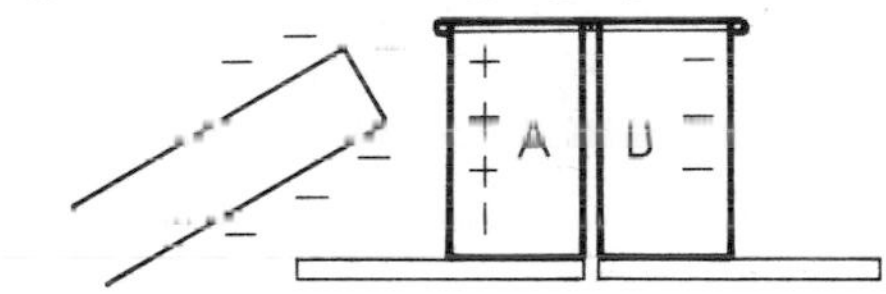

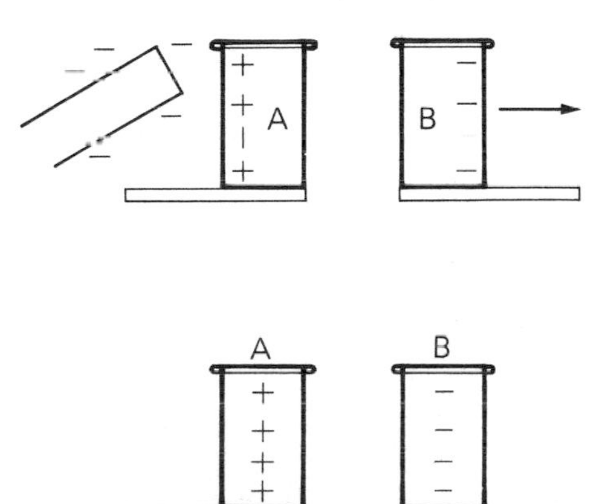

Charging an electroscope by induction

We have already seen that a negatively charged rod brought near to an uncharged electroscope causes a deflection due to the repelled electrons from the electroscope disc. We can provide these electrons with an escape route along a conductor which is our finger to the earth. We stand on the earth and are therefore in electrical contact with it, Figure 16.15.

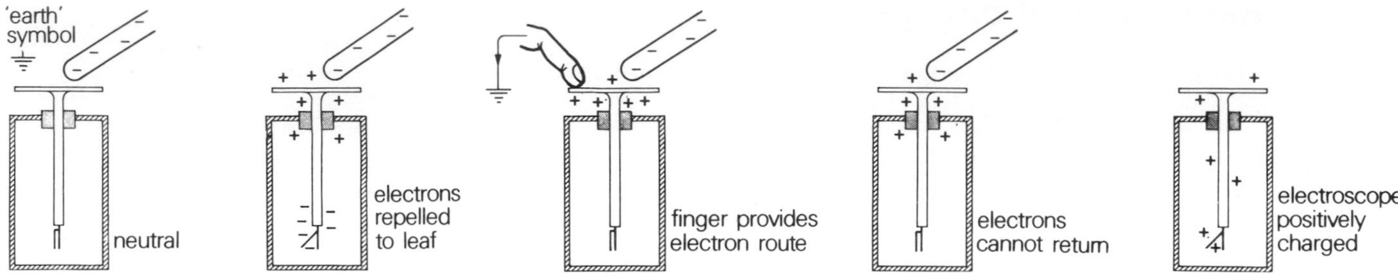

Figure 16.15 Charging an electroscope by induction

The earth can be considered as an object to which we can add as many electrons as we like or from which we can remove as many electrons as we like. The symbol for the earth is given above.

When we remove our finger from the disc of the electroscope the electrons cannot return from earth since there is now no conducting path. The electroscope is now without some of its electrons and is therefore **positively** charged.

Notice that the object which has been charged by induction always has the **opposite** charge to that of the object which is brought near it.

16.7 The electrophorus

An electrophorus consists of a metal disc with insulating handle and a separate polythene square, Figure 16.16.

If we rub the polythene square with a cloth, it becomes negatively charged. When we place the disc on the square, electrons in the metal which are free to move are repelled by the negative charge on the square and so move to the top of the disc. Touching the top of the disc momentarily, we are earthing the disc and providing an escape route from the negatively charged square. The disc has lost some electrons and is therefore **positively** charged and can be removed, Figure 16.17.

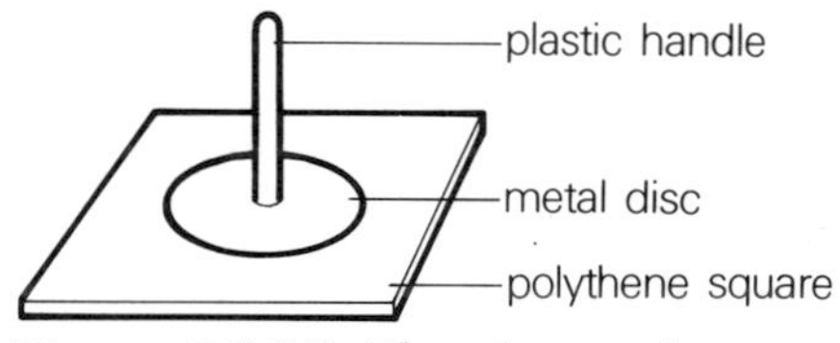

Figure 16.16 The electrophorus

Figure 16.17 Charging an electrophorus disc

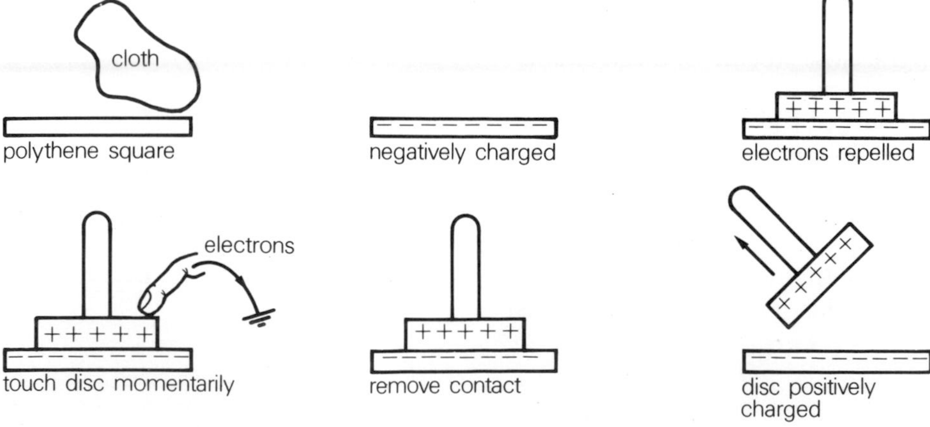

The metal disc does not remove electrons from the polythene square. The metal and polythene surfaces, while apparently smooth, make contact in relatively few places, Figure 16.18.

Polythene is an insulator and so the electrons cannot flow to the points of contact. We have used charging by induction to charge an electrophorus disc. Using a positively charged square we would obtain a negatively charged disc.

The electrophorus disc can be used for transferring charge to objects. When a charged electrophorus disc is brought in contact with the top of an uncharged electroscope, the charge is **shared** and the electroscope shows a deflection after the electrophorus disc has been removed.

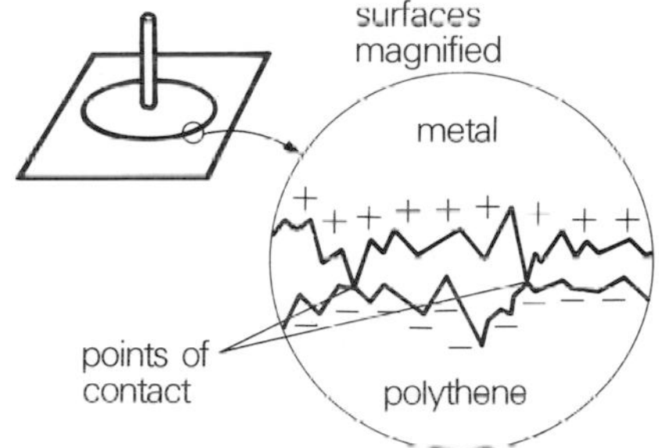

Figure 16.18

16.8 Electric field patterns

Around a charged rod there is a region where an electric force can be detected. We call this region an **electric field**. A charged object placed in an electric field experiences a force.

Using a high voltage supply we can investigate the regions near to a charged metal object which is placed in a dish of oil, Figure 16.19.

When the high voltage supply is switched on and a little semolina is sprinkled on the surface of the oil, an electric field pattern is formed, Figure 16.20.

Figure 16.19 Electric fields experiment

Figure 16.20 Electric field patterns

a) Field between positive and negative rod

b) Field between parallel plates

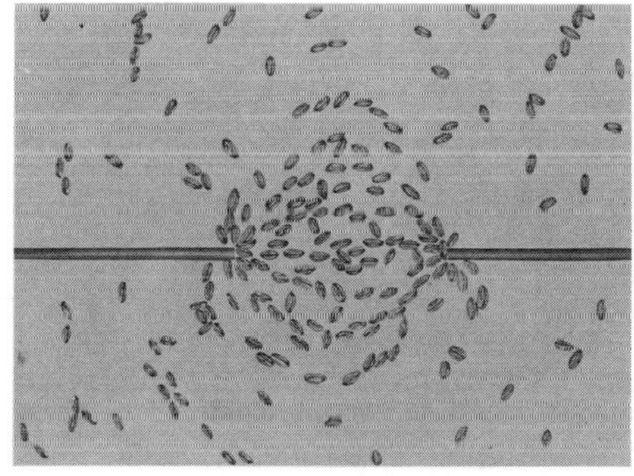

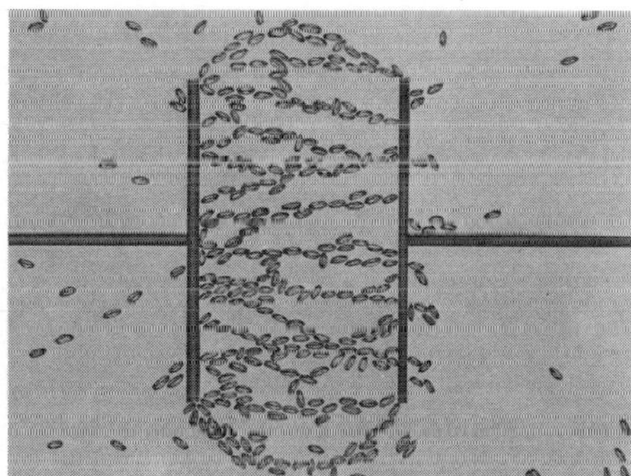

We define electric field lines as lines along which a positive charge would move and they represent the direction of the electrical force on the charge. These are drawn as in Figure 16.21 with the direction from positive to negative.

Figure 16.21 Electric field lines

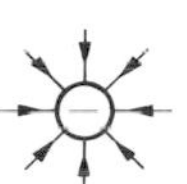

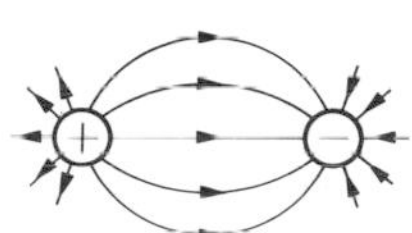

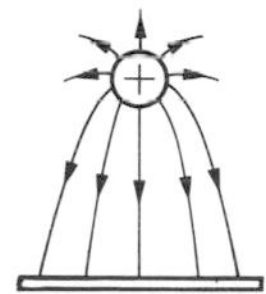

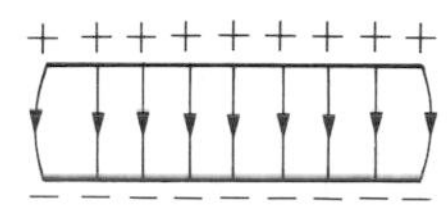

Summary

Objects can be charged by rubbing. Electrons can be rubbed off one object and on to another.

There are two types of charge: positive and negative.

A positive charge will repel a positive charge and attract a negative charge.
A negative charge will repel a negative charge and attract a positive charge.

Materials through which charge can flow are called conductors.
Materials through which charge cannot flow are called insulators.

Charge can be detected by an electroscope. Either a positive or a negative charge will cause a deflection of the electroscope leaf.

We can find out the type of charge using an electroscope.
If the charge increases the deflection of a positively charged electroscope, it is positive.
If the charge increases the deflection of a negatively charged electroscope, it is negative.

The total charge in an isolated system is conserved.

An object being charged by induction always has the opposite charge of the object which does the charging.

An electrophorus disc is charged by induction; we can use it to transfer charge to another object.

The region near an electric charge is called an electric field. Electric field lines are drawn from positive to negative.

Problems

1 Explain how an object becomes charged. What type of material is most likely to keep its charge?

2 Describe how you would use an electroscope to find the charge of an object. What else would you need to use?

3 Explain what is meant by the term 'charging by induction'.

4 Describe and explain the working of an electrophorus.

5 If a metal sphere on an insulating stand has a charge of 12 units, and it is brought in contact with an identical insulated metal sphere and then removed, what will be the charge on each sphere?

6 A negatively charged polythene strip is brought close to the disc of a gold-leaf electroscope. The diagrams show successive steps in charging the gold-leaf electroscope.

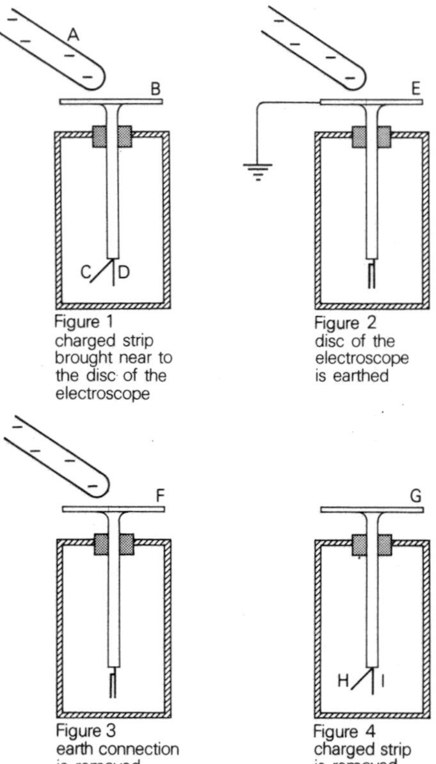

Figure 1 charged strip brought near to the disc of the electroscope

Figure 2 disc of the electroscope is earthed

Figure 3 earth connection is removed

Figure 4 charged strip is removed

Copy and complete this table, inserting the sign of the charge (if any) at the positions labelled. (The sign of the charge at A is given as an example.)

Position	A	B	C	D	E	F	G	H	I
Sign of charge	—								

SCEEB

7 When a boy rubbed a brass rod on a duster and then held it near a few small pieces of paper it did not pick them up or even move them. He concluded that there was no transfer of electric charge between the duster and the brass rod. He was wrong.

7·1 Suggest why the pieces of paper did not move.

7·2 Explain how you would alter this experiment to show that he was wrong. *SCEEB*

8 **8·1** A positively charged rod is brought near an uncharged gold-leaf electroscope but does not touch it.

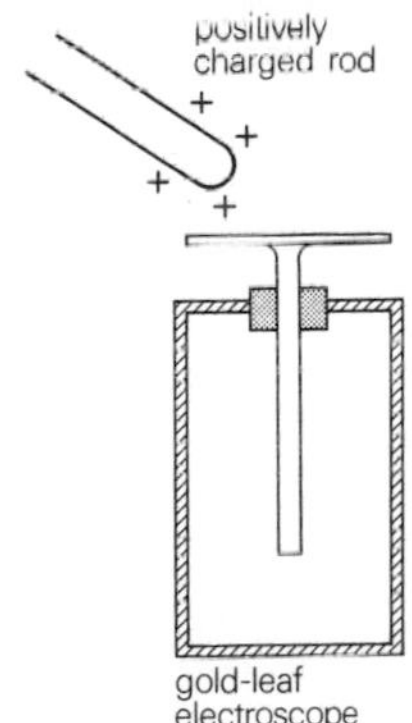

Copy the diagram of the gold-leaf electroscope into your answer book and complete it showing on the electroscope the position of the gold leaf (or leaves).
Indicate the charge distribution over the electroscope.

8·2 A jet of water from a tap is attracted by either a positively charged rod or a negatively charged rod. Suggest an explanation. *SCEEB*

9 The following diagrams show a negatively charged electroscope alone, and then with different plastic rods brought near to it.

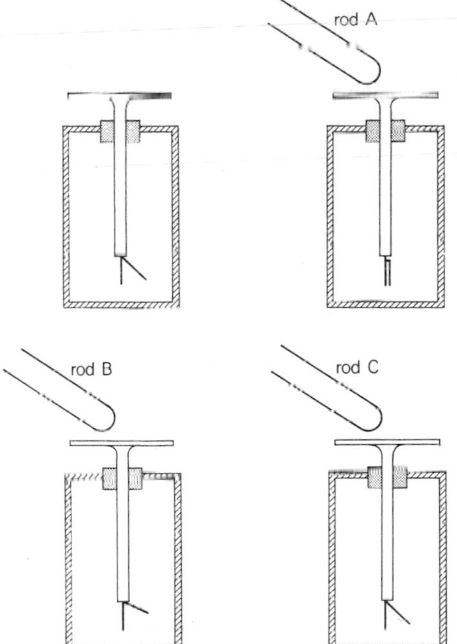

What type of charge (if any) is on each of the three rods? *SCEEB*

17 Electric circuits

17.1 Current

Introduction

We live in an age in which the daily life of many people depends on electricity, e.g. for heating, lighting, transport, etc. Computers, television sets, pocket calculators, digital watches, Pioneer spacecraft, and Concorde aircraft, all contain electric circuits. We shall begin by trying to find out what electric current is.

We have studied electrical charges at rest, but under suitable conditions charges can move. Robert J. Van de Graaff, an American scientist working at Princeton (1931) developed a machine which continuously charged a metal dome. A smaller version of the machine, called a Van de Graaff Generator, is shown in Figure 17.1. A rubber belt is charged by friction as it passes over rollers which are driven by an electric motor. The charge is carried by the belt to a hollow metal dome on an insulating perspex column. The charge gathers on the dome and cannot escape since the dome is insulated from its surroundings. If, however, we connect a conductor (wire) from the dome of the Van de Graaff generator to a suitable sensitive electric meter and then from the meter to the base of the generator, charge on the dome can escape through the wires producing a meter deflection, Figure 17.1. Charge flowing in the wire produces a meter deflection.

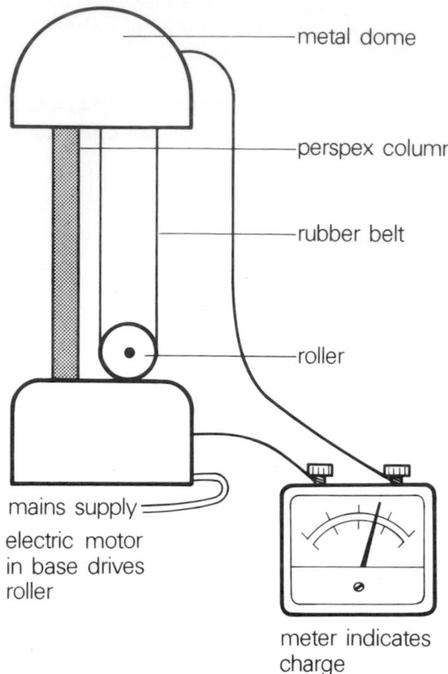

Figure 17.1 Charge flow

When, instead, we connect a cell to the meter we also obtain a deflection due to the electric current, Figure 17.2.

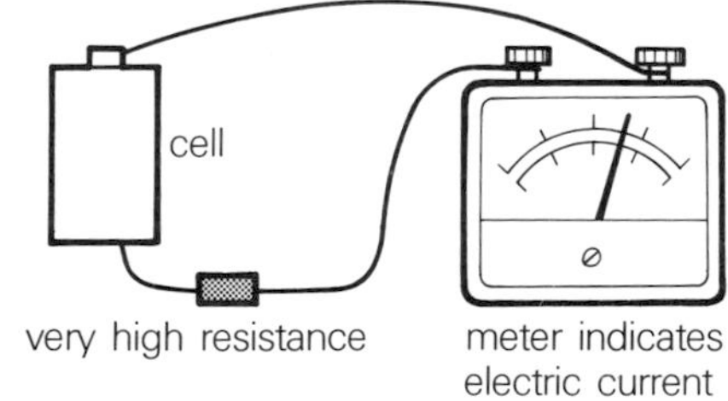

Figure 17.2 Current

We conclude that flow of charge and electric current are the same. An electric current is a flow of electrons and will therefore flow from negative to positive.

We have, in fact, set up two electric circuits in which there is a continuous path of electrical conductors around which charges can flow.

The electric force or **electromotive force** (e.m.f.) needed to make the charge flow around a circuit is usually provided by a battery or power supply unit. A chemical reaction takes place in the battery, releasing energy which causes the charges to flow.

Defining the unit of current

We have been able to define most quantities in terms of three basic units, the metre, kilogram and second. All electrical quantities can be defined in terms of these three units plus a fourth basic unit. In the SI system, the unit chosen is the unit of current called the **ampere** after the French scientist André Ampère. The following experiments indicate the way in which a definition of the ampere can be arrived at.

Forces due to currents

When charge flows up an aluminium tape and down another held very close to it (Figure 17.3), the tapes are found to repel each other. A current can produce a force. This is in fact called a magnetic force, which we shall study in Chapter 18.

Figure 17.3 Force due to currents

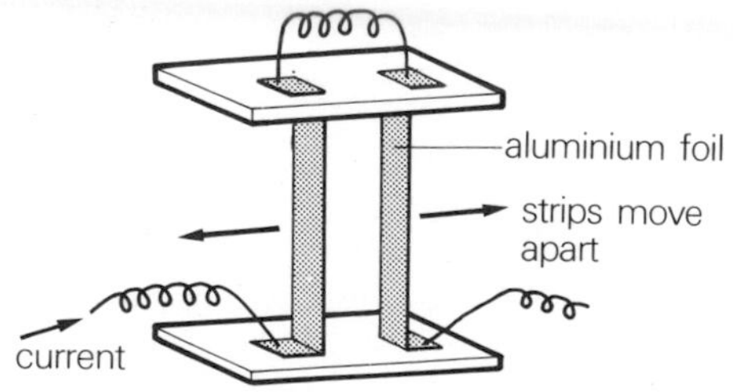

Current balance

Using the apparatus in Figure 17.4, we can balance this force due to the current by a gravitational one.

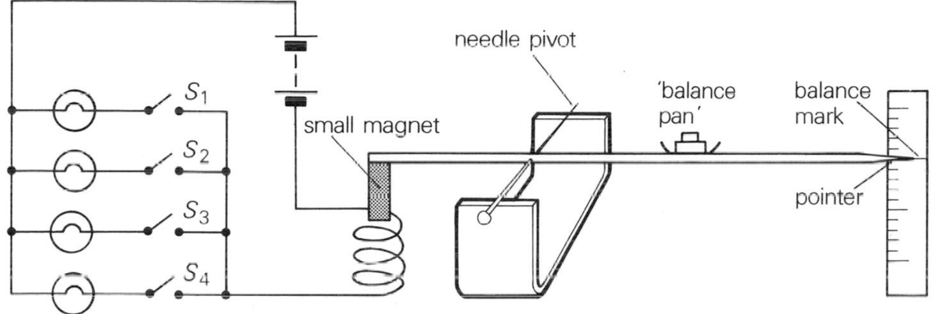

Figure 17.4 Current balance

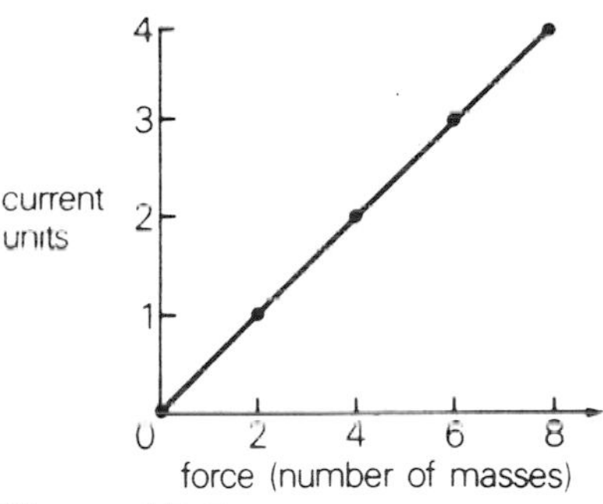

Figure 17.5

This is called a current balance. When charge flows through the coil, a force is exerted on the magnet which tends to attract it into the coil. We can balance this force by the weight of some small masses placed in the balance pan so that the lever is balanced, i.e. when the pointer is level with the balance mark. Each lamp in the circuit in Figure 17.4 draws the same current. By switching on lamps 1, 2, 3, and 4, we obtain 1, 2, 3, and 4 units of current. When the current is increased by equal amounts each time, the force increases and therefore more masses are needed each time to balance it. Typical results are shown as a graph in Figure 17.5.

This experiment shows that the magnetic force produced by an electric current is directly proportional to (varies directly as) the current. This force can be used as a measure of the size of the current.

The ampere is defined as:

'the constant current which, flowing in two infinitely long, straight, parallel conductors of negligible cross section, placed in a vacuum 1 metre apart, produces between them a force of 2×10^{-7} N per metre of wire.'

Apparatus in the National Physical Laboratory (Figure 17.6) which works on the same principle, is used to define and measure current to a high degree of accuracy.

Figure 17.6 Current balance

Units

The SI unit of current, the ampere, is abbreviated to A. Smaller currents are often described in terms of sub-multiples of this unit.

1 milliampere = 1 mA = 0·001 A = 10^{-3} A;
1 microampere = 1 μA = 0·000 001 A = 10^{-6} A.

Having defined the ampere, we can now describe another electrical quantity in terms of it.

The size of the current depends on the amount of charge passing any point in a circuit every second.

$$I = \frac{Q}{t}$$

where I is the electric current,
Q is the electric charge,
t is the time.

Multiplying both sides of this equation by t

$I \times t = Q$
$\Rightarrow Q = I \times t$

We arrive at a definition of the unit of charge.

If a current of 1 ampere flows for 1 second, then 1 unit of charge has flowed. The unit of charge is called the **coulomb**, after Charles Augustin Coulomb, another notable French scientist. One coulomb of charge is abbreviated to 1 C.

1 coulomb = 1 ampere second
1 C = 1 A s

One electron has a negative charge of $1{\cdot}6 \times 10^{-19}$ C, which means that about 6×10^{18} electron charges make up one coulomb. A current of 1 ampere in a wire means that about 6×10^{18} (six million million million) electrons pass a point on the wire every second.

Example 1

A current of 3 amperes flows for 120 seconds in a circuit containing a lamp. How much charge has flowed through the lamp?

$I = 3$ A; $t = 120$ s; $Q = I \times t$.
$\Rightarrow Q = 3 \times 120$
$\Rightarrow Q = 360$
360 coulombs of charge has flowed through the lamp.

Symbols used in circuit diagrams

lamp	L
switch	S
cell	
battery	
ammeter	A
voltmeter	V
conductors crossing with no electrical connection	
junction of conductors	

Current in circuits

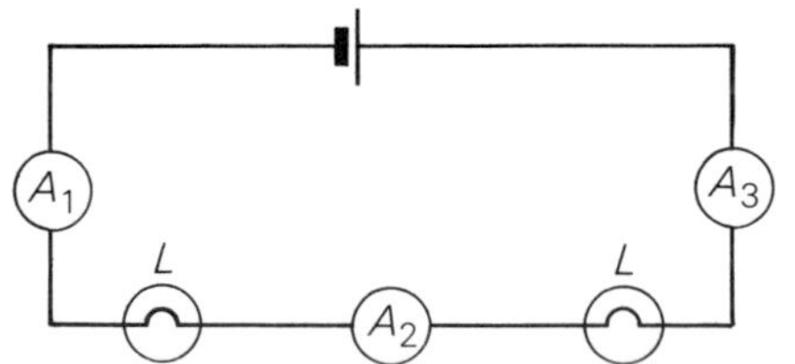

Figure 17.7 Current in a series circuit

Figure 17.7 shows a series circuit diagram. The ammeter readings A_1 A_2 and A_3 are the same.
The current is the same all the way round the circuit.
Figure 17.8 shows a parallel circuit diagram. The ammeter readings are $A_1 = 4$ A, $A_2 = 1$ A, $A_3 = 3$ A. The sum of the currents A_2 and A_3 in the branches is equal to the supply current A_1 (or A_4).
The total current leaving a circuit junction is equal to the total current entering.

When we describe current in our electric circuits, we will be referring to the movement of electrons and this direction will be from negative to positive. In some textbooks the current is described as flowing from positive to negative. It is then often referred to as 'conventional current'.

It is important to remember whether you are using electron flow or conventional current. Throughout this book we shall be using **electron flow**.

Figure 17.8 Current in a parallel circuit

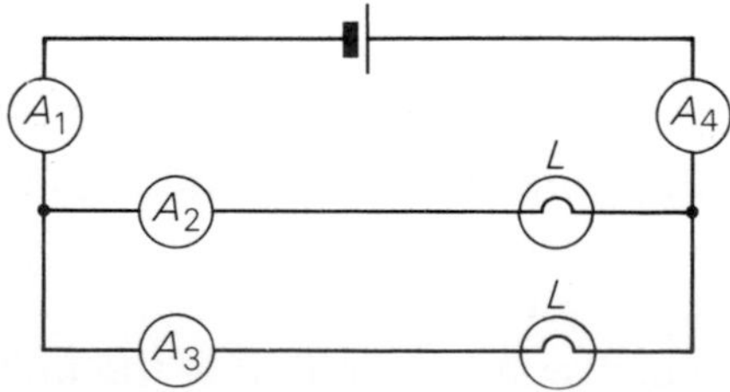

17.2 Potential difference

Having defined charge, we can now define the quantity which causes the charge to flow, called **potential difference** (often referred to as p.d.).

Let us consider a situation in which we have a pair of charged metal plates, Figure 17.9. If we wish to transfer negative charge from the left-hand plate to the right-hand one, we need to overcome the attraction of the positive plate and the repulsion of the negative one: we must therefore do work.

If we require 1 joule of energy to transfer 1 coulomb of charge between the plates, we say that there is a potential difference (p.d.) of 1 volt.

Moving 1 coulomb of charge through a potential difference of 1 volt requires 1 joule of energy.

Moving Q coulombs of charge through a potential difference of 1 volt requires Q joules of energy.

Moving Q coulombs of charge through a potential difference of V volts requires $Q \times V$ joules of energy.

Electrical energy $E = Q \times V$ joules.

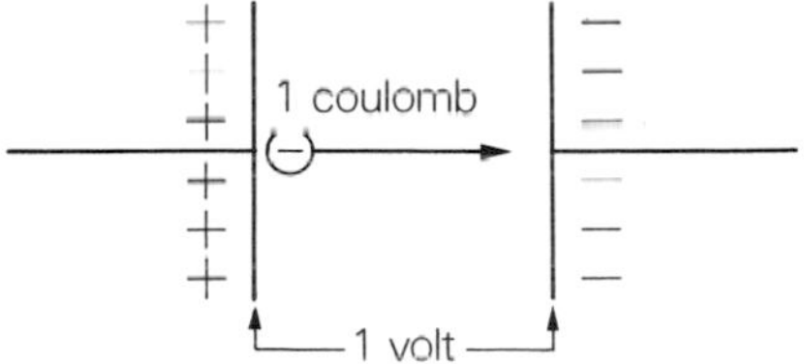

Figure 17.9 Defining the potential difference

Units

The SI unit of electrical energy is the joule and the unit of charge is the coulomb.

Since $E = Q \times V$

$$\Rightarrow \quad V = \frac{E}{Q}$$

The unit of potential difference is therefore given by

$$1 \text{ volt} = \frac{1 \text{ joule}}{1 \text{ coulomb}}$$

$$1 \text{ V} = 1 \text{ J C}^{-1}$$

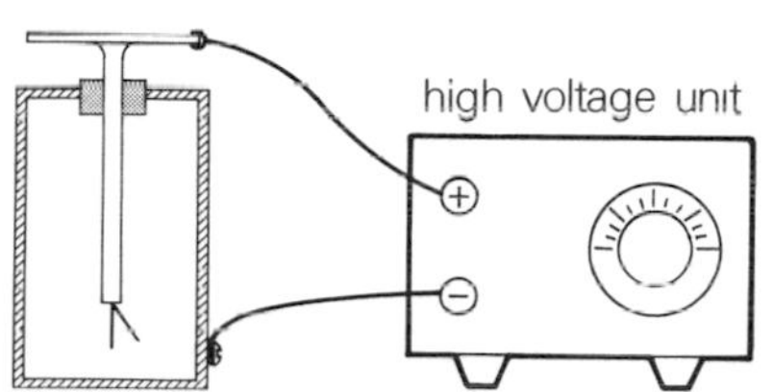

Figure 17.10 Gold-leaf electroscope indicating p.d.

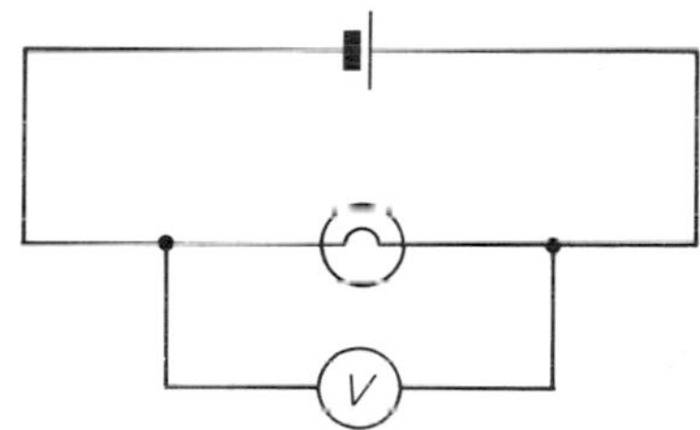

Figure 17.11 Voltmeter connection

Potential difference and the gold-leaf electroscope

When the electroscope disc is connected to the positive terminal of a high voltage supply and the case is connected to the negative terminal, we find that the leaf is deflected. The electroscope is indicating **potential difference**, Figure 17.10. The greater the potential difference, the greater is the deflection.

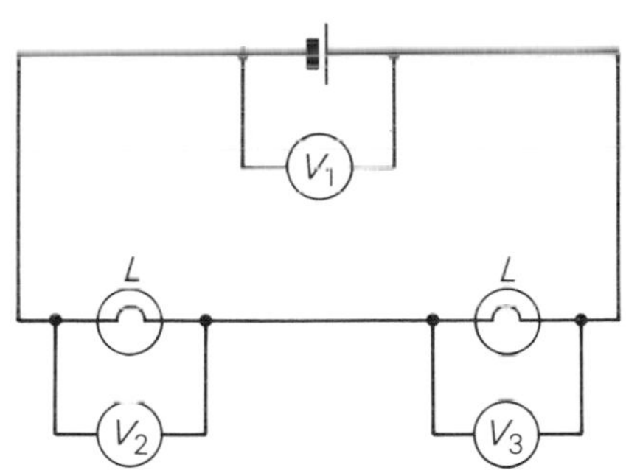

Figure 17.12 p.d. in a series circuit

Potential difference in circuits

We shall use a voltmeter to measure the potential difference between two points in a circuit. The voltmeter must be placed in parallel with the component in order to measure the p.d. across it, Figure 17.11.

Figure 17.12 shows a series circuit. The voltmeter readings are $V_1 = 6$ V, $V_2 = 2$ V and $V_3 = 4$ V, so $V_1 = V_2 + V_3$.

The potential difference across the supply is equal to the sum of the potential differences across the individual components.

Figure 17.13 shows a parallel circuit. The voltmeter readings are $V_1 = 1{\cdot}5$ V, $V_2 = 1{\cdot}5$ V, $V_3 = 1{\cdot}5$ V. $V_1 = V_2 = V_3$. The potential difference across the supply is equal to the potential difference across all the individual components. The p.d. across all the components is the same.

Figure 17.13 p.d. in a parallel circuit

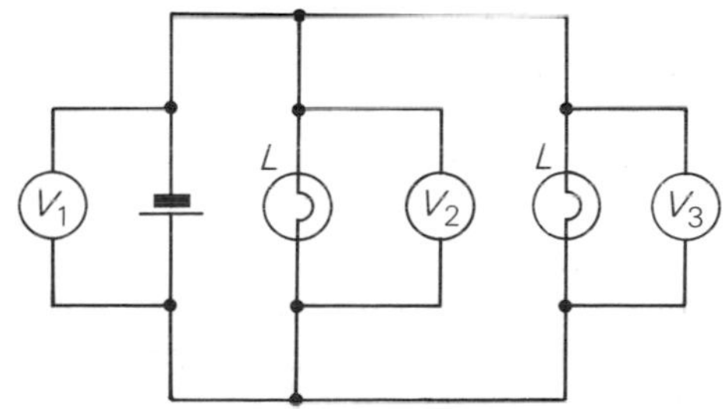

17.3 Electrical power

We can combine the two electrical quantities, current and potential difference, to give us information about the electrical power in a circuit.

Electrical energy = charge × potential difference

$$E = Q \times V$$

As $Q = I \times t$

$\Rightarrow E = I \times t \times V$

Multiply both sides by $\frac{1}{t}$

$$\frac{E}{t} = I \times V$$

As $\frac{\text{energy}}{\text{time}}$ = power P

$$P = I \times V$$

1 watt = 1 ampere × 1 volt

Summary

electrical energy $E = I \times t \times V$

electrical power $P = I \times V$

Example 2

What is the power of the lamp L in the circuit below if the ammeter reading is 4 A and the voltmeter reading is 12 V?

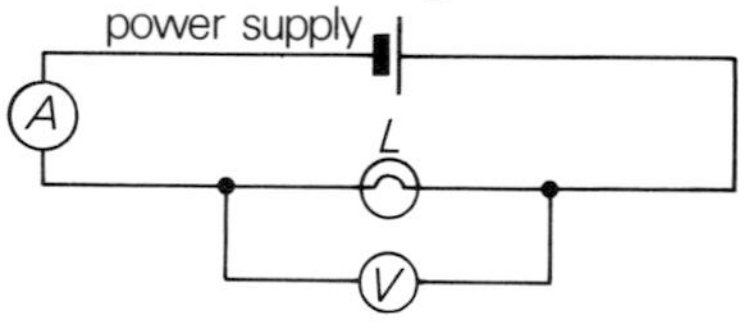

Potential difference across lamp L = 12 volts

Current flowing through lamp L = 4 amperes

Power of lamp $P = I \times V$

$= 4 \times 12$

$= 48$

Power of lamp is 48 watts

Example 3

What current flows through a 250-volt 100-watt lamp operating at the correct voltage?

Power $P = I \times V$

$100 = I \times 250$

Multiply both sides by $\frac{1}{250}$

$$\frac{100}{250} = \frac{I \times 250}{250}$$

$$I = \frac{100}{250}$$

$$= 0{\cdot}4$$

Current through the lamp is 0·4 A

Checking a voltmeter calibration using an immersion heater

Using the apparatus in Figure 17.14, we can heat a known mass of water for a measured time and note the temperature rise. We also take the voltmeter and ammeter readings. The electrical energy supplied equals the heat gained by the water, assuming no energy is lost. We can calculate the energy using two different equations:

Heat energy supplied $= c \times m \times \Delta T$

Electrical energy $= I \times t \times V$

The method of calculation is given below.

mass of water m	= 0·200 kg
initial temperature of water	= 20·0 °C
final temperature of water	= 30·0 °C
temperature rise ΔT	= 10·0 °C
current I	= 4·00 A
voltmeter reading	= 12·0 V
time of heating t	= 175 s

$$c \times m \times \Delta T = I \times t \times V$$
$$4\,200 \times 0{\cdot}200 \times 10{\cdot}0 = 4{\cdot}00 \times 175 \times V$$
$$8\,400 = 700 \times V$$
$$V = 12{\cdot}0$$

Calculated potential difference = 12·0 volts
Actual voltmeter reading = 12·0 volts
So the meter reads correctly at this point on the scale.
The experiment can be repeated for various p.d.'s so that many calibration points can be checked.

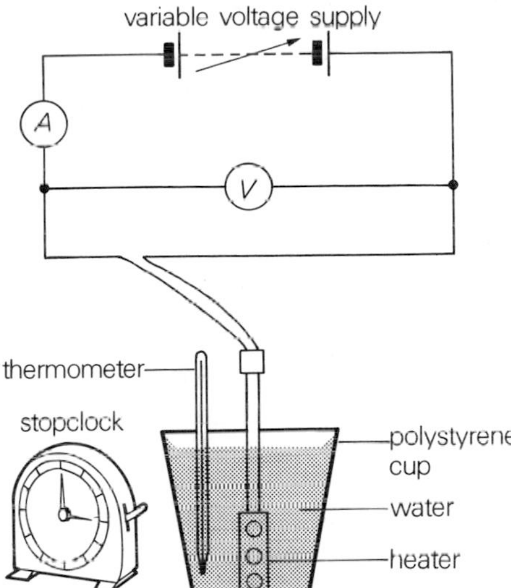

Figure 17.14 Voltmeter calibration check

Example 4

An electrical heater is placed in 200 g of water at 15 °C and increases the temperature to 25 °C in 5 minutes. If the current is 4 A, what is the potential difference across the heater, assuming no heat loss?

Electrical energy = $I \times t \times V$; heat energy = $c \times m \times \Delta T$
Electrical energy supplied = Heat energy gained

$$I \times t \times V = c \times m \times \Delta T$$
$$\Rightarrow 4 \times (5 \times 60) \times V = 4\,200 \times 0{\cdot}2 \times 10$$
$$\Rightarrow \quad V = \frac{4\,200 \times 0{\cdot}2 \times 10}{4 \times 5 \times 60}$$
$$\Rightarrow \quad V = 7{\cdot}0$$

Potential difference across heater is 7·0 volts.

17.4 Ohm's Law and resistance

Ohm's Law

Using the circuit in Figure 17.15, we can investigate how varying the potential difference across a component called a **resistor** R affects the current flow.
The current through R is measured by an ammeter and the p.d. across R by the voltmeter. Typical results are given in Table 1.

V (volts)	1·5	3·0	4·5	6·0
I (amperes)	0·5	1·0	1·5	2·0
V/I	3·0	3·0	3·0	3·0

Table 1 Ohm's Law experimental results

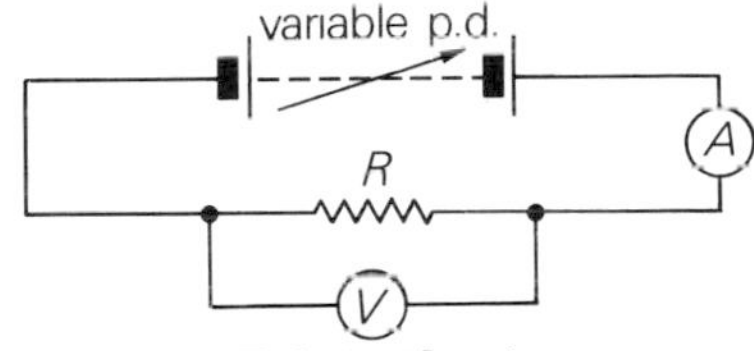

Figure 17.15 Ohm's Law experiment

A graph of the results is shown in Figure 17.16.
The graph of V against I is a straight line through the origin. This implies that the applied p.d. is directly proportional to (varies directly as) the current. $V \propto I$.

From the last row of Table 1, the ratio V/I is constant. When repeated for different conductors it is found that V/I is a constant for each conductor. This is called **Ohm's Law** after a German scientist G. S. Ohm.

Figure 17.16 Ohm's Law graph

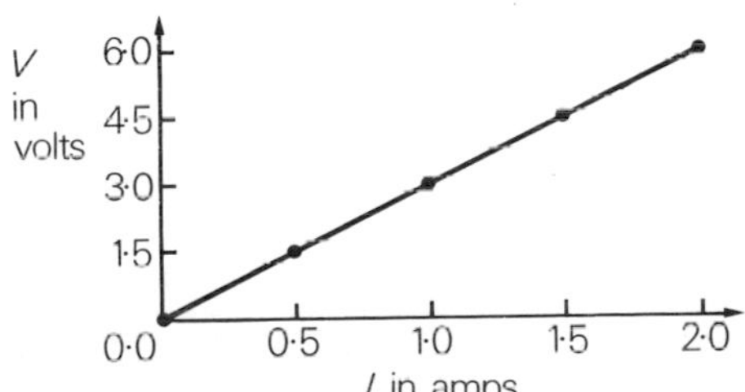

For any conductor, the **resistance** R is defined as

$$\frac{\text{potential difference across the conductor}}{\text{current through the conductor}}; \quad R = \frac{V}{I}$$

Units

As the unit of p.d. (V) is the volt and the unit of current (I) is the ampere, the unit of resistance (R) is the volt per ampere ($V\ A^{-1}$). This unit is called the ohm and is abbreviated to the Greek letter omega (Ω).
The resistance of a conductor is affected by a number of factors all of which were investigated by G. S. Ohm. Ohm made wires of different materials, lengths and thicknesses. He investigated each one of these factors while keeping all the others constant.

Components which obey Ohm's Law are called resistors and they come in many shapes, sizes and values, Figure 17.17.

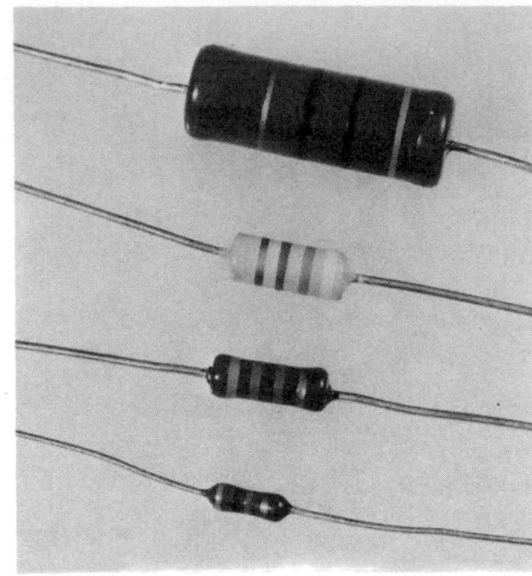

Figure 17.17 Resistors

Symbols

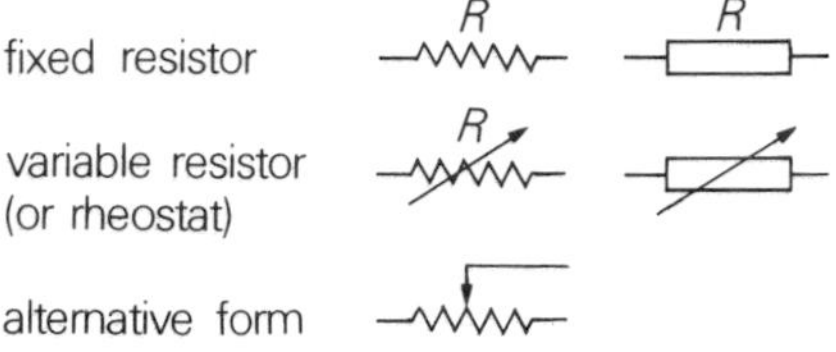

The values of resistors cover a very wide range and so the values are abbreviated in the usual manner:

1 000 ohms = 1 kΩ = 1 kilohm
1 000 000 ohms = 1 MΩ = 1 megohm

The simplest resistor is a length of wire.
The longer the wire, the greater is the resistance.
The thinner the wire, the greater is the resistance.
The resistance of a wire also depends on the material.

Table 2 shows the resistance of 1 metre of various types of wire of the same thickness.

Type of wire	*Resistance of 1 metre*
Constantan	3·5Ω
Manganin	3·0Ω
Iron	0·75Ω
Tungsten	0·42Ω
Copper	0·14Ω

Table 2 Resistance for the same thickness of wire of various metals.

Figure 17.18 Variable resistors

Measuring resistance

Using the circuit in Figure 17.19, we can take a range of convenient p.d. and current readings by adjusting the variable resistor in order to find the value of the unknown resistor.
A typical set of results is given in Table 3.

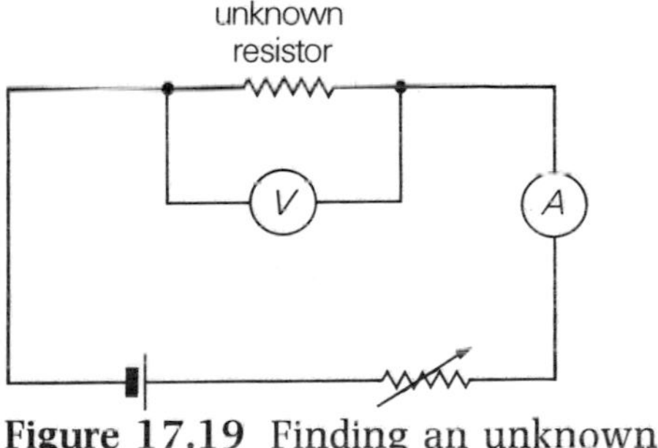

Figure 17.19 Finding an unknown resistance

V (volts)	1·0	1·2	1·4	1·6	1·8
I (amperes)	0·49	0·62	0·70	0·81	0·93
V/I (ohms)	2·04	1·94	2·00	1·97	1·93

Table 3 Measuring resistance

The value of the unknown resistor obtained by averaging the values of the ratio V/I is 1·98. The resistance of the unknown resistor is approximately 2 ohms.

Example 5

What current will flow through a 1·0 kΩ resistor when a potential difference of 12 V is applied across it?

Using the equation $\frac{V}{I} = R$

$\Rightarrow \quad \frac{12}{I} = 1\,000$

Multiplying both sides by I

$\Rightarrow 12 = 1\,000 \times I$

Multiplying both sides by $\frac{1}{1\,000}$

$\Rightarrow \quad \frac{12}{1\,000} = I$

$\Rightarrow 12{\cdot}0 \times 10^{-3} = I$

A current of 12·0 × 10⁻³ amperes (12 mA) flows through the resistor.

Example 6

A resistor has a p.d. of 6 volts applied across it and the current flowing through it is 0·1 ampere. What is the resistance of the resistor?

Using the equation $R = \frac{V}{I}$

$\Rightarrow \quad R = \frac{6}{0{\cdot}1}$

$\Rightarrow \quad R = 60$

Resistor has resistance of 60 ohms

Effect of temperature on resistance

We can investigate how resistance depends on temperature with the apparatus in Figure 17.20. The resistor in this experiment is a lamp filament.
The variable resistor R is used to vary the current. As the current through the lamp is increased, the filament glows more brightly until it is at normal brightness. This means that its temperature is gradually increasing. Typical readings obtained in this type of experiment are given in Table 4.

Figure 17.20 Resistance and temperature

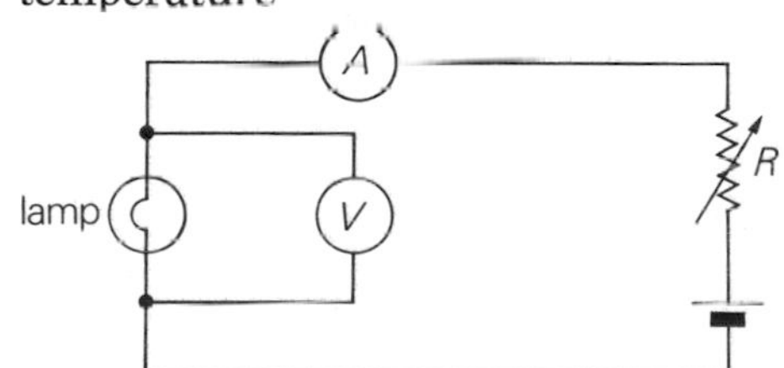

V (volts)	1·3	2·5	3·8	4·8
I (amperes)	0·14	0·20	0·24	0·28
V/I (ohms)	9·3	12·5	15·8	17·1

Table 4 Change of resistance for various temperatures

The table clearly indicates that, as the lamp filament becomes hotter, its resistance increases.

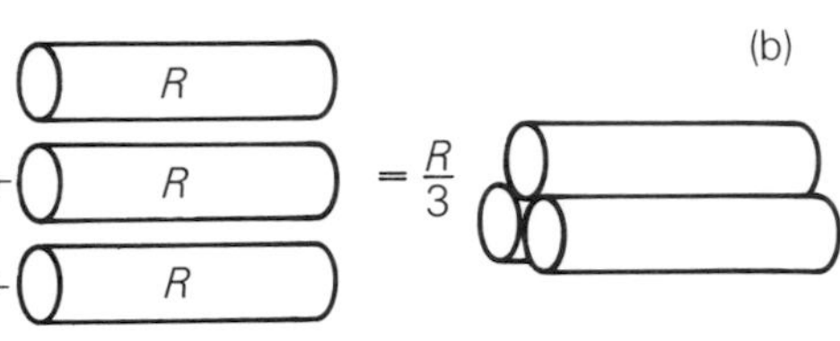

Figure 17.21

Combination of resistors

We have already mentioned two factors which relate to resistor combinations.

a) The resistance of a wire depends on its length, and therefore resistors in series would probably combine by adding, Figure 17.21a.

b) The resistance of a wire depends on its cross section (the thinner the wire the more resistance), and so the more wires we bundle together (in parallel) the less resistance we would expect, Figure 17.21b.

We shall look at these situations in detail, using our knowledge of Ohm's Law and circuits.

Resistors in series

Two resistors R_1 and R_2 are joined in series with a power supply of voltage V volts and the current is I amperes. We wish to find the combined resistance R_s, Figure 17.22.

$$\frac{V}{I} = R_s$$

where V is the supply voltage, I is the current, R_s is the total resistance in the series circuit.

$\Rightarrow V = I \times R_s$

We know that $V = V_1 + V_2$

By applying Ohm's Law to resistors R_1 and R_2 in turn:

$V_1 = I \times R_1$ and $V_2 = I \times R_2$

$$\Rightarrow V = I \times R_1 + I \times R_2$$
$$= I(R_1 + R_2)$$
$$\Rightarrow \frac{V}{I} = R_1 + R_2$$
$$\Rightarrow R_s = R_1 + R_2$$

When resistors are in series, their total resistance is the sum of the individual resistances.

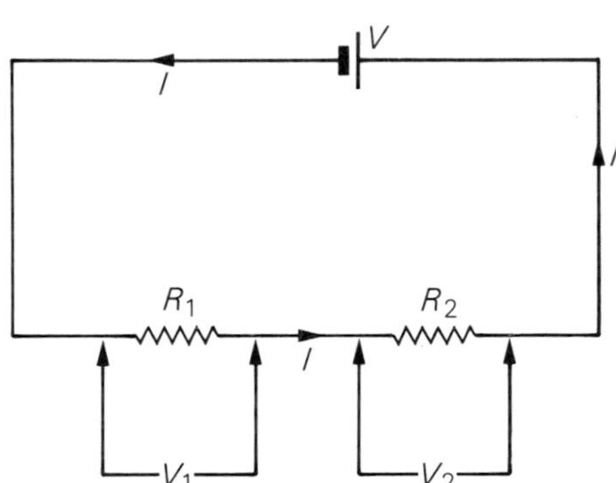

this is equivalent to

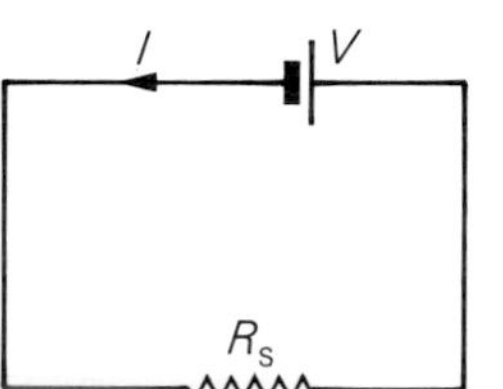

Figure 17.22 Series circuit

Example 7

What is the total resistance in the circuit shown? What current flows through the 2 ohm resistor if the power supply is 20 V?

$$R_s = R_1 + R_2 + R_3 + R_4$$
$$= 1 + 2 + 3 + 4$$
$$= 10$$

Total resistance in the circuit is 10 ohms

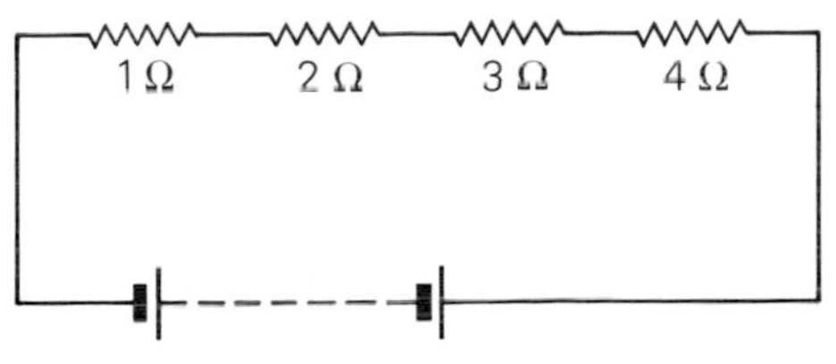

Using Ohm's Law $R = \frac{V}{I}$

$$10 = \frac{20}{I}$$

Multiplying both sides by I,

$\Rightarrow 10 \times I = 20$

Multiplying both sides by $\frac{1}{10}$,

$\Rightarrow I = 2$

The current is the same all round a series circuit.

Therefore the current through the 2 ohm resistor is 2 A.

Example 8

One metre of copper wire has resistance 1·2 ohms. How much resistance has 3·0 metres of the same wire?

1·2 Ω 1·2 Ω 1·2 Ω

1 metre 1 metre 1 metre

$R_s = R_1 + R_2 + R_3$

$= 1{\cdot}2 + 1{\cdot}2 + 1{\cdot}2$

$R_s = 3{\cdot}6$

Three metres of copper wire has a resistance of 3·6 ohms.

Resistors in parallel

Two resistors R_1 and R_2 are joined in parallel with a power supply of voltage V volts and the supply current is I amperes. We wish to find the combined resistance R_p.

The current through R_1 is I_1 and through R_2 is I_2, Figure 17.23.

$$\frac{V}{I} = R_p$$

where V is the supply voltage, I is the supply current, R_p is the total resistance in the parallel circuit.

$$\Rightarrow \quad I = \frac{V}{R_p}$$

The supply current splits up so that $I = I_1 + I_2$

Applying Ohm's Law to resistors R_1 and R_2

$$I = \frac{V_1}{R_1} + \frac{V_2}{R_2}$$

But since the resistors are in parallel, $V_1 = V_2 = V$

$$\Rightarrow \quad I = \frac{V}{R_1} + \frac{V}{R_2} = V\left(\frac{1}{R_1} + \frac{1}{R_2}\right)$$

$$\Rightarrow \frac{V}{R_p} = V\left(\frac{1}{R_1} + \frac{1}{R_2}\right)$$

Multiply both sides by $\frac{1}{V}$

$$\Rightarrow \frac{1}{R_p} = \frac{1}{R_1} + \frac{1}{R_2}$$

When resistors are in parallel, the reciprocal of the total resistance is the sum of the reciprocals of the individual resistances.

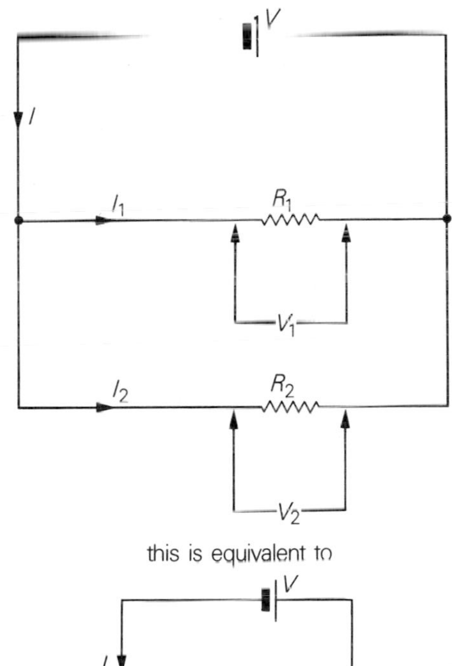

Figure 17.23 Parallel circuit

Example 9

Find the total resistance in the circuit below.

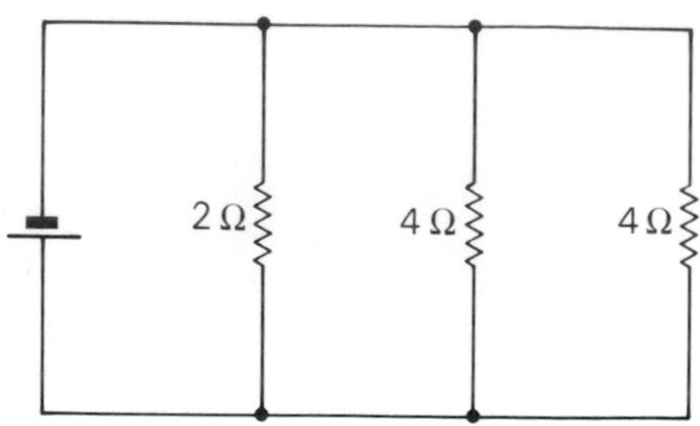

$$\frac{1}{R_p} = \frac{1}{R_1} + \frac{1}{R_2} + \frac{1}{R_3}$$
$$= \frac{1}{2} + \frac{1}{4} + \frac{1}{4}$$
$$= \frac{2+1+1}{4}$$
$$\Rightarrow \frac{1}{R_p} = \frac{4}{4}$$
$$\Rightarrow R_p = 1$$

The total resistance is 1 ohm.

Example 10

What is the total resistance of four equal lengths of identical resistance wire connected in parallel if each wire has a resistance of 4·8 ohms?

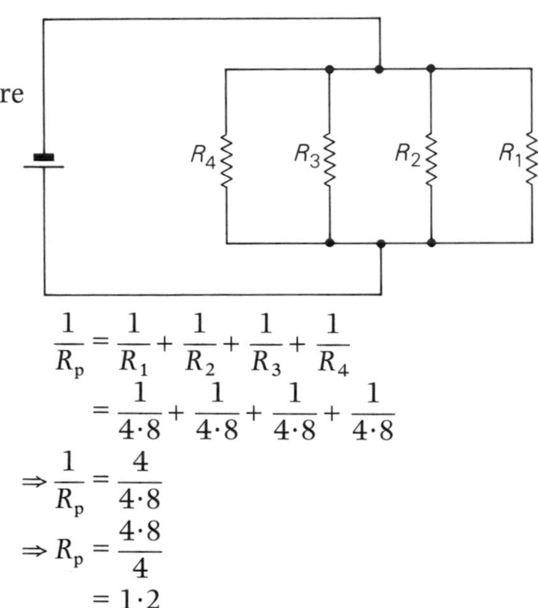

$$\frac{1}{R_p} = \frac{1}{R_1} + \frac{1}{R_2} + \frac{1}{R_3} + \frac{1}{R_4}$$
$$= \frac{1}{4{\cdot}8} + \frac{1}{4{\cdot}8} + \frac{1}{4{\cdot}8} + \frac{1}{4{\cdot}8}$$
$$\Rightarrow \frac{1}{R_p} = \frac{4}{4{\cdot}8}$$
$$\Rightarrow R_p = \frac{4{\cdot}8}{4}$$
$$= 1{\cdot}2$$

The combined resistance of the four resistors is 1·2 ohms.

Example 11

Find the total resistance in the circuit below.

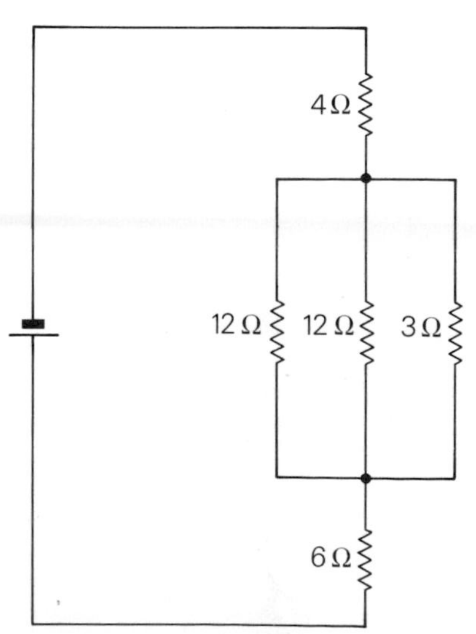

If we first combine the three resistors which are in parallel:

$$\frac{1}{R_p} = \frac{1}{3} + \frac{1}{12} + \frac{1}{12}$$
$$= \frac{4+1+1}{12}$$
$$= \frac{6}{12}$$
$$\Rightarrow \frac{1}{R_p} = \frac{1}{2}$$
$$\Rightarrow R_p = 2 \text{ ohms}$$

Now combining this with the other two resistors in series:

$$R_s = 4 + 2 + 6$$
$$\Rightarrow R_s = 12$$

Total resistance in the circuit is 12 ohms.

Example 12

In the circuit below, the potential difference across the 4 ohm resistor is 6 volts. What is the potential difference across the 3 ohm resistor?

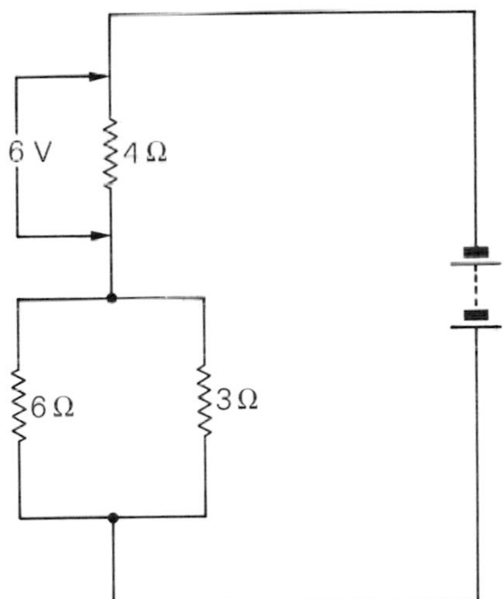

Applying Ohm's Law to the 4-ohm resistor:

$$R = \frac{V}{I}$$

$$\Rightarrow 4 = \frac{6}{I}$$

Multiplying both sides by I

$$\Rightarrow 4 \times I = 6$$

Multiplying both sides by $\frac{1}{4}$

$$\Rightarrow I = \frac{6}{4}$$

$$\Rightarrow I = 1{\cdot}5 \text{ amperes}$$

Combining the pair of resistors in parallel:

$$\frac{1}{R_p} = \frac{1}{R_1} + \frac{1}{R_2}$$

$$= \frac{1}{3} + \frac{1}{6}$$

$$\Rightarrow \frac{1}{R_p} = \frac{2+1}{6}$$

$$= \frac{3}{6}$$

$$\Rightarrow R_p = \frac{6}{3} = 2 \text{ ohms}$$

The circuit therefore simplifies to:

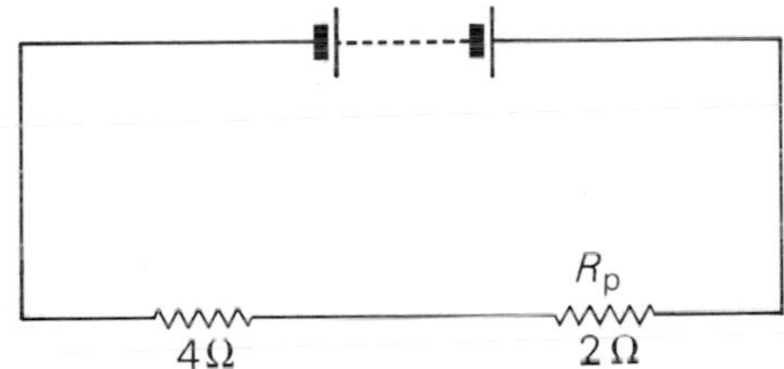

Potential difference across R_p is given by

$$V = I \times R_p = 1{\cdot}5 \times 2 = 3{\cdot}0$$

potential difference across R_p is 3 volts,
$\Rightarrow$ potential difference across each of the resistors in parallel is 3 volts,
Potential difference across the 3-ohm resistor is 3 volts.

Summary

An electric current is a flow of electrons from negative to positive.

The SI unit of current is the ampere (A). It is defined in terms of the force exerted between two wires carrying a current.

The SI unit of charge is the coulomb (C). It is the amount of charge involved when a current of 1 ampere flows for 1 second: 1 C = 1 A s.

$Q = I \times t$ where Q = charge, I = current, and t = time.

The current is the same all the way round a series circuit.

The total current leaving a circuit junction is equal to the total current entering.

If one joule of energy is required to transfer one coulomb of charge between two points, then the potential difference is one volt: $1\ \mathrm{V} = 1\ \mathrm{J\ C^{-1}}$.

$$E = Q \times V \text{ or } V = \frac{E}{Q}$$

In a series circuit, the potential difference across the supply is equal to the sum of the potential differences across the individual components.

In a parallel circuit, the potential difference across all the parallel components is the same.

Electrical power (watts) equals current (amperes) times potential difference (volts): $P = I \times V$.

Electrical energy (joules) equals current (amperes) times time (seconds) times voltage (volts): $E = I \times t \times V$.

Ohm's Law: $\frac{V}{I}$ = constant.

Resistance (ohms) equals potential difference (volts) divided by current (amperes): $1\Omega = 1\ \mathrm{V\ A^{-1}}$.

$$R = \frac{V}{I}$$

The resistance of a piece of wire depends on its length, thickness, and composition. The resistance of metals increases with temperature.

Combining resistors in series: $R_s = R_1 + R_2$.

Combining resistors in parallel: $\frac{1}{R_p} = \frac{1}{R_1} + \frac{1}{R_2}$.

Problems

1 A pupil wishes to insert a 5·5-ohm resistor into a circuit but has only a packet of 2-ohm resistors. He uses a combination of eight 2-ohm resistors. How did he arrange them?

2 What value of resistor must be inserted in a circuit which contains a 6-volt, 36-watt lamp connected to a 12-volt power supply, in order that the lamp should operate normally?

3 Draw a circuit diagram to illustrate three lamps connected in parallel with a power supply so that two of the lamps may have their brightness varied independently, using rheostats.

4 Which of the lamps in the following circuit will be the brightest, assuming that all the lamps are identical?

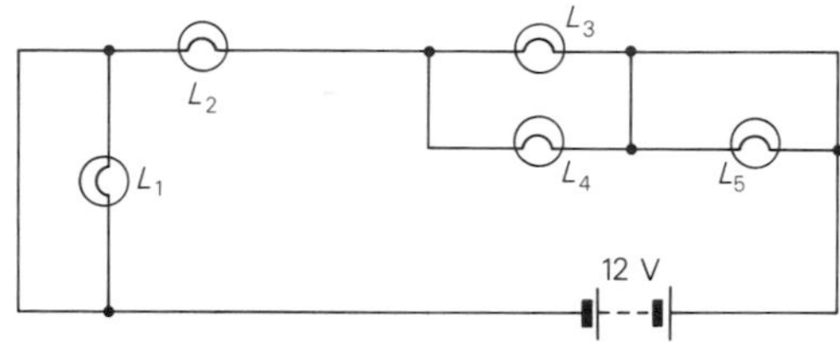

5 This is a light circuit with two similar lamps operated from one battery.

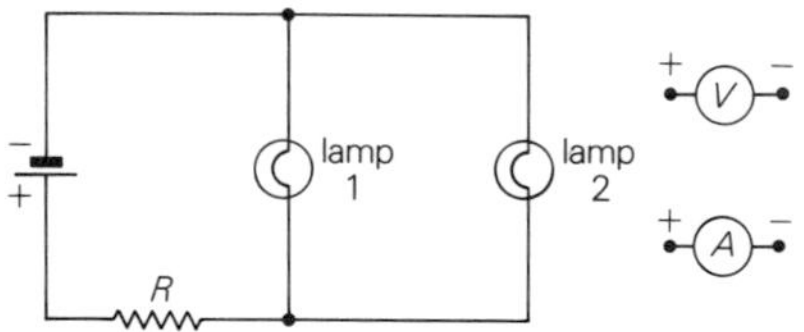

5·1 If a short circuit developed in this circuit which caused both bulbs to 'blow', where is it most likely to have occurred?
5·2 Redraw the above circuit to include all of the following:
a) a switch which will switch on and off only one of the lamps;
b) an ammeter to measure the total current taken from the battery;
c) a voltmeter to measure the potential difference across the resistor R.
(Use the symbols shown above for the ammeter and the voltmeter.) *SCEEB*

6 The circuit shown below could be used to light a 4-volt bulb from a 12-volt supply.

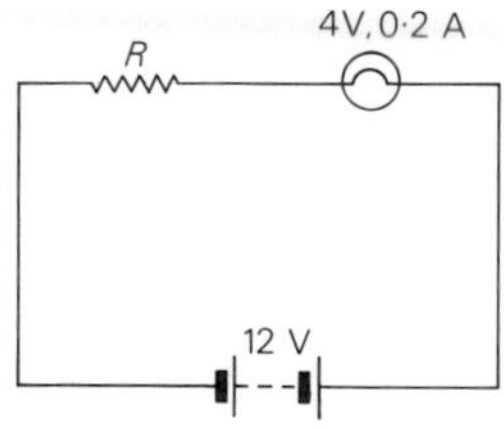

6·1 What should be the voltage drop across the resistor R so that the bulb operates at its normal rating?
6·2 What should be the resistance of R so that the bulb operates at its normal rating?

6·3 At what rate is electrical energy being converted into heat energy in the resistor R?
6·4 What percentage of the electrical power given out by the supply is wasted as heat in the resistor R?
6·5 If the resistor R were replaced by a series arrangement of 2 bulbs, both identical to the one shown, what would be the effect on the brightness of the original bulb? Explain your answer. *SCEEB*

7 A potential difference of 12 volts is applied across a resistance wire.
7·1 What quantity of electric charge must flow through the wire so that 1 680 joules of heat energy may be produced in it?
7·2 If it takes one minute for this charge to pass, what is the current through the wire?
7·3 If all the heat being produced is used to raise the temperature of 1 kg of water, what will be the rise in temperature of the water in one minute?
Specific heat capacity of water $= 4200$ J kg^{-1} °C^{-1} *SCEEB*

8 This is part of the wiring diagram of a section of the electrical system of a car.

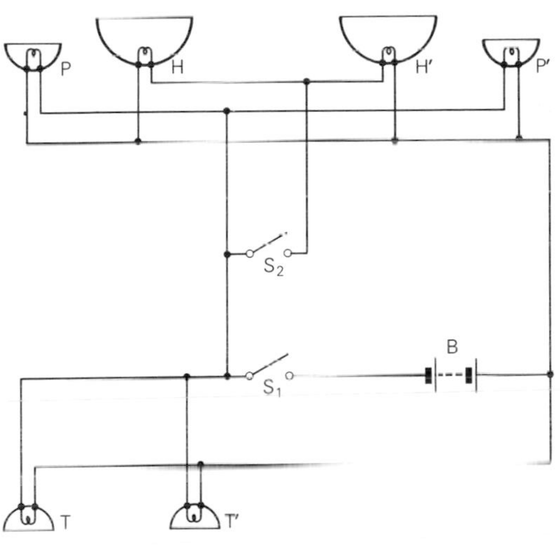

┬ represents wires connected electrically
┼ represents wires crossing but *not* connected electrically

H H′ headlights
P P′ parking lights
T T′ tail lights
B car battery
S_1 S_2 switches

8·1
a) How would you switch *all* the lights *on*?
b) Now, how would you switch *off* the headlights *only*?
8·2 What would happen to the lights if they were *all on* and tail light T′ were broken? Give a reason for your answer.
8·3 All lights have the same battery B as their source of energy. How can the lights be of different brightness in spite of this? *SCEEB*

9 An electrical component is contained in a closed box. Two terminals from the component are connected in the circuit at X and Y as shown.

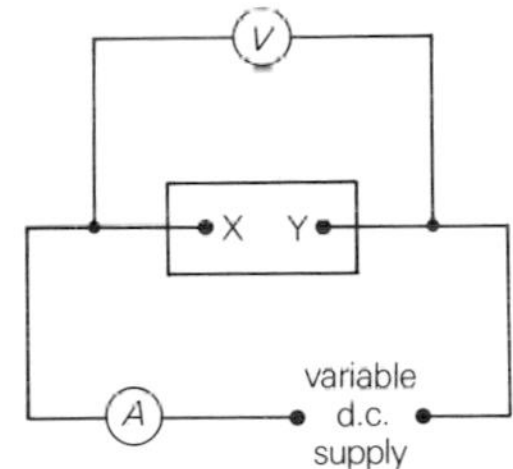

A pupil is asked to examine the relationship between the voltage across XY and the current in the circuit. He gradually increases the voltage and obtains the following results:

V volts	1·0	2·0	3·0	4·0	5·0
I amperes	0·20	0·27	0·34	0·36	0·37

9·1 Calculate the value of the resistance of the box from the fourth set of results, namely $V = 4{\cdot}0$ V, $I = 0{\cdot}36$ A.
9·2 Draw a graph of *I* against *V*.
9·3 What conclusion regarding the resistance of the contents can be drawn from the graph?
9·4 In what way does the shape of the graph differ from what you might have been led to expect from the law connecting current and voltage?
9·5 On looking inside the box, the pupil discovers it contains a torch bulb. Use your knowledge of a bulb to explain the conclusion in part **9·3**. *SCEEB*

10 A pupil uses the circuit shown below to find the resistance of wire AB.

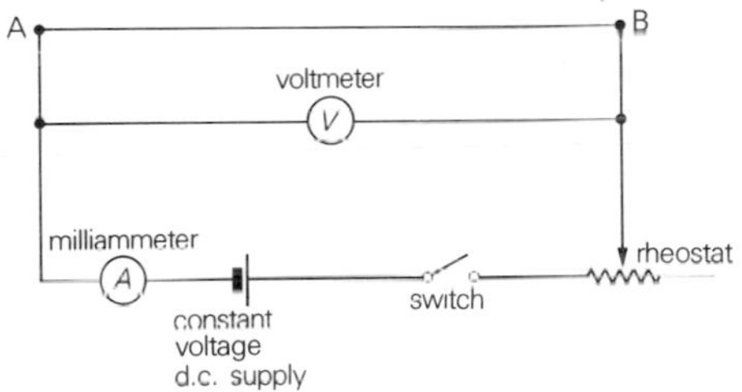

The following readings are obtained.

POTENTIAL DIFFERENCE (in volts)	1·05	1·40	1·80	2·20
CURRENT (in amperes)	0·15	0·20	0·25	0·30

10·1 State how the changes in the meter readings are produced.
10·2 Draw a graph of the potential difference against the current.
10·3 From these experimental results, what is the value of the resistance of wire AB?
10·4 To check his result, the pupil now selects a fifth value of 3·50 volts for the voltmeter reading. Copy the milliammeter dial shown on the next page and indicate on it the new needle position he should expect.

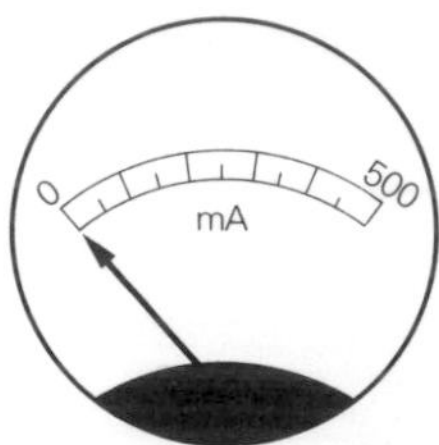

10·5 If the pupil keeps the switch closed for a long time, he notices that this fifth current reading decreases slightly. Explain this observation. *SCEEB*

11 In this circuit the two lamps, X and Y, are operated at their rated voltage and power.

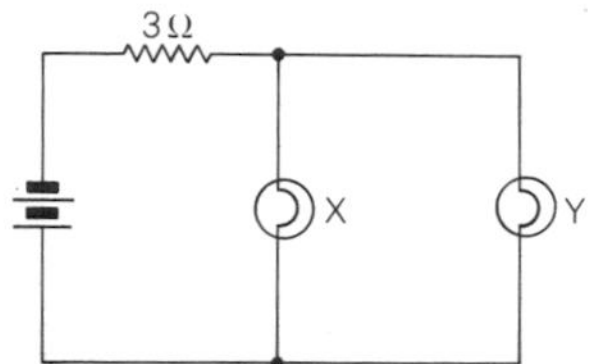

Lamp X is rated 6 V 12 W; lamp Y is rated 6 V 24 W.
Calculate:

11·1 the current in the 3 Ω resistor,

11·2 the potential difference across the battery. *SCEEB*

12 When switch S is open, the voltmeter reads 6·0 volts and the ammeter reads 1·0 ampere. When switch S is closed, the voltmeter reads 3·0 volts and the ammeter reads 1·5 amperes.

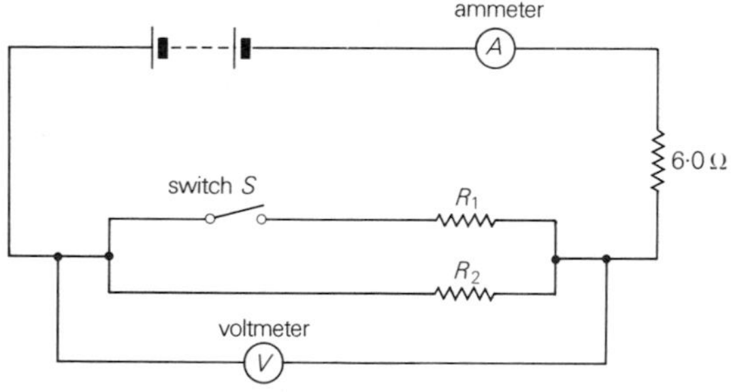

Calculate the values of R_1 and R_2. *SCEEB*

13 **13·1** A pupil uses the circuit shown to investigate resistors in series. He has in addition one ammeter, one voltmeter and some connecting wire.

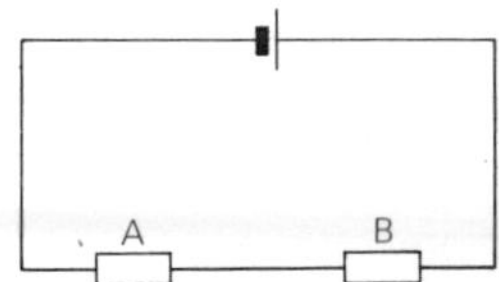

a) Draw separate diagrams to show how these meters could be connected to this circuit to find the resistance of A only; the resistance of B only; and the combined resistance of A and B in series.

b) State the relationship which the pupil should find between the resistance of A, the resistance of B and the combined resistance of A and B in series.

13·2 The pupil decides to build an instrument for measuring resistance directly. He bases his design on the following circuit diagram.

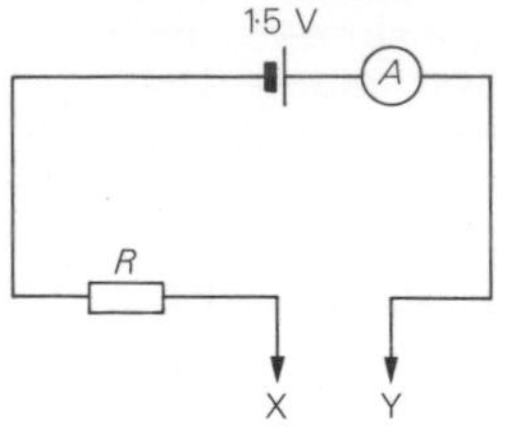

The resistance to be measured is connected between X and Y.

a) Describe how this instrument may be calibrated so that the meter indicates resistance directly in ohms.

b) Give one reason why resistor R is included in the circuit of the instrument. *SCEEB*

14 **14·1** Two resistors of resistance $R_1 = 10\ \Omega$ and $R_2 = 15\ \Omega$ respectively were connected in series across the 20 volt terminals of a d.c. supply.

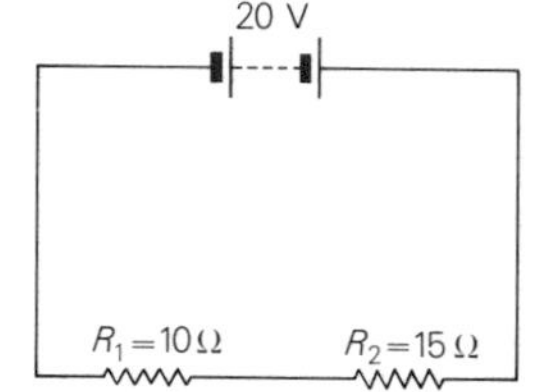

a) What is the combined effective resistance of R_1 and R_2 in series?

b) Calculate a value for the current through each of the resistors.

c) Deduce the potential difference across each resistor.

Suppose you wanted to make an experimental test of your calculations. Redraw the above circuit with voltmeter(s) and ammeter(s) in place to test the calculated values of current and potential difference.

14·2 'Fairy lights' were designed to have ten similar 1·2 W lamps in series across a 240 V supply. When connected in this way what is:

a) the voltage across each lamp,

b) the current through each lamp?

Suggest what would happen if one of the fairy lights were connected directly across the 240 V supply. *SCEEB*

15 Copy down the words in column A. Select, from column B opposite, the word or group of words which is equivalent to the one in column A.

	Column A	*Column B*
15·1	milliampere	1 000 A; 100 A; $\frac{1}{100}$ A; $\frac{1}{1\,000}$ A.
15·2	joule	kilogram metre; watt/second; newton metre; newton/second.
15·3	coulomb	ampere second; ampere/second; volt ohm; joule/second.
15·4	watt	joule/second; joule second; ampere second; coulomb/second.
15·5	ohm	ampere/volt; ampere/second; volt/ampere; volt ampere.

SCEEB

18 Magnetic effect of current

18.1 Magnets

Men have been aware of the existence of magnets for over two thousand years. Lodestone, a naturally occurring mineral, will attract pieces of iron, and men have used magnets as simple compasses for many centuries.

We recognize a **magnet** by its property of attracting certain metals, notably iron, cobalt and nickel. These metals, which are attracted by a magnet, are said to be magnetic.

A **magnetic substance** is a substance which can be attracted by a magnet.

Magnets come in various shapes and sizes. The simplest shape of magnet is that of a bar magnet. If a bar magnet is allowed to move freely, it will swing round until it points in a north–south direction. Figure 18.1 shows how to support a magnet on two watch glasses so that it can rotate freely.

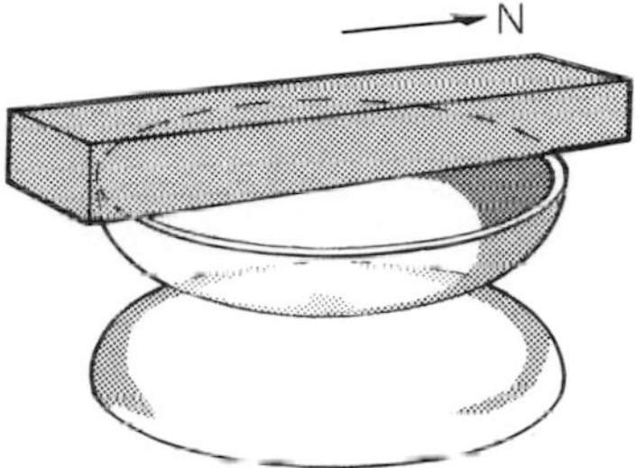

Figure 18.1

It is this important direction-seeking property of magnets which enables us to use them as compasses for navigation. The end of the bar magnet which points north is called the north pole of the magnet, and the end which points south is called the south pole of the magnet.
The **north** pole of a magnet is the **north-seeking** pole
The **south** pole of a magnet is the **south-seeking** pole.

When a magnet is dipped into iron filings, the iron filings cling to it, particularly around the poles of the magnet, Figure 18.2. The magnetic effect is strongest at the poles of the magnet.

Figure 18.2 Poles of a bar magnet

When a single bar magnet is allowed to rotate freely, it settles down in a north–south direction. If another magnet is brought up to it, both exert a force on each other (Figure 18.3) and line up as follows:

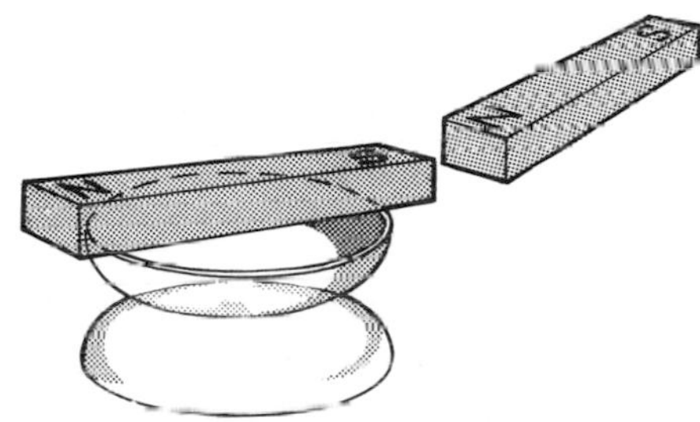

Figure 18.3 Attraction between magnets

A north pole attracts a south pole and repels a north pole.
A south pole attracts a north pole and repels a south pole.

Thus we find that two **like** poles **repel** each other and two **unlike** poles **attract** each other.

18.2 Magnetic fields

If we sprinkle iron filings on to a card, placed over a bar magnet (Figure 18.4) and tap the card gently, the iron filings settle into a pattern as shown in Figure 18.5.

Figure 18.4 Obtaining a field pattern

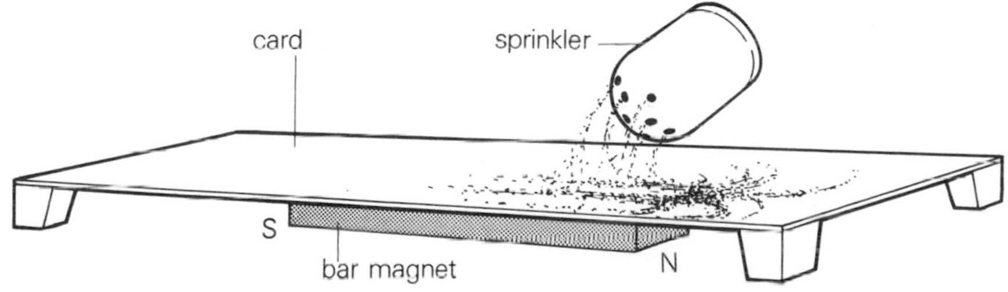

Figure 18.5 Bar magnet field pattern

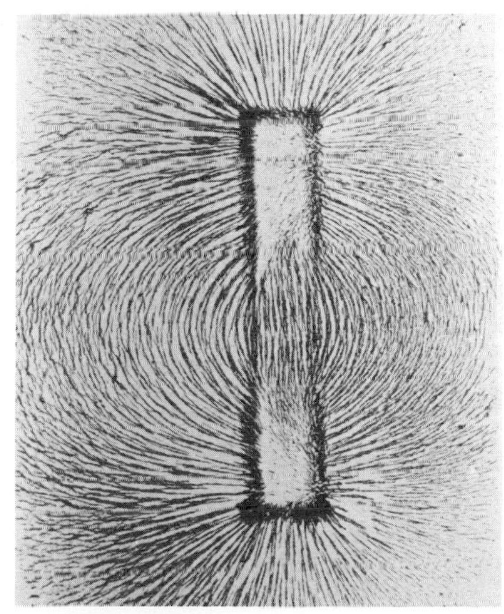

The iron filings give a rough picture of the pattern of the magnetic field. A compass is a small magnet. If we place a small compass in the magnetic field, it will point along the direction of one of the lines making up the magnetic field pattern.

When there are two magnets close together, their magnetic fields affect each other. Figure 18.6 shows two photographs of magnetic field patterns between (a) two like poles and (b) two unlike poles.

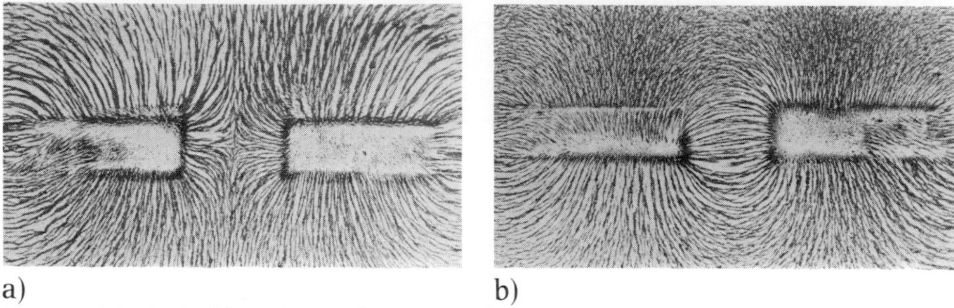

Figure 18.6 Field patterns for two poles

Figure 18.7 Using a plotting compass

We may use a small plotting compass to plot the magnetic field about a magnet as follows.

Place the plotting compass near the magnet resting on a piece of paper, and mark a dot on the paper at the tip (north pole) of the compass needle. Then move the compass so that its tail (south pole) points to the dot and again mark a dot on the paper at the tip of the compass needle. By continually repeating this process we map out a line of the magnetic field, Figure 18.7. Then join up these dots to form a line. We can plot another line in the pattern by starting from another point near the magnet.

In this way we build up the pattern of the magnetic field as a series of lines known as the **magnetic field lines**. The direction of a field line is away from a north pole and towards a south pole. Figure 18.8 shows the magnetic field lines for a bar magnet. Where the magnetic field is strongest, the field lines are closest together.

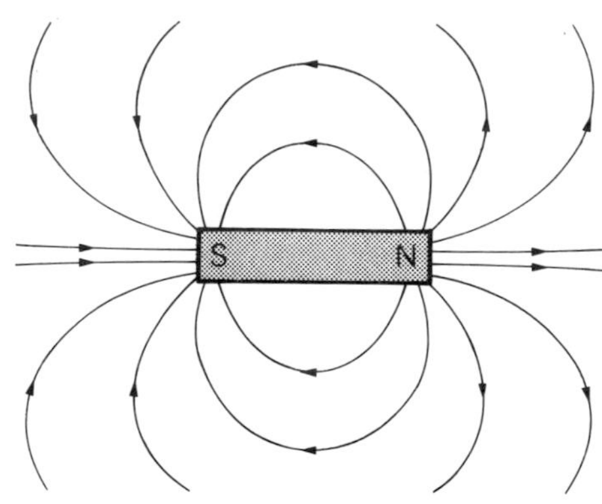

Figure 18.8 Field lines for a bar magnet

18.3 Magnetic field about a wire

If a current flows through a wire passing vertically through a card, we can show that there is a magnetic field pattern about the wire. We sprinkle iron filings on the card and tap the card gently. This shows that the magnetic field pattern is circular, Figure 18.9.

We may find the direction of the field lines by placing plotting compasses around the wire and switching on the current, Figure 18.10.

Figure 18.9 Obtaining a field pattern about a wire

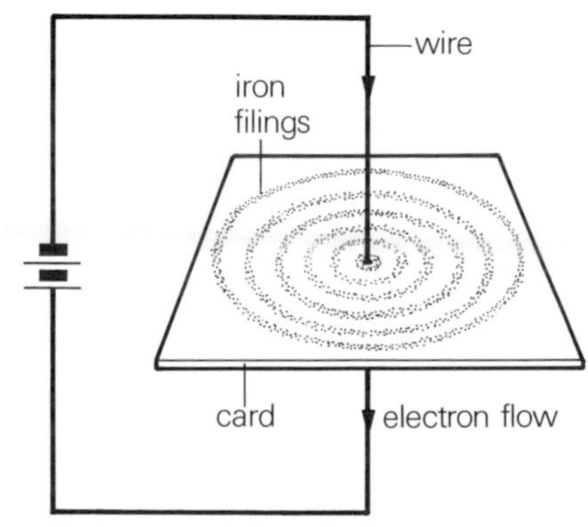

Figure 18.10 Direction of field lines about a wire.

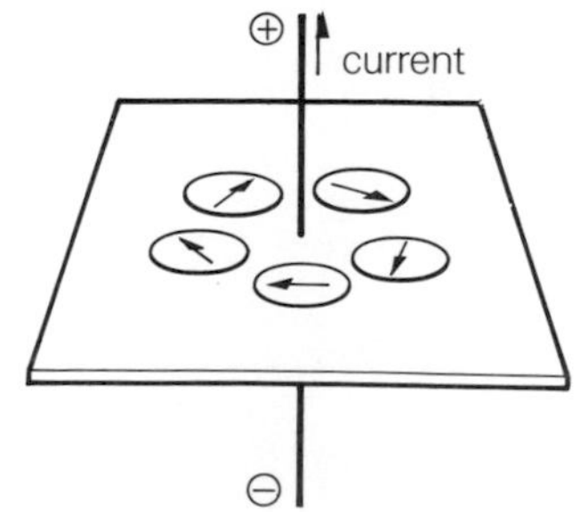

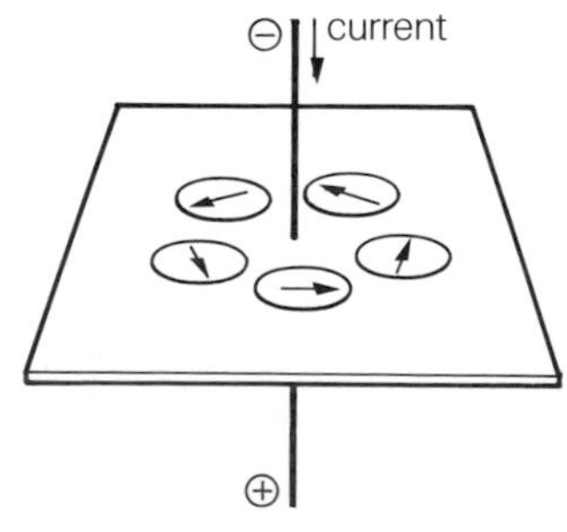

We may remember the direction of the magnetic field lines by using the left-hand rule for a straight current-carrying conductor:

Clench the left hand with the thumb extended. Point the thumb along the wire in the direction of the current and the fingers curl round in the direction of the magnetic field lines, Figure 18.11.

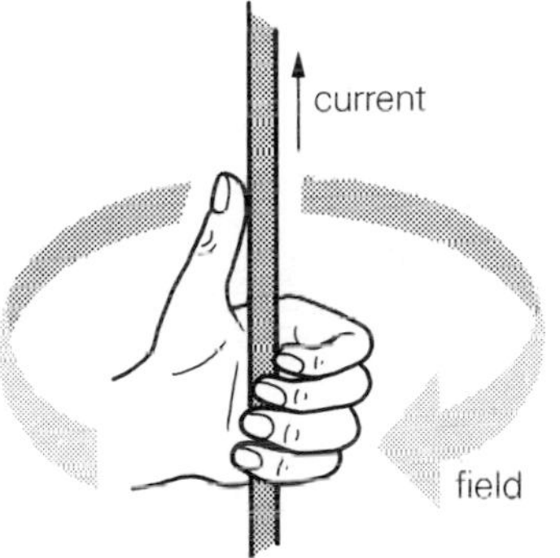

Figure 18.11 Left-hand rule

The conventional way of drawing a cross section of a wire, viewed end on, is as follows:

⊕ current away from you

⊙ current towards you

It is easy to remember this if you think of the cross as being the tail of an arrow pointing away from you and the dot as being the tip of an arrow pointing towards you.

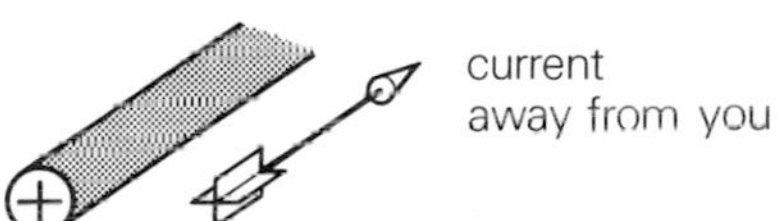

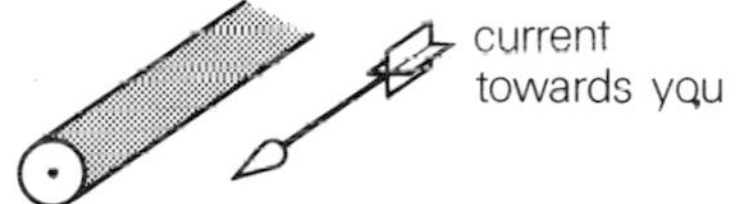

The magnetic field around a straight current-carrying wire is stronger nearer to the wire. So when we draw in the magnetic field lines we draw them closer together near the wire. We can find the direction of the field lines using the left-hand rule Figure 18.12.

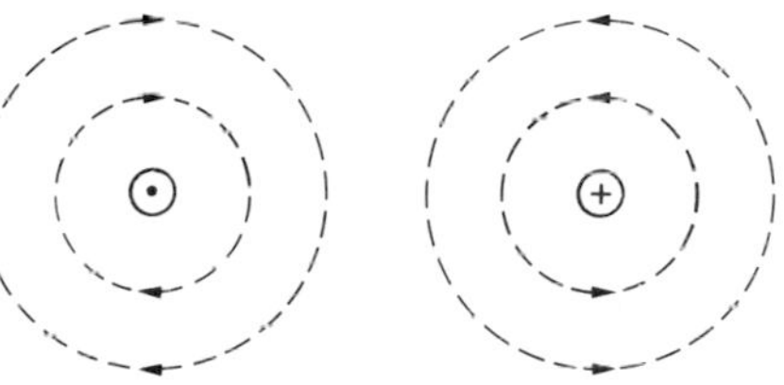

Figure 18.12 Direction of field lines around a straight wire

18.4 Magnetic field about a solenoid

A solenoid is a long coil of wire. We can make a solenoid by winding twenty or thirty turns of wire closely on to a pencil. If a current flows through a solenoid resting on a card, we can show a magnetic field pattern about the solenoid. Sprinkle iron filings on the card and tap the card gently. The iron filings tend to cluster around the end of the coil in a way similar to their clustering around the poles of a bar magnet, Figure 18.13 shows the magnetic field lines of a solenoid. Figure 18.14 shows how these field lines can be demonstrated using iron filings.

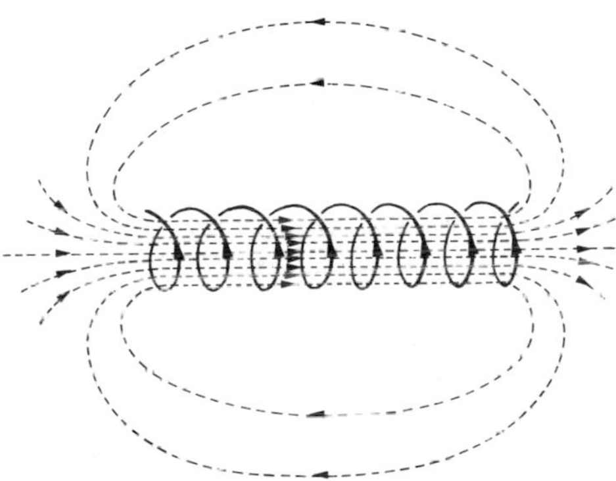

Figure 18.13 Solenoid as a magnet

Figure 18.14 Magnetic field lines for a solenoid

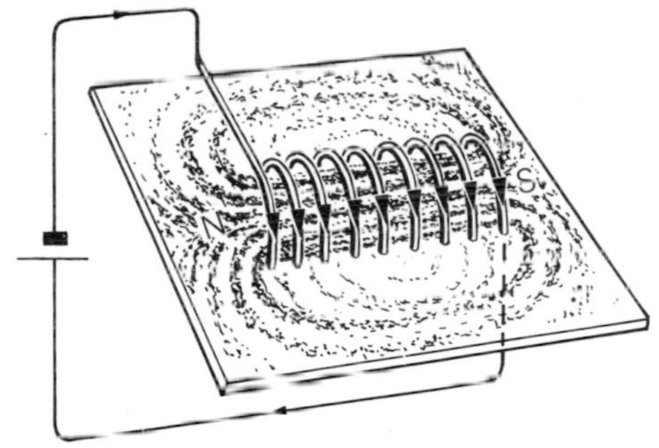

The magnetic field lines inside the solenoid are equally spaced, showing that the magnetic field is of uniform strength.

The magnetic field lines outside the solenoid form a pattern which is the same as that round a bar magnet. For the bar magnet, the direction of the field lines is away from the north pole and towards the south pole. In the case of the solenoid, the direction of the field lines depends on the direction of the current in the solenoid. Experiment shows that the end of the solenoid which acts as the north pole is given by another version of the left-hand rule: the rule for a current-carrying solenoid.

Curl the fingers of the left hand in the direction of the current and the extended thumb points to the north pole, Figure 18.15.

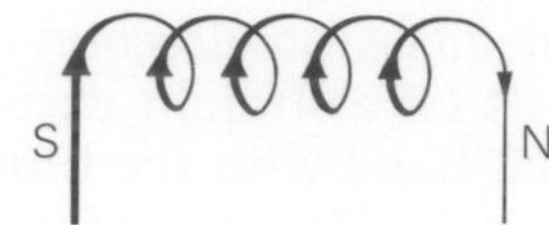

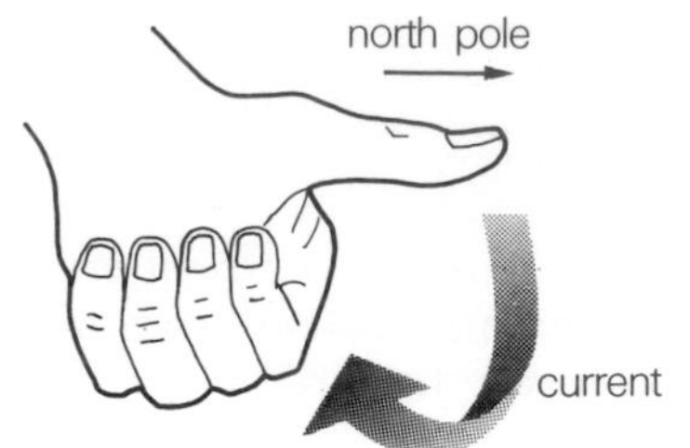

Figure 18.15 Left-hand rule for a solenoid

18.5 Electromagnets

While the magnetic field pattern of a solenoid is similar to that of a bar magnet, the solenoid does not normally have a very strong magnetic field. If, however, the solenoid is formed by coiling the wire round a piece of 'soft' iron, the magnetic field produced when a current flows in the solenoid is much stronger. ('Soft' iron is iron which cannot become permanently magnetized.)

An **electromagnet** consists of wire coiled around a soft iron core. They can range in size from those small enough to fit in the ear-piece of a telephone to those large enough to lift large quantities of scrap iron, Figure 18.16.

Figure 18.16

One advantage of an electromagnet over a permanent magnet for moving scrap iron is that it can be switched off. A permanent magnet cannot be switched off and it would be difficult to remove any iron that had become attached to it.

The uses of electromagnets are many and varied. One important use is in a relay switch. This is a circuit containing an electromagnet which is used to switch another circuit on and off. Figure 18.17 shows a circuit of a car starter-motor in which a relay is used to switch on the circuit containing the car battery and the starter motor.

The advantage of having a relay switch in the circuit is that it only needs a small current and therefore does not require thick wiring. The starter motor circuit requires heavy wiring to carry the large current involved and it would not be desirable to lead these thick wires to a switch on the dash-board. The G.P.O. uses relays extensively in telephone exchanges.

Figure 18.17 Use of a relay

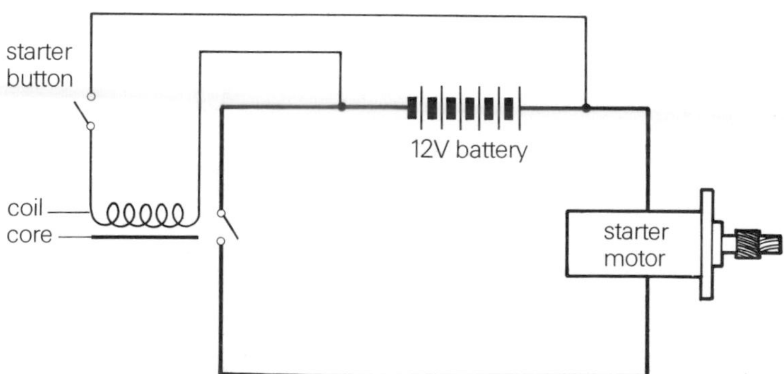

18.6 Force on a current-carrying conductor

In the last chapter we saw that the ampere was defined in terms of the force between two straight current-carrying wires. This force exists because each wire is in the magnetic field of the other wire. This is an example of a current-carrying conductor experiencing a force when it is in a magnetic field.

Further examples of this occur when a current-carrying conductor is in the magnetic field of a permanent magnet. We can investigate these effects using magnadur magnets which have poles on the flat faces, Figure 18.18.

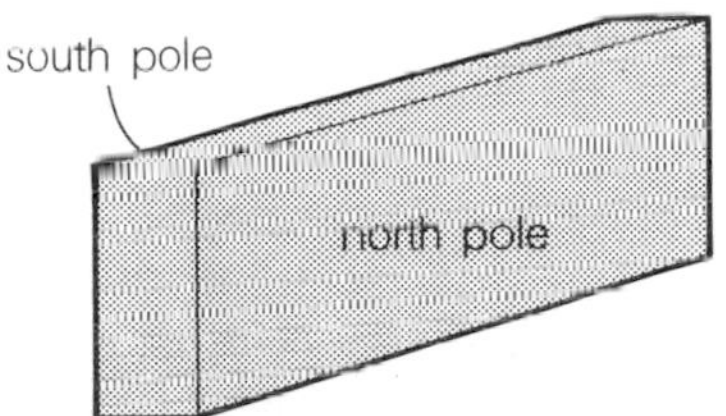

Figure 18.18 Magnadur magnet

When these are placed on a steel yoke, with opposite poles facing each other, a uniform magnetic field is formed between the magnets, Figure 18.19.

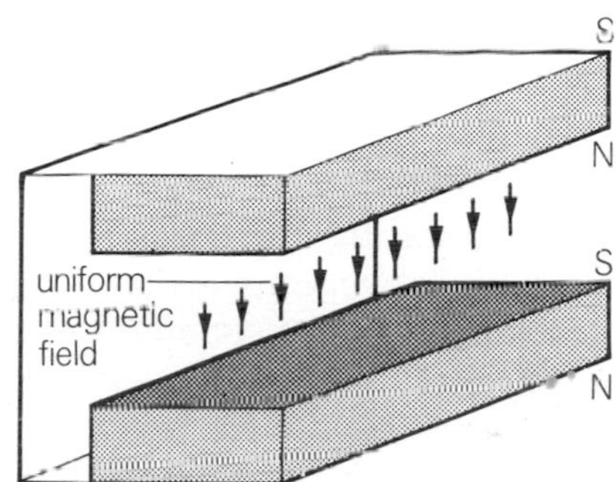

Figure 18.19 Producing a uniform magnetic field

Figure 18.20 shows the apparatus that we use to investigate the force on a straight wire in a magnetic field.

Two wires from the power supply project horizontally and support a third wire that is free to slide along them. All three wires are un-insulated so that electrical contact is maintained. When the power supply is switched on, the moveable wire either slides away from the steel yoke or into it, depending on the direction of the field and of the current. When we try different current and field directions, we obtain the results shown in Figure 18.21.

Figure 18.20 Investigating the force on a straight wire

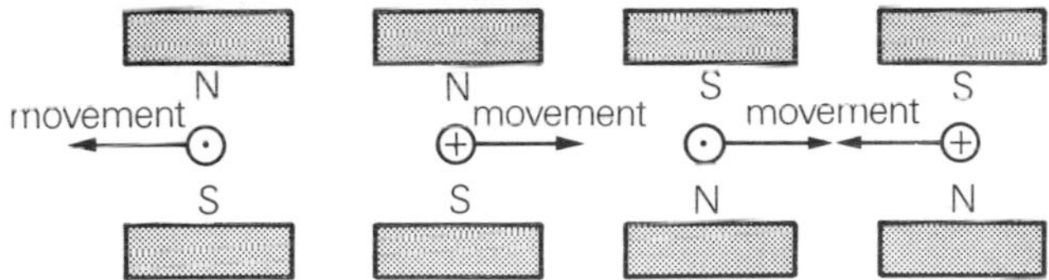

Figure 18.21 Movement of straight wire in magnetic field

The direction of the magnetic field is from the north pole to the south pole. Each current shown is at right angles to the magnetic field and the resulting movement is at right angles to both the current and the field. The current, field and movement are at right angles to each other.

You can remember the direction of the current, the magnetic field and the movement relative to each other by using the right-hand motor rule. Hold the right hand with the thumb, first finger and second finger at right angles to each other, Figure 18.22:

if the *F*irst finger points in the direction of the *F*ield,
and the se*C*ond finger points in the direction of the *C*urrent,
then the thu*M*b points in the direction of the *M*ovement.

Figure 18.22 Right-hand motor rule

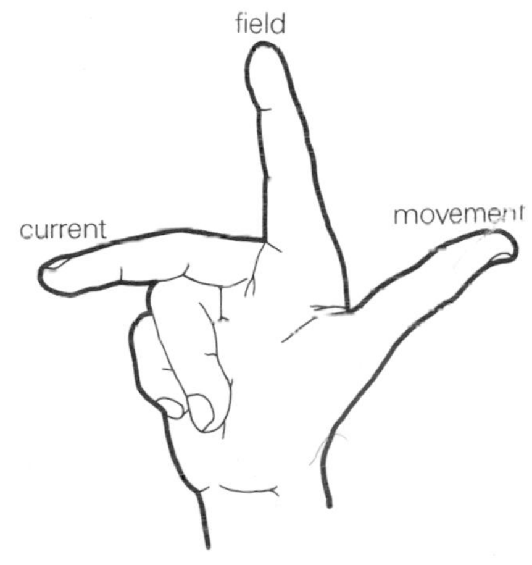

18.7 Loudspeaker

One practical application of the fact that a current-carrying conductor experiences a force in a magnetic field is in the loudspeaker, Figure 18.23. The cardboard cone is fixed at its edges to a rigid metal frame, but the centre is free to move. The coil attached to the cone is surrounded by a permanent magnet which is shaped so that one pole is inside the coil and the other pole surrounds the coil. This creates a magnetic field which is at right angles to the wire which makes up the coil.

Since the force on a current-carrying conductor in a magnetic field is at right angles to both the current direction and the magnetic field direction, the coil will move to the right or to the left, depending on the current direction. If the current direction is reversed, the force on the coil is reversed.

If an alternating current is passed through the coil, the coil will vibrate with a frequency equal to that of the alternating current. The coil is attached to the cone and the vibrations of the coil are transmitted to the cone. Sound is produced by vibrating objects. The pitch of the sound depends on the frequency of the vibrations and the loudness of the sound depends on the amplitude of the vibrations. The amplitude of the vibrations of the loudspeaker can be controlled by controlling the size of the current through the coil.

A loudspeaker converts electrical energy into sound energy. The pitch of the note produced is controlled by the frequency of the alternating current in the speaker coil and the loudness is controlled by the size of the alternating current.

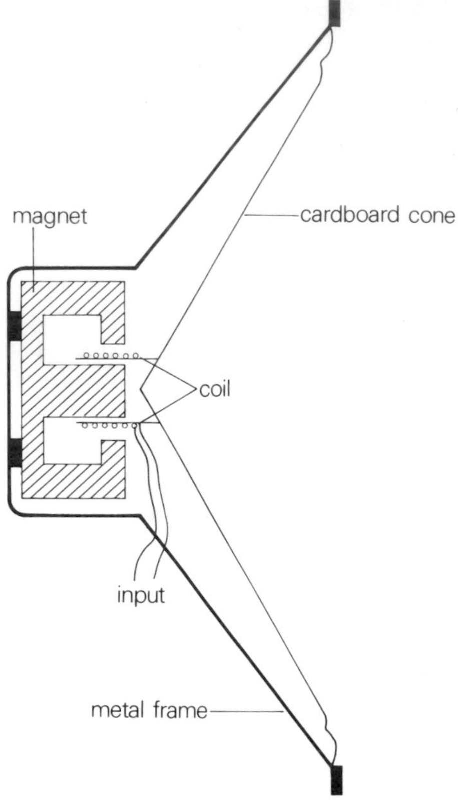

Figure 18.23 Cross section of a loudspeaker.

18.8 Electric motor

The movement of a current-carrying conductor in a magnetic field has an obvious application in an electric motor. However, the movement that we have so far investigated is a single movement at right angles to the magnetic field, and not the continuous movement that we want from an electric motor.

Figure 18.24 shows two wires in a magnetic field. Because the currents are in opposite directions, the directions of movement produced (found by the right-hand motor rule) are opposite, one wire moving up and one wire moving down.

A similar situation exists if, instead of two separate wires, we use one wire made into a loop, Figure 18.25. The upward force on one side of the loop and the downward force on the other side result in a rotation of the loop.

If we want this loop to keep turning we must arrange that the turning force is always in the same direction. With the arrangement shown in Figure 18.25, this means that the current in the side of the loop nearest the south pole must always be flowing **away** from us while the current in the side of the loop nearer the north pole must always be flowing **towards** us. To ensure this, we arrange that the side of the loop nearest the south pole is in contact with the negative terminal of the power supply, while the side of the loop nearest the north pole is in contact

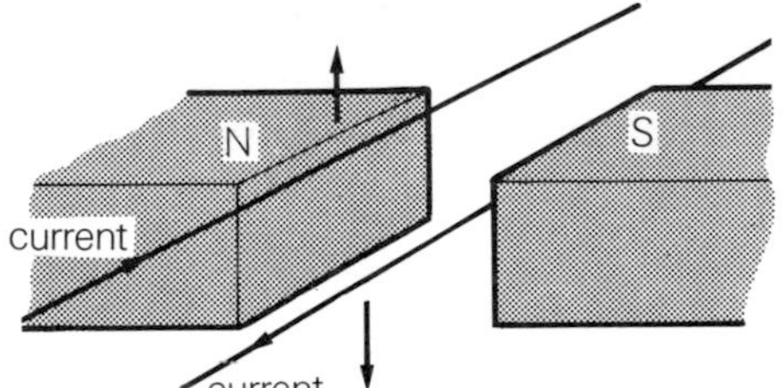

Figure 18.24 Two wires in a magnetic field

Figure 18.25 Loop in a magnetic field

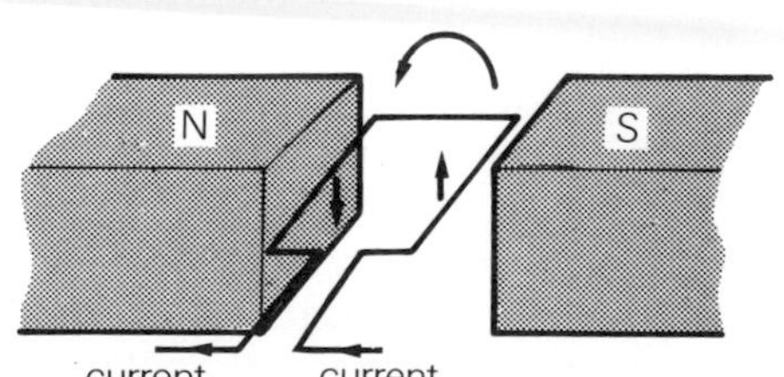

with the positive terminal. To do this we use a split-ring commutator, a simple example of which is shown in Figure 18.26.

The two halves of the split ring are connected to the opposite sides of the loop. Contact between the power supply and the halves of the split ring are made through the carbon brushes (carbon is a conductor), which are held against the commutator by springs. The loop and the commutator rotate, but the brushes remain stationary. As the loop rotates, the part of the loop on the right is always in contact with the negative terminal, while that on the left is always in contact with the positive terminal.

In practice a motor with a single loop would not be very effective: a coil of many turns increases the turning force of the motor, and makes it more useful.

Figure 18.27 shows a model electric motor constructed by winding a coil around a wooden core. The split ring consists simply of the two ends of the wire of the coil stripped of their insulation and placed in contact with the bared ends of wire from the power supply.

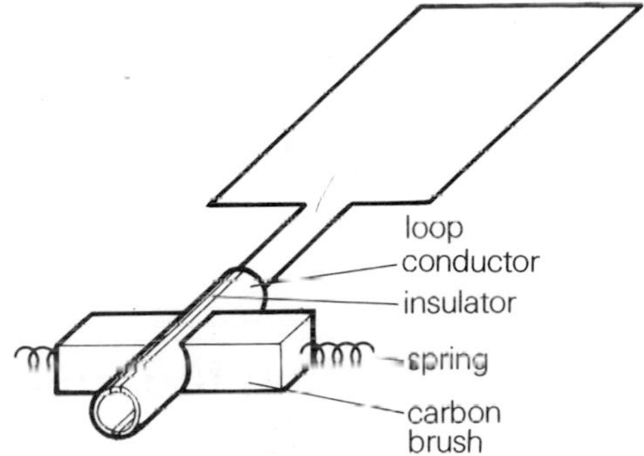

Figure 18.26 Split-ring commutator

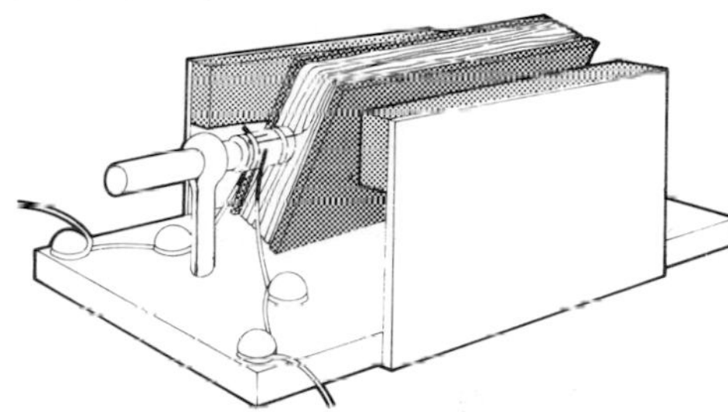
Figure 18.27 Model motor

Figure 18.28 Model moving-coil meter

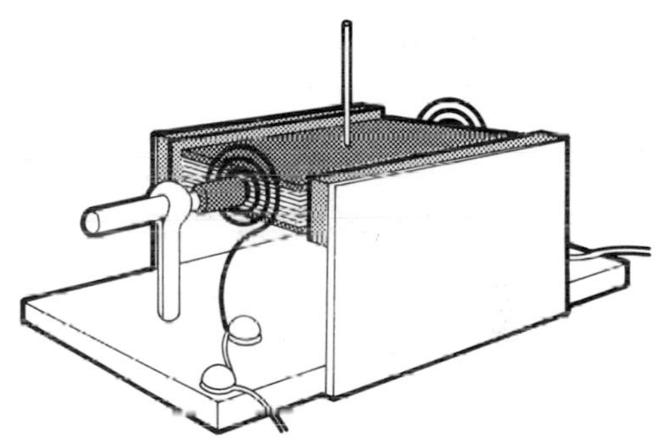

18.9 Moving-coil meter

In the construction of an electric motor we were concerned with arranging the coil and its connections so that the coil rotated continuously when a current flowed through it. The forces that turned the coil were produced as a result of a current flowing in the magnetic field.

If we arrange our coil so that it is restrained by a spring (Figure 18.28), the turning forces will move the coil against the spring. When a current passes through the coil, it turns against the spring. If the current increases, the coil turns further. If a pointer is attached to the coil so that the pointer moves over a scale, we can calibrate the scale to measure the current in the coil. This is the principle on which the moving coil meter is based.

18.10 Full-scale deflection

The deflection of the pointer on a moving coil depends on the current flowing through the coil. The current which will produce a full-scale deflection is called the full-scale deflection current I_{fsd} for the basic meter.

As the coil of a meter consists of a length of wire it has a resistance which we shall call r. This resistance is usually referred to as the resistance of the meter and it does not change with the current through the meter.

Whenever a current I flows through a resistance R a potential difference V exists across it, and these three quantities are related by the equation

$$R = \frac{V}{I}$$

$$\Rightarrow V = I \times R.$$

The potential difference that exists across the meter when there is a full-scale deflection is called the full-scale deflection voltage V_{fsd}. It is related to I_{fsd} and r by the equation

$$V_{fsd} = I_{fsd} \times r$$

Example 1

A meter has a full-scale deflection voltage of 100 mV and a full-scale deflection current of 10 mA. What is the meter resistance?

$V_{fsd} = 100 \text{ mV} = 0{\cdot}1 \text{ V}$

$I_{fsd} = 10 \text{ mA} = 0{\cdot}01 \text{ A}$

$V_{fsd} = I_{fsd} \times r$

$\Rightarrow 0{\cdot}1 = 0{\cdot}01 \times r$

$\Rightarrow \dfrac{0{\cdot}1}{0{\cdot}01} = r \qquad 10 = r$ **The meter resistance is 10Ω.**

18·11 Adapting the meter to measure higher currents

The conditions for full-scale deflection of the meter are dependent on the basic meter, Figure 18.29.

If we wish to use the meter to measure currents greater than I_{fsd} we must adapt the meter so that only a maximum current of I_{fsd} actually flows through the basic meter and the rest of the current bypasses the meter through a **shunt**, Figure 18.30.

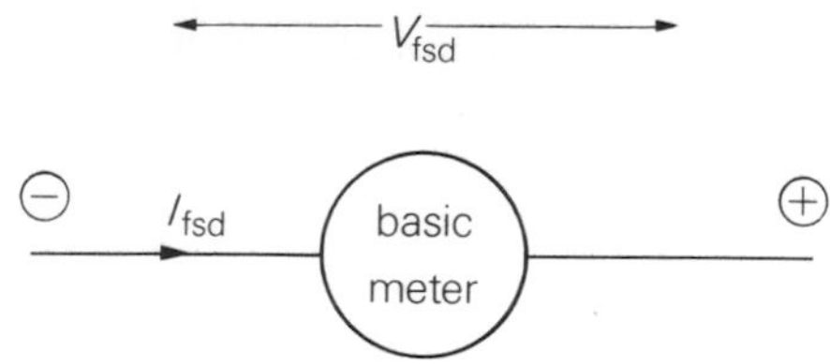

Figure 18.29 Meter characteristics

Calculation of the appropriate shunt value

Let the maximum current to be measured = I
Let the resistance of the shunt = R_s

The circuit of the adapted meter giving full-scale deflection is shown in Figure 18.31

Since the total current flowing into a junction is equal to the total current flowing out of a junction,

$I = I_{fsd}$ + current flowing through the shunt
$\Rightarrow I - I_{fsd}$ = current flowing through the shunt.

For full-scale deflection, the potential difference across the basic meter is V_{fsd}.

The potential difference across two resistors in parallel is the same, so the potential difference across the shunt is V_{fsd}.

Using the equation $V = I \times R$
p.d. across the shunt = current through the shunt × shunt resistance

$$V_{fsd} = (I - I_{fsd}) \times R_s$$

$$\Rightarrow \frac{V_{fsd}}{I - I_{fsd}} = R_s$$

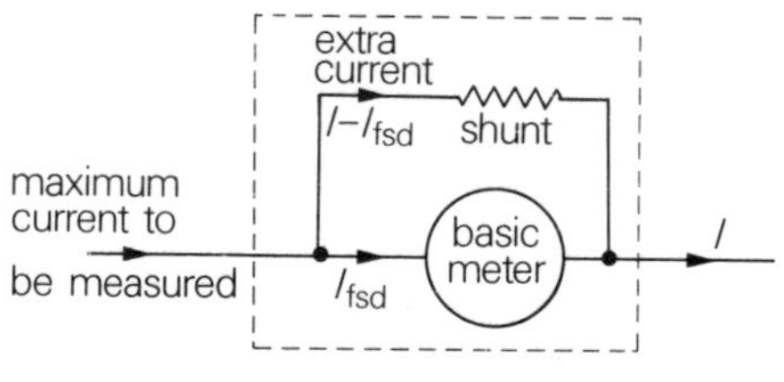

Figure 18.30 Meter and shunt

R_s, $I-I_{fsd}$, I, r, I_{fsd}, V_{fsd}

Figure 18.31 Calculation of shunt value

Example 2

A 0–10 mA meter has a full-scale deflection when the potential difference across it is 100 mV. How would you adapt the meter to read 0–1 A?

For full-scale deflection:

p.d. across the basic meter = 100 mV = 0·1 V
current through the basic meter = 10 mA = 0·01 A
total current through the meter and the shunt = 1 A
current through the shunt = 1 − 0·01 = 0·99

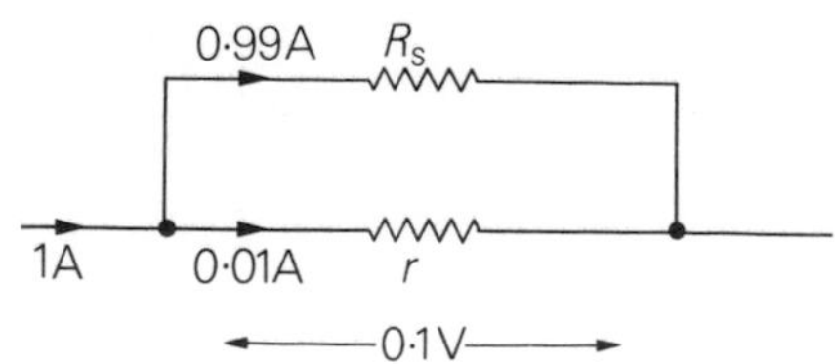

Using $V = I \times R$

$0{\cdot}1 = 0{\cdot}99 \times R_s$

$\Rightarrow \dfrac{0{\cdot}1}{0{\cdot}99} = R_s$

$\Rightarrow 0{\cdot}10 = R_s$

The resistance of the shunt is 0·10 Ω.

18·12 Adapting the meter to measure higher p.d.'s

Again it is important to remember that, for full-scale deflection of the basic meter, we have the conditions shown in Figure 18.29.

If we wish to use the meter to measure potential differences greater than V_{fsd} we must adapt the meter so that, when the maximum potential difference is across the adapted meter, the potential difference across the basic meter is V_{fsd}.

The meter is adapted by connecting a resistor of appropriate value in series with the basic meter so that the excess potential difference is dropped across this resistor, Figure 18.32. Such a resistor is called a **multiplier**.

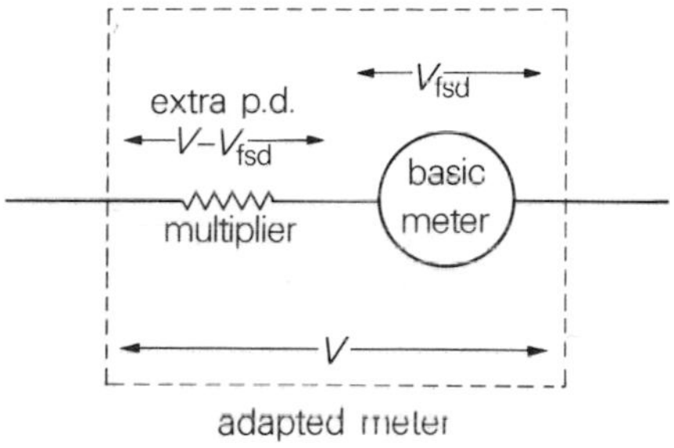

extra p.d. $V-V_{fsd}$ · V_{fsd} · multiplier · basic meter · V · adapted meter

Figure 18.32 Meter with multiplier

Calculation of the appropriate multiplier value

Let the maximum potential difference to be measured $= V$
Let the resistance of the multiplier $= R_m$

The circuit of the adapted meter giving full-scale deflection is shown in Figure 18.33.

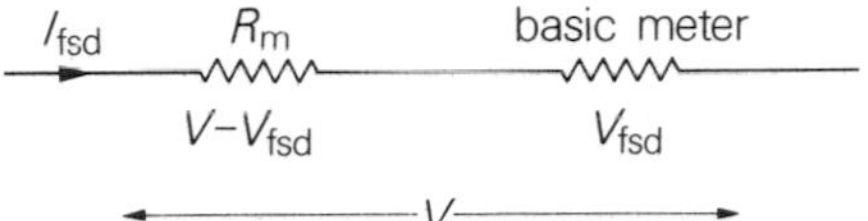

Figure 18.33 Calculation of multiplier value

For two resistors in series, the total potential difference across both of them is equal to the sum of the potential differences across each of them. Therefore

$V = V_{fsd}$ + potential difference across the multiplier
$\Rightarrow V - V_{fsd}$ = potential difference across the multiplier.

For full-scale deflection, the current through the basic meter is I_{fsd}. The current through two resistors in series is the same and so the current through the multiplier is I_{fsd}.
By the application of Ohm's Law,

p.d. across the multiplier = current through the multiplier × resistance of the multiplier.

$$V - V_{fsd} = I_{fsd} \times R_m$$
$$\Rightarrow \frac{V - V_{fsd}}{I_{fsd}} = R_m$$

Example 3

A 0–10 mA meter has a full-scale deflection when the potential difference across it is 100 mV. How would you adapt the meter to read 0–5 V?

For full-scale deflection:
current through the basic meter = 10 mA = 0·01 A
p.d. across the basic meter = 100 mV = 0·1 V
total p.d. across the meter and the multiplier = 5 V
⇒ p.d. across multiplier = 5 − 0·1 = 4·9 V

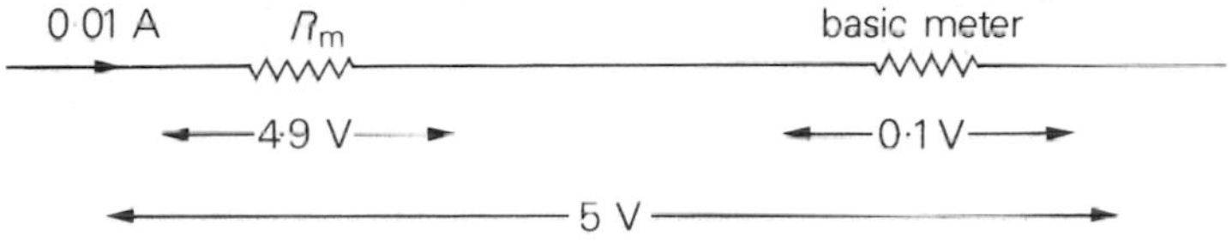

0·01 A · R_m · basic meter · 4·9 V · 0·1 V · 5 V

$$4{\cdot}9 = 0{\cdot}01 \times R_m$$
$$\Rightarrow \frac{4{\cdot}9}{0{\cdot}01} = R_m$$
$$\Rightarrow 490 = R_m$$

The resistance of the multiplier is 490 Ω.

18·13 Meters in circuits

An ammeter measures the current in a circuit. It is connected in series with the circuit so that the current to be measured passes through the ammeter. The resistance of the ammeter should be as low as possible so that, when it is connected into the circuit it does not increase the total resistance and thus reduce the current. When a meter is adapted with a shunt, which is a low resistance connected in parallel with the meter, the overall resistance of the meter is reduced.

A voltmeter measures the potential difference between two points in a circuit. Connecting the voltmeter between two points in a circuit means that it forms a parallel branch in the circuit. Current flows through this parallel branch, but this current should be kept very small by ensuring that the resistance of the voltmeter is as high as possible. The current between the two points in the main circuit then remains practically unchanged. When a multiplier (which is a high resistance) is connected in series with the meter, the overall resistance of the meter is increased.

A good measuring instrument should have little effect on the quantity that it measures. Figure 18.34 shows how connecting a voltmeter into a circuit changes the current by I_V, and how connecting an ammeter into a circuit changes the voltage by V_A.

A meter has a positive (red) terminal and a negative (black) terminal. The positive terminal must always be connected to the positive side of the circuit and the negative terminal to the negative side of the circuit.

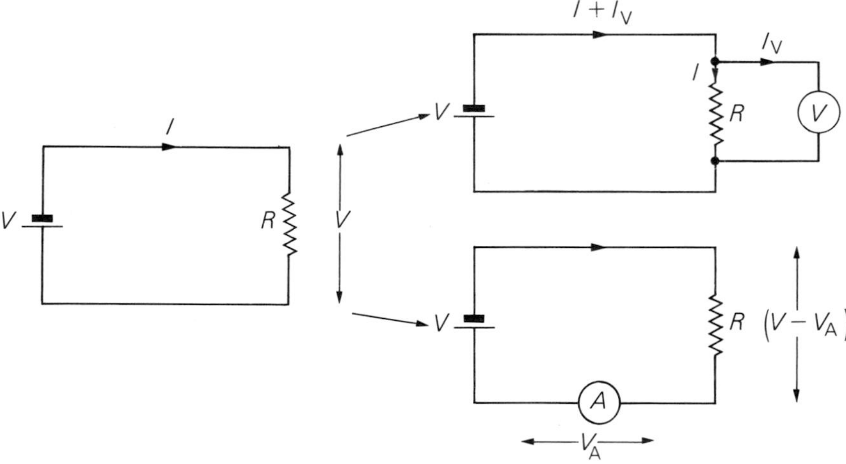

Figure 18.34 Effect of meters on circuits

Example 4

Show how you would connect the ammeter and voltmeter in the circuit to measure the current through the lamp and the potential difference across it.

The ammeter is connected in series with the lamp and the voltmeter in parallel with it. Note that the positive terminal of each meter is nearest to the positive terminal of the supply.

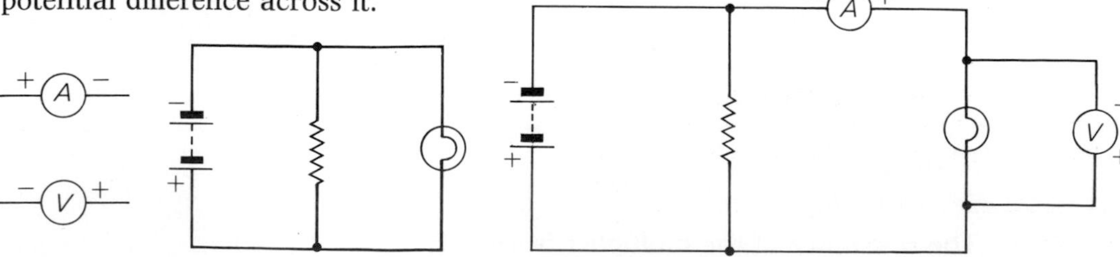

Summary

The north pole of a magnet is the north-seeking pole; the south pole of a magnet is the south-seeking pole.

Like poles repel and unlike poles attract.

The magnetic field about a straight current-carrying wire is circular with its centre at the wire.
The direction of the field lines is given by the left-hand rule: if the left hand is clenched with the thumb extended, the thumb points in the direction of the current and the fingers curl in the direction of the magnetic field.

The magnetic field of a current-carrying solenoid is similar to that of a bar magnet.
The north pole of a current-carrying solenoid is given by the left-hand rule for a solenoid: if the left hand is clenched with the thumb extended, the thumb points to the north pole and the fingers curl in the direction of the current.

An electromagnet is a solenoid with a soft iron core.
A relay is an electromagnet used to operate a switch in a second separate circuit.

A straight current-carrying wire at right angles to a magnetic field experiences a force at right angles to both the wire and the field.
The direction of the movement that this force tends to produce is given by the right-hand motor rule: if the right hand is held with the thumb and first two fingers at right angles to each other, the first finger points in the direction of the field; the second finger points in the direction of the current; and the thumb points in the direction of the motion.

An electric motor uses the force on a current-carrying conductor to produce rotation. Contact to the supply is made through a split-ring commutator.

A moving-coil meter is similar to a motor, but the coil rotates against the tension of a spring. The larger the current, the greater is the angle of rotation.

At full-scale deflection, the basic meter resistance r of a moving-coil meter is given by $V_{fsd} = I_{fsd} \times r$.

A moving-coil meter may be adapted to measure higher currents by connecting a shunt resistor of the appropriate value in parallel with it.

A moving-coil meter may be adapted to measure higher potential differences by connecting a multiplier resistor of the appropriate value in series with it.

An ammeter is connected in series in a circuit. A voltmeter is connected in parallel in a circuit.

A good ammeter has a low resistance. A good voltmeter has a high resistance.

Problems

1 Describe how you would use a bar magnet to separate iron filings from sand.

2 If you had a mixture of powdered iron, nickel and carbon, which one could you separate out? Explain your answer.

3 If you had an unmarked magnet, describe how you could discover which was the south pole. You are not allowed to use another magnet.

4 Draw the magnetic field patterns in the following cases.
 4·1 a single bar magnet;
 4·2 two south poles near to each other;
 4·3 a north pole and a south pole when they are near to each other.

5 Show the symbol for the cross section of a wire with the current flowing towards you, and draw in the magnetic field pattern around the wire. Indicate the direction of the magnetic field lines.

6 When magnetic field lines are used to describe a magnetic field pattern, how do they indicate the strength of the field? What determines the direction of the field lines?

7 A meter has the following characteristics:
Full-scale deflection current = 10 mA.
Full-scale deflection voltage = 100 mV.
Show how you would adapt the meter to measure over the following ranges (give the values of any components used, and show how they are connected):
 7.1 0–100 mA.
 7.2 0–1 V.

8 Draw a diagram of the cross section of a current-carrying solenoid and draw the magnetic field pattern. Mark the end of the solenoid which acts in the same way as the north pole of a bar magnet.

9 Copy the three diagrams of the cross section of current-carrying wires in magnetic fields, and show in each case the movement (if any) that is produced if the wires are free to move.

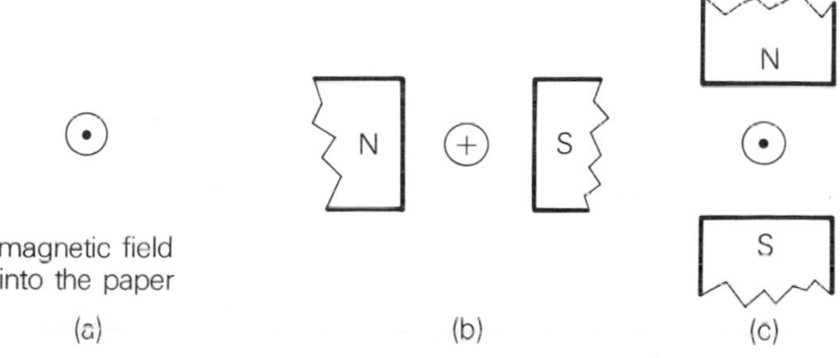

19 Electromagnetic induction

In the last chapter we saw that electrical energy could be changed into kinetic energy. We shall now investigate whether the reverse is possible.

19.1 Production of current

When current flows through a straight wire at right angles to a magnetic field, the wire tends to move in the direction indicated by the right-hand motor rule. When a bar magnet is brought up to a current-carrying solenoid, the solenoid is attracted or repelled. These are two cases in which the magnetic field produced by a current flowing in a conductor interacts with another magnetic field, tending to produce movement. We are now going to consider what happens when we move a conductor in a magnetic field.

If we connect a solenoid to a centre-zero milliammeter and bring the solenoid up to a magnet, we obtain a deflection on the milliammeter while the solenoid is moving towards the magnet. When the solenoid and magnet are stationary, there is no deflection on the milliammeter. When we move the solenoid away from the magnet a deflection on the milliammeter is again obtained, but in the opposite direction, Figure 19.1.

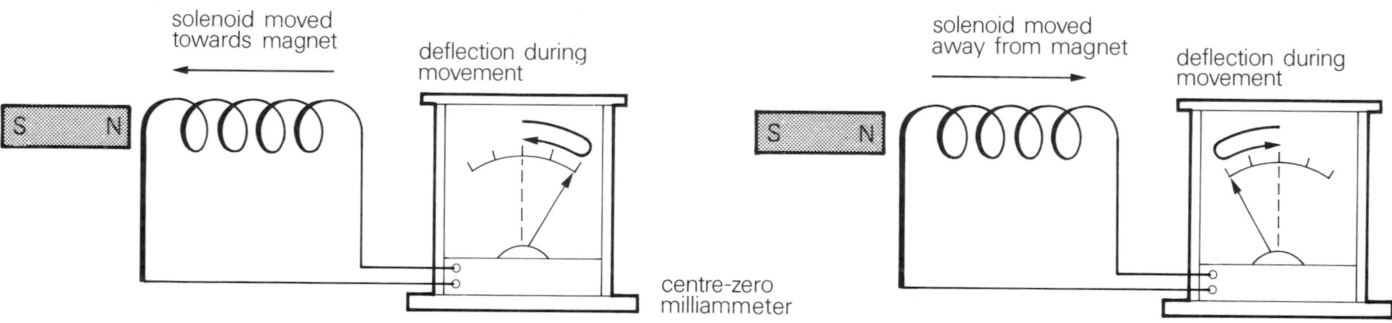

Figure 19.1 Induction in a solenoid

The same results are obtained if we keep the coil stationary and move the magnet towards and away from the coil.

Using this apparatus we observe that

1. A current flows only while there is a changing magnetic field around the solenoid.
2. Movements of the magnet towards and away from the solenoid produce deflections in opposite directions.
3. The direction of current flow depends on which pole of the magnet is used; a change of poles produces reversal of current direction.

In order to produce a current flow, we must have a potential difference. The movement of the solenoid relative to the magnetic field results in a potential difference called an electromotive force (e.m.f.). This process is an example of **electromagnetic induction**. An e.m.f. produced in this way is called an **induced electromotive force**.

The induction of an e.m.f. by the movement of a bar magnet relative to a solenoid was one of the first discoveries made by Michael Faraday (Figure 19.2), one of the most distinguished scientists of the

Figure 19.2 Michael Faraday

nineteenth century. Figure 19.3 shows a photograph of the solenoid and magnet used by Michael Faraday.

It was Faraday who, in 1831, first discovered electromagnetic induction. This discovery must rate as one of the most important, as it forms, among its many applications, the basis of our whole system of electricity generation and transmission.

If a straight wire is moved through a magnetic field (Figure 19.4), an e.m.f. is induced across the ends of the wire, and we observe that a current flows through a microammeter connected to the ends of the wire.

Again, a flow of current occurs only when the conductor is moving in the magnetic field. As in the case of a current flowing through a conductor in a magnetic field and producing movement, the current, magnetic field and movement are all at right angles to each other.

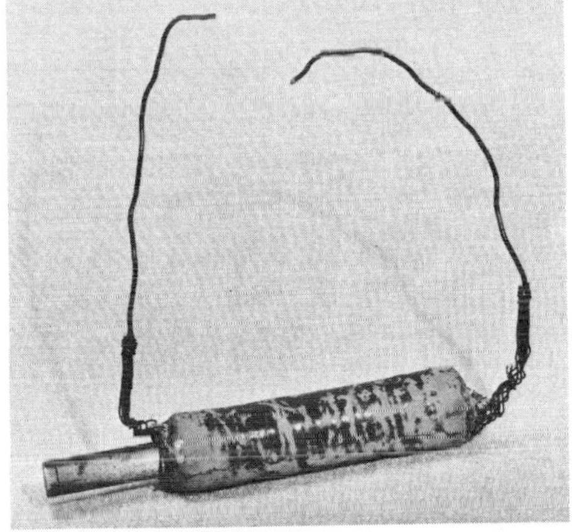

Figure 19.3 Faraday's solenoid and magnet

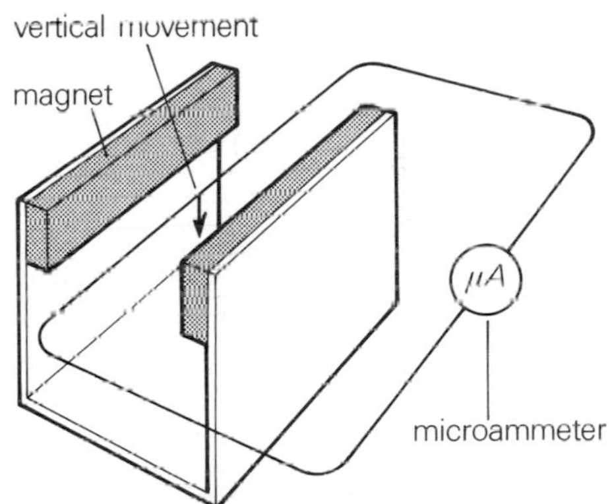

Figure 19.4 Induction in a straight wire

19.2 Magnitude of induced e.m.f.

The following factors affect the magnitude of the induced e.m.f.

1. **Length of conductor** We have already seen that we can induce an e.m.f. across the ends of a solenoid by moving a magnet towards the solenoid, and that the e.m.f. will drive a current through a milliammeter connected to the ends of the solenoid. The size of deflection on the milliammeter is a measure of the size of the induced e.m.f. If we repeat the experiment using the same magnet moving in the same way towards a solenoid with a greater number of turns, we get a larger deflection on the meter, showing that the induced e.m.f. is greater. This is because the moving magnetic field is cutting through more turns of the conductor.
2. **Strength of the magnetic field** If we repeat the experiment using a stronger magnet, the induced e.m.f. is greater. When the changing magnetic field is stronger, the induced e.m.f. is greater.
3. **Speed of movement** In our experiments on induction, we produced an induced e.m.f. by moving a conductor in a magnetic field or by moving a magnetic field past a conductor. When we increase the speed of movement in these experiments, we obtain a larger meter deflection, showing that the induced current and hence the induced e.m.f. is greater.
4. **Direction of movement** In order to induce an e.m.f. across a straight wire we have moved the wire at right angles to the magnetic field. Figure 19.5 shows a wire being moved an equal distance

a) at right angles to the magnetic field;
b) at an angle to the magnetic field;
c) parallel to the magnetic field.

From this diagram we can see that, for equal distance travelled, the wire moving at right angles to the magnetic field cuts through the greatest number of field lines, while that moving parallel to the field cuts through no field lines. As a result of this, the induced e.m.f. is greatest when the movement is at right angles to the magnetic field and is zero when the movement is parallel to the magnetic field. For angles in between, the nearer the angle between the direction of movement and the magnetic field is to 90°, the greater is the induced e.m.f.

Figure 19.5

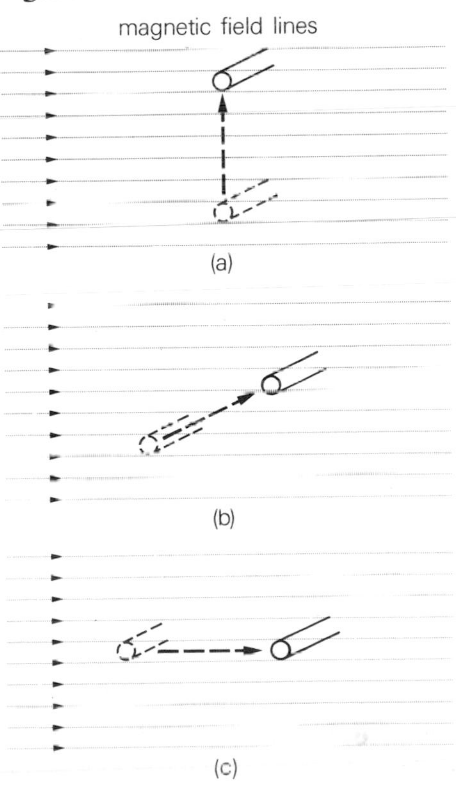

19.3 Direction of induced e.m.f.

If we suspend a coil so that it can swing freely, and induce a current in it by moving a magnet to and fro, we obtain the results shown in Figure 19.6.

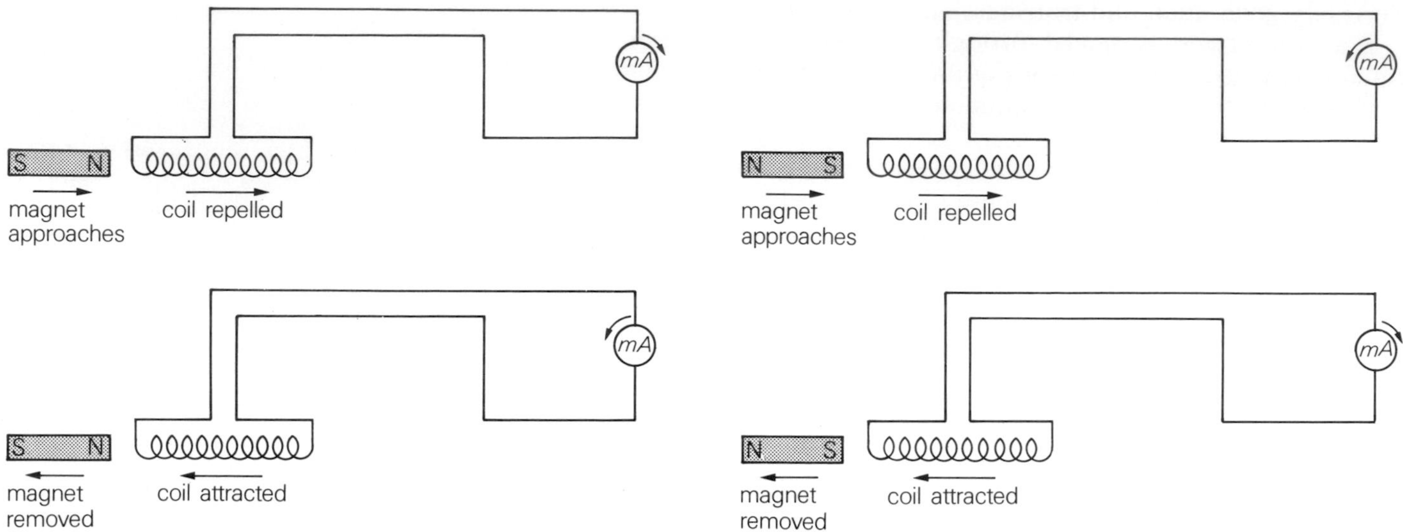

Figure 19.6

We see from these results, that, whenever the magnet is approaching the coil, there is repulsion and, whenever the magnet is moving away from the coil, there is attraction. The induced current flows in such a way as to cause repulsion of an approaching magnet and attraction of a departing magnet. This illustrates **Lenz's Law** which states:

'The direction of the induced current is such that it opposes the change producing it.'

Example 1

Figure 19.7 shows a straight wire being moved down through a magnetic field. Will the induced current flow from A to B or from B to A?

Figure 19.7

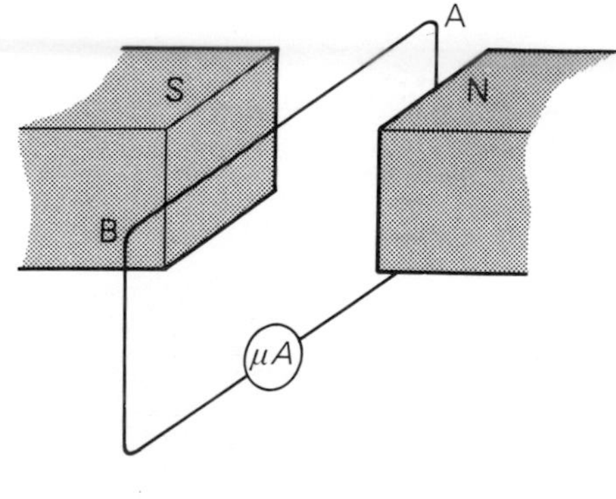

The wire is being moved down through the magnetic field. Therefore, by Lenz's Law, we know that the induced current will flow in such a way as to oppose the downward movement. This means that the current direction is such as to try to make the wire move up through the magnetic field.

Using the right-hand motor rule (Figure 19.8), if we point the first finger in the direction of the field and the thumb in the direction of the motion that the induced current tends to produce, the second finger shows that the direction of the induced current is from A to B.

Figure 19.8

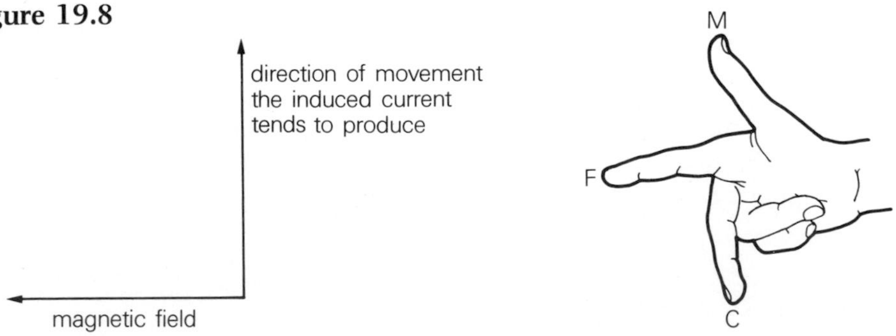

direction of movement the induced current tends to produce
magnetic field
M
F
C

Example 2

Figure 19.9 shows a resistor XY connected across a solenoid. When a magnet is moved towards the solenoid, an e.m.f. is induced and this causes a current to flow through the resistor.
In which direction does the current flow through the resistor?

Figure 19.9

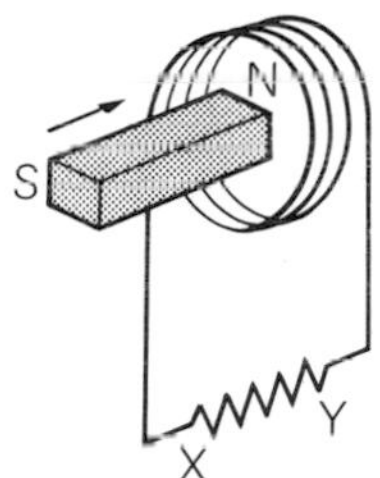

When the magnet is moved towards the coil, an induced current will flow in such a way as to oppose the motion (Lenz's Law).

Therefore the induced current will flow so that the end of the solenoid nearest the magnet will act as a north pole.

Using the left-hand rule, if we point the thumb towards the north pole of the solenoid, the fingers curl round in the direction of the current, Figure 19.10.

Thus the current flows in a clockwise direction in the end of the coil facing the magnet. Since the current must flow throughout the complete circuit, it must be flowing from Y to X through the resistor, and Y must be more negative than X.

Figure 19.10

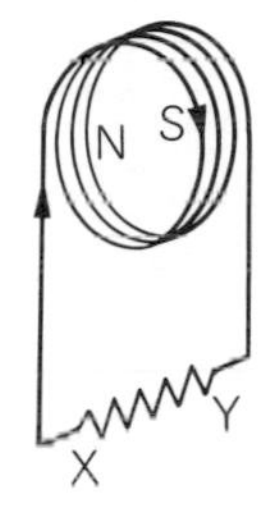

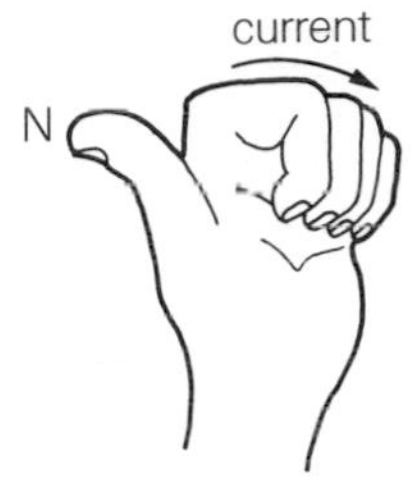

19.4 Generation of electricity

Generation of d.c.

If we take a model electric motor, attach it to a milliammeter and turn the coil by hand, a reading is obtained on the milliammeter, Figure 19.11.

When the motor is used in reverse in this way, it is acting as a **dynamo**. The electric motor converts electrical energy into kinetic energy. The dynamo converts kinetic energy into electrical energy.

Since the connections to the coil in the dynamo are through a split-ring commutator as in the motor (Figure 19.12), one terminal is always in contact with the side of the coil that is moving down through the magnetic field, and the other terminal is always in contact with the side of the coil that is moving up.

This produces an e.m.f. across the two terminals that varies but that is always in the same direction. An e.m.f. that is always in the same direction is called d.c. The term d.c. actually stands for 'direct current' but the term is also used to describe a potential difference.

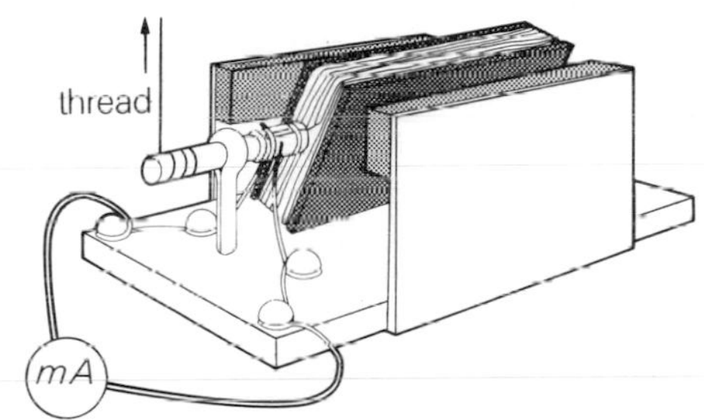

Figure 19.11 Model dynamo

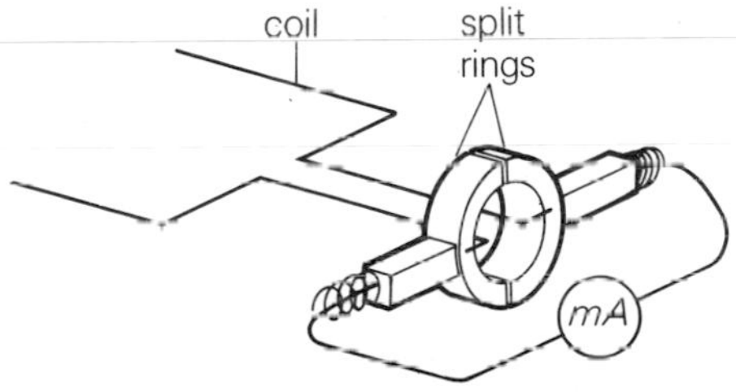

Figure 19.12 Split-ring commutator

Generation of a.c.

Another type of connection between a rotating coil and a circuit uses slip rings, Figure 19.13.

With slip rings, one terminal is always connected to one side of the coil, whether that side is moving up or down. This means that the e.m.f. between the terminals completely changes direction every half revolution of the coil.

A generator with slip rings produces alternating current (a.c.) and it is called an alternator or a.c. generator.

Figure 19.13 Slip rings

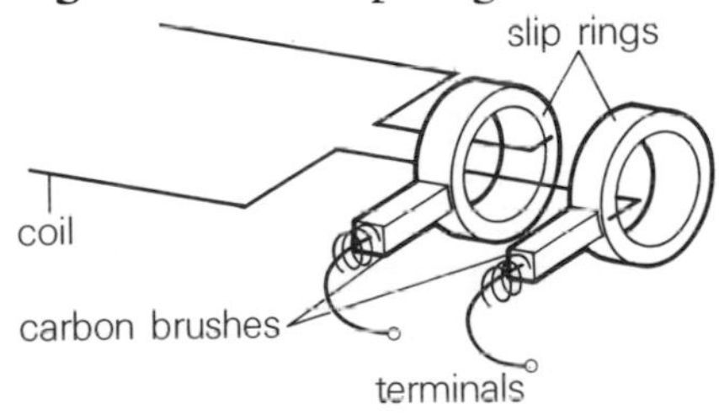

Generators exist in a wide range of sizes, from the bicycle dynamo used to supply d.c. to bicycle lights, to the large a.c. generators used to supply electricity to the national grid. These a.c. generators are usually driven by steam turbine. The steam is produced using heat from burning coal, gas or oil, or from atomic power. Figure 19.14 shows a photograph of an a.c. generator at a power station.

Figure 19.14 a.c. generator at a power station

19.5 C.R.O. display of generated e.m.f.

We can use an oscilloscope to give a trace which shows how the e.m.f. varies with time. The shape of the trace is the same as that for a graph of e.m.f. against time: e.m.f. is on the vertical axis and time on the horizontal axis.

Figure 19.15 shows cathode ray oscilloscope traces of the output of a dynamo and an alternator.

In both cases the value of the e.m.f. varies between 0 V and a maximum value, known as the peak e.m.f., but in the case of the dynamo the e.m.f. is always in one direction, while in the case of the alternator it continually changes direction.

We can deduce how the e.m.f. varies between 0 V and the peak e.m.f. if we consider the coil as it rotates. Figure 19.16 shows the cross section of a loop which is being rotated in a magnetic field and the direction of flow of the induced current at each stage.

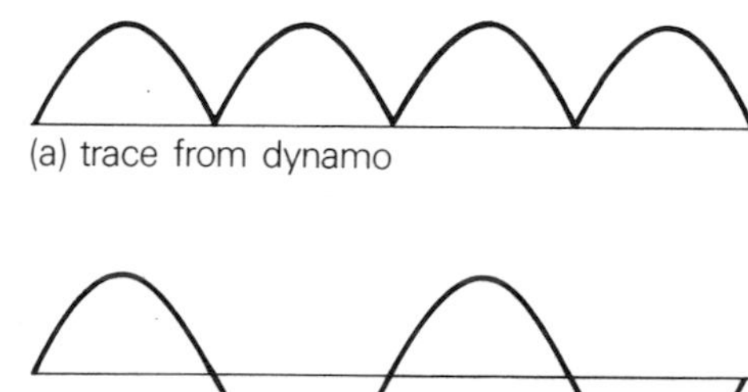

Figure 19.15

Figure 19.16 Loop rotating in a magnetic field

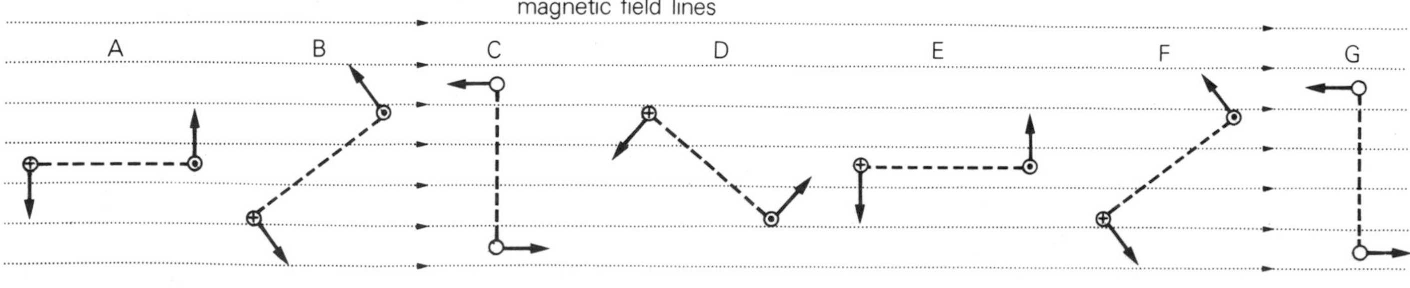

Let us consider the magnitude of the induced e.m.f. at each of the stages shown in Figure 19.16.

A and E The sides of the loop are moving at right angles to the magnetic field and the induced e.m.f. is at a maximum.

C and G The sides of the loop are moving parallel to (are not cutting through) the magnetic field, and the induced e.m.f. is zero.

Figure 19.17 relates these stages to the graph of the induced e.m.f. against time across the terminals of loops with each type of connection.

Mains electricity is of the wave form illustrated for an a.c. generator output.

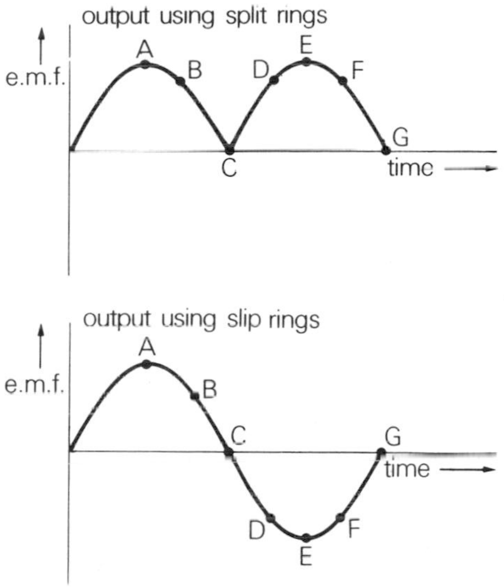

Figure 19.17 Variation of induced e.m.f.

19.6 Microphone

In Chapter 18 we saw how a loudspeaker converted a varying electric current into vibrations of the cone of the loudspeaker, and these vibrations produced sound waves in the surrounding medium. A microphone reverses this process. The sound vibrations received by the microphone are converted into an alternating electric current of the same frequency as the sound.

One common type of microphone is the moving-coil microphone. Figure 19.18 shows a cross section of one type of moving-coil microphone.

When the diaphragm is vibrated by the sound waves, the coil vibrates in the magnetic field between the central north pole and the surrounding south pole of the magnet. When the coil is moving in, the induced current will be in one direction and when it is moving out the induced current will be in the opposite direction. Thus the frequency of the output current from the microphone will be the same as that of the vibrations of the coil. In this way the sound signal is converted into an electrical output of the same frequency.

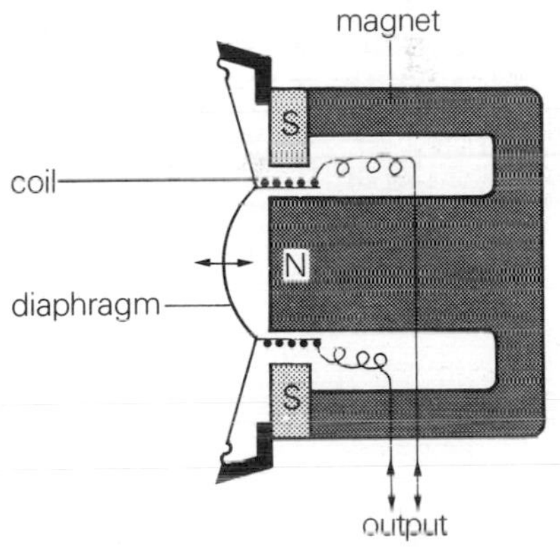

Figure 19.18 Moving-coil microphone

19.7 Record player

The signal that is recorded on a record consists of a groove cut in the surface of the record. When the stylus (or needle) runs along the groove, the groove makes the stylus vibrate, Figure 19.19.

These vibrations of the stylus are then converted into an electrical signal. In the case of a magnetic pick-up, there are small coils attached to the stylus and these coils vibrate between the poles of the permanent magnet, inducing an a.c. current of the same frequency as the vibrations of the stylus. This induced current is the signal that is then amplified and played through a loudspeaker.

Figure 19.19 Magnetic pick-up

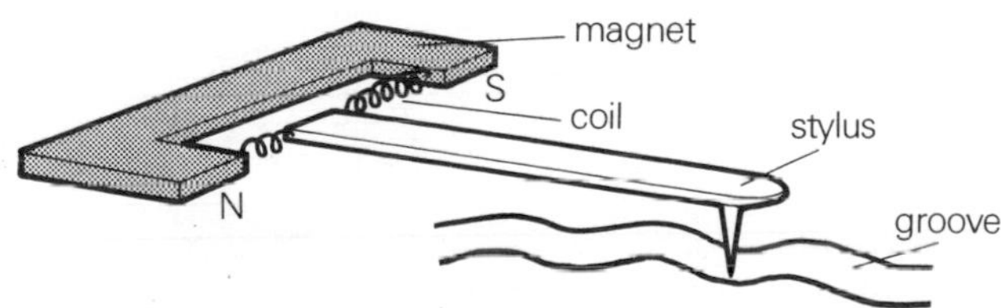

In the case of stereophonic records, two separate signals are both recorded in a single groove. This groove makes the stylus vibrate both horizontally and vertically, and two sets of coils pick up the signals from these two vibrations. These two signals are amplified separately and fed into separate loudspeakers.

19.8 Tape recorder

The tape recorder employs the principles of electromagnetism to record a signal on a magnetic tape. The tape is made of plastic which contains very fine particles of a magnetic powder. The powder is usually an oxide of iron or chromium. The pick-up head of the tape recorder is a specially shaped iron core through a coil, Figure 19.20. When the tape passes the end of the core, recorded variations in the magnetic field of the powder on the tape produce a varying magnetic field between the ends of the core. This induces a varying e.m.f. across the coil. This signal is then amplified and played through the loudspeaker.

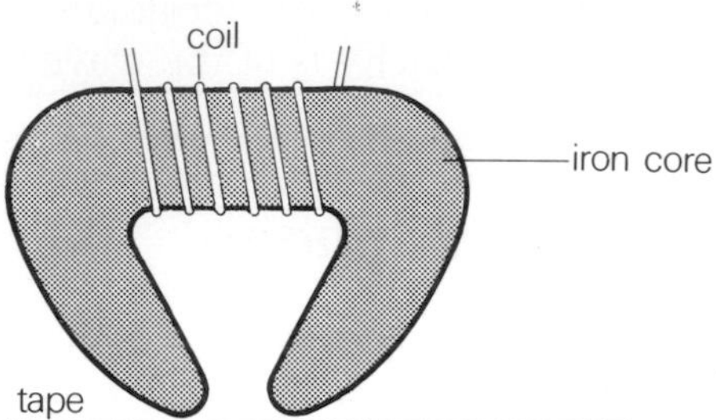

Figure 19.20 Tape recorder pick-up head

A recording is made on the tape by the reverse process. An amplified signal from a microphone, record player or radio, is passed through the coil in the head. The varying current in the head magnetizes the magnetic powder in the tape that is moving past the head. Since the magnetizing current varies, it produces variations in the recorded magnetic field on the tape.

Figure 19.21 illustrates the steps in the recording and play-back of a tape recorder.

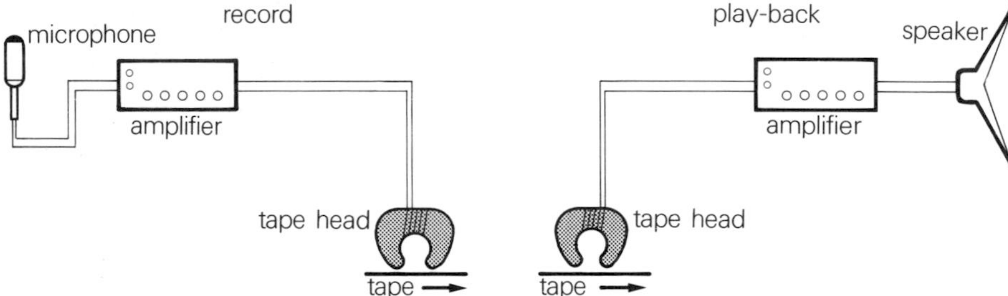

Figure 19.21 Recording and play-back in a tape recorder

19.9 Transformer

In 1831 Michael Faraday first demonstrated the principle on which the transformer operates. He wound a **primary coil** of many turns of copper wire around half of an iron ring, and connected it in series to a power supply and a switch. This formed the primary circuit. On the other half of the iron ring he wound a **secondary coil**, also consisting of many turns of copper wire. The ends of the secondary coil were connected by a long piece of wire. This coil and connecting wire made up the secondary circuit. Figure 19.22 shows a photograph of the original iron ring used by Faraday, and a diagram of the circuit that he used.

Figure 19.22

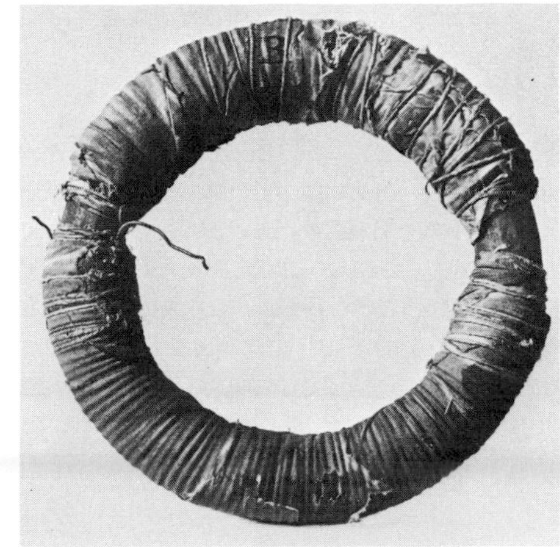

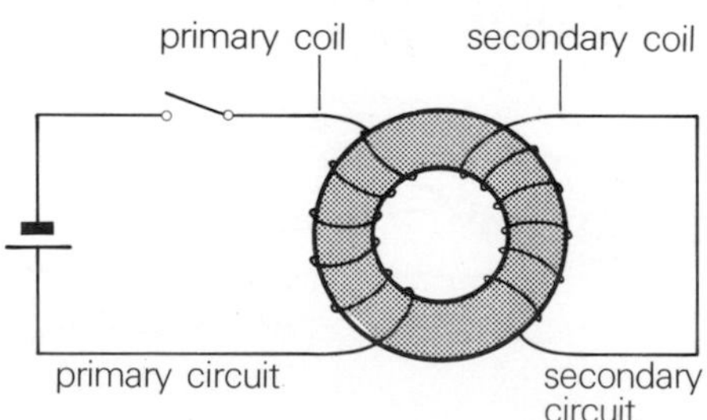

Faraday showed that there was a momentary flow of current in the secondary circuit at the instant when the switch was closed, and another momentary flow of current at the instant when the switch was opened. When the current in the primary coil was steady, no current flowed in the secondary circuit.

In this experiment Faraday had demonstrated the transfer of electrical energy from the primary circuit to the secondary circuit. He had built the first simple transformer. A transformer consists of two separate coils wound on a core. The symbol for a transformer is shown in Figure 19.23. In this symbol, the vertical line represents the core.

We can easily repeat Faraday's experiment using the circuit shown in Figure 19.24.

When we close the switch in the primary circuit, the meter gives a kick in one direction. When we open the switch, the meter gives a kick in the opposite direction. Again, the current only flows momentarily at the instant of opening and closing the switch.

When the switch is closed, the primary circuit is completed and there is a rapid build-up of the current through the primary coil from zero to its maximum value. This produces a rapid growth of magnetic field in the primary coil. The secondary coil is in this rapidly changing magnetic field and so an e.m.f. is induced across it. When the current reaches its maximum value in the primary circuit, the steady current produces a steady magnetic field. Since an e.m.f. is only induced when a conductor is in a changing magnetic field, no e.m.f. is induced across the secondary coil when the primary current is constant. When the switch is opened, the rapid decrease in current in the primary coil causes a rapid collapse of the magnetic field which results in another induced e.m.f. across the secondary coil.

The iron core increases the strength of the magnetic field through the coils when a current flows.

With a d.c. supply in the primary circuit, we produce a changing magnetic field by opening and closing a switch. A more convenient way of producing a changing current, and hence a changing magnetic field, is to use an a.c. supply in the primary circuit. We can study the induced e.m.f. by connecting the secondary coil to a cathode ray oscilloscope, Figure 19.25.

Using this apparatus we find that an alternating current in the primary coil produces an induced e.m.f. in the secondary coil of the same frequency as the supply.

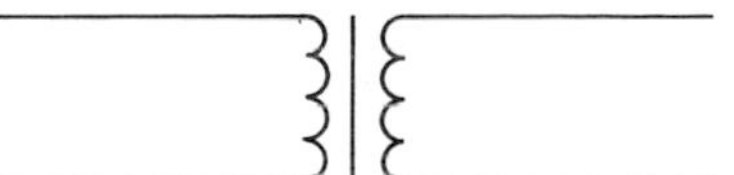

Figure 19.23 Symbol for transformer with core

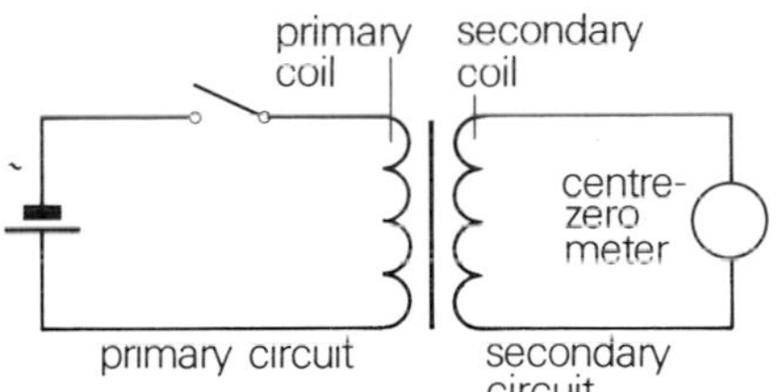

Figure 19.24

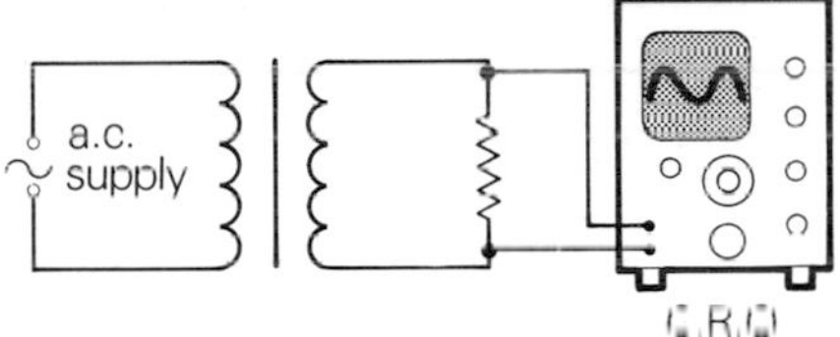

Figure 19.25 Transformer output

19.10 p.d. in a transformer

We use the circuit shown in Figure 19.26 to investigate the relationship between the potential differences across the primary and secondary coils. The a.c. voltmeter used is one with a number of ranges.

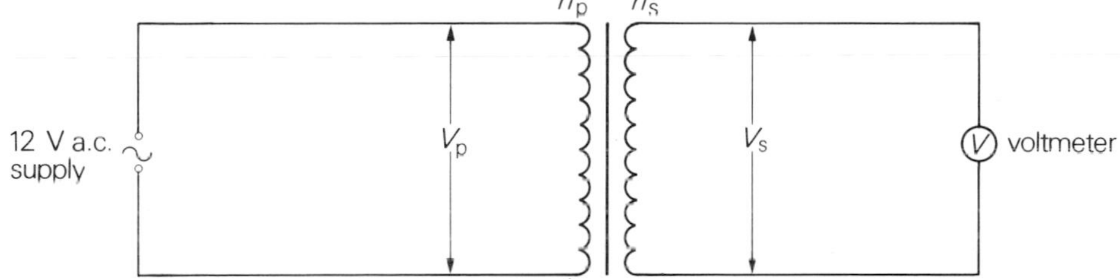

Figure 19.26 Potential differences in a transformer

n_p = number of turns in primary coil
n_s = number of turns in secondary coil
V_p = p.d. across primary coil
V_s = p.d. across secondary coil.

We keep the primary circuit the same throughout the experiment. Various secondary coils of different numbers of turns are used and the potential difference across the secondary coil is measured in each case. A set of sample results is shown in Table 1.

number of turns in primary coil n_p	*number of turns in secondary coil* n_s	*p.d. across primary coil* V_p	*p.d. across secondary coil* V_s
500	125	12 V	3 V
500	250	12 V	6 V
500	500	12 V	12 V
500	2 500	12 V	60 V
500	5 000	12 V	120 V

Table 1

These results illustrate the following facts about a transformer with an a.c. supply.

1. When the number of turns in the secondary coil equals the number of turns in the primary coil, the potential difference across the primary coil is equal to the potential difference across the secondary coil.

$n_s = n_p \Rightarrow V_s = V_p$

2. When the number of turns in the secondary coil is greater than the number of turns in the primary coil, the potential difference across the secondary coil is greater than the potential difference across the primary coil.

$n_s > n_p \Rightarrow V_s > V_p$

Such a transformer is called a **step-up** transformer.

3. When the number of turns in the secondary coil is less than the number of turns in the primary coil, the potential difference across the secondary coil is less than the potential difference across the primary coil.

$n_s < n_p \Rightarrow V_s < V_p$

Such a transformer is called a **step-down** transformer.

The general rule illustrated by these results is $\frac{V_s}{V_p} = \frac{n_s}{n_p}$

19.11 Transformer currents and power

A transformer with an a.c. supply transfers energy from the primary circuit to the secondary circuit. In all transformers there are some energy losses, but we shall first consider an ideal transformer (a theoretical transformer) in which there is no loss of energy. This means that the power output is equal to the power input.
In an ideal transformer: power output = power input.

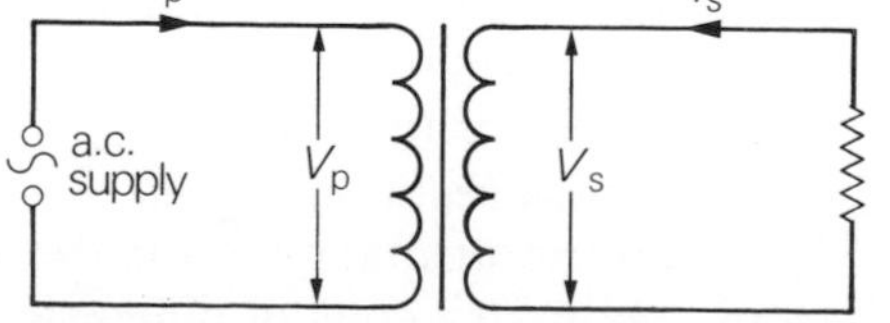

I_p = current in primary coil
I_s = current in secondary coil

For the transformer circuit shown in Figure 19.27

power input = $I_p \times V_p$
power output = $I_s \times V_s$

Figure 19.27 Currents in a transformer

If the transformer is an ideal transformer

$I_p \times V_p = I_s \times V_s$

$\Rightarrow \quad \frac{I_p}{I_s} = \frac{V_s}{V_p}$

Thus the ratio of the currents is the reciprocal of the ratio of the potential differences.

Example 3

A transformer steps up the potential difference from 12 V to 48 V. If the current flowing in the primary circuit is 2 A, what current is flowing in the secondary circuit, assuming that the transformer is an ideal transformer?

$V_p = 12$ V; $I_p = 2$ A

$V_s = 48$ V

$$\frac{I_p}{I_s} = \frac{V_s}{V_p}$$

$$\Rightarrow \quad \frac{2}{I_s} = \frac{48}{12}$$

$$\Rightarrow \quad 2 = \frac{48}{12} \times I_s$$

$$\Rightarrow 2 \times \frac{12}{48} = I_s$$

$$\Rightarrow \quad 0{\cdot}5 = I_s$$

Current in secondary circuit is 0·5 A.

A transformer which steps up the potential difference, steps down the current.

A transformer which steps down the potential difference, steps up the current.

Example 4

Figure 19.28 shows a circuit in which a transformer is being used to operate a 12 V, 24 W lamp from a 240 V a.c. supply.

Assuming that there is no power loss in the transformer, and that the lamp is operating at its correct rating, calculate

a) the number of turns in the secondary coil n_s;

b) the current in the secondary circuit I_s;

c) the current in the primary circuit I_p.

Figure 19.28

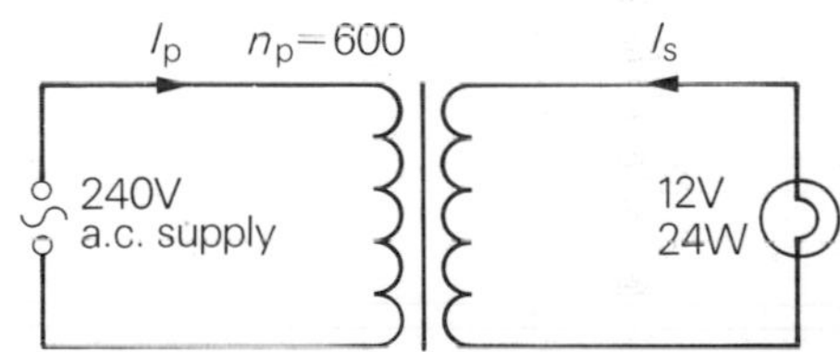

a) p.d. across the secondary coil V_s = p.d. across the lamp

p.d. across the lamp = 12 V

$V_s = 12$ V

p.d. across the primary coil $V_p = 240$ V

$$n_p = 600$$

$$\frac{n_s}{n_p} = \frac{V_s}{V_p}$$

$$\Rightarrow \quad \frac{n_s}{600} = \frac{12}{240}$$

$$\Rightarrow \quad n_s = \frac{12}{240} \times 600$$

$$\Rightarrow \quad n_s = 30$$

Number of turns in the secondary coil is 30.

b) If the lamp is operating at its correct rating 24 W, 12 V,

the power of the lamp = 24 W

the p.d. across the lamp = 12 V

power = current × potential difference

power of the lamp = (current through the lamp) × (p.d. across the lamp).

$$\Rightarrow \quad 24 = I_s \times 12$$

$$\Rightarrow \quad 2 = I_s$$

Current in the secondary circuit is 2 A

c) The power output of the transformer = power of the lamp = 24 W

power input = (current through the primary coil) × (p.d. across the primary coil)

power input = $I_p \times 240$

power input of the transformer = power output of the transformer

$\Rightarrow \quad 24 = I_p \times 240$

$\Rightarrow \frac{24}{240} = I_p$

$\Rightarrow \quad \frac{1}{10} = I_p$

$\Rightarrow \quad 0{\cdot}1 = I_p$

Current in the primary circuit is 0·1 A

19.12 Transformer energy losses

In a real transformer some energy is lost in the transformer, which means that the power output is less than the power input. There are several causes of power loss in the transformer and we shall consider the three most important causes and how they may be reduced.

1. Heating effect in the coils

When there is a current in a resistor, heat is produced. The power transformed in heating the resistor is given by

power = (current through resistor) × (p.d. across the resistor)
power = $I \times V$
$V = I \times R$ from Ohm's Law
power = $I \times I \times R = I^2R$

The coils are made of copper which is a good conductor, but the wire required for a coil of many turns has a great length and has an appreciable resistance. The 2 400 turn coil shown in Figure 19.29 is made of approximately 300 m of wire and has a resistance of 63 Ω.

When there is a current in the wire in the coil, power is transformed in heating the coil. This loss can be reduced if the coil is made of wire of lower resistance by using thicker wire.

Figure 19.29 2 400 turn coil

2. Eddy currents in the core

The soft iron core is a conductor and is in a changing magnetic field. This results in an induced e.m.f. in the core and this in turn produces currents in the core. Such currents are called **eddy currents**, Figure 19.30.

The eddy currents in the core produce heating of the core and result in further loss of power.

Eddy currents are reduced by using a laminated iron core. This consists of a number of thin sheets of iron glued together, with thin layers of insulation between them, Figure 19.31. Since there is no current through the layers of insulation, eddy currents are reduced.

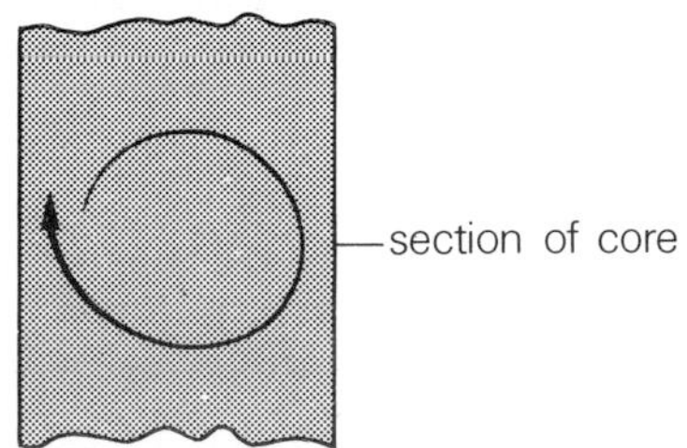

Figure 19.30 Eddy currents in the core

Figure 19.31 Laminated iron

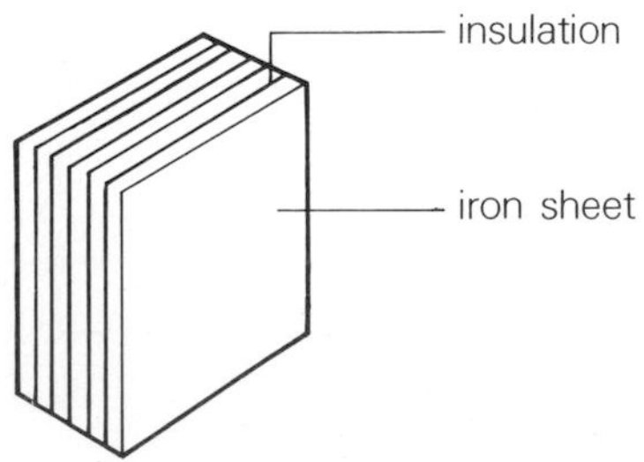

3. Magnetization and demagnetization of the core

The iron core is continually magnetized and demagnetized by the alternating current, and this transforms energy into heat in the core.

This energy loss is reduced by using soft iron for the core. Soft iron does not retain its magnetism and requires little energy to be magnetized and demagnetized.

Efficiency of a transformer

$$\text{percentage efficiency} = \frac{\text{useful power output}}{\text{power input}} \times 100$$

The efficiency of a transformer can be very high and it is possible to achieve an efficiency of 99%.

19.13 Use of transformers

The most important use of transformers is for stepping up and stepping down voltages. A single appliance such as a TV set has components which operate at different potential differences ranging from a few volts to several kilovolts. The TV set is operated from a 240 V mains supply and uses transformers to step the potential difference up or down for the individual components. The power supply which you use in the laboratory steps down the 240 V of the mains supply to the lower and safer potential difference that we require. By tapping the secondary coil at various points we can obtain different outputs from a single transformer. Figure 19.32 shows the circuit of such a transformer. The required potential difference is obtained by connecting between the 0 V terminal and the appropriate terminal.

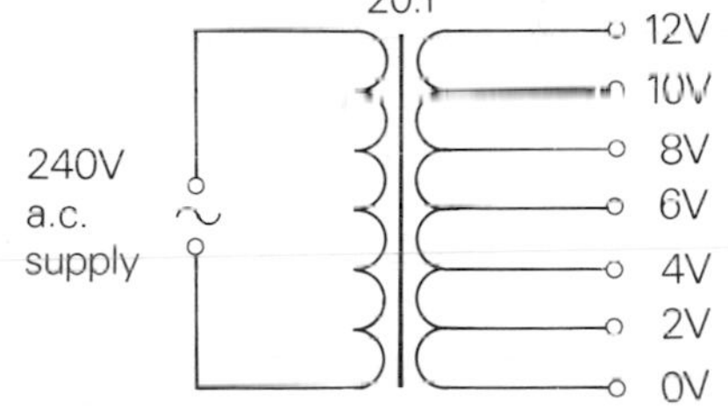

Figure 19.32 Tapped transformer

One very important use of transformers is in the National Grid by which electrical power is transmitted throughout the country. The power is transmitted through many miles of cable. The resistance of the cable is kept as low as possible, but it is still appreciable. As we have seen, power is lost by the heating effect of a current through a resistor. This power loss is given by

$$\text{power} = I^2R$$

where I = current through the resistor
R = resistance of the resistor.

Figure 19.33 shows an experiment to demonstrate how the power loss in transmission lines can be reduced by using transformers.

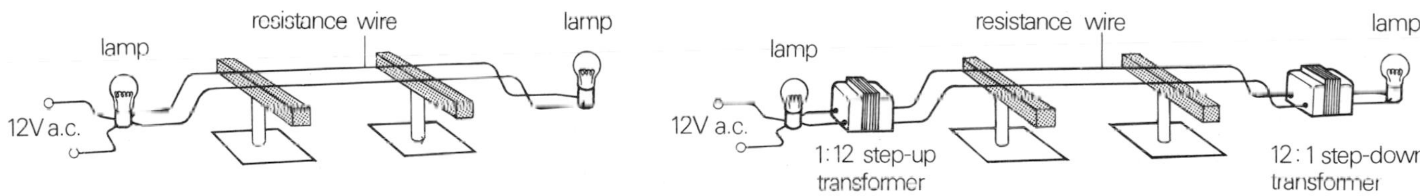

Figure 19.33

When power from the supply is transmitted directly along the model power lines, the lamp at the far end of the lines is less bright than that at the supply end. This is because there is a loss of power in the lines. However, when the supply potential difference is stepped up before transmission, and then stepped down at the far end of the lines, the lamp at the far end of the line is much brighter. When the power is transmitted at higher potential, there is less loss of power in the power line. This is done with an a.c. supply since a transformer will not operate with a d.c. supply.

The loss of power in a transmission line is given by

power loss = (current in line)^2 × (resistance of line).

The resistance of the line is constant, so the power loss is reduced by transmitting power at lower currents. This means stepping up the potential difference (with the resulting step down of current), and stepping down the potential difference to a safe level at the supply end. There is a practical limit to the potential difference that can be used, because a high potential requires very good insulation.

In the National Grid, power from generating stations is generated at 11 kV. This potential difference is stepped up for transmission. Different parts of the grid transmit at different potentials of 132 kV, 275 kV and 400 kV. Step-down transformers are then used to step down the potential difference in stages at the supply end to various potential differences used in industry and to the 240 V for the domestic supply.

Figure 19.34 shows two transformers, one at either end of the large size range that can be obtained.

Figure 19.34

Summary

When a conductor experiences a changing magnetic field, an e.m.f. is induced across it. The magnitude of the induced e.m.f. can be increased by

increasing the strength of magnetic field used;
increasing the rate of change of the magnetic field;

or using a longer conductor.

Lenz's Law states that: the direction of the induced current is such as to oppose the change that produces it.

A dynamo with a split-ring commutator produces a changing d.c. electromotive force.

An alternator (a.c. generator) with slip rings produces an alternating electromotive force.

Dynamos and alternators convert kinetic energy into electrical energy.

A moving-coil microphone converts sound energy into electrical energy by allowing the sound vibrations to vibrate a coil in a magnetic field.

In a transformer the electrical energy is transferred from the primary coil to the secondary coil.
In a step-up transformer there are more turns in the secondary coil than in the primary coil. A step-up transformer steps up the voltage; this is accompanied by a step-down of the current.

In a transformer, the ratio of the secondary potential difference V_s to the primary potential difference V_p equals the ratio of the number of turns in the secondary coil n_s to the number of turns in the primary coil n_p.

$$\frac{V_s}{V_p} = \frac{n_s}{n_p}$$

For an ideal transformer the power output equals the power input. In a real transformer there are power losses because of

the heating effect of the current in the coils;
the heating effect of eddy currents in the core;
the heating effect of the magnetization and demagnetization of the core.

Problems

1 50 kW of power are supplied at the end of power lines of total resistance 10 Ω.

1·1 If the power is transmitted at a potential difference of 250 V between the power lines, what is the power loss?
1·2 If the power is transmitted at a potential difference of 2·5 kV between the power lines, what is the power loss?

2 **2·1** Draw a circuit diagram to show how a 240 V supply and a transformer can be used to operate a 12 V lamp at the correct rating. Assume that the transformer is ideal.
2·2 Is the supply a.c. or d.c.?
2·3 If the transformer has 2 000 turns in the primary coil, how many turns will it have in the secondary coil?
2·4 If the lamp is rated at 24 W, what is the current in the primary coil of the transformer?

3 A bar magnet is attached to one end of a long wooden rod AB which is hung from a pivot at X so that the magnet can swing in and out of the coil PQ.

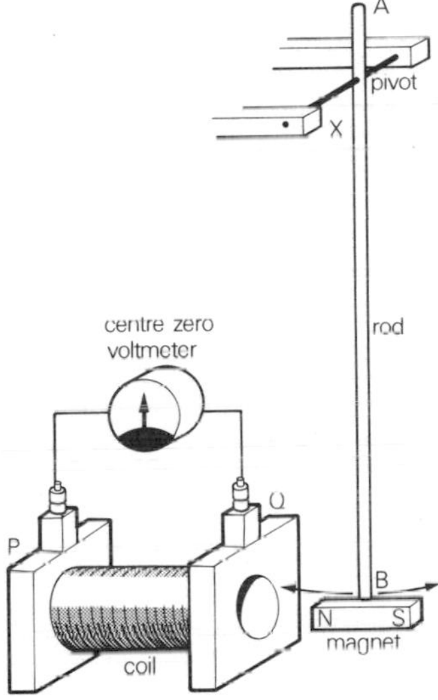

The magnet is pulled to the right and released. It makes ten swings before coming to rest. Describe the changes which take place in the voltmeter reading during this time.
What difference will there be in the voltmeter readings if the experiment is repeated but with

3·1 the coil PQ replaced by one of more turns of the same wire,
3·2 the magnet reversed? *SCEEB*

4 **4·1** Electric power is transmitted from a generating station to distant consumers by high voltage overhead power lines.

a) State the advantage of transmitting electric power at high voltage.
b) Why is a.c. used in preference to d.c.?
c) Explain why at least two transformers are necessary.

4·2 A classroom model of electric power transmission uses two transformers, a 2 V a.c. supply, two 6 m lengths of resistance wire, a 2 V lamp and the necessary connecting wire.

a) Draw a fully labelled circuit diagram for the model.
b) What does the resistance wire represent? *SCEEB*

5 A long pendulum is set swinging between the opposite poles of two flat magnets and in a direction parallel to the long faces as shown. A and C represent the two extreme positions of swing and B the centre position. Instead of a cord the ball is supported by a thin wire whose ends are connected to a centre zero microammeter. The bottom connection is very flexible so that the pendulum will keep swinging.

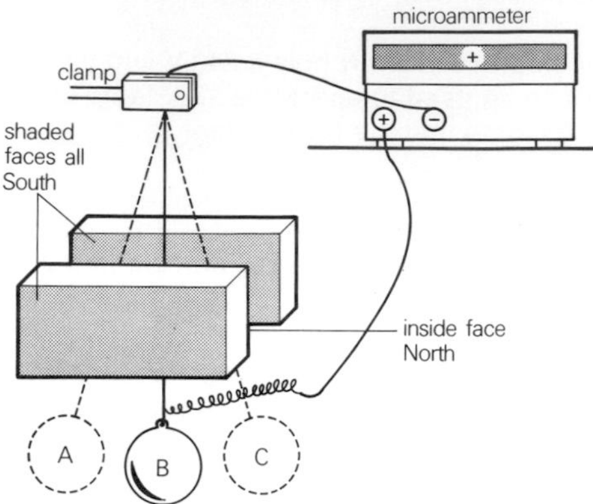

5·1
a) Why does the microammeter show a maximum reading when the bob is at B and moving to the right?
b) Why does it show a zero reading when it is at C?
5·2 What will the microammeter read when the bob is at
a) B, moving to the left;
b) A?
5·3 If the pendulum is set swinging from the south face to the north face and back again, so that its movement is at right angles to that shown in the diagram, explain what effect this will have on the readings.
5·4 How could you increase the maximum meter readings using the same apparatus? *SCEEB*

6 **6·1** In the circuit shown below, the soft iron C cores are separated by a thin card and the bulb B_2 is just lit.

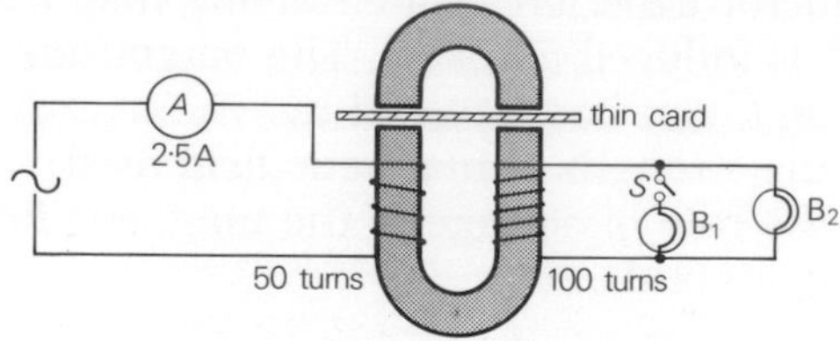

If the card is now removed and the two parts of the C core are clamped together, explain what happens to
a) the ammeter reading and
b) the brightness of the bulb B_2.
If S is now closed, what happens to
c) the ammeter reading
d) the brightness of bulb B_2?
6·2 This is a model of an electrical distribution system.

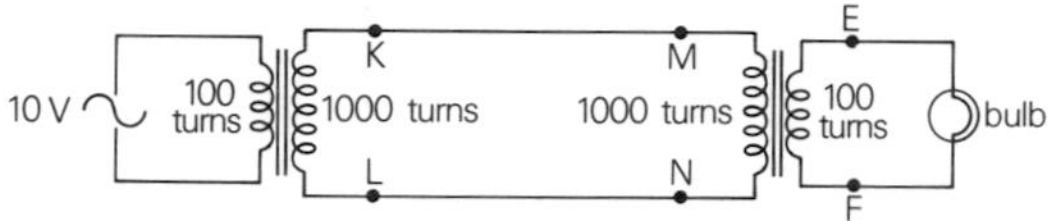

If the transformers are considered ideal, what p.d. would occur across KL?
KM and LN each represent a long length of cable, each having a resistance of 5 ohms. If the current flowing in KM is 0·2 A, what would be the:
a) p.d. across KM,
b) p.d. across MN and across EF,
c) current through the bulb? *SCEEB*

20 Domestic electricity

Electrical appliances play a very large part in our lives. We have washing machines, vacuum cleaners, etc. to help us with the housework, TV's, radios, etc. for home entertainment and a wide range of heaters to keep us warm. Electricity is taken for granted and often not treated with the respect that it deserves. Electricity can kill by shock or an electrical fault can result in a fire. In this chapter we shall consider the electricity supply in the home, the cost of electricity and some of the precautions taken to protect us from the possible consequences of electrical faults in the home.

20.1 Electrical appliances

Electrical household appliances in this country are designed to operate from a 240 V a.c. supply. Each appliance normally has a power rating marked on it. This indicates the electrical power that it transforms when operating at the correct potential difference. Table 1 lists some typical power ratings for household appliances.

Electrical appliance	*Power rating*
immersion heater	3 000 W (3 kW)
electric fire	2 000 W (2 kW)
electric kettle	2 000 W (2 kW)
electric iron	1 000 W (1 kW)
hair dryer	400 W
colour television set	300 W
black and white television set	135 W
refrigerator	120 W
light bulb	60 W–100 W

Table 1

When we pay for electricity we are paying for the electrical energy that is transformed. The power is the energy that is transformed in unit time.

$$\text{power} = \frac{\text{energy}}{\text{time}}$$

$$\text{energy} = \text{power} \times \text{time}$$

Let us consider the electrical energy transformed by an electric fire in one hour.

$$\begin{aligned} \text{power} &= 2\,000 \text{ W} \\ \text{time} &= 1 \text{ hour} \\ &= 3\,600 \text{ s} \\ \text{energy} &= \text{power} \times \text{time} \\ \Rightarrow \text{energy} &= 2\,000 \times 3\,600 \\ &= 7\,200\,000 \text{ J} \end{aligned}$$

The SI unit for energy is the joule, but this is a small unit. If a 2 kW fire consumes 7 200 000 J in one hour, we can see that the number of joules used in a house in the two months between meter readings would be huge. This means that the joule is too small a unit to use for recording the electrical energy used in a house. The unit used by the Electricity Board is the kilowatt hour (kW h) which is the energy used by a 1 kW appliance operating for 1 hour.

energy (kW h) = power (kW) × time (h)

$$1\ \text{kW} = 1\,000\ \text{J s}^{-1}$$
$$1\ \text{h} = 3\,600\ \text{s}$$
$$\Rightarrow \quad 1\ \text{kW h} = 1\,000 \times 3\,600\ \text{J}$$
$$\Rightarrow \quad 1\ \text{kW h} = 3\,600\,000\ \text{J}$$

Generally speaking, household appliances with the highest power rating are those that are used for heating, and these are the most expensive to run.

The electrical energy used by a household is measured by a kilowatt-hour meter, an example of which is shown in Figure 20.1.

Figure 20.1

20.2 Fuses in the house

A fuse is a simple device for ensuring that the current in a circuit does not rise too high when a fault occurs. Figure 20.2 shows different types of fuses, and the symbol that is used to represent a fuse in circuit diagrams.

The fuse itself is simply a thin wire of low melting point. The different types of fuse shown are merely different ways of supporting this wire so that it can easily be inserted in a circuit. Using the circuit shown in Figure 20.3, we can demonstrate how a fuse protects a circuit.

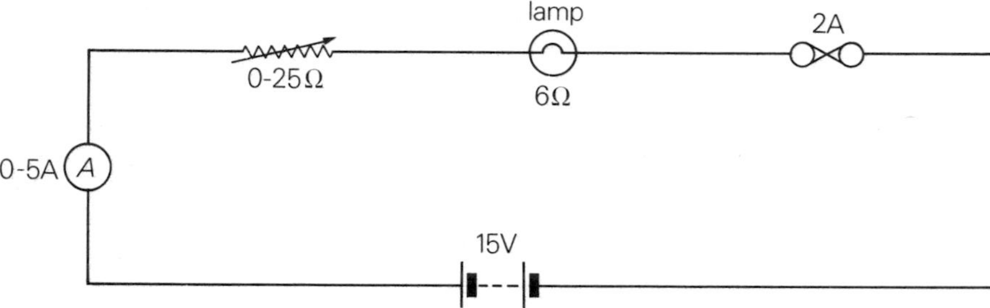

Figure 20.3

symbol

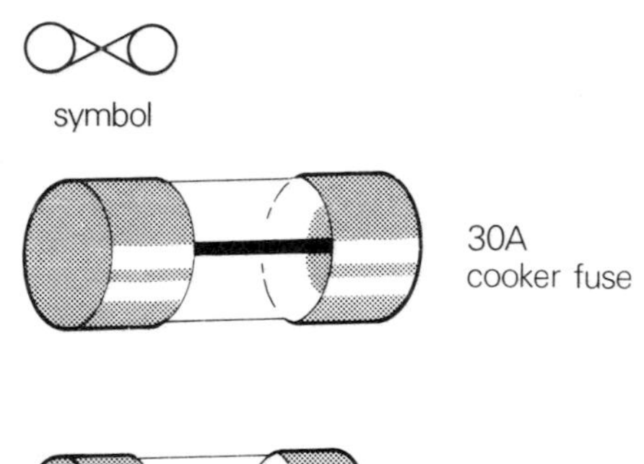

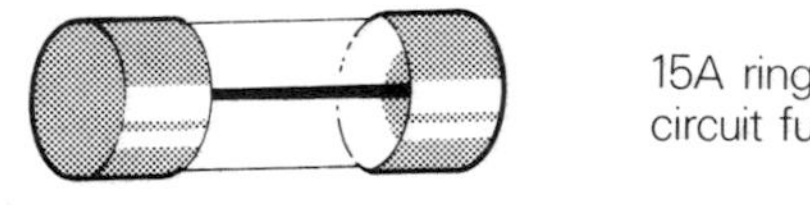

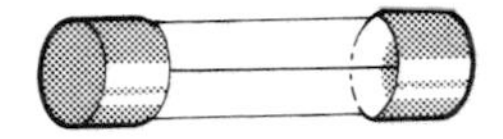

Figure 20.2 Fuses

If we set up the circuit with the variable resistor set to its maximum value, a current reading of about 0·5 A is obtained. As the resistance in the circuit is reduced, the meter reading increases until the current reaches just over 2 A. At this point the fuse wire melts and the current instantly falls to zero because there is no longer a complete circuit. Fuses are made with different ratings by varying the thickness (and hence the resistance) of the wire. The fuse rating is the estimated maximum current that can flow through the fuse without the wire melting. When the current rises above the fuse rating, the heating effect of the current through the fuse melts it. When this happens we usually say that the fuse blows.

Modern household circuits are usually well protected by fuses. The electricity enters the house through the Electricity Board fuse which usually has a rating of 60 or 80 A. If this fuse blows, the electricity supply to the whole house is cut off. It is rare for this fuse to blow, and if it does so the Electricity Board must be contacted. They will investigate to find the cause of the fuse blowing, and carry out the necessary repair. The householder is not allowed to break the Electricity Board's seal on the box containing this fuse.

The mains fuse box contains a number of fuses protecting wiring to various parts of the house. In a modern house, there are the following separate circuits, containing mains fuses of the given rating.

circuit	*fuse rating*
fixed lighting	5 A
water heating	15 A
cooking	30–45 A
ring circuit (wall outlets)	30 A

Electric central heating would have a separate circuit.

These fuses are to protect the wiring from excess current. If the current through the wires is too great, it can result in overheating of the wires and is a potential fire risk.

In addition to the protection of these fuses, appliances which are plugged into the ring circuit are further protected by fuses in the plugs, and often in the appliance itself. The rating of the fuse for an appliance must be higher than the maximum current for the normal use of that appliance, for otherwise the normal current will blow the fuse. It is usual to choose the fuse rating which is just above the maximum current for the appliance. If the fuse rating is much higher than this, it is possible for a fault to result in a current high enough to damage the appliance, but not high enough to blow the fuse.

Since the power rating for an appliance is known and it operates at mains voltage, the current through the appliance can be calculated as follows:

power P = (current I) × (potential difference V)

$$P = I \times 240$$

$$\frac{P}{240} = I$$

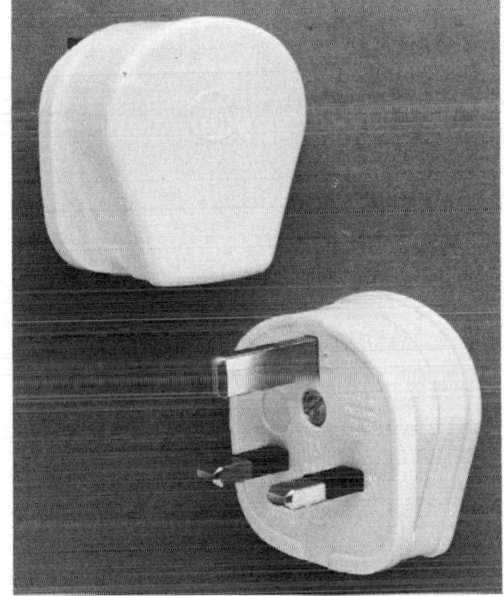

Figure 20.4

In a modern house the sockets are usually designed to take 13 A plugs, Figure 20.4. Older wiring systems may have round-pin sockets and plugs. A 13 A plug is one which is designed to be safe for currents up to 13 A: no household appliance which is plugged in to a mains socket will allow a current greater than 13 A unless the appliance is faulty.

Figure 20.5 shows how a cable is connected in a 13 A plug.

Figure 20.5 Wiring inside a 13 A plug

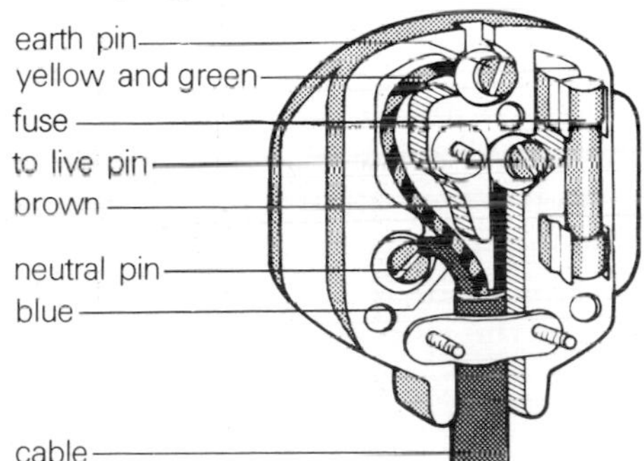

The colour coding indicated is that which is in modern use. Older appliances may have a red wire for live, a black wire for neutral and a green wire for earth.

A range of fuse ratings, including 2 A, 5 A, 10 A and 13 A, is available for these plugs. For simplicity, the Electricity Board recommend the use of 3 A fuses for appliances rated below 750 W and 13 A fuses for appliances rated between 750 W and 3 000 W. Failure to use the correct fuse can result in an excess current and expensive damage to an appliance. The excess current could cause overheating and also damage to other parts of the circuit.

The fuse is always connected into the live wire leading to a circuit. The neutral wire is at an approximate potential of 0 V. Because the potential difference of the supply is between the live wire and the neutral, it is the live wire that is at high potential and this is the wire that should be disconnected when a fault develops. In a similar way, switches in household circuits are in the live wire from the supply. This ensures that the appliance is isolated from the high potential when the switch is open.

The earth pin in the plug makes contact with house wiring which is connected to a metal water pipe leading to the earth, and is always at the same potential as the earth. If the casing of the appliance is connected to this pin, it will be kept at earth potential. Earth potential is usually taken to be 0 V. Since we are earthed, we too are at earth potential. Thus by earthing the case, we ensure that there can be no potential difference between us and the case and we cannot receive a shock from the casing. If a fault develops in the circuit which results in the live wire making contact with the casing, a fuse would blow but the casing would remain at earth potential.

20.3 The ring main

Household appliances in this country are designed to operate at 240 V. This means that, whatever the number of appliances that we wish to connect into the mains supply, the potential difference across each one should still be 240 V. This is done by connecting them in parallel. The modern circuit for doing this is the ring main. A diagram illustrating a ring main circuit is given in Figure 20.6. The earth wire has been left out for simplicity.

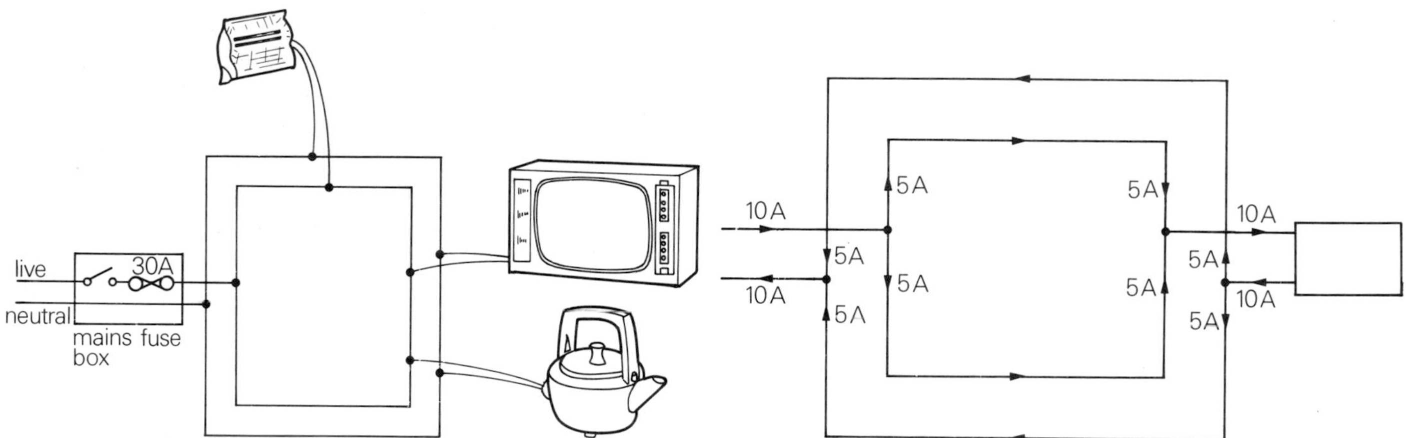

Figure 20.6 Ring circuit

Figure 20.7 Current in a ring circuit

The wiring in the ring circuit is designed to safely carry 15 A, yet it is protected by a 30 A fuse in the mains fuse box. The reason for this can be seen if we look at Figure 20.7 which illustrates a single appliance drawing a current of 10 A from a ring circuit. We can see that there are two paths to the appliance in a ring circuit: the current in any part of the ring circuit is half the current to the appliance.

Summary

An electrical appliance has a power rating; this indicates the power that it consumes when it operates at the correct potential difference.

The unit for electrical energy used by the electricity boards is the kilowatt hour (kW h).
1 kilowatt hour is the energy used by a 1-kilowatt appliance operating for 1 hour.

A fuse will melt if the current in the circuit rises above the value indicated by the fuse rating. A fuse is used to prevent too large a current flowing in the circuit and damaging components or causing fire.

A fuse is connected in series in the live wire of the circuit.

The modern wiring in a house is a ring main for the wall sockets. There are separate circuits for the cooker, the lights, the water heater, and central heating.

Problems

1 The following table gives details of some household electrical appliances. Each appliance is fitted with either a 3 A fuse or a 13 A fuse. Copy and complete the table.

appliance	*power rating*	*operating voltage*	*current drawn*	*most suitable fuse rating*
food mixer		240 V	0·3 A	
lamp	100 W	240 V		
heater	3 kW	240 V		
stereo unit		240 V	1·5 A	

2 A consumer pays 2·5 p for each kilowatt hour of electrical energy used. What will he pay to run

2·1 a 3 kW immersion heater for 3 hours?
2·2 a 200 W lamp for 5 hours?

3 **3·1** A battery charger has a 240 V a.c. input and a 12 V d.c. output. The manufacturer recommends a maximum charging current of 2 amperes.

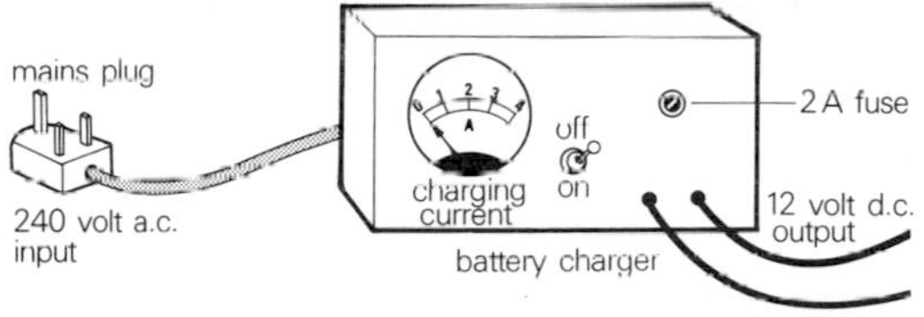

a) Name the **two** *vital* circuit components inside the battery charger and state the purpose of each component.
b) A 1 A fuse is fitted into the mains plug. Show by calculation whether this will allow a maximum output current of 2 amperes. Assume that the charger is 100% efficient.

3·2 The wiring diagram of a modern mains appliance is shown below.

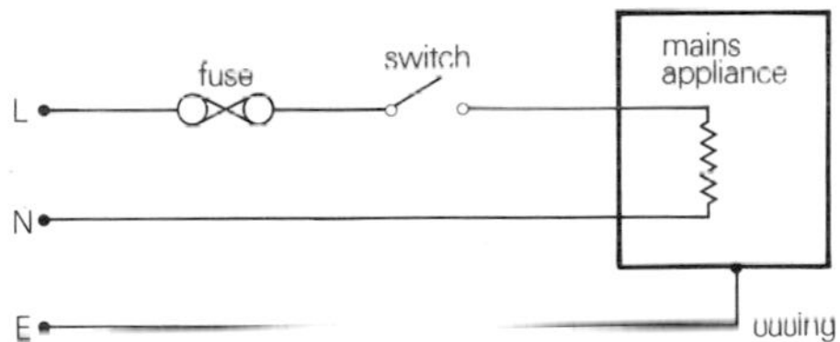

a) What is the purpose of lead E?
b) Explain why it is important when wiring this appliance to a mains plug, that the wire L is connected to the LIVE pin and wire N to the NEUTRAL pin, rather than the other way round. *SCEEB*

4 A pupil experimenting with fuses sets up the following circuit, including two 3 A fuses.

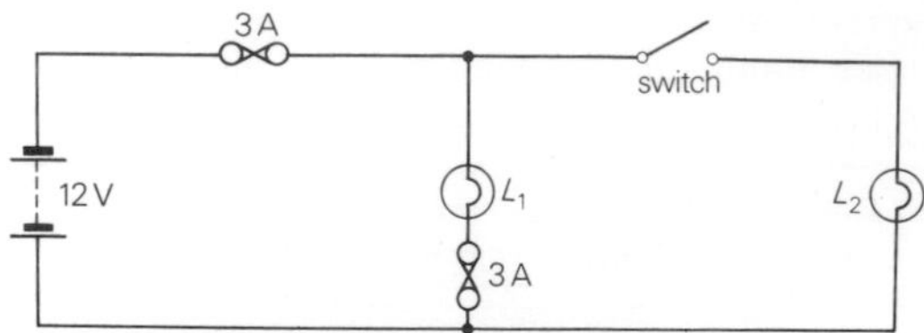

4·1 Before closing the switch, which lamp(s), if any, will light?
4·2 After closing the switch, which lamp(s), if any, will light?

5 The diagram shows part of a house-wiring system.

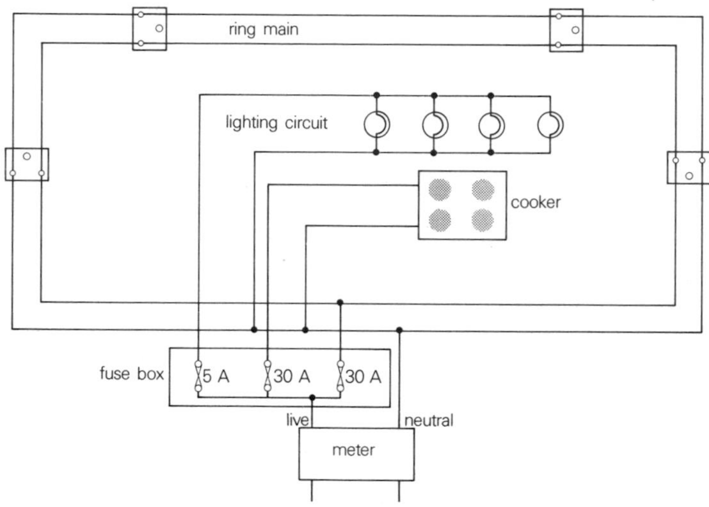

5·1 There are usually about ten 13 A points in the ring main. Why is a 30 A fuse considered suitable for this circuit?
5·2 If the 30 A cooker fuse 'blew', what would happen to the components in the other circuits, e.g. lighting, ring main?
5·3 No earth connections are shown in the diagram. Why are such connections required?
5·4 What physical quantity does the meter measure?
5·5 If a heater rated at 240 V, 2 kW were to be connected in the ring main, what value of fuse (2 A, 5 A, 10 A, or 13 A) would it be advisable to fit in the plug?
5·6 What difference would you expect to find between the wire used in the lighting and the ring main circuits? Explain your answer. *SCEEB*

21 Electronics

21.1 Thermionic emission

When we heat a high melting point metal to a very high temperature, there is enough energy supplied to some of the electrons in the metal to enable them to escape completely from its surface. Electrons which are emitted under these circumstances are said to be produced by **thermionic emission.** A thin filament of metal can be heated by passing a current of a few milliamperes through it.

We can show that these thermionically emitted electrons exist by enclosing a filament (C) in a glass container from which most of the air has been removed. The air is removed so that there will be no collisions between the electrons and air particles. If we include in the glass container another piece of metal (A) a short distance away we can attract electrons across to it, Figure 21.1.

Since there are two pieces of metal (electrodes) inserted into the container, we call it a **diode**. When we attach electrode A to the positive terminal of a battery and electrode C to the negative terminal, we detect a current with the milliammeter, Figure 21.2. Since the battery drives electrons from negative to positive, the electrons must have travelled from the filament C to the other electrode A.

The filament is often made from a lower melting point metal coated with an oxide which emits electrons at a much lower temperature and thus less energy is needed.

The filament is called the **cathode** and the metal plate A the **anode** of the diode.

When the diode is connected in the reverse way (Figure 21.3), with the anode connected to the negative terminal of the battery and the cathode to the positive, no current flows. The electrons are not attracted to the negative anode.

In both cases there is no current when the filament supply is switched off since there is no thermionic emission.

Electrons will only flow in the one direction from the cathode to the anode. We call this device a **diode valve** because (like a valve on a bicycle tyre) it only allows flow in one direction.

In the physics laboratory we can investigate this behaviour using a large demonstration diode as in Figure 21.4.

Figure 21.1 Diode

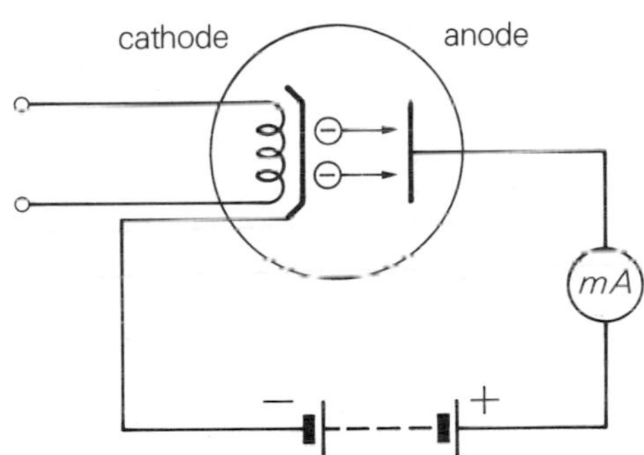

Figure 21.2 Diode current

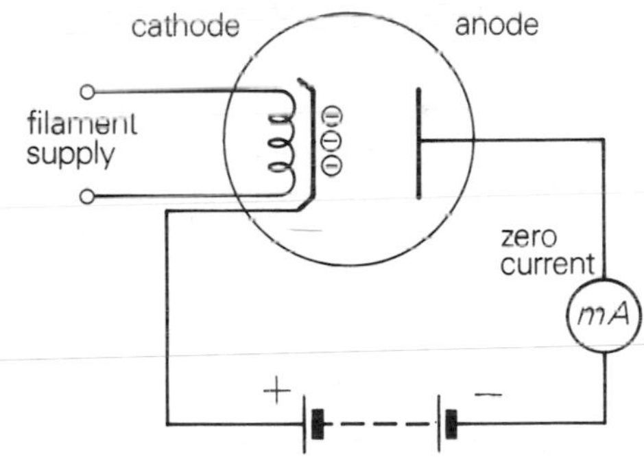

Figure 21.3 No current flow

Figure 21.4 Demonstration diode

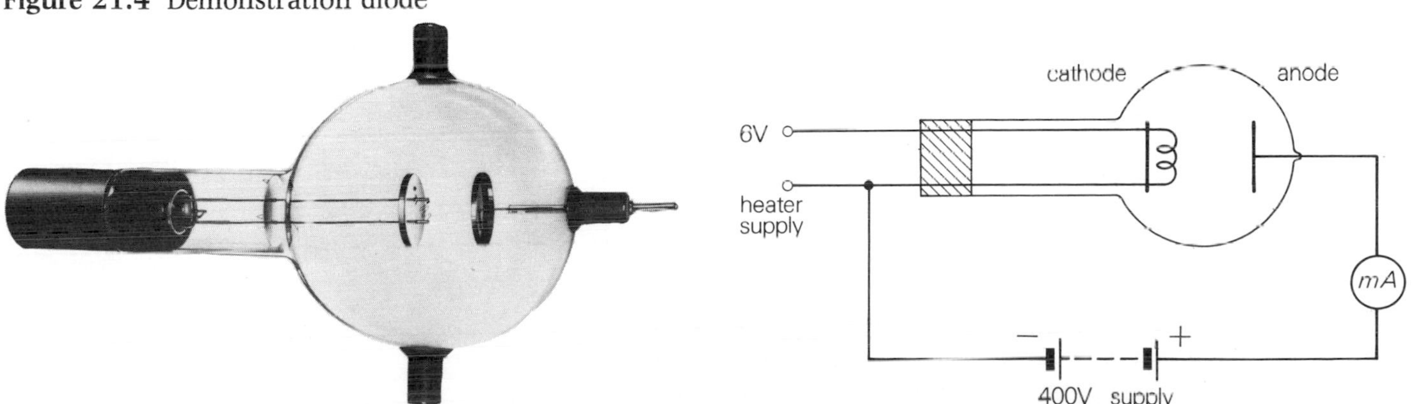

Half-wave rectification

One use of a diode valve is to convert alternating current to direct current. Remember that alternating current involves a varying current which changes direction, Figure 21.5.

When this type of supply replaces the battery in Figure 21.4 we obtain a different current, Figure 21.6.

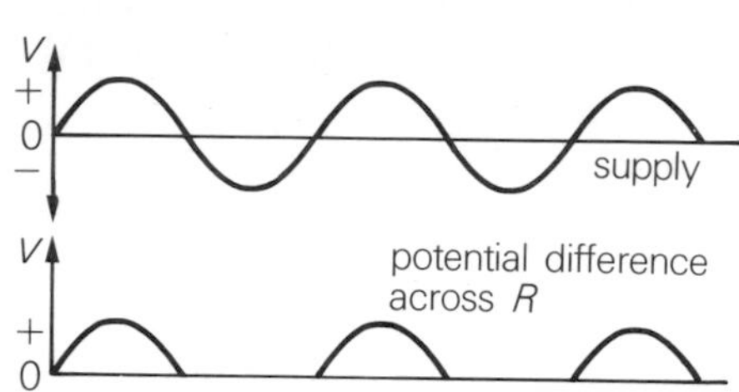

Figure 21.5 Alternating current

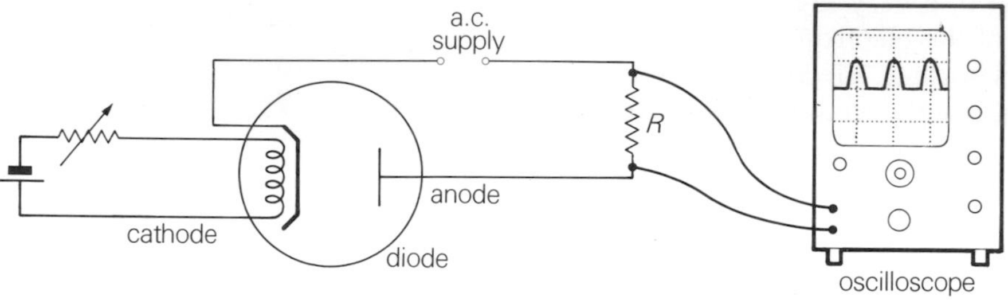

Figure 21.6 Circuit for half-wave rectification

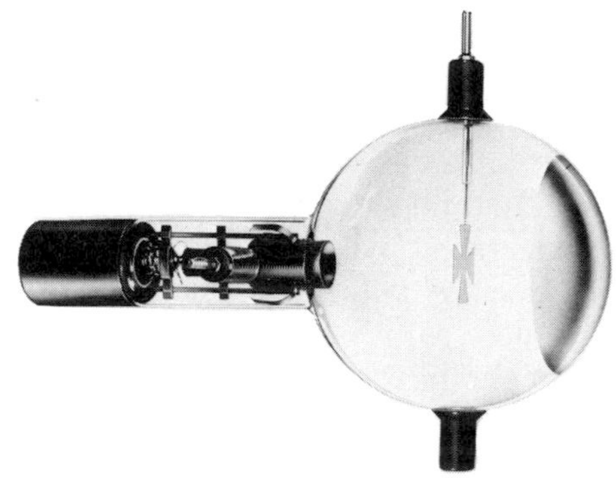

Figure 21.7 Graph for half-wave rectification

The electrons are able to flow from the cathode to the anode of the diode but not from the anode to cathode, so that there is a current for only half the time to give half waves. The current through the resistor R results in the p.d. variation shown on the oscilloscope, Figure 21.7. We have changed a.c. to d.c. This process is called **half-wave rectification**. Because V varies directly as I, the current through the resistor also changes with time in exactly the same way as the potential difference across the resistor.

21.2 Cathode rays

An important piece of equipment which makes use of thermionic emission is called an oscilloscope or **cathode ray oscilloscope**, often abbreviated to C.R.O. It is really a rather sophisticated diode valve and it is so named because the flow of electrons from cathode to anode was originally thought to be a type of radiation called cathode rays. We shall look at three experiments which confirm that cathode rays are electrons and at the same time we shall introduce the more complex C.R.O.

Figure 21.8 Maltese Cross tube

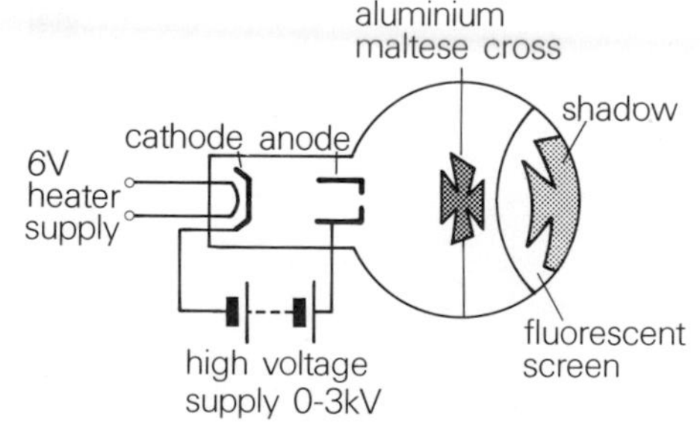

1. Energy of cathode rays

We shall use the Maltese Cross tube. This is basically a diode with an obstacle and a fluorescent screen, Figure 21.8. When the heater supply is switched on, the filament glows and a shadow of the Maltese Cross is cast on the screen. When the 3 kV supply is connected, electrons produced at the cathode are accelerated towards the anode which is a hollow metal cylinder. The electrons pass through the anode and on towards the metal Maltese Cross. The electrons which strike the Maltese Cross are stopped, but those which travel past it strike the fluorescent screen where their kinetic energy is converted to light energy (green light). A shadow is cast by the Maltese cross which corresponds exactly to the light shadow produced by the cathode filament alone. We can see therefore that 'cathode rays' possess energy and travel in straight lines to produce a shadow of any obstacle in their path.

If we bring a bar magnet near to the tube, the cathode ray shadow moves but the light shadow does not. Cathode rays are not, therefore, a form of electromagnetic radiation but they do in fact move in the direction indicated by the motor rule for moving electrons, Figure 21.9.

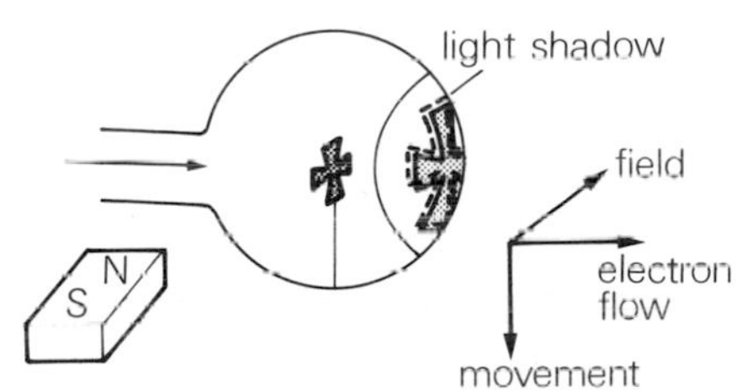

Figure 21.9 Magnetic deflection of cathode rays

2. Deflection of cathode rays by electric field

By inserting two extra electrodes in the form of parallel metal plates with a fluorescent screen between them we form an electrostatic deflection tube, Figure 21.10.

When we connect a large potential difference across the two metal plates we create an electric field between the two plates and the cathode rays are attracted towards the upper positive plate in the same way as electrons.

If we apply a continuously varying p.d. to the plates (an a.c. voltage) the electron beam scans upwards and downwards as the p.d. varies, Figure 21.11.

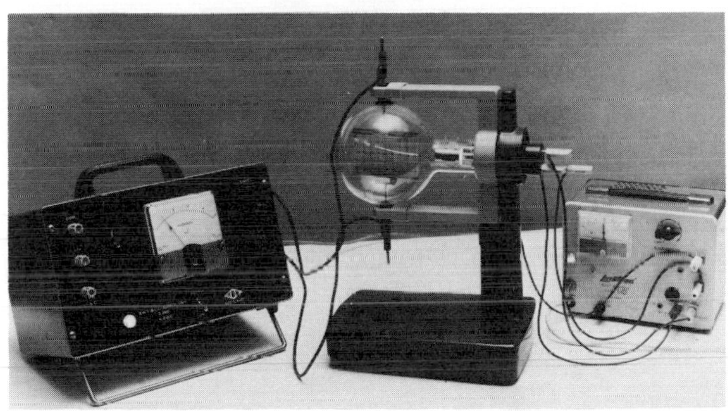

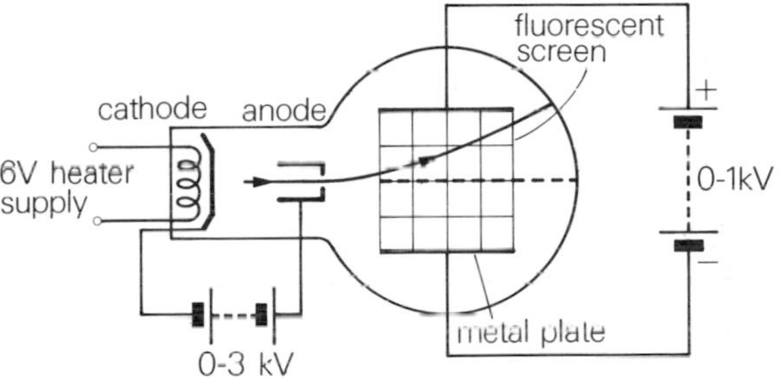

Figure 21.10 Electrostatic deflection tube

3. Collection of cathode rays

In 1895 a French scientist named Jean Perrin performed an experiment which was designed to convince people at that time that cathode rays were not a type of radiation, but rather, a flow of electrons through the evacuated tube.

We can demonstrate this experiment using an up-to-date version of the apparatus called a Perrin tube, Figure 21.12.

If we accelerate the cathode rays towards a cylindrical anode with a small pinhole in the end of it, a narrow beam emerges and travels on to strike the fluorescent screen and produce a spot of light.

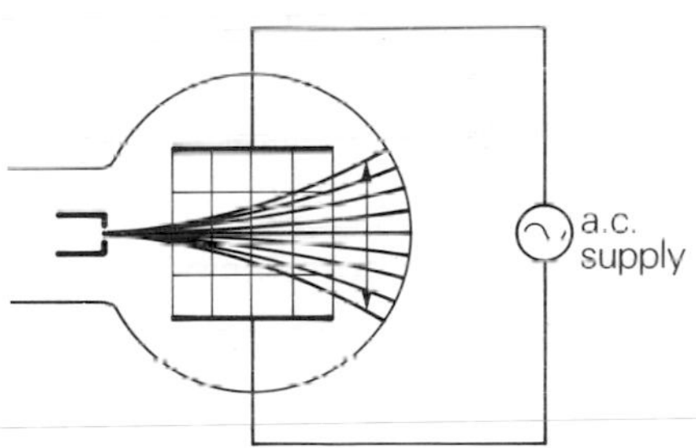

Figure 21.11 Scanning

Figure 21.12 Perrin tube

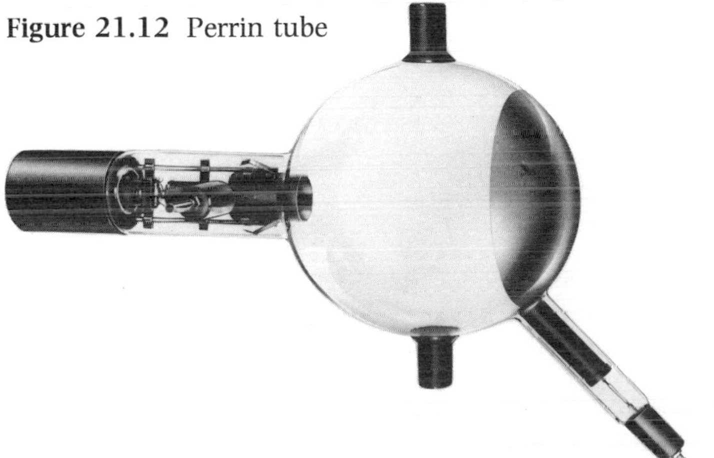

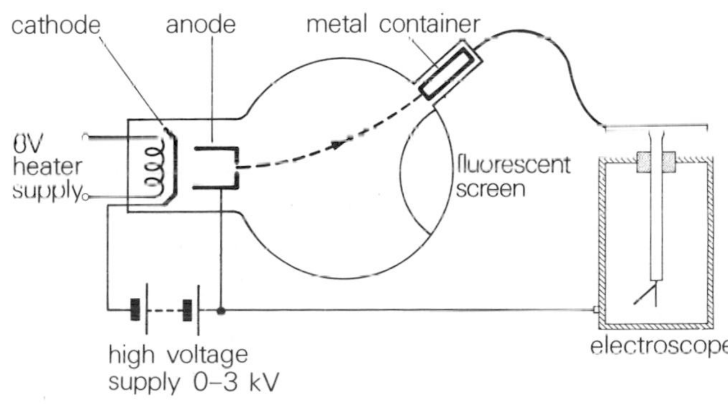

We can deflect the beam up into a small cylindrical metal container using the magnetic field of a bar magnet. When the beam enters the metal container, the gold leaf electroscope which is attached to it shows a deflection. Using a negatively charged polythene rod held near to the electroscope disc, we find that the leaf rises further. This means that the electroscope is negatively charged. Thus we conclude that cathode rays have the same charge as electrons.

21.3 The cathode ray oscilloscope

Construction

We have now looked at the basic factors which when combined with some complex electronic circuits, make up a cathode ray oscilloscope, Figure 21.13.

Figure 21.13 Cathode ray oscilloscope

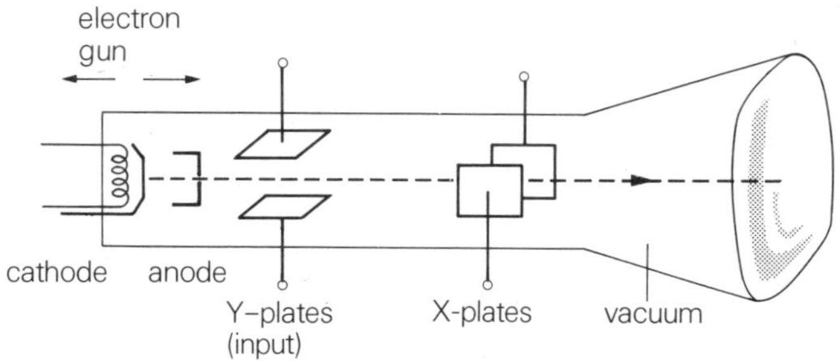

Figure 21.14 C.R.O. tube

A cathode ray oscilloscope can be as simple as the one in Figure 21.13 or as complicated as the one in a colour television set.

Figure 21.14 shows the basic construction of the C.R.O. tube. The electron gun (cathode and anode system) fires a beam of electrons towards the screen. This beam then passes between two pairs of electrodes called Y-plates and X-plates. The electrical input signal is usually connected across the Y-plates.

Figure 21.15 C.R.O. Y-plate deflection

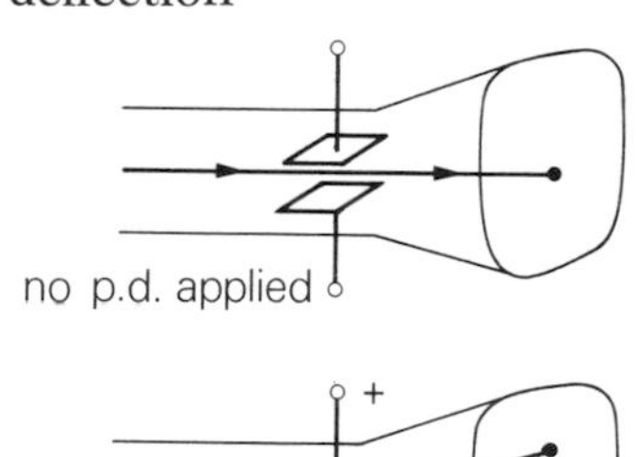

d.c. applied + −

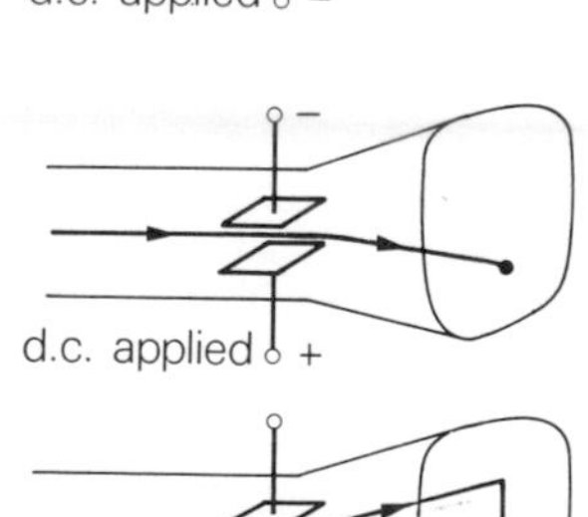

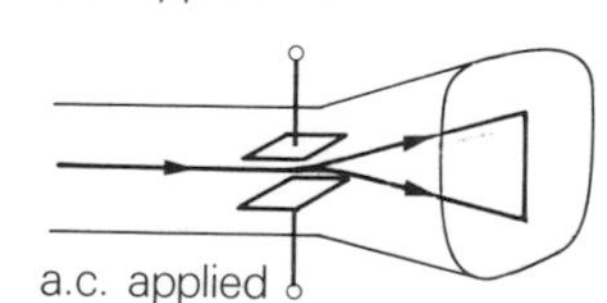

Brightness and focus

A C.R.O. is used to produce a trace (or picture) on a fluorescent screen. The picture is produced by a beam of electrons and we can vary the brightness and sharpness of the trace by means of controls called the 'brightness' and 'focus' controls.

Y-plates

If the electron beam is focussed on the centre of the screen, we obtain a spot of light. When we apply a potential difference to the Y-plates, the electron beam is deflected due to the electric field and the electrons strike the screen at a different point. The deflection is directly proportional to the p.d. applied to the Y-plates, Figure 21.15.

If the top Y-plate is positive, then the beam of electrons is deflected upwards in the Y-direction; if this plate is negative, the beam is deflected in the opposite direction. When an a.c. potential is applied to the Y-plates, we obtain a vertical line of light on the screen.

Gain

If the voltage applied to the Y-plates is too small to produce any detectable deflection of the beam, it can be amplified by means of a built-in amplifier controlled by the gain knob on the C.R.O., Figure 21.16.

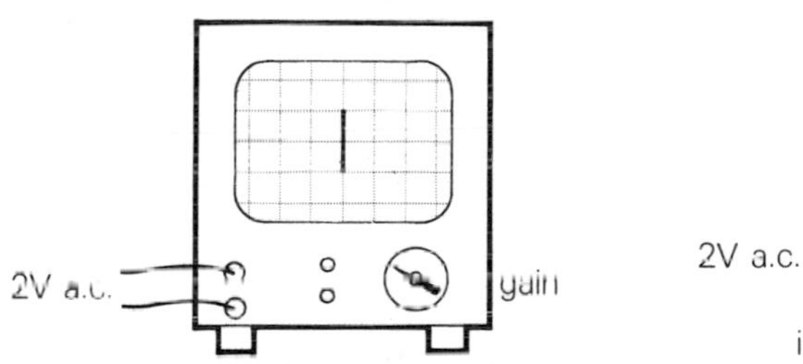

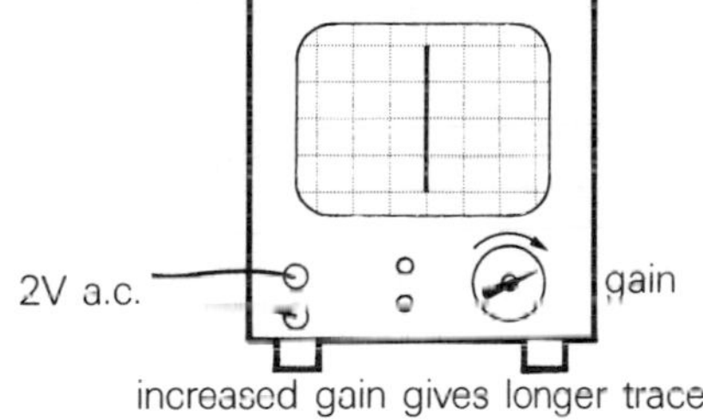

Figure 21.16 Gain control

X-plates

The second pair of metal plates inserted in the C.R.O. tube are called the X-plates; when we apply a p.d. to them, we obtain a deflection of the electron beam in the X-direction, Figure 21.17.
When the value of the p.d. applied to the Y-plates is varying with time, we obtain a vertical line trace on the screen of the C.R.O. If the beam is also deflected to the right at a constant speed, we can see the variation in the p.d. on the Y-plates.

Imagine that your pencil point is producing the up-and-down vertical trace on the screen; if you then move your hand to the right at a constant speed, you will see how the trace is produced, Figure 21.18.

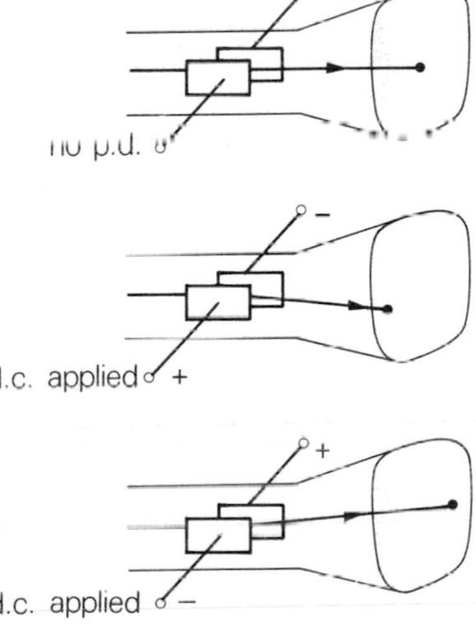

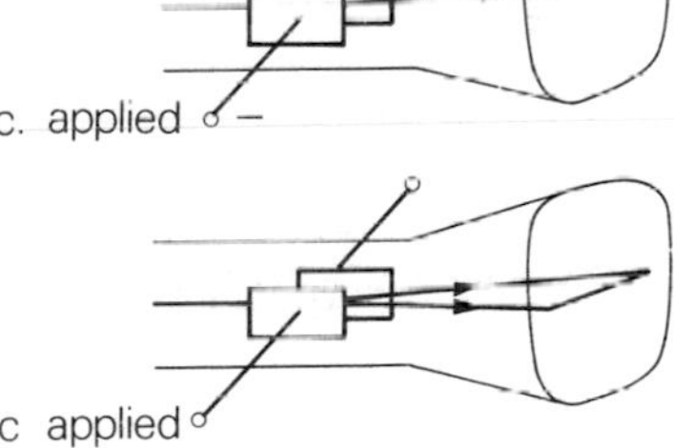

Figure 21.17 C.R.O. X-plate deflection

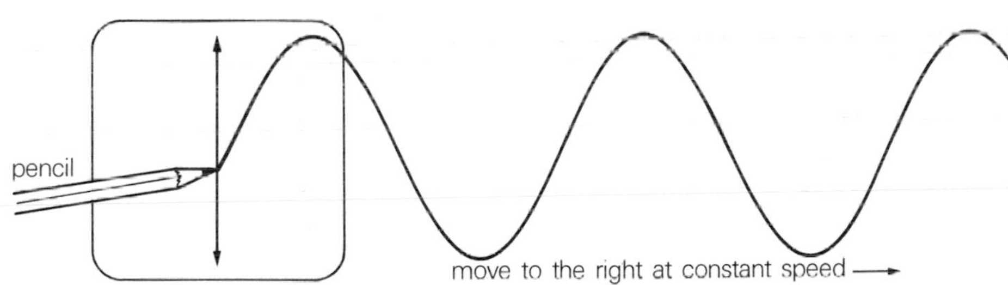

Figure 21.18 Making an a.c. trace

The C.R.O. screen is of limited size however and we cannot keep on deflecting the beam to the right. The p.d. applied to the X-plates is therefore of the form shown in Figure 21.19.

Figure 21.19 Sawtooth p.d.

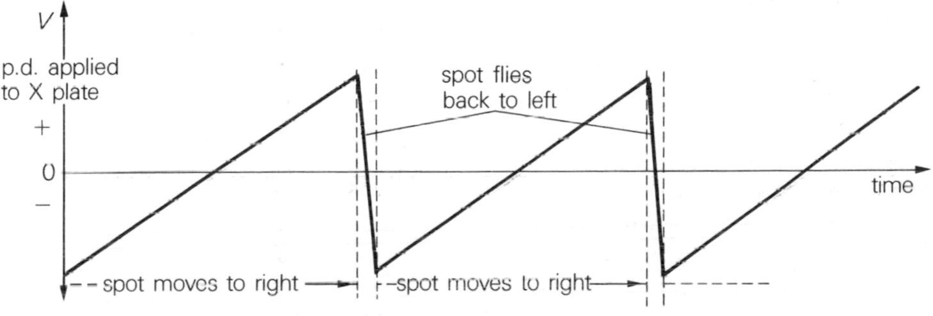

The graph of p.d. variation is often called a 'saw tooth' wave because of its shape. This wave form is produced by a circuit called the **time-base circuit** which is connected to the X-plates. It can be adjusted to vary the speed at which the electron beam is made to scan across the C.R.O. screen to produce the trace.

If we sweep the beam across slowly, we obtain a trace with a certain number of complete waves, Figure 21.20.

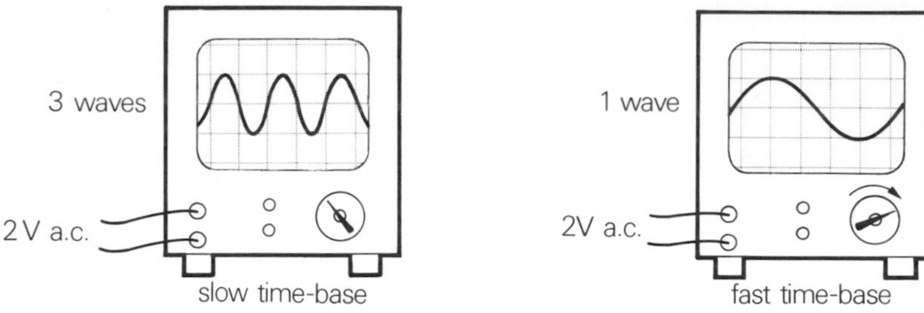

Figure 21.20 C.R.O. time base

When the time-base speed is increased, then there is only time to display a trace of one wave before the beam flies back to begin making another wave from the left to coincide with the first one.

21.4 Use of the C.R.O.

1. Measurement of potential difference

We can measure potential difference using the C.R.O. The C.R.O. is a very good voltmeter because of its very high resistance. The apparatus in Figure 21.21 can be used to demonstrate this use.

We first connect one cell (1·5 V) to the Y-plates and adjust the gain control until the trace indicating the p.d. is 1 cm above the zero volts line, Figure 21.22 (a). Then if we connect one further cell in series, we can measure the combined p.d., Figure 21.22 (b).
Figure 21.22 (c) shows the result obtained with three 1·5 V cells in series, while Figure 21.21 (d) shows what happens when we connect these cells the reverse way round.

Sometimes the gain control is already calibrated in volts per centimetre (V cm^{-1}) so that every centimetre of deflection represents a known p.d. This means that we can measure an unknown p.d.

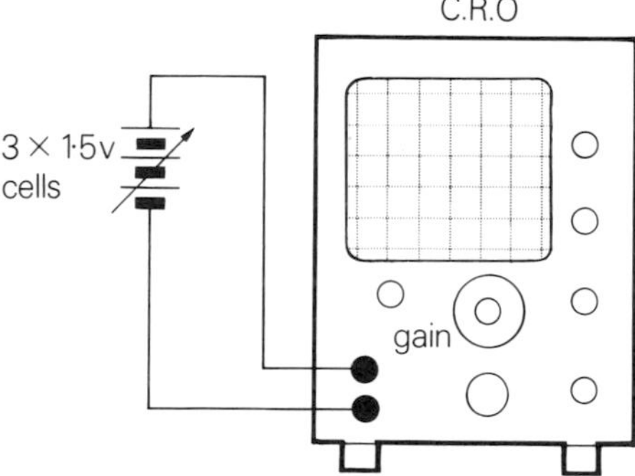

Figure 21.21 Using a C.R.O. to measure p.d.

Figure 21.22 p.d. measurement

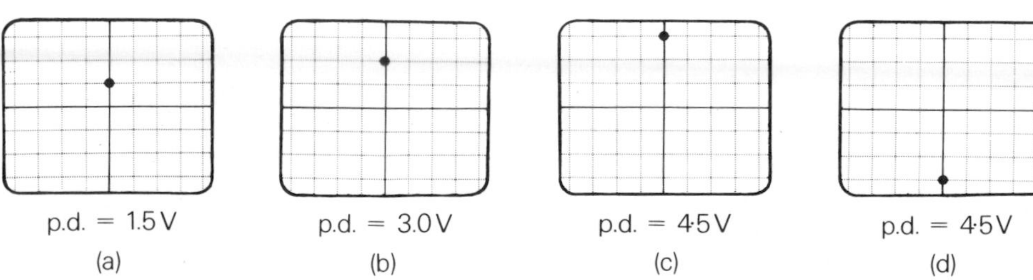

Example 1

The C.R.O. gain control is set at 2 V cm^{-1}. What is the unknown p.d. applied to the Y-plates?

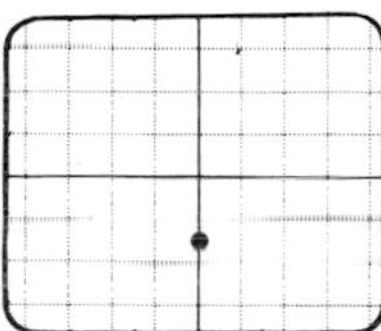

1 cm is equivalent to 2 volts
$\Rightarrow$ 1·5 cm is equivalent to 1·5 × 2 volts
$\Rightarrow$ p.d. = 3 volts
The unknown p.d. is 3 volts.

2. Waveform studies

We can use the C.R.O. to study the voltage waveforms from a low voltage a.c. power pack, Figure 21.23.

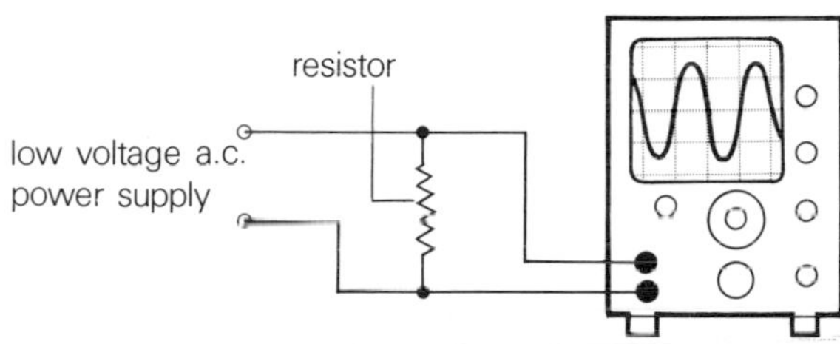

Figure 21.23 a.c. waveform trace

The C.R.O. can also display a sound-wave trace, as we have seen in Figure 5.19, page 50. When we connect a small microphone to the Y-plate input terminals, a p.d. variation is produced as the sound energy is converted to electrical energy and is shown as a trace on the C.R.O. screen.

Waveforms may appear rather different depending on whether the time-base is on or off. Figure 21.24 gives two such examples.

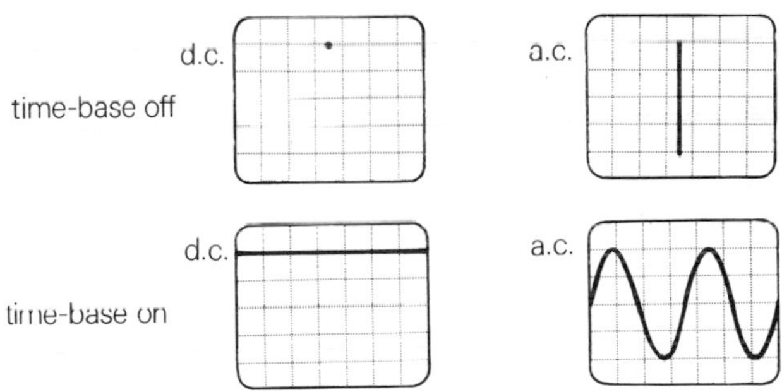

Figure 21.24 Effect of time base

3. Frequency measurement

We can use the C.R.O. to measure frequency by comparing a waveform of unknown frequency with one of known frequency.

We can adjust the time-base of the C.R.O. until we have one complete wave on the screen, Figure 21.25 (a). Without altering the controls of the C.R.O. if we then apply the signal of unknown frequency instead, we can compare frequencies by counting the number of complete waves, Figure 21.25 (b). There are two waves produced in the same time and therefore the unknown frequency is twice the known one. In this case, if the known frequency is 50 Hz, the unknown frequency is 100 Hz.

Figure 21.25 Frequency measurement

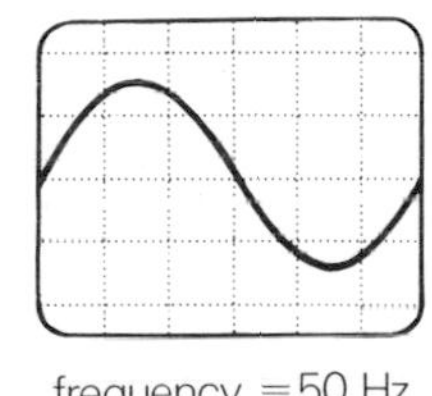

frequency = 50 Hz
one wave
(a)

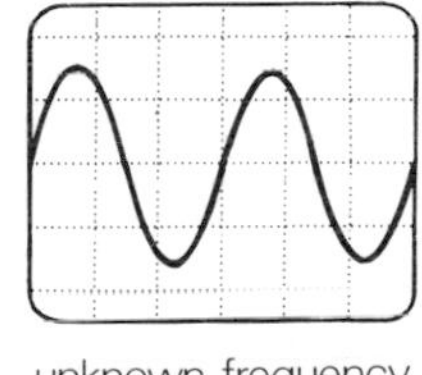

unknown frequency
two waves
(b)

Example 2

The frequency of the signal on the left is 100 Hz. What is the frequency of the signal which produces the trace on the right without adjusting the C.R.O. controls?

The frequency of three complete waves on the screen is 100 Hz. Therefore the frequency of one complete wave that occupies the same space on the screen is $100 \times \frac{1}{3} = 33{\cdot}3$ Hz.

$\therefore$ unknown frequency = 33·3 Hz.

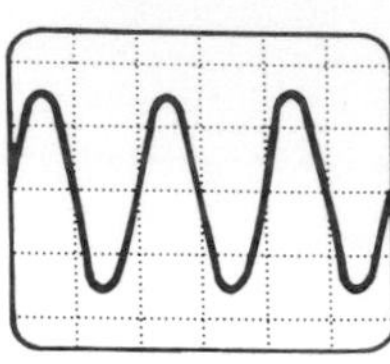

100 Hz

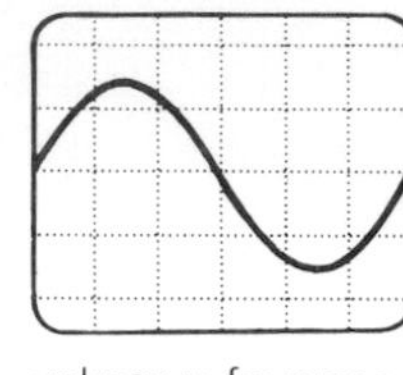

unknown frequency

Often the C.R.O. time-base is calibrated in milliseconds per centimetre (ms cm^{-1}). This tells us how long the trace takes to travel 1 cm to the right of the screen.

Example 3

The time-base of a C.R.O. is calibrated to 2 ms cm^{-1}. If one complete wave occupies 10 cm, what is the frequency of the wave?

The time-base moves 1 cm in 2 ms

$\Rightarrow$ the time-base moves 10 cm in $10 \times 2 = 20$ ms $= \frac{1}{50}$ s

$\Rightarrow$ the wave travels one complete wavelength in $\frac{1}{50}$ s

$\Rightarrow$ the frequency of the wave = 50 s^{-1} = 50 Hz.

The frequency of the wave is 50 Hz.

21.5 Semiconductors

We have divided materials into two categories called non-conductors (insulators) and conductors. There is however an 'in-between' category because the high insulation properties of some materials can be changed to reasonable conduction by the addition of impurities. Table 1 gives examples of all three categories. The 'in-betweens are called **semi-conductors.**

Conductors	*Semiconductors*	*Insulators*
gold	germanium	air
silver	silicon	nylon
copper	selenium	polythene

Table 1

For example if we add atoms of arsenic, we obtain impure germanium which is then able to conduct electricity. This is known as n-type germanium. The addition of atoms of indium instead results in p-type germanium.

This impure germanium ('doped' germanium) is referred to as a **semiconductor** material. Silicon is another material which can be treated in a similar manner to give n-type silicon or p-type silicon.

Semiconductors are the basis of all the sophisticated electronics which we have today. They are used to make solid-state diodes, transistors, and many other complex components.

Semiconductors must be prepared to a very high degree of purity and the amounts of the impurity atoms with which we 'dope' the material must be controlled very accurately. The impurities are often diffused into the pure crystals of silicon or germanium.

21.6 Solid state diodes

It is possible for n-type and p-type semiconductor to be together as one single continuous crystal. This is achieved by diffusing impurities from both sides of the crystal. Such an arrangement is called a **p-n junction**.

We can investigate the conduction properties of the p-n junction using the simple circuit shown below, Figure 21.26.

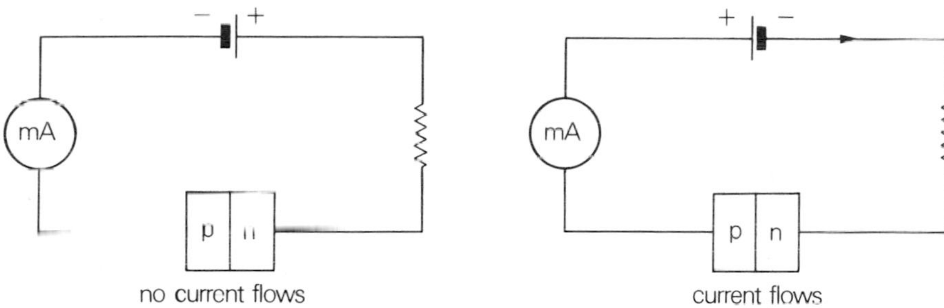

Figure 21.26 p-n junction

When we connect the p-type side of the junction to the negative side of the cell, no current flows.
When we connect the p-type to the positive side of the cell a current flows.

A p-n junction behaves like a thermionic diode; it allows current flow in one direction only. This means that electron flow is from n-type to p-type. The symbol is shown in Figure 21.27.
Using the symbol for the p-n junction (known as a diode), we can draw circuits to represent Figure 21.26. When current flows, the diode is said to be **forward biased**; if no current flows, it is **reverse biased**, Figure 21.28.

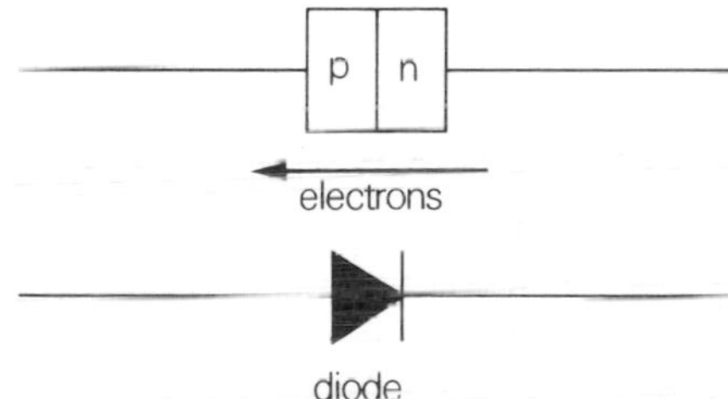

Figure 21.27 p-n junction (diode) symbol.

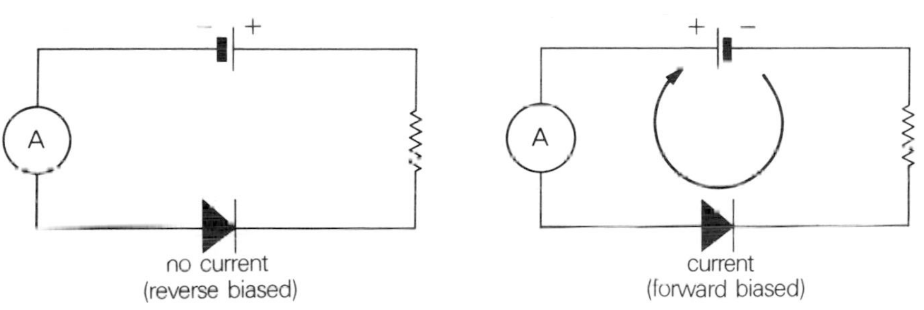

Figure 21.28 Biasing of diode

A p-n junction diode acts in the same way as a diode valve when a p.d. is applied to it.

Solid state diodes vary in size; a selection of diodes is shown in Figure 21.29.

Figure 21.29 Solid state diodes

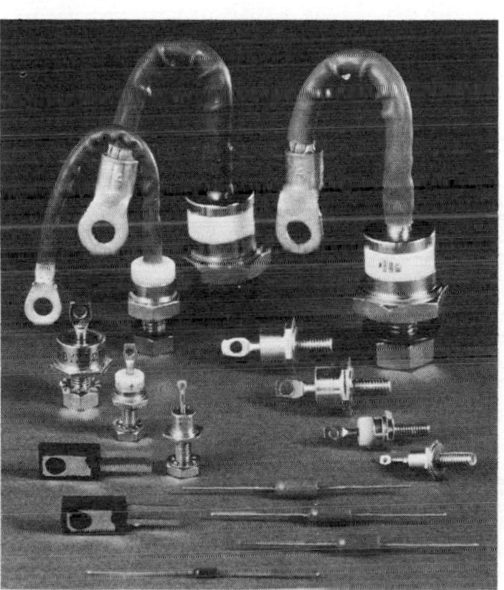

21.7 Rectification

When we convert a.c. to d.c. we are carrying out the process called **rectification.**
We can demonstrate how a solid state diode performs as a rectifier using the circuit below, Figure 21.30.

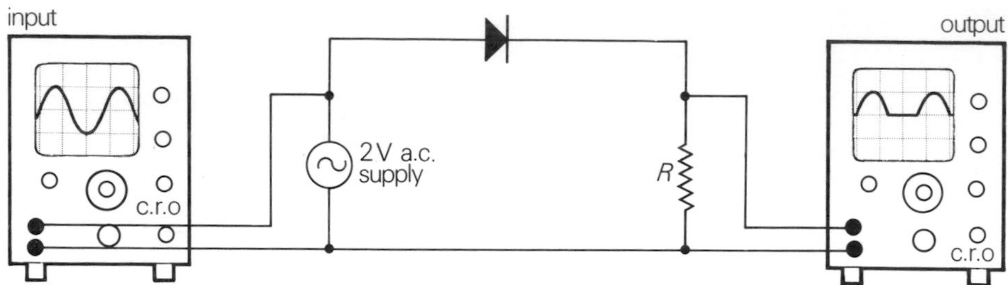

Figure 21.30 Half-wave rectification

The diagram compares the C.R.O. trace obtained from the power supply alone with the trace of the p.d. across the resistor when the diode is in the circuit. This shows that the diode is indeed only allowing current flow in one direction. We obtain current flow and hence a p.d. across resistor *R* only when the supply p.d. is 'forward-biasing' the diode. This is called **half-wave rectification.**

We can obtain **full-wave rectification** by using four diodes connected to form a circuit as in Figure 21.31, which is often called a rectifier bridge circuit.

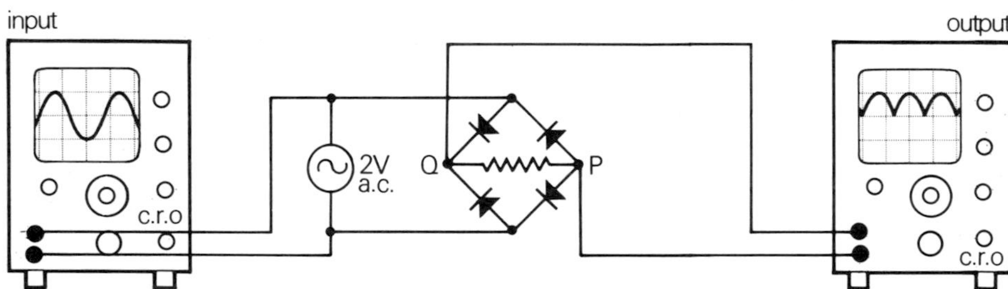

Figure 21.31 Full-wave rectification

We can follow how full-wave rectification takes place if we consider in detail what happens to the flow of electrons when the applied p.d. causes electron flow in the two different directions, Figure 21.32.

Figure 21.32 Full-wave rectification

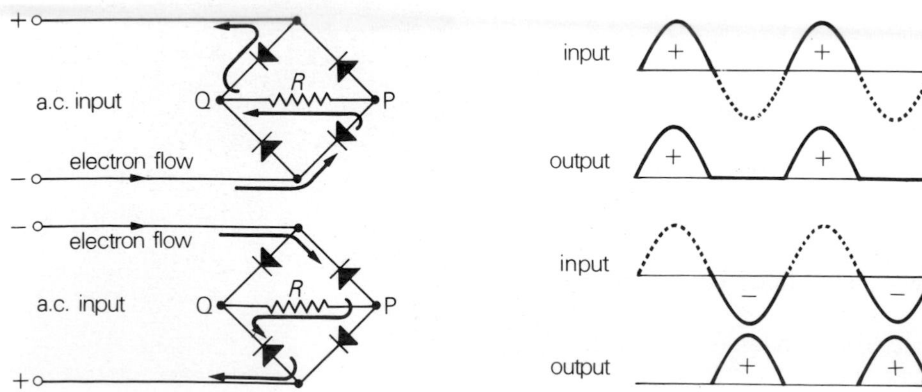

We note that the change in polarity of the input p.d. results in a change in the path taken by the electrons. Two different diodes are used for each half wave, but the electron flow is always in the direction PQ through the resistor. Since electrons flow from P to Q, Q is the positive terminal of the d.c. output, Figure 21.33.

In a power supply where a transformer is used to step down the p.d., we need only two diodes to produce full-wave rectification. An example of such a system using a centre-tapped transformer to obtain a low voltage from the mains supply is given in Figure 21.34.

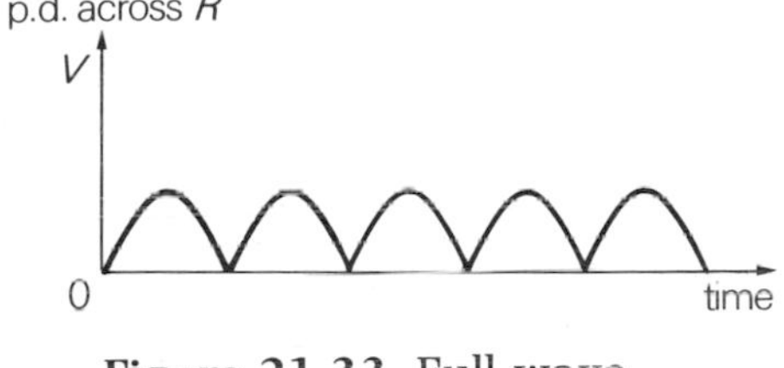

Figure 21.33 Full-wave rectification output

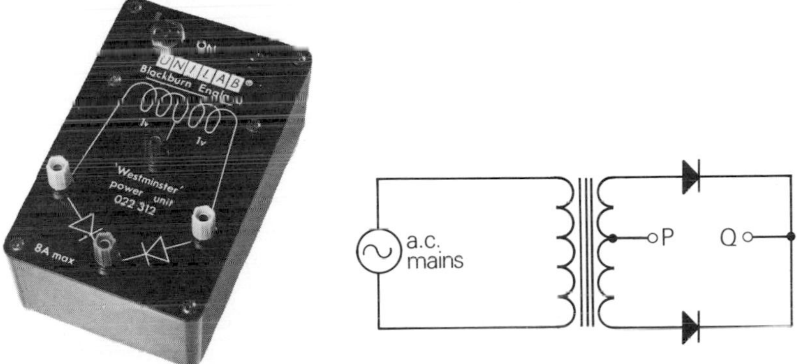

Figure 21.34 Low voltage d.c. unit

Figure 21.35 shows the direction of electron flow for two successive half waves.
Once again the electron flow is from P to Q through the resistor connected to the power supply and we obtain a graph of potential difference against time as in Figure 21.33.

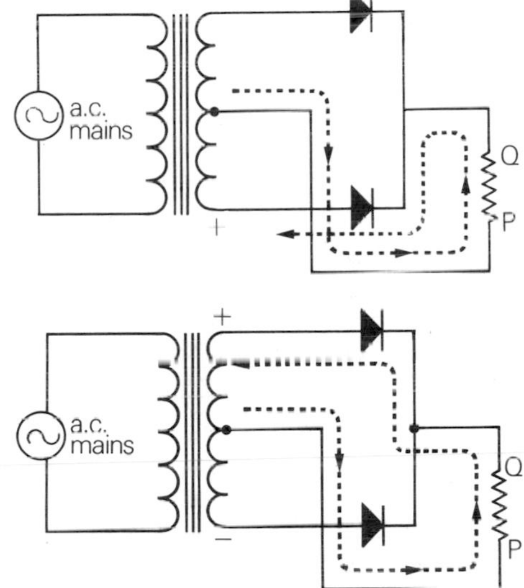

Figure 21.35 Alternative full-wave rectification circuit

21.8 Transistors

The transistor is not a little box from which we can hear the pop music of the day; the box is a transistor radio which of course contains transistors, Figure 21.36.

Figure 21.36 Transistors and transistor radio

The transistor is a semiconductor device which consists of two p-n junctions back to back. There are two possible arrangements: a p-n-p sandwich or an n-p-n sandwich, Figure 21.37. The different regions of semiconducting material in the transistor are given the names **collector**, **base** and **emitter**. They are often contained inside a small metal or plastic case.

In a p-n-p transistor, the centre layer (called the base) is made of n-type material and is very thin. The two outer layers are thicker and are called the collector and emitter. Figure 21.38 shows how the transistor is usually connected.

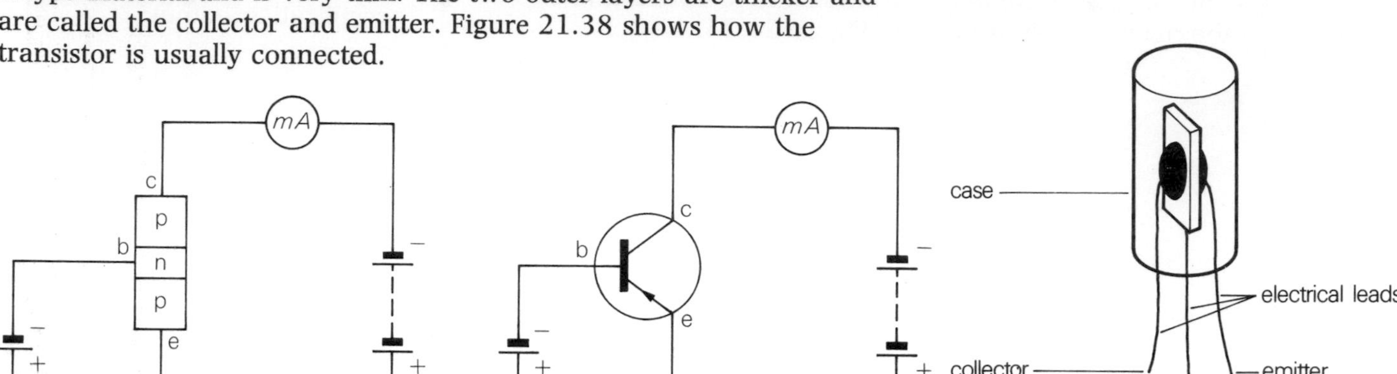

Figure 21.38 Transistor circuit

Figure 21.37 p-n-p transistor construction and symbol

Although the emitter-base junction is forward biased, the base-collector junction is not: it is reverse biased.

We can use the circuit in Figure 21.39 to investigate how the transistor behaves.

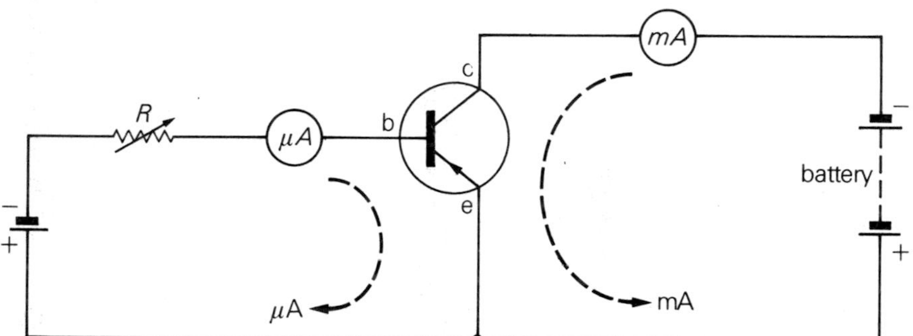

Figure 21.39 Transistor characteristics

If we alter the current flowing to the base of the transistor by adjusting the rheostat R, we find that a change of a few microamperes in this part of the circuit results in a change in the milliammeter reading in the collector-emitter part of the circuit. A very small change (μA) in the base current results in a much larger change (mA) in the collector current. We say that the transistor **amplifies** the change in the base current. The increased current is of course supplied by the battery on the right.

We can use the transistor as a current amplifier to amplify the very small signal produced by a microphone to a suitable size to produce an audible sound in an earphone, Figure 21.40.

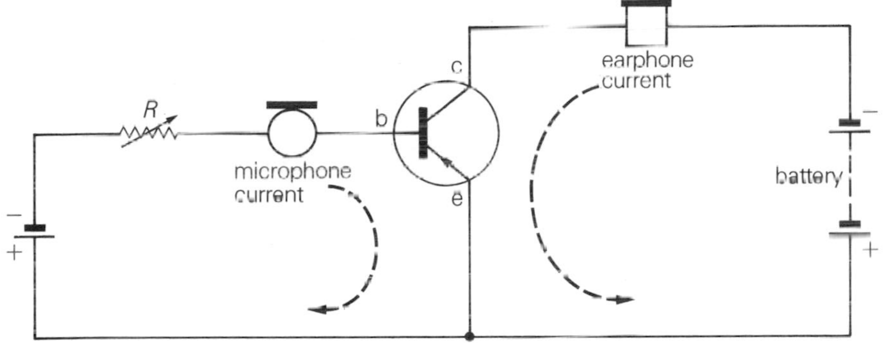

Figure 21.40 Amplification

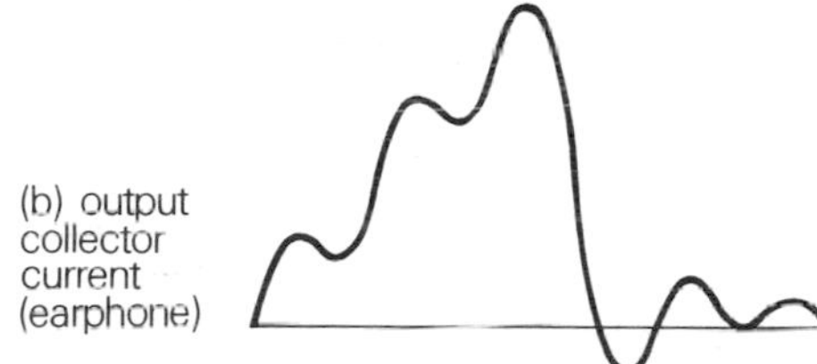

Figure 21.41 Amplifier input and output

In practice the circuits employed with transistors in them are rather more complicated than this. Figure 21.41 shows how the variations of the input (microphone) signal are copied exactly by the output current (earphone).

The increased use of semiconductor devices instead of valve-type components has resulted in great savings in size, in power consumption (since there are no heater supplies required), and in cost due to high reliability. The potential differences required are also much lower. But it is the development of the integrated circuit (Figure 21.42) which has contributed most to the advance of electronics. In such a circuit thousands of components, including semiconductor devices can be included on a chip no larger than the head of a pin.

Figure 21.42 Integrated circuits

Summary

When a thin filament of metal is heated, electrons may be emitted due to thermionic emission.

In a diode valve, electrons flow from the negative cathode to the positive anode. There is a one-way current.

A diode valve can convert a.c. to half-wave rectified d.c.

Cathode rays are fast moving electrons. A cathode ray oscilloscope (C.R.O.) can be used to measure p.d. and frequency.

A solid state diode is composed of n-type and p-type semiconductor material.

A rectifier bridge circuit may be used to convert a.c. to full-wave rectified d.c.

A transistor has three connections: collector, base, and emitter. A transistor can be used to amplify current.

Problems

1 Describe a diode valve and how it works.

2 Draw a diagram to illustrate how to obtain full-wave rectification using solid-state diodes.

3 Explain what each of the following C.R.O. controls does: brightness, focus, gain, time-base, X-plates, Y-plates.

4 Describe how to use a C.R.O. to measure
4·1 potential difference
4·2 frequency.

5 Draw the symbol for a transistor. What does a transistor do in a circuit when it is connected as in the diagram below?

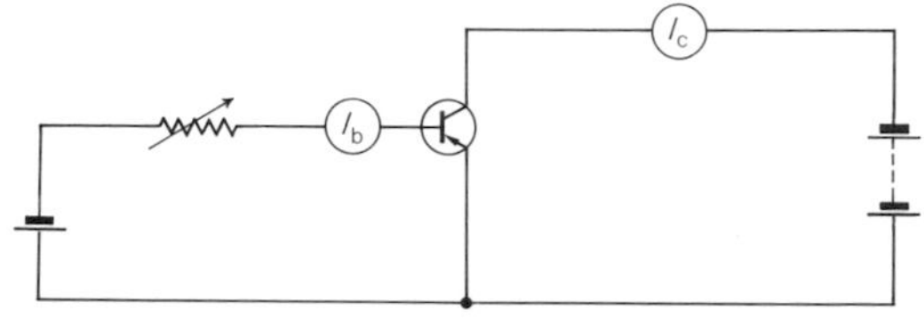

6 A cathode ray tube was modified by sealing a small metal can inside the glass envelope as shown.

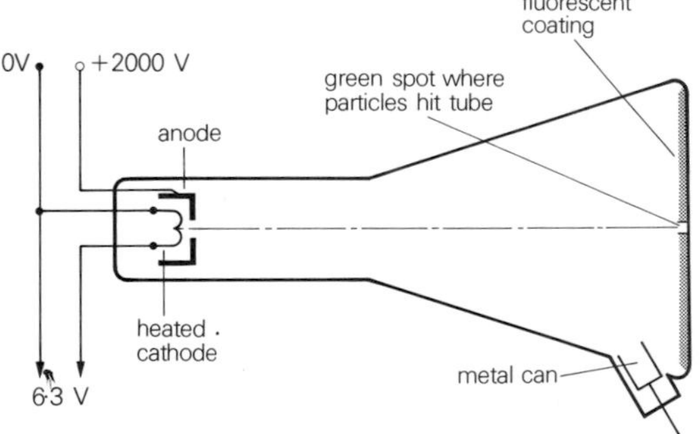

How could you
6·1 deflect the particles into the metal can;
6·2 tell if the particles were charged;
6·3 decide whether the charge was positive or negative?
SCEEB

7 This is a test circuit for a p-n-p transistor.

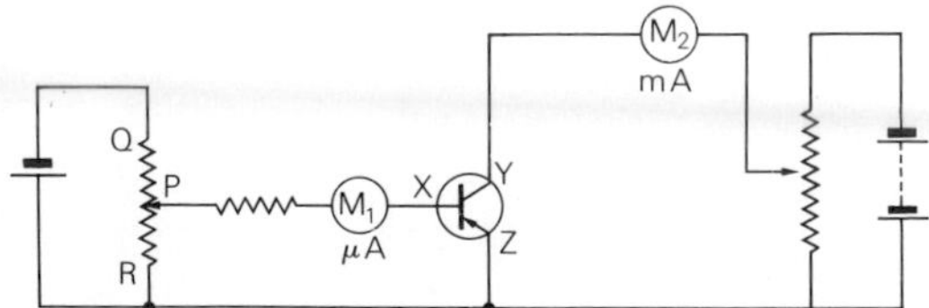

7·1 Which of the terminals X, Y and Z correspond to the emitter, base and collector?
7·2 If the sliding contact P is moved towards end Q, what would happen to the reading on:
a) the microammeter M_1,
b) the milliammeter M_2?
7·3 How do the corresponding changes of current through M_1 and M_2 compare in size? *SCEEB*

8 The diagrams show experiments in which important components have been omitted, and are indicated by capital letters.

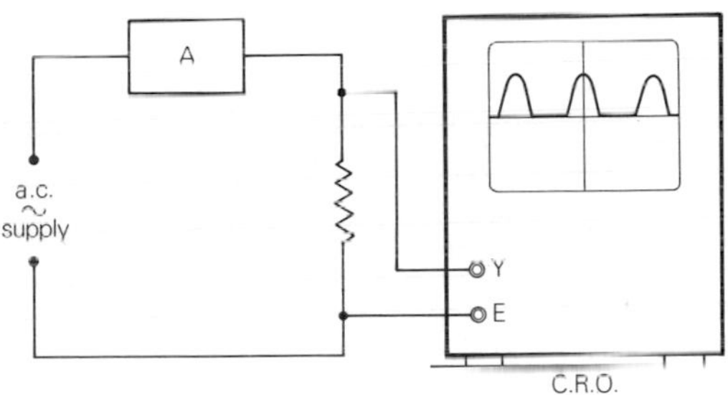

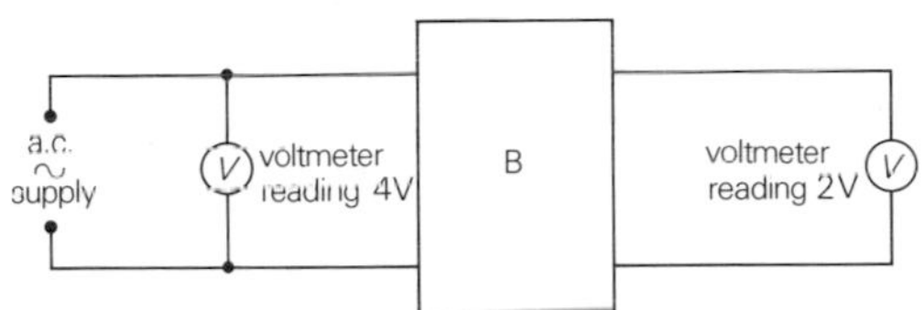

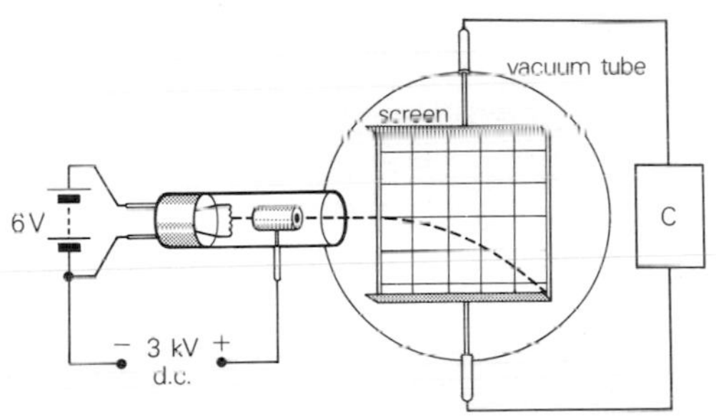

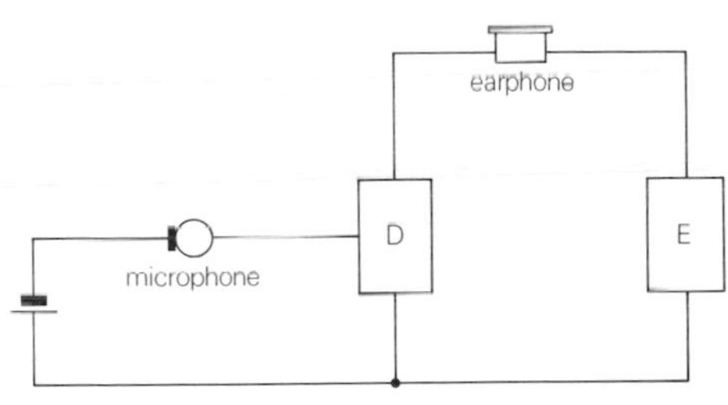

State what each of the items A, B, C, D and E is, and **briefly** describe its purpose in its circuit.

SCEEB

9 **9·1** The symbol for a thermionic diode valve is shown below.

a) What are the parts of this valve which are labelled A, C and H?
b) What is the function of H?
c) What is the function of C?

9·2 In the circuit shown below a reading was obtained on the milliammeter.

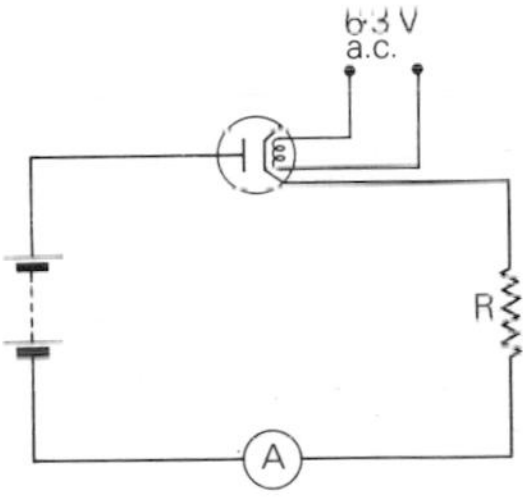

What would be the effect (if any) on the milliammeter reading when each of the following alterations was made in turn to the above circuit?
a) Reversing the resistor R.
b) Disconnecting the 6·3 V a.c., supply.
c) Reversing the battery.

9·3 The following circuit can be used to convert alternating current to direct current. A cathode ray oscilloscope was connected across the terminals X and Y.

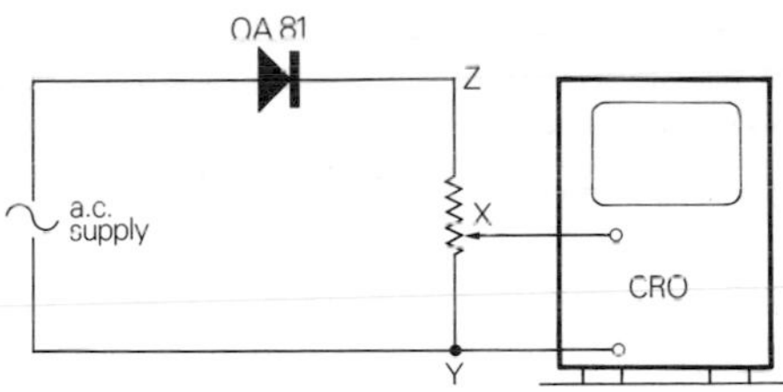

a) What is the component labelled OA81?
b) Sketch what would be seen on the screen of the C.R.O. with a suitable time-base switched on.
c) How would the trace be affected if the contact X were moved closer to Z?
d) Which terminal, Z or Y, would you mark positive?

SCEEB

10 Answer this question on a page of graph paper. It is arranged that this trace is seen on the screen of a cathode ray oscilloscope before each of the following adjustments is made.

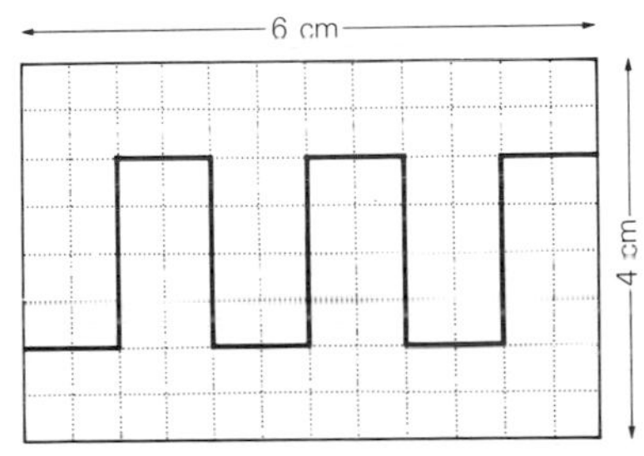

10·1 The Y-shift control altered slightly;
10·2 the Y-gain control altered slightly;
10·3 the time-base frequency increased.
Draw patterns in separate rectangles (6 cm by 4 cm) on graph paper to illustrate the effect of **each** of these alterations.

SCEEB

22 Radioactivity I

22.1 Discovery

Natural radioactivity was discovered by Henri Becquerel in Paris in 1896. He was at the time investigating the properties of uranium salts, in particular their fluorescence (glowing) in the presence of light. He thought that X-rays might be connected with this fluorescence and to test this he wrapped a photographic plate in black paper, put a silver coin on top and then some uranium salt on top of that. When the crystal of uranium salt was exposed to sunlight it glowed. On development, the photographic plate turned black except for the area below the coin, so it looked as if the uranium salt was emitting X-rays which were penetrating the paper but not the coin.

Being a good scientist, Becquerel decided to repeat his experiment. However the weather was now cloudy and he simply set the experiment up again with a new photographic plate, and put the apparatus in a drawer so that it would be ready when the sun came out. After several days the sun was still not shining and Becquerel decided to develop the plate even though there had been no sunshine and so no fluorescence. To his astonishment, the plate turned as black as before.

However, Becquerel did not discard this result as an accident but instead investigated it properly by repeating the experiment in total darkness and again obtained the same result. Becquerel had discovered what Madame Curie later called **radioactivity**.

Becquerel's experiment can be repeated using sealed photographic film with a radioactive source held above it, Figure 22.1.

A metal key is placed on the film and the apparatus left overnight. The resulting radioactivity photograph is shown in Figure 22.2.

The fact that the key can be seen shows that something has penetrated the wrapping of the film, but not the key.

We call this 'something' **radiation** and the source a **radioactive** source

Becquerel did not know that this radiation was dangerous and was fortunate to avoid any harmful after-effects. These radiations can cause ionization of molecules in living cells (see section 22.5), and this can lead to the whole cell being destroyed.

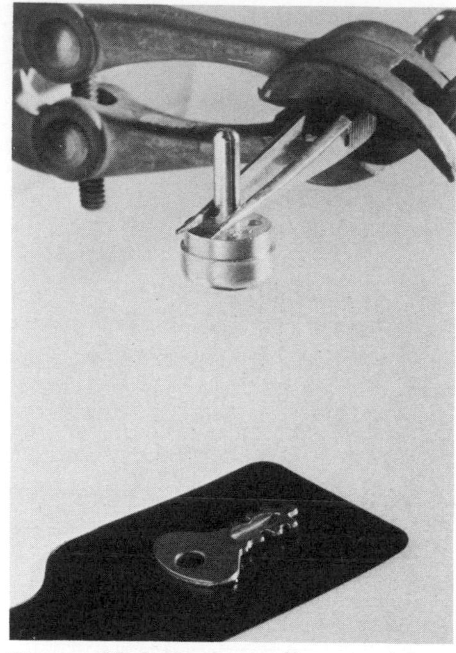

Figure 22.1 Becquerel's experiment

Figure 22.2 Radioactivity photograph

22.2 Absorption and deflection of radiation

When we repeated Becquerel's experiment, we saw that the radiation was stopped by the key although it could penetrate the wrapping of the film. We will now look at the ways in which radiation can be stopped.

When radiation penetrates matter, it loses energy by collisions with particles. It is then said to have been **absorbed**. By looking at this property of absorption it can be demonstrated that there are three distinct types of radiation.

Absorption of radiation

If we take a radioactive source which emits all the types of radiation, we can detect them using a Geiger tube connected to a ratemeter (explained in section 22.5). The apparatus used is shown in Figure 22.3.

Figure 22.3 Investigating absorption

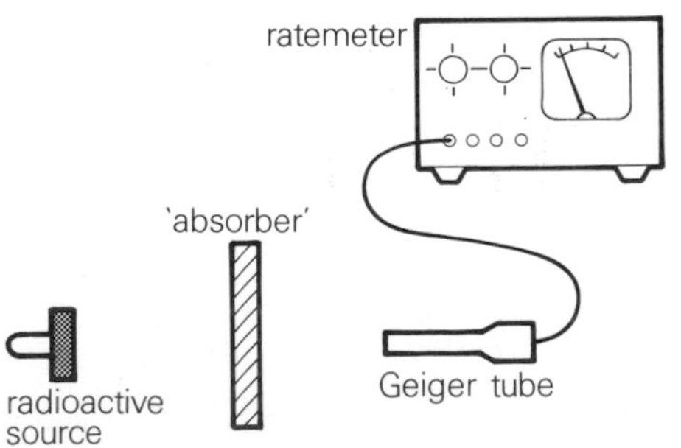

When a sheet of thin paper is inserted between the source and the detector, the reading on the ratemeter drops. This indicates that some of the radiation has been absorbed. The radiation that is absorbed by thin paper is called **alpha radiation** (α radiation). A radioactive source which emits only alpha radiation is called a pure alpha source.

When a sheet of aluminium is now inserted between the source and detector, the reading drops further. This indicates that some more of the radiation has been absorbed. The extra radiation that is absorbed by the sheet of aluminium is called **beta radiation** (β radiation). A radioactive source which emits only beta radiation is called a pure beta source.

When a thick sheet of lead is inserted between the source and detector, the reading drops even further. This indicates that there is a third type of radiation which is being absorbed. The radiation that is absorbed by a thick sheet of lead is called **gamma radiation** (γ radiation). A radioactive source which emits only gamma radiation is called a pure gamma source. Some sources emit more than one type of radiation.

The results of this experiment are summarized by Figure 22.4.

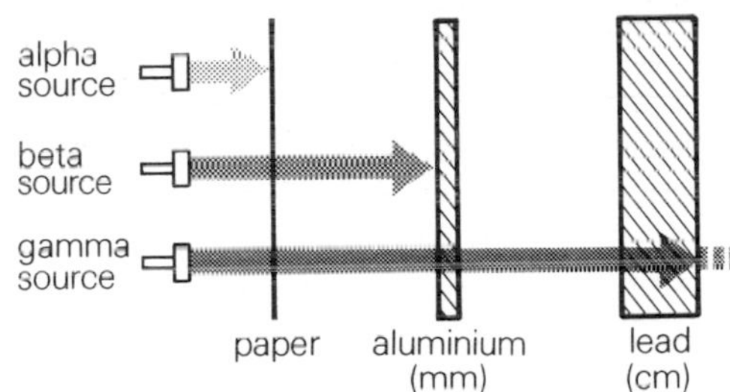

Figure 22.4 Absorbing α, β and γ radiations

Range in air

The range of radiation in air is the distance it can penetrate air before it becomes totally absorbed. It can be investigated by changing the separation of the radioactive source and detector.

When a source of alpha radiation is used with a thin-window Geiger tube and ratemeter, the following results are obtained.

As the separation is increased, the reading decreases until (beyond a few centimetres) the reading drops suddenly and only a low count remains.

This indicates that the range of alpha radiation in air is a few centimetres.

Background radiation

In the above experiment, the low count rate that remains, even though all radiations from the radioactive source have been absorbed, is called **background radiation**. Background radiation is present all the time and originates from outer space and from natural radioactivity around us.

Deflection of radiation

If radiation can be deflected by magnetic or electric fields, it must consist of charged particles. We can use this to test whether alpha, beta and gamma radiations consist of charged particles.

Deflection by a magnetic field

The deflection of radiation by a magnetic field can be demonstrated by inserting a strong magnet between the source and the detector.

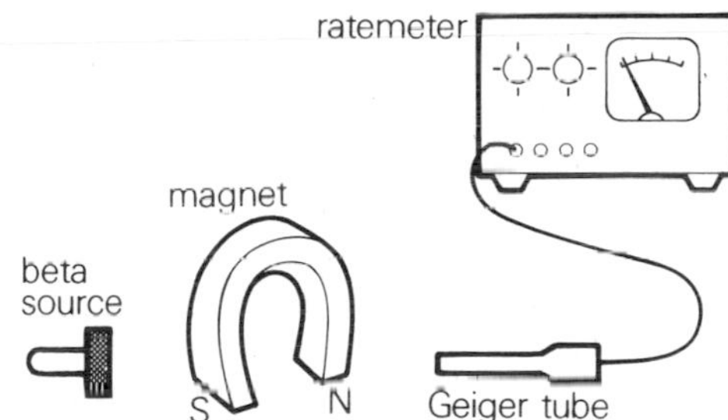

Figure 22.5 Deflection of beta radiation by a magnetic field

Figure 22.5 shows a Geiger tube connected to a ratemeter which is detecting radiation from a beta source. When a magnet provides a magnetic field across the path of the radiation, the reading on the ratemeter drops. This means that some beta radiation is being deflected by the magnetic field and is no longer reaching the detector. We conclude that beta radiation consists of charged particles.

Figure 22.6 Nature of the charge carried by beta particles

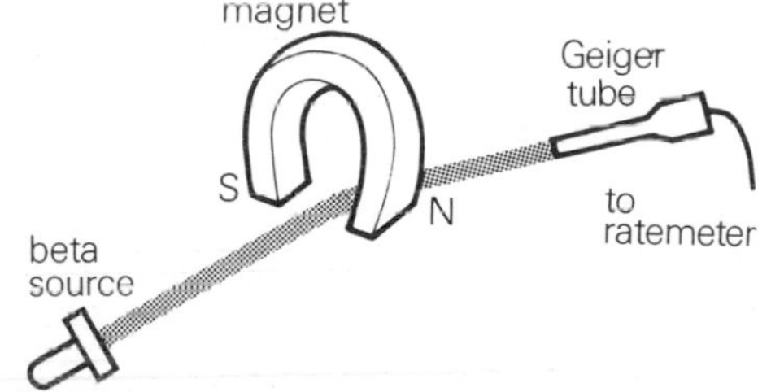

We can also show that beta particles carry a negative charge, Figure 22.6. The source and Geiger tube are tilted slightly so that without the magnet present, the detector indicates a low count. When

the magnet is inserted between the source and the detector, the reading increases showing that the direction of the beta particles is being altered and they are now reaching the detector. It is important to note that the direction of the magnetic field, the original direction of the beta particles, and the direction of the force which is changing their direction, are all at 90° to each other. From the direction in which the particles are bent, it can be deduced (using the right-hand motor rule: see page 201) that beta particles carry a negative charge.

This experiment is not so convincing for alpha radiation because it is not so strongly deflected and also its range in air is too small. However, the deflection of alpha radiation can be demonstrated using a sealed plastic tube from which the air has been removed, Figure 22.7. Because the range of alpha radiation is much greater in the absence of air, a small deflection of the alpha radiation can be made more noticeable. When the magnet provides a magnetic field across the path of the radiation, the reading on the detector drops, showing that, like beta radiation, alpha radiation also consists of charged particles.

If the plastic tube is bent, the reading also drops. Using the magnet to deflect the radiation so that it again reaches the detector, it is seen that the direction of the magnetic field, the original direction of the alpha particles, and the direction of the force which is changing their direction, are again all at 90° to each other. From the direction in which they are bent it can be deduced that alpha particles carry a positive charge.

Figure 22.8 shows a cloud chamber photograph of alpha and beta particle tracks bent in a magnetic field. The working of the cloud chamber is explained in section 22.5.

If the original experiment is carried out with a source of gamma radiation, no change in the reading occurs. This indicates that gamma radiation does not consist of charged particles. In fact, other experiments have proved that it is a form of electromagnetic radiation.

The deflections of alpha, beta, and gamma radiations in magnetic fields are summarized in Figure 22.9.

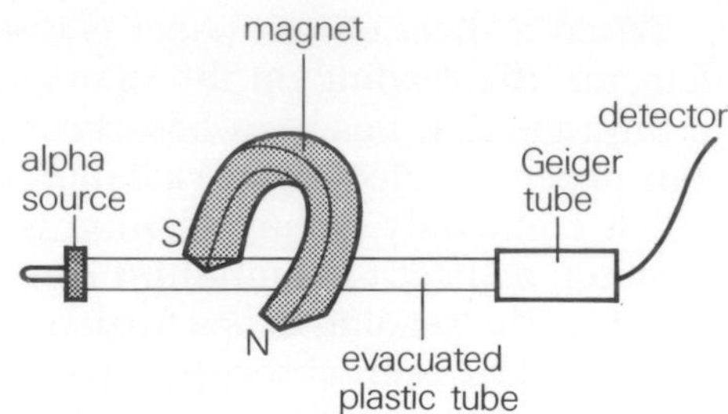

Figure 22.7 Nature of the charge carried by alpha radiation

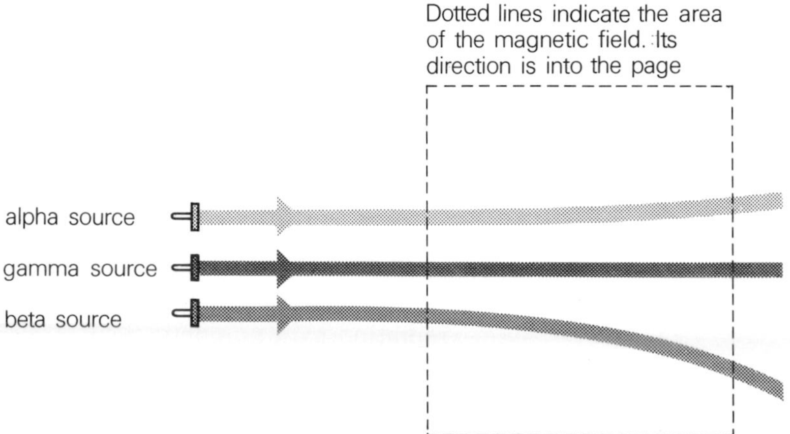

Figure 22.9 Deflection by a magnetic field

Figure 22.8 Cloud chamber photographs

Alpha tracks in a strong magnetic field.

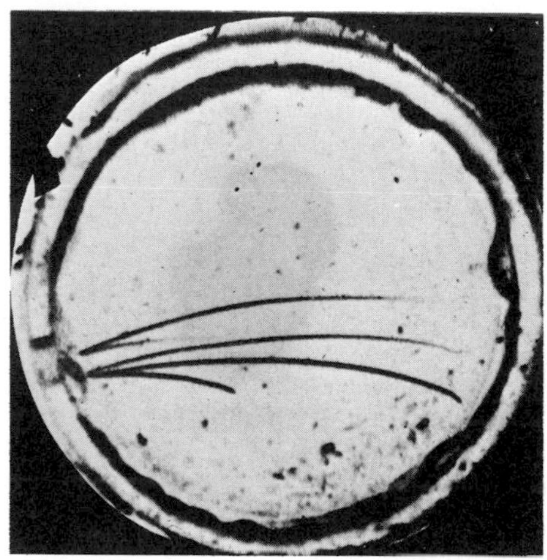

Beta tracks in a weak magnetic field.

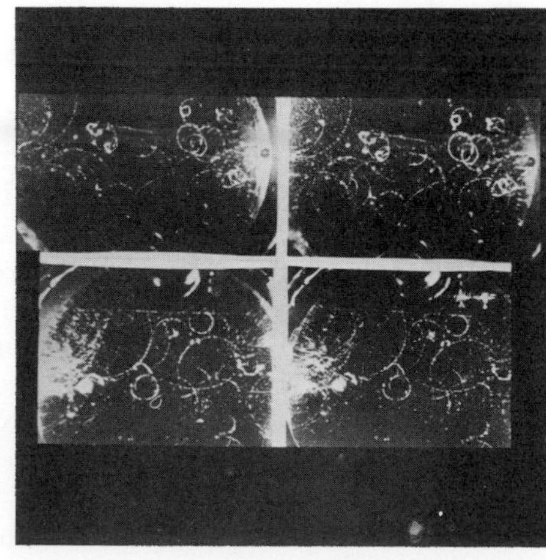

Deflection by an electric field

Using magnetic fields we have shown that,

a) alpha particles are positively charged.
b) beta particles are negatively charged, and
c) gamma radiation carries no charge.

If these three radiations are passed through an electric field between two parallel plates, they are deflected as shown in Figure 22.10.

The alpha radiation is deflected towards the negative plate, the beta radiation is strongly deflected towards the positive plate, but the gamma radiation remains undeflected.

Because of their much lower mass, beta particles are more strongly deflected than alpha particles by both magnetic and electric fields.

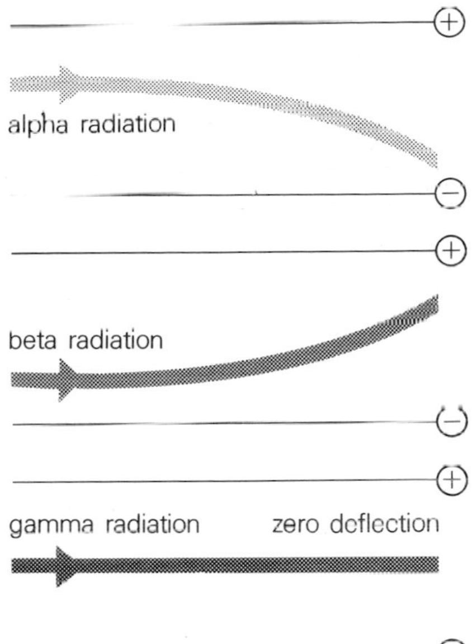

Figure 22.10 Deflection in an electric field

22.3 The nuclear atom

An atom consists of a small positively charged **nucleus** with negatively charged **electrons** around the nucleus. It is the nucleus that emits the radiation, Figure 22.11.

The nucleus consists of **protons** and **neutrons** which have similar mass, while the electrons have a much smaller mass. The neutron is uncharged, while the proton and electron have equal but opposite charges (positive and negative respectively). As the protons and neutrons are in the nucleus, it is the nucleus that carries all the positive charge and practically all the mass of an atom. However the electrons carry an equal amount of negative charge and so the atom as a whole is electrically neutral. These properties are summarized in Table 1 which takes the unit of mass as that of a proton and the unit of charge also as that of a proton.

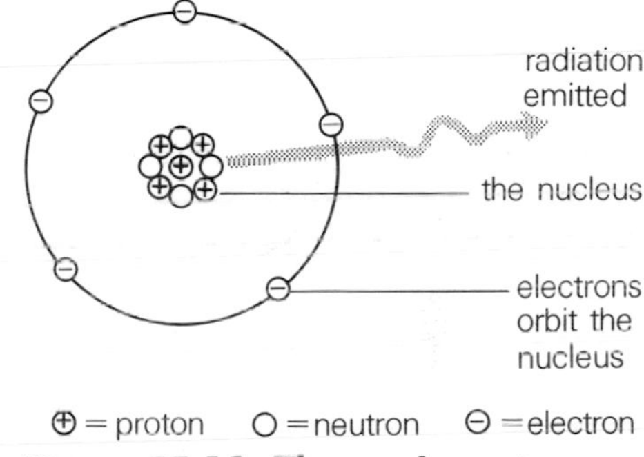

Figure 22.11 The nuclear atom

Particle	*Approximate mass (proton = 1)*	*Charge (proton = +1)*	*Is it in the nucleus?*
Proton (p)	1	+1	Yes
Neutron (n)	1	0	Yes
Electron (e)	$\frac{1}{1840}$	−1	No

Table 1 Properties of atomic particles

The number of protons in the nucleus of an atom is its **atomic number** Z. The atomic number is therefore the number of positive charges in the nucleus. For example, an atomic number of 2 means that the atom has a charge of +2 in its nucleus. Each different element has a different atomic number. The element with an atomic number of 2 is helium. All the elements can be arranged in a list in increasing order of atomic number starting with Z = 1 (hydrogen). This list makes up the Periodic Table.

The total number of protons and neutrons in the nucleus of an atom is called its **mass number** A. The mass number is therefore the total number of nuclear particles (or nucleons) and indicates the total mass of the nucleus, where the unit of mass is that of one proton. For example, the most common form of helium has a mass number of 4. This means that a helium nucleus has 4 nucleons or that its mass is 4 units.

Table 2 shows the atomic and mass numbers of the first four elements.

Different elements have different chemical symbols. An element with

Atomic number		*Contents of the nucleus*	*Element*	*Mass number*
1	⊕	1 p	hydrogen	1
2		2 p + 2 n	helium	4
3		3 p + 4 n	lithium	7
4		4 p + 5 n	beryllium	9

+ or p = proton O or n = neutron

Table 2 Atomic and mass numbers

the chemical symbol X, mass number A, and atomic number Z is described more simply by writing it as ${}^{A}_{Z}X$.

Helium has an atomic number of 2 and a mass number of 4. As the chemical symbol for helium is He, it is written as ${}^{4}_{2}He$.

It is possible for nuclei with the same atomic number to have different mass numbers. Atoms with the same atomic number (i.e. the same number of protons), have the same chemical properties and are therefore given the same name and symbol. For example, a nucleus with an atomic number of 2 and a mass number of 3 is a form of helium and is written as ${}^{3}_{2}He$.

Nuclei of the same element (i.e. same atomic number) but with different mass numbers, are called **isotopes**. This means that ${}^{4}_{2}He$ and ${}^{3}_{2}He$ are isotopes of the element helium. Both have 2 protons in the nucleus, but the first isotope has 2 neutrons (giving a mass number of $2 + 2 = 4$), while the second isotope has 1 neutron (giving a mass number of $2 + 1 = 3$).

It is also possible for different *elements* to have the same mass numbers. For example, ${}^{212}_{82}Pb$ and ${}^{212}_{83}Bi$ represent forms of the elements lead and bismuth respectively. Although they have different atomic numbers, they have the same total number of nucleons in the nucleus. A nucleus of ${}^{212}_{82}Pb$ consists of 82 protons and 130 neutrons, giving a mass number of 212. A nucleus of ${}^{212}_{83}Bi$ consists of 83 protons and 129 neutrons, giving the same mass number of 212.

22.4 Properties of α, β and γ radiations

Alpha radiation

Alpha radiation is the emission of alpha particles by a radioactive source. An alpha particle is the nucleus of a helium atom which has a mass of 4 units since it contains 4 nuclear particles and a charge of +2 units since it contains two protons. Alpha radiation has a range of a few centimetres in air. Alpha particles travel slowly compared with beta particles because of their large mass. Alpha particles can be deflected by electric or magnetic fields, as can any moving charged particle.

Beta radiation

Beta radiation is the emission of beta particles. A beta particle is an electron moving at high speed (approaching the speed of light). It has a charge of −1. Beta radiation is more penetrating than alpha radiation and has a range in air of many centimetres. As it has a very small mass and is a moving charged particle, it is strongly deflected by a magnetic field. It is also strongly deflected by an electric field.

Gamma radiation

Gamma radiation is a form of electromagnetic wave with a very high frequency and very short wavelength. It travels at the speed of light as do all electromagnetic waves and obviously has no mass or charge. It is extremely penetrating and its range even in lead is many centimetres. Electric or magnetic fields have no effect on gamma radiation.

The properties of alpha, beta, and gamma radiations are summarized in Table 3.

Type	*Nature*	*Mass* (*proton* = 1)	*Charge* (*proton* = +1)	*Speed*	*Absorbed by*	*In an electric or magnetic field is*
Alpha particle α	Helium nucleus 2 p + 2 n	4	+2	Up to $\frac{1}{10}$ of the speed of light	A sheet of paper or a few centimetres of air	Deflected
Beta particle β	Fast moving electron	$\frac{1}{1840}$	−1	Up to $\frac{9}{10}$ of the speed of light	A few millimetres of aluminium	Strongly deflected
Gamma ray γ	High frequency electromagnetic radiation	0 Waves have no mass!	0	Speed of light	Many centimetres of lead (but not totally absorbed).	Not deflected

Table 3 Properties of α, β and γ radiations

22.5 Detection of α, β and γ radiations

Ionization

When an electron is added to or taken away from an atom, an **ion** is formed. An ion is therefore negatively or positively charged. The process of forming ions is called ionization.

If a lighted match is held near a charged electroscope, the leaf falls, Figure 22.12.

The flame has produced positive and negative ions in the air around the disc of the electroscope. If the electroscope is negatively charged, the disc will attract the positive ions. The positive ions (which are short of electrons) remove electrons from the disc to become neutral atoms. Thus the disc loses its extra electrons and so the leaf falls. You should be able to work through a similar argument for a positively charged electroscope. Both positively and negatively charged electroscopes discharge when a flame is brought near the disc of the electroscope.

By this time you may be puzzled as to where radioactivity fits into

Figure 22.12 Flame discharging an electroscope

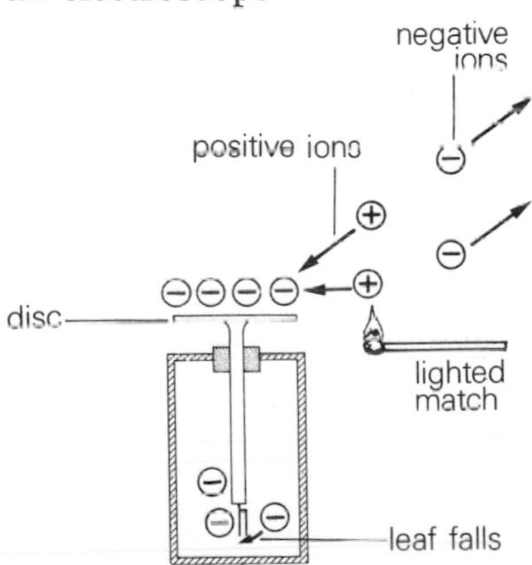

this discussion! In fact, many detectors of radioactivity use the ability of radiation to produce ionization.

When an alpha source is held near the disc of a charged electroscope, the leaf falls because an alpha source produces a great deal of ionization in air. A charged electroscope can therefore be used as a detector of alpha radiation.

Detectors

1 Gold-leaf electroscope

We have seen that a charged electroscope will detect radiation. The strength of the source can be estimated by the rate at which the leaf falls. This detector works well only for alpha radiation (strongly ionizing).

2 Spark counter

A spark counter consists of a wire gauze below which is a thin wire. A high voltage supply is used to provide a potential difference of a few thousand volts between the wire and the gauze, Figure 22.13.

When an alpha source is brought near the gauze the resulting ionization of the air allows a charge to flow. This current is indicated by sparks between the wire and the gauze. The rate at which sparks are produced indicates the strength of the alpha source. This detector only works for alpha radiation.

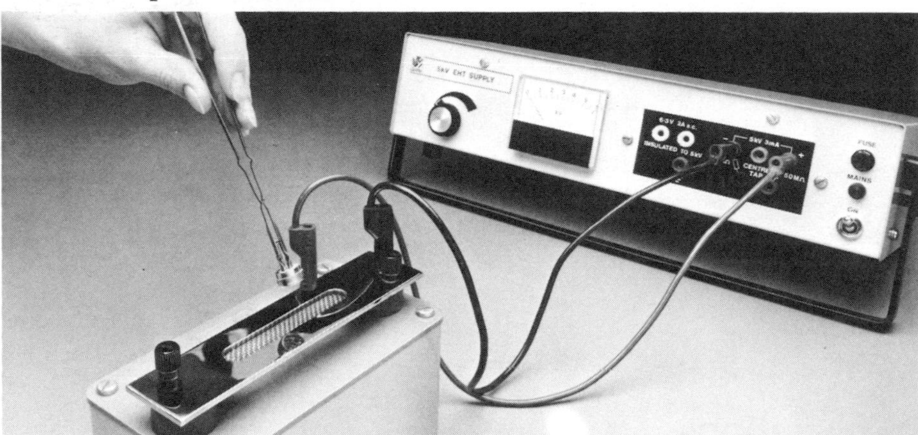

Figure 22.13 Spark counter

3 Geiger-Müller tube

A Geiger-Müller tube contains two electrodes across which a potential difference of around 400 V is applied, Figure 22.14. One electrode consists of a central wire while the second electrode is a cylindrical tube. The tube is filled with gas at low pressure and has a thin mica window at one end through which radiation can enter. When radiation passes through the window, it causes ionization of the gas and as a result an electrical pulse travels between the two electrodes.

The output of the Geiger tube can be registered on a scaler (Figure 22.15 a) which counts individual pulses and registers them on dials, or on a ratemeter (Figure 22.15 b) which indicates the average number of pulses per second on a meter. A scaler gives a measure of the total amount of radiation entering the Geiger tube in a given time. A rate-meter indicates the rate at which radiation is entering the Geiger tube.

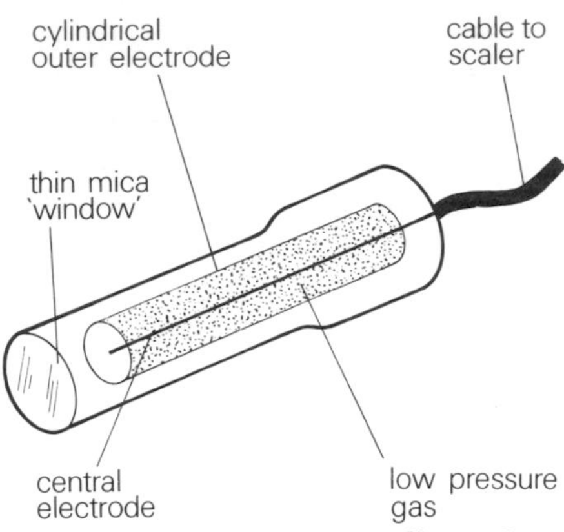

Figure 22.14 Geiger-Müller tube

Figure 22.15

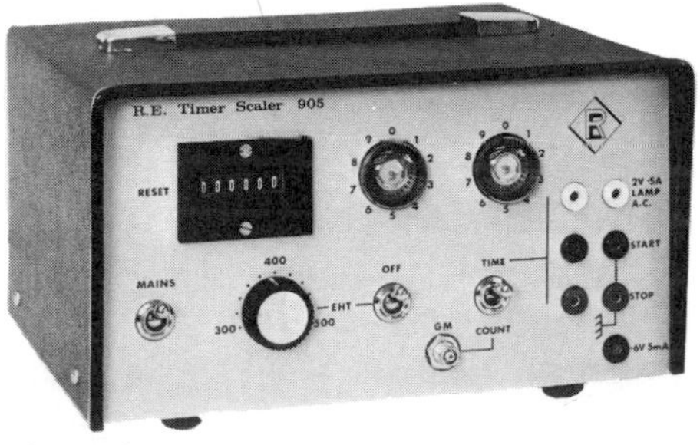

a) Scaler

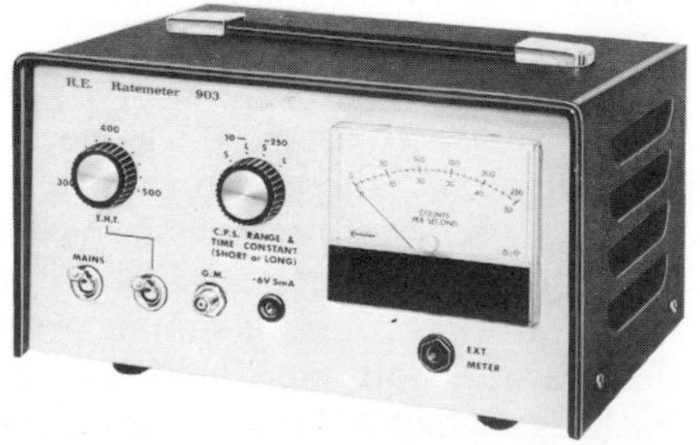

b) Ratemeter

A Geiger tube with a scaler or ratemeter will detect beta and gamma radiation. It will also detect alpha radiation provided the window is thin enough to allow it to enter the tube.

4 Ionization chamber with a d.c. amplifier

An ionization chamber consists of a metal can with wire gauze forming a lid. Inside the box is a metal rod and a plate which is just below the gauze, Figure 22.16. The metal plate is insulated from the box and connected to the d.c. amplifier by the rod. When an alpha source is brought near the gauze, it causes ionization of the air and a very small current between the gauze and the plate. A battery provides a potential difference between the gauze and the plate.

The ionization chamber therefore works in a similar way to the spark counter. In the case of the ionization chamber, there is a current without any visible spark.

The small current is amplified by the d.c. amplifier to give a reading on a milliammeter that is proportional to the activity of the source. This detector works well for alpha radiation (strongly ionizing) and can also detect beta radiation (less strongly ionizing).

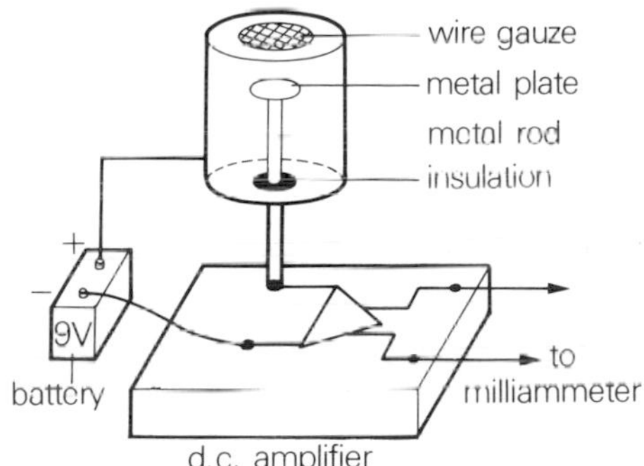

Figure 22.16 Ionization chamber and d.c. amplifier

5 Cloud chamber

A type of cloud chamber you may meet in school is the diffusion cloud chamber, Figure 22.17.

The floor of the chamber is cooled using solid carbon dioxide. A felt ring at the top of the chamber is soaked with alcohol. The alcohol evaporates and cools as it diffuses downwards. After about a minute, conditions are right for the alcohol vapour to condense. However, the alcohol will only condense on a particle of dust or a charged particle. As a radioactive source produces ions in air, the alcohol will condense on these charged particles.

When the chamber is illuminated, the alcohol droplets reflect the light and so can be seen. In this way the passage of radiation through the chamber causes a track to be seen. Note that what you are seeing is not the actual radiation but alcohol droplets which have condensed on the ions produced by the radiation. When the perspex top of the chamber is rubbed with a cloth to charge it, the resulting electric field removes any unwanted ions.

The cloud chamber shows up the tracks of alpha and of beta radiations, Figure 22.18.

Figure 22.17 Diffusion cloud chamber

Figure 22.18 Cloud chamber photographs

a) alpha tracks

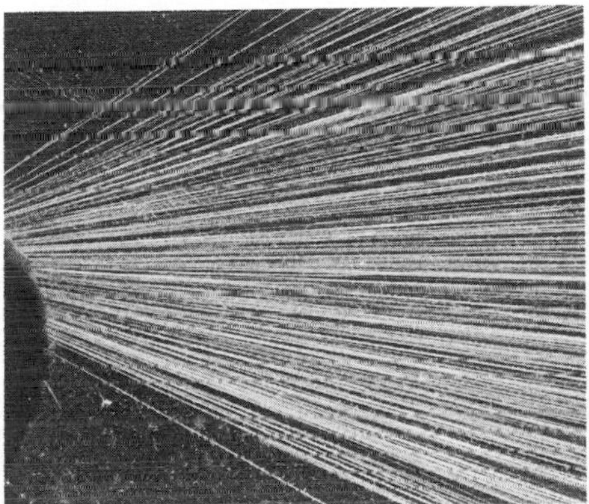

b) beta tracks

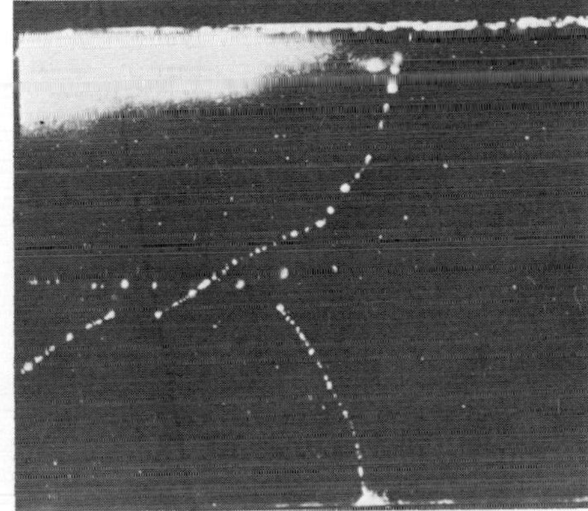

In photograph (a), note that the tracks are thick because alpha particles produce many ions in air. They are found to be roughly the same length showing they have about the same range.

In photograph (b), note that the tracks are more 'wispy' and are not straight. This is because beta particles do not ionize air as strongly as alpha particles, while the straggly nature of the tracks is because beta particles, being smaller than alpha particles, are jolted each time they collide with an air molecule.

The original cloud chamber was invented in 1911 by C. T. R. Wilson. His chamber showed up the paths of radiation in moist air, Figure 22.19.

When the pressure in the chamber was reduced by pumping out some air, the air became cooler and conditions were favourable for water particles to condense on the ions produced by the passage of radiation. In a similar way to the diffusion cloud chamber, tracks were produced by the passage of radiation through the chamber.

One of the most important detectors used in research today is the **bubble chamber**, Figure 22.20.

The bubble chamber produces tracks as a result of bubbles forming in a liquid. The bubbles are formed on the ions produced by radiation in a similar way to vapour condensing on ions in a diffusion cloud chamber.

Figure 22.19 Wilson cloud chamber

Figure 22.20 Bubble chamber

6 Photographic film

Radiation affects photographic film in a similar way to light. Figure 22.2 on page 244 shows a radioactivity photograph that was obtained from the Becquerel-type experiment. Photographic film can detect alpha, beta, and gamma radiation.

7 Solid state detector

Many types of solid state detector have been developed. One you may meet in school consists of a circuit containing a p-n semiconductor junction, Figure 22.21.
Radiation falling on the junction causes a pulse in the circuit which can be detected using a scaler or ratemeter. Other solid state detectors can be used to detect alpha, beta, and gamma radiations.

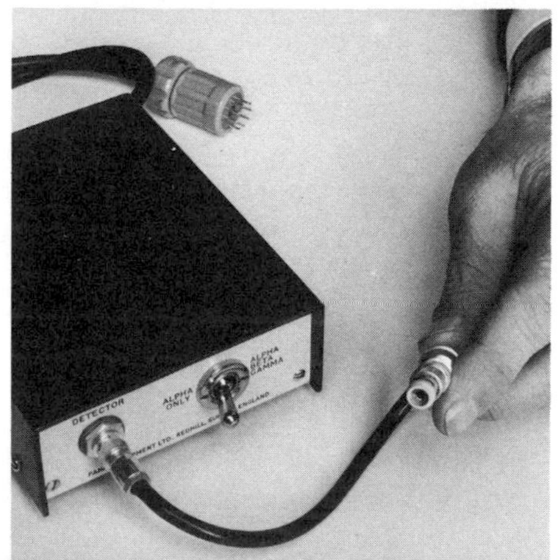

Figure 22.21 p-n junction solid state detector

22.6 Half-life

Radioactive decay is totally irregular (or **random**). This means that it is impossible to predict when one particular atom will disintegrate. For example, when a spark counter is detecting alpha radiation, the sparks do not occur at a regular rate but at random time intervals. Similarly in a cloud chamber both alpha and beta tracks appear at irregular intervals, and with a Geiger tube/scaler the pulses are seen to occur in a random way.

A practical radioactive source contains many millions of atoms. Although it is impossible to know when any one atom will disintegrate, it is possible to determine the average number of disintegrations per second when large numbers of atoms are present. The average number of disintegrations per second is called the **activity** of the radioactive source. The unit of activity is the **curie**, written as Ci. In practice the curie turns out to be a rather large unit so the millicurie (mCi) or microcurie (μCi) is used.

The activity of a radioactive source will decrease with time because as atoms are disintegrating all the time, there will gradually be fewer and fewer atoms left which have still to disintegrate. The time taken for half the total number of atoms to disintegrate is called the **half-life** of the source. It follows that after one half-life there will only be half the number of the original nuclei which have still to disintegrate.
Hence in a time of one half-life, the activity will have fallen to half its original value, Figure 22.22.

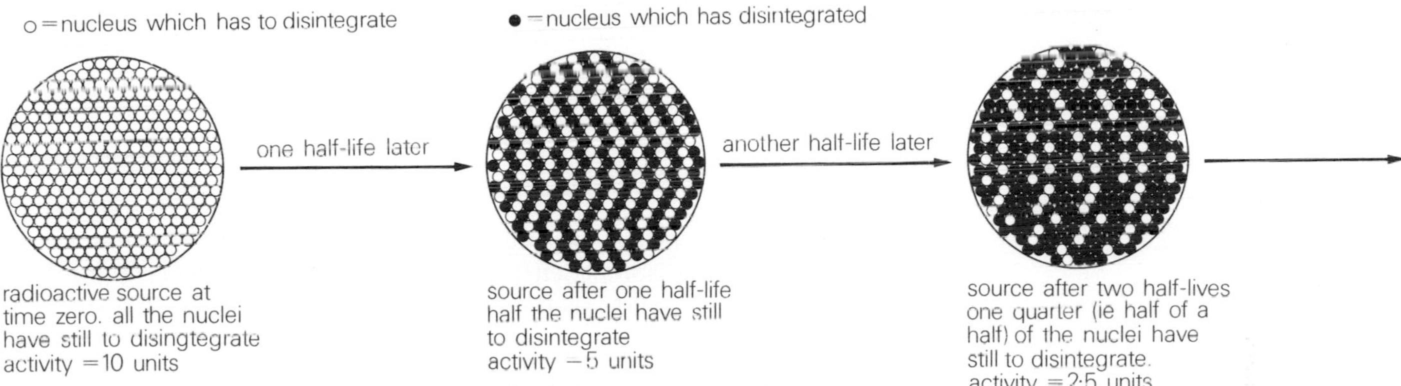

Figure 22.22 Disintegration over two half-lives

After a second half-life, there will only be one quarter (i.e. half of a half) of the original nuclei which have still to disintegrate. Hence in a time of two half-lives, the activity will have fallen to one quarter of its original value. You should be able to understand that after three half-lives the activity will have fallen to one eighth (i.e. half of a quarter) of its original value.

Note that the diagram shows a relatively small number of nuclei, and so is only an approximation. A practical source contains many millions of atoms and it is only when this very large number is present that it is possible to determine half-life accurately. Note also that the total mass hardly alters at all; the material does not disappear.

Half-lives of radioactive sources vary from less than a millionth of a second to more than a thousand million years. It is useful to know the half-life of a radioactive source to be able to predict how long it will remain active. A measurement of half-life is also a useful method of identifying an unknown radioactive isotope, since each isotope has its own half-life.

Example 1

A radioactive source has a half-life of 14 days. On 1st May it is labelled 10 mCi. How should it be labelled 28 days later?

After one half-life (i.e. 14 days) its activity will have fallen to half its original value. Therefore, on the 15th May its activity will be 5 mCi.

After a further 14 days its activity will have fallen to half its activity on the 15th May. Therefore, on the 29th May, i.e. 28 days later, the radioactive source should be labelled 2·5 mCi.

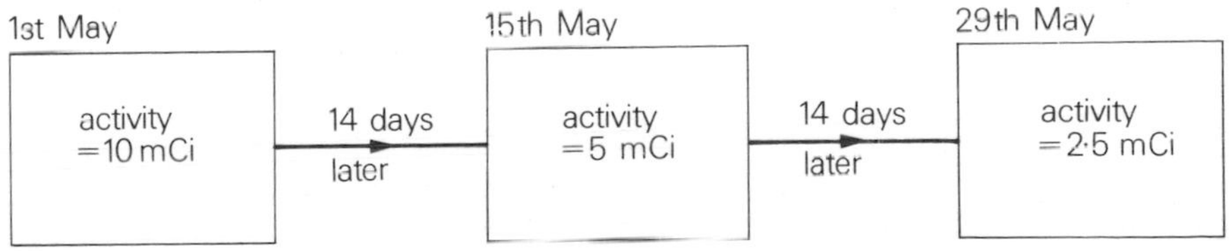

Example 2

a) The half-life of iodine-131 ($^{131}_{53}I$) is 8 days. Starting with 12 g of iodine-131, how much iodine will have decayed in 24 days?
b) Where has this quantity of iodine gone?
c) Why does this radioisotope present less problems in disposal than one of half-life 5 years?

a) After one half-life (i.e. 8 days) only half the original iodine will have still to decay. Therefore, after 8 days, the undecayed mass will be 6 g. After a further 8 days only half of a half (i.e. a quarter) of the original iodine will have still to decay. Therefore after 16 days the undecayed mass will be 3 g. After a further 8 days only half of a half of a half (i.e. an eighth) of the original iodine will have still to decay. Therefore after 24 days the undecayed mass will be 1·5 g. Hence in 24 days (12 − 1·5) = 10·5 g of iodine will have actually decayed.

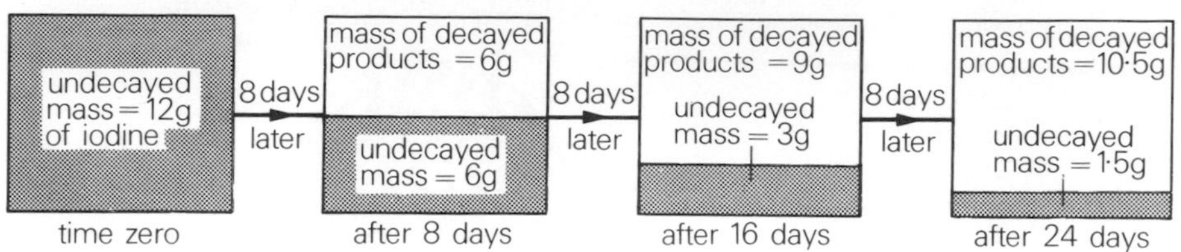

b) The iodine which has decayed has not vanished. Most of it has become the product isotope left behind after the decay. The original isotope is often referred to as the **parent**, while the product isotope is called the **daughter**.

c) This radioisotope is easily disposed of because its activity will fall very quickly. After 32 days its activity is only one-sixteenth of its former value. However, a radioisotope with a half-life of 5 years would take many years before its activity was low enough to allow easy disposal. A further problem is that the daughter product may also be radioactive!

Half-life of radon-220 ($^{220}_{86}Rn$)

Radon-220 is a radioactive gas formed from the decay of thorium. Radon-220 will in turn decay with a half-life of around 54 s. This half-life can be measured by squeezing radon-220 gas from a plastic bottle into a sealed ionization chamber. The small current produced can be amplified by a d.c. amplifier to give a meter reading proportional to the activity of the gas, Figure 22.23.

The activity of the gas in the chamber decreases with time and so the meter reading gradually falls. Meter readings are taken every 10 seconds and the half-life of radon-220 can be found from a graph of activity against time, Figure 22.24.
The activity is 0·8 at 17 s. The activity has fallen to 0·4 at 71 s. i.e. (71 − 17) = 54 s later. The activity has halved in 54 s. Therefore the half-life of radon-220 is 54 seconds.

Figure 22.23 Half-life of radon apparatus

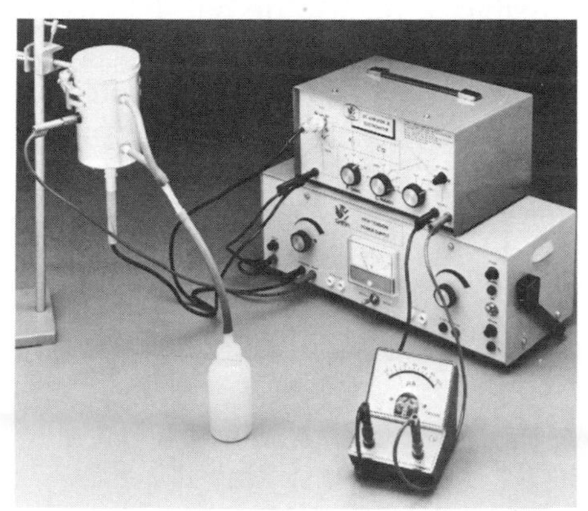

Figure 22.24 Activity against time for radon-220

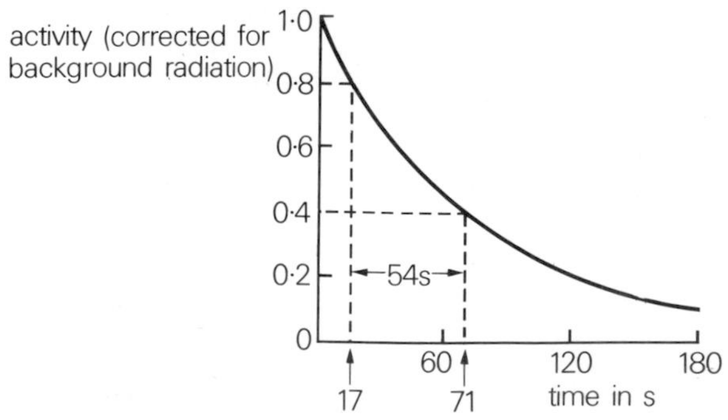

Example 3

A source of beta radiation is held near a Geiger tube which is connected to a ratemeter. The table below shows how the count rate recorded by the ratemeter varies with time.

If the count rate due to background radiation is 60 counts per minute, find the half-life of the source.

Time (minutes)	0	5	10	15	20
Count-rate (counts per minute)	1 660	1 100	750	510	350

Firstly, we must subtract the count rate due to the background radiation from all the ratemeter readings. This give the corrected count rate, i.e. the count rate actually due to the source.

Time (minutes)	0	5	10	15	20
Corrected count-rate (counts per minute)	1 600	1 040	690	450	290

To find the half-life of the source we construct a graph of the corrected count rate against time.

From the graph, the time taken for the count-rate to fall from 1 600 to 800 is 8 minutes. This means that the activity of the source falls to half its original value in 8 minutes. Therefore, the half-life of the source is 8 minutes.

22.7 Nuclear reactions

Radioactivity occurs as a result of instability of the nucleus. When radiation is emitted, the structure of the nucleus changes and it becomes more stable. If an alpha or beta particle is emitted, the atomic number of the so-called parent nucleus changes and a new element (called the daughter product) is formed. This daughter product may itself be unstable and emit radiation, the process continuing in a **decay chain** until a stable isotope is formed. Any excess energy in the nucleus as the result of alpha or beta emission can be emitted as gamma radiation.

Alpha emission

An alpha particle is the nucleus of a helium atom and consists of 2 protons and 2 neutrons. The atomic number of an alpha particle is therefore 2 and its mass number is 4. As the chemical symbol for helium is He, it can be represented by ${}^4_2He$.

Consider a nucleus X which emits an alpha particle to become the daughter nucleus Y. First, its atomic number decreases by 2 because the total charge must remain the same. Secondly, its mass number decreases by 4 because the total mass must remain the same. This can be represented by the equation:

$$\underset{\text{parent}}{{}^A_ZX} \rightarrow \underset{\text{daughter}}{{}^{A-4}_{Z-2}Y} + \underset{\text{alpha particle}}{{}^4_2He}$$

A diagrammatic representation of this is shown in Figure 22.25.

Figure 22.25 Alpha emission

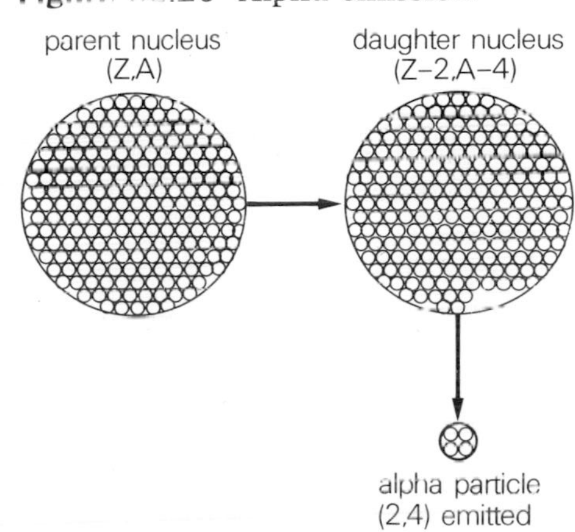

Example 4

Radium 226 ($^{226}_{88}Ra$) is unstable and emits an alpha particle. Find the new nucleus formed, stating its mass number and atomic number.

The decay can be represented as:

$^{226}_{88}Ra \rightarrow ^{A}_{Z}? + ^{4}_{2}He$

The mass number before must equal the sum of the mass numbers after.

i.e. $226 = A + 4$

$\Rightarrow \quad A = 222$

The atomic number before must equal the sum of the atomic numbers after, because the total charge does not change.

i.e. $88 = Z + 2$

$\Rightarrow \quad Z = 86$

The element which has an atomic number Z = 86 is radon, Rn. Hence the new nucleus is $^{222}_{86}Rn$.

Note that the emission of an alpha particle results in a new atom two places lower in the list of atomic numbers.

Beta emission

A beta particle is a negatively charged electron. A beta particle is produced as the result of a neutron disintegrating and leaving a proton in the nucleus with the emission of an electron. It is important to realize first that there are no electrons actually in the nucleus, and secondly that the electron which is emitted as a beta particle is not one of the electrons orbiting the nucleus.

A neutron can be represented by $^{1}_{0}n$ as it has no charge and a mass of 1 unit. A proton can be represented by $^{1}_{1}p$ as it has a charge of +1 unit and a mass of 1 unit. The emission of a beta particle can be represented by the following equation:

$^{1}_{0}n \rightarrow ^{1}_{1}p + ^{A}_{Z}e,$

where e is the symbol for an electron. As the top numbers in the equation must balance (by conservation of mass), we can assign A = 0 to the beta particle. As the bottom numbers in the equation must balance (by conservation of charge), we can assign Z = −1 to the beta particle.

Hence the beta particle is represented by $^{0}_{-1}e$ and the emission of a beta particle is represented by:

$^{1}_{0}n \rightarrow ^{1}_{1}p + ^{0}_{-1}e$

Remember that the numbers assigned to an electron are purely for convenience in solving this kind of equation.

Consider a nucleus X which emits a beta particle to become the daughter nucleus Y. The change in the nucleus is that a neutron has disintegrated and a proton has been formed. This means that although the mass number remains unchanged, the formation of a proton results in the atomic number increasing by 1.

Thus the emission of a beta particle can be represented by the equation:

$^{A}_{Z}X \rightarrow {}_{Z+1}^{\;\;A}Y + ^{0}_{-1}e$

A diagrammatic representation of this is shown in Figure 22.26.

Figure 22.26 Beta emission

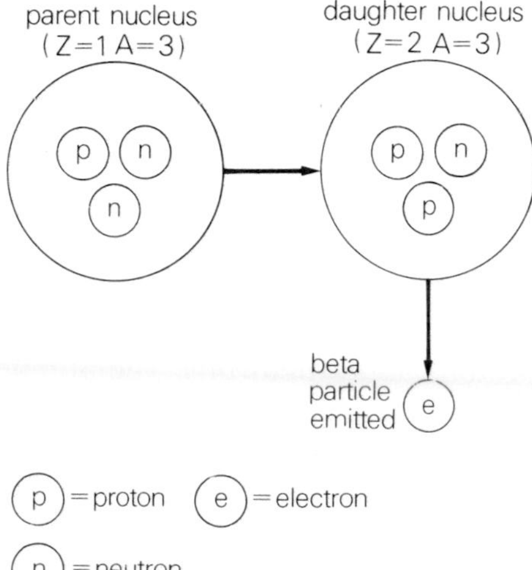

Example 5

Carbon 14($^{14}_{6}C$) decays by emitting a beta particle. Find the new nucleus formed, stating its atomic number and mass number.

The decay can be represented as:

$$^{14}_{6}C \rightarrow {}^{A}_{Z}? + {}^{0}_{-1}e$$

The mass number before must equal the sum of the mass numbers after.

i.e. $14 = A + 0$

$\Rightarrow \quad A = 14$

The atomic number before must equal the sum of the atomic numbers after.

i.e. $6 = Z - 1$

$\Rightarrow \quad Z = 7$

From a list of atomic numbers, the element with an atomic number of 7 is nitrogen, N. Hence the new nucleus is $^{14}_{7}N$.

Gamma emission

Gamma radiation is emitted as a result of instability in the nucleus. When the particles of the nucleus rearrange to become more stable, the extra energy can be emitted as gamma radiation.

As gamma radiation is high frequency electromagnetic radiation, it has no mass or charge. Hence the emission of gamma radiation leaves the atomic number and mass number unchanged.

Summary

Natural radioactivity was discovered by Henri Becquerel.

When radiation penetrates matter, its energy is absorbed by collisions with particles. The range of radiation is the distance it can penetrate before it becomes totally absorbed.

Background radiation is always present and originates from outer space and from natural radioactivity around us.

The atomic number Z of an atom is the number of positive charges (protons) in the nucleus.
The mass number A of an atom is the number of protons plus neutrons in the nucleus.
An element with chemical symbol X, mass number *A*, and atomic number Z is written as ${}^{A}_{Z}X$.
Isotopes are nuclei of the same element (same atomic number) but with different mass numbers.

An alpha particle is a helium nucleus (charge +2).
A beta particle is a fast moving electron (charge −1).
A gamma ray is a high frequency electromagnetic radiation (no charge) and travels at the speed of light.

Ionization is the formation of ions from atoms by the addition or removal of electrons. Many detectors of radioactivity depend on the ability of the radiation to produce ionization.

Radiation detectors:
The gold-leaf electroscope or the spark counter only detect alpha radiation.
The Geiger-Müller tube with a scaler or a ratemeter detects alpha, beta and gamma radiation.
The ionization chamber with d.c. amplifier detects alpha and beta radiation.
The cloud chamber detects alpha and beta radiation.
Photographic film or solid-state detectors detect all three types of radiation.

Radioactive decay is totally irregular (random). The activity of the source is the number of disintegrations per second.
The unit of activity is the curie (Ci).
The half-life of a source is the time taken for its activity to fall to one half of its original value.

Alpha emission can be represented by
$${}^{A}_{Z}X \rightarrow {}^{A-4}_{Z-2}Y + {}^{4}_{2}He$$
The mass number falls by 4 and the atomic number falls by 2.

Beta emission can be represented by
$${}^{A}_{Z}X \rightarrow {}^{A}_{Z+1}Y + {}^{0}_{-1}e$$
The mass number is unchanged and the atomic number increases by 1.

Gamma emission does not affect either the mass number or the atomic number.

Problems

1 Becquerel's experiment is repeated using photographic film and a radioactive source. Should an α, β or γ source be used to produce a shadow of a key on the resulting photograph? Explain your answer.

2 Explain whether alpha, beta or gamma radiations can be deflected by a magnetic field. Describe an experiment which verifies that *one* of these radiations can be deflected by a magnetic field.

3 Explain how you could prove that silver-113 emits only beta radiation.

4 Describe how you could show absorption of
4·1 alpha radiation,
4·2 beta radiation, and
4·3 gamma radiation.

5 Explain why a positively charged electroscope is discharged when a flame is brought near the disc of the electroscope.

6 Describe three different detectors, one which detects α radiation, one which detects β radiation, and one which detects γ radiation. For each one explain how it detects the radiation.

7 A radioisotope has a half-life of 3 days. If the original activity is 6·4 μCi, what will the activity be after 9 days?

8 A radioisotope has a half-life of 600 years. If the original activity is 36 μCi, how long will it take for the activity to fall to 4·5 μCi?

9 The activity of a radioactive source is measured using a Geiger-tube and ratemeter. The graph shows how the count-rate recorded by the ratemeter varies with time.
9·1 From the graph, estimate the half-life of the source.
9·2 Explain why the count-rate never falls below 5 counts per second.

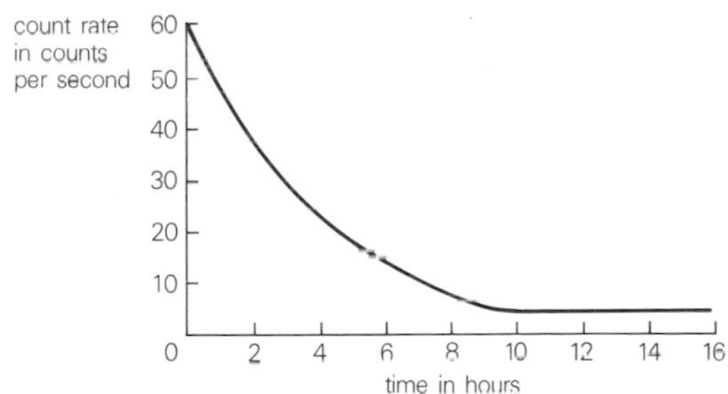

10 An isotope of radon is represented by $^{222}_{86}Rn$.
10·1 State how many protons and neutrons are contained in a nucleus of this isotope.
10·2 If this isotope decays by emitting an alpha particle, find the new atomic number and mass number of the daughter product.

11 The isotope actinium-226, $^{226}_{89}Ac$, decays by the emission of a beta-particle. Find the atomic number and mass number of the daughter product.

12 The following represents part of the decay chain for thorium.

$^{232}_{90}Th \rightarrow {}^{228}_{88}Ra \rightarrow {}^{228}_{89}Ac \rightarrow {}^{228}_{90}Th \rightarrow {}^{224}_{88}Ra$

State what particles are emitted at each part of the chain.

13 Identify the following radiations (not all of which come from radioactive sources) from the evidence supplied.

Radiation A absorbed in a few centimetres of air; deflected by a magnetic field.
Radiation B can be detected by certain photographic plates, some phototransistors and a thermometer with a blackened bulb; not visible to the human eye.
Radiation C very penetrating rays; not deflected by a magnetic field; harmful to living things.
Radiation D mostly absorbed by a few millimetres of aluminium; deflected by a magnetic field.
Radiation E has a wavelength of several metres; an aerial is required for the transmission of these waves.
Radiation F appears yellow to the naked eye; not deflected by a magnetic field; absorbed by black paper.

SCEEB

14 When a scientist in a nuclear power station examined some sea water which had become radioactive, he found that the emission of beta particles and gamma rays was due to a radioactive isotope which had a half-life of 15 hours.
14·1 In the above statement:
a) what are 'beta-particles',
b) what are 'gamma-rays',
c) what is meant by 'half-life of 15 hours'?
14·2 Why was the count-rate reduced when he placed a 1 cm thick aluminium sheet between the sea water and the detecting instruments? *SCEEB*

15 **15·1** An experiment was performed to determine the correct working voltage for a Geiger-Müller tube. A radioactive source was placed in front of the tube and the d.c. voltage across the tube was set at different values. The count-rate was measured for each value of the voltage used, and the values obtained are shown in the table below.

Count-rate (counts/s)	0	10	40	65	70	71	69	70	75	90
Voltage	250	300	350	400	500	600	700	800	850	900

a) Plot these points on a graph with a suitable scale and with the count-rate on the vertical axis.
Sketch the best line which represents the graph.
b) Explain why it is not necessary for the curve to pass through all the points.
c) Indicate a region on the graph where the count rate is independent of the voltage. (Write CONSTANT COUNT RATE at this region and label the ends of the region A and B.)

15·2
a) A radioactive isotope has a half-life of 14 days. On 1st May 1970 it is correctly labelled 10 millicuries. How should it be labelled 28 days later?
b) Why does this radioactive isotope present less problems in disposal than one of half-life 5·3 years?
c) If the isotope in part (a) is a β emitter, how would you produce a narrow beam of β particles from it? *SCEEB*

16 **16·1** An artificial radioisotope emits α particles. Immediately after its manufacture, a sample of this radioisotope has a measured activity of 800 counts per minute, after allowing for background radiation. The graph shows how its activity varies with time thereafter.

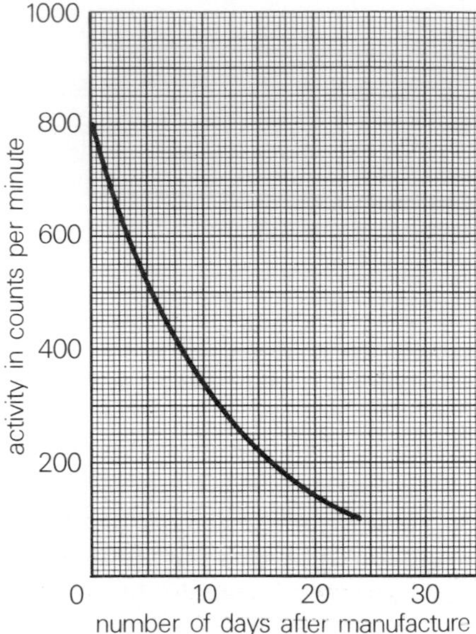

a) What is the activity of this α source when it is first used in the laboratory three days later?

b) What is the half-life of this source?

c) When its activity is below the average background count of 50 counts per minute, the source can be considered suitable for disposal. After how many days can the source first be considered suitable for disposal?

d) Describe a laboratory experiment to confirm that the radioisotope emits α particles only.

16·2 Explain why α radiation from an external source is likely to be less harmful to the internal body organs than β or γ radiation. *SCEEB*

17 A Geiger-Müller tube with an automatic counter is used by a pupil to examine a radioactive source which emits one type of radiation only.

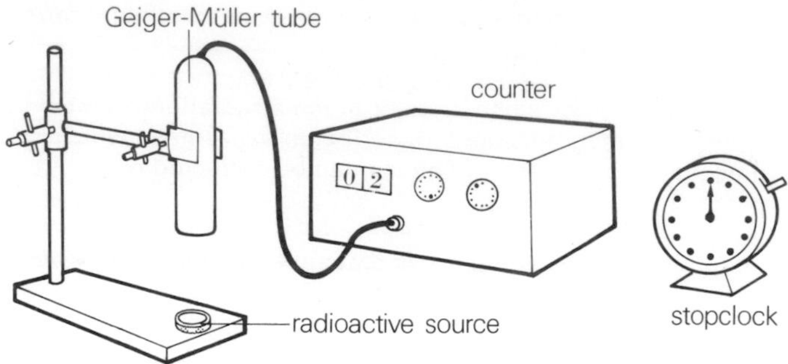

When the Geiger-Müller tube is more than 10 cm from the source, an average count rate of 5 counts per minute is obtained. The Geiger-Müller tube is now moved until it is 2 cm from the source and the results in the following table are taken over a period of 80 minutes. The source is then removed.

TIME (minutes)	0	20	40	60	80
COUNT-RATE (number of counts in one minute)	101	65	43	29	21

17·1 What type of radiation is emitted by the source?

17·2 Suggest what the count-rate would be after the source is removed.

17·3 Plot a graph of count-rate for the source against time, using square-ruled paper. Hence estimate the half-life of the source.

17·4 During the experiment, the pupil reported that 'the counter might be sticking since it appeared to be counting irregularly during the experiment'.
Comment on the pupil's interpretation of the irregular counting. *SCEEB*

18 A nuclear power station supplies an average power of 150 MW to the National Grid. The power station's reactor consumes 0·3 kg of uranium fuel per hour.

18·1

a) Calculate the energy supplied to the National Grid in 1 hour.

b) If the energy extracted from 1 kg of uranium fuel in the nuclear reactor is $5{\cdot}4 \times 10^{12}$ J, calculate the efficiency of the power station.

18·2

a) Part of the radioactive waste from the reactor has a half-life of 30 years. A drum of this waste material is encased in a concrete block before storage. The measured radiation level outside the concrete block is 512 units.
If the maximum safe allowable level is 1 unit, how long must the concrete block be kept in secure storage before it is safe to approach it?

b) How does the concrete casing help to make the radioactive waste safe during storage? *SCEEB*

23 Radioactivity II

23.1 The Rutherford atom

We now consider an atom as consisting of a small central nucleus which contains almost all the mass, and electrons around the nucleus, Figure 23.1.

Over 99 per cent of the mass is concentrated in the nucleus and most of the atom consists of empty space.

This 'model' of the atom was proposed in 1911 by Lord Rutherford (Figure 23.2) as a result of experiments, known as Rutherford's Scattering Experiments, which were carried out by Geiger and Marsden. Until then, the subject was rather confused due to various conflicting ideas.

The most well known of atomic 'models' existing before 1911 was that proposed by Sir J. J. Thomson. He considered the atom as consisting of negatively charged particles embedded in a sphere of uniformly distributed positive charge. This was Thomson's 'plum pudding' model of the atom. Rutherford's Scattering Experiments, which we will now consider in more detail, led to the downfall of Thomson's theory.

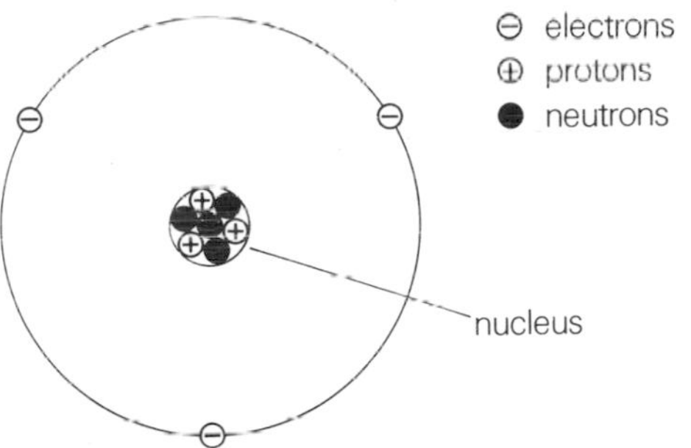

Figure 23.1 The nuclear atom

Figure 23.2 Lord Ernest Rutherford

Rutherford's scattering experiments

Geiger and Marsden fired alpha particles at a thin gold foil and measured the angles through which the alpha particles were deflected. The apparatus used is shown in Figure 23.3.

The experiment was conducted in a partial vacuum to reduce collisions of the alpha particles with air particles. A radioactive source emitted alpha particles which were formed into a narrow beam by passing them through a slit.

This narrow beam struck a thin gold foil. Any alpha particles which passed through the foil were detected by a zinc sulphide screen which emitted a flash of light when an alpha particle struck it. The individual flashes of light were counted by an observer looking at the screen through a microscope which could be rotated along with the screen.

Geiger and Marsden soon found that most of the particles passed straight through the gold foil without deflection, although even a thin gold foil is still several hundred atoms thick. This implied that an atom is not solid and is mostly empty space.

Figure 23.3 Rutherford's scattering experiment

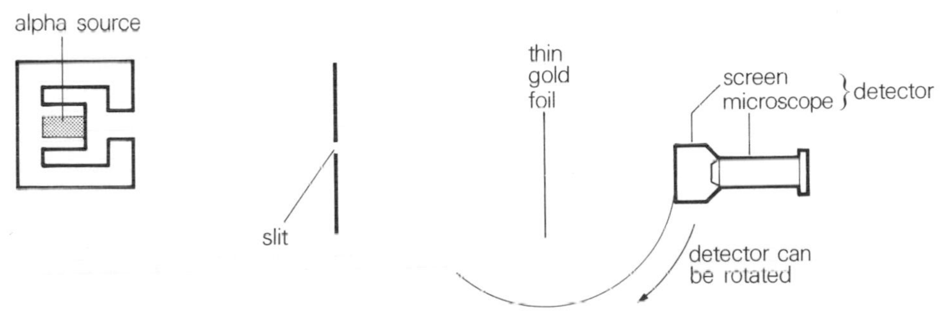

However, a much more significant event occurred a few days later when they rotated the detector and looked for alpha particles which had been scattered through large angles. Rutherford himself had expected this search to be fruitless because, as the foil was very thin, a large deflection could not conceivably be produced even from several successive collisions.

Much to Rutherford's astonishment, they found that some alpha particles actually came backwards after hitting the foil. To quote Rutherford himself, 'It was quite the most incredible event that has ever happened to me in my life. It was almost as incredible as if you fired a 15-inch shell at a piece of tissue paper and it came back and hit you!' Table 1 shows Geiger and Marsden's results. Remember that every single particle was counted by an observer!

Angle of deflection	*Experimental count*
5°	8 289 000
10°	502 700
30°	7 800
60°	477
105°	69
150°	33

Table 1 Angles of scatter of alpha particles

These large deflections could not be explained by collisions with electrons because an alpha particle is approximately 7 300 times more massive than an electron. The only way of explaining this result was to consider practically all the mass and positive charge as being concentrated in a tiny volume called the **nucleus**.

When a positively charged alpha particle came very near to the concentration of positive charge in the nucleus, there would be a large repulsive force between them. This would result in a one-in-a-million chance of an alpha particle being deflected through a very large angle. As most of the particles would miss the tiny nucleus, they would go straight through the atom, Figure 23.4.

The answer may seem fairly obvious, but in fact this was a brilliant deduction and was a great turning point in physics that laid the foundation for all modern atomic and nuclear science.

These experiments also produced the first estimate of the diameter of a nucleus of around 10^{-14} metres or a millionth of a millionth of a centimetre! As an atomic diameter is around 10^{-10} metres, ten thousand nuclear diameters could fit into one atomic diameter!

Years later, the development of the cloud and bubble chambers allowed photographs to be taken of the alpha scattering process. Figure 23.5 shows alpha particle tracks in helium gas.

Note that most of the tracks are straight but that one alpha particle (and this happens only very occasionally) has been deflected through a large angle. Elastic collisions between equal masses always result in a 90° angle of separation, so the photograph confirms that an alpha particle is a helium nucleus.

In conclusion, not only did Rutherford's experiment lead to a new model of the atom, but also for the first time an atomic 'missile' was being used as an investigating probe. Many new particles have since

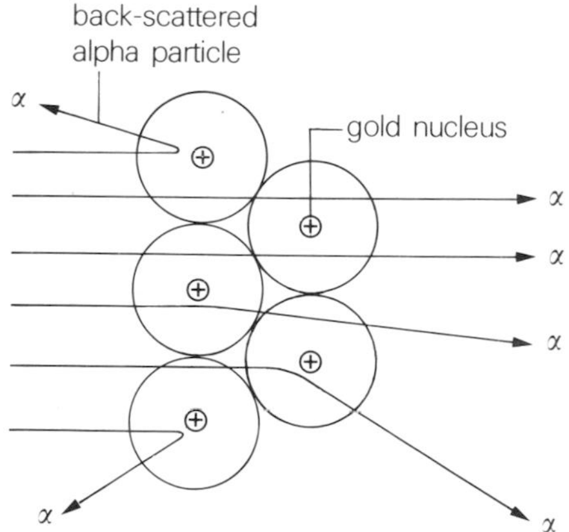

Figure 23.4 Scattering of alpha particles

Figure 23.5 Alpha tracks in a helium cloud chamber

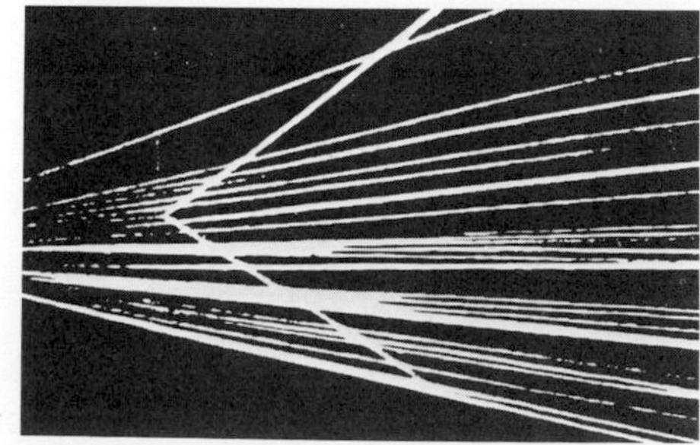

been discovered by bombarding nuclei with other nuclear particles in particle accelerators (see section 23.3).

A scattering analogue

An analogue is a demonstration designed to help us understand a physical process.

Figure 23.6 shows one example of an analogue of Rutherford's scattering experiment which may help you to understand the scattering process.

A steel ball is rolled down a ramp on to a curved 'hill'. Unless the ball is aimed directly at the centre of the hill is it not deflected very much. However, when the ball is aimed at the centre, it is deflected through a large angle and may even come back along its original path. This is similar to the deflection of an alpha particle by a nucleus where the ball represents the alpha particle and the hill represents the repulsive field of the nucleus.

Figure 23.6 Alpha-scattering analogue

23.2 Applications of radioactivity

The properties of α, β, and γ radiation are used in an increasing number of ways in medicine, agriculture and industry. Provided adequate safety regulations are obeyed, radioactivity methods can be very helpful.

The uses of radioactive sources include:

1 using the fact that they are radioactive and therefore **detectable**, to trace their progress, e.g. through a human body. This use involves adding them in extremely small amounts to the area to be studied. They are then called **tracers**;

2 using the **properties** of the radiation that they emit e.g. absorption;

3 using the **energy** that they emit to produce electrical power.

1. Uses of tracers

Medicine

One example of the use of tracers in medicine is in the treatment of thrombosis (bloodclots). The position of a bloodclot can be determined by adding radioactive sodium to the body and using detectors to find where the flow stops. Other uses of tracers in medicine include studying the uptake of certain elements by organs of the body. For example, the function of the thyroid gland (which normally takes up any iodine that enters the body) can be investigated by injecting radioactive iodine. Figure 23.7 is a photograph of a thyroid 'scan'. The now radioactive thyroid can be clearly seen.

Figure 23.7 Thyroid scan

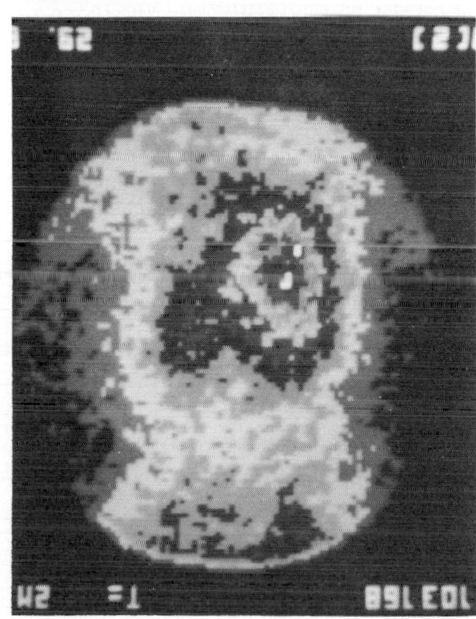

Industry

One example of the industrial use of tracers is in making the dredging of harbours and canals more efficient. This is done by adding radioactive silicon to the mud and measuring the way in which the mud moves and builds up.

Agriculture

One example of the use of tracers in agriculture is in studying the uptake of certain elements by plants and animals, e.g. radioactive phosphorus added to the fertilizer for a wheat field can enable the amount of phosphorus taken up from the fertilizer to be determined.

Other uses of tracers in agriculture include making the vitamins to be added to animal food radioactive. This enables the proportion of vitamins in the food to be subsequently checked by measuring its activity. Radioisotopes with a short half-life are used so that the food is safe by the time it is eaten.

2. Uses of the properties of radiation

Before discussing the applications of radioactive sources, we will first consider the possible **safety hazards**. Alpha, beta and gamma radiations can all produce ionization of the molecules in living cells. As this can lead to the whole cell being destroyed, the radiation is extremely harmful to all living things, including us.

Alpha sources are very dangerous if taken into the human body because they produce a great deal of ionization, but are relatively safe outside the body because of their short range in air.

Beta and gamma sources do not produce so much ionization, but nevertheless they are much more penetrating than alpha sources. So these two types of radiation can damage internal cells, even from outside the body.

These effects are complicated by the fact that
(i) different sources remain active for different lengths of time,
and (ii) certain radioactive elements are taken up by specific organs.

There is concern about the radioactive source plutonium-239 which can be used as a fuel for nuclear reactors. Its half-life is 25 thousand years, so it would remain active for a very long time if there was an accident involving it. Plutonium is also produced as a by-product in certain nuclear reactors and there are great problems in its transport and disposal.

Medicine

One example is in the treatment of cancer. A radioactive cobalt-60 source will kill off malignant cancer cells in the human body. As the source is always directed at the cancerous area, the cancer cells receive radiation all the time. However, as the source is moving, the surrounding healthy cells are only receiving radiation for a short time and are therefore not damaged. Figure 23.8 shows the unit and how it rotates.

Other uses of radioactivity in medicine include the sterilization of medical instruments.

Figure 23.8 Cobalt therapy unit used in the treatment of cancer

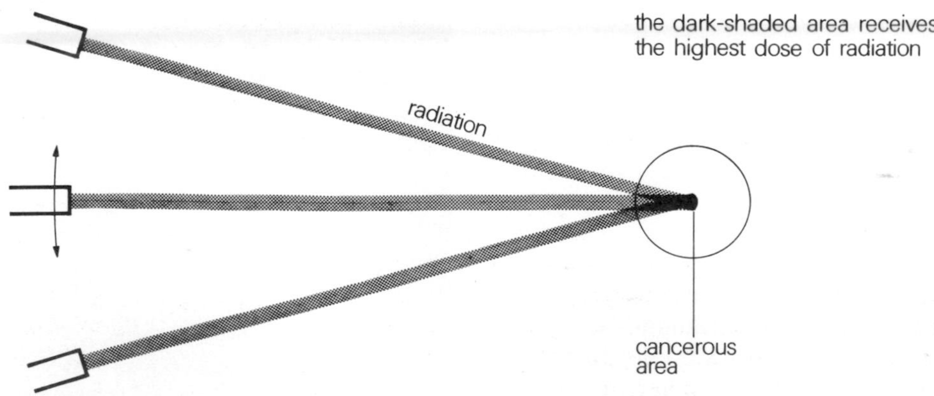

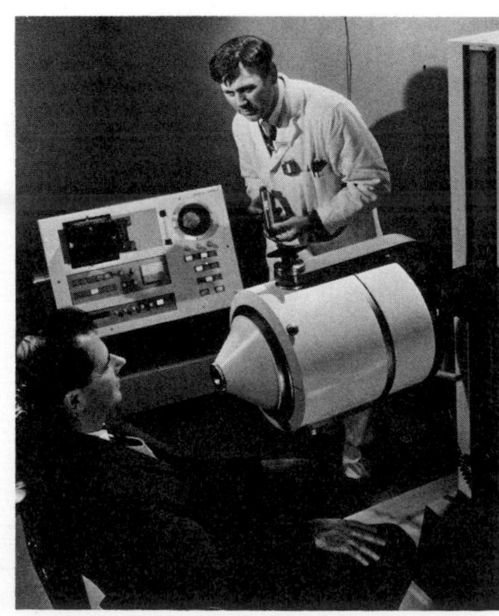

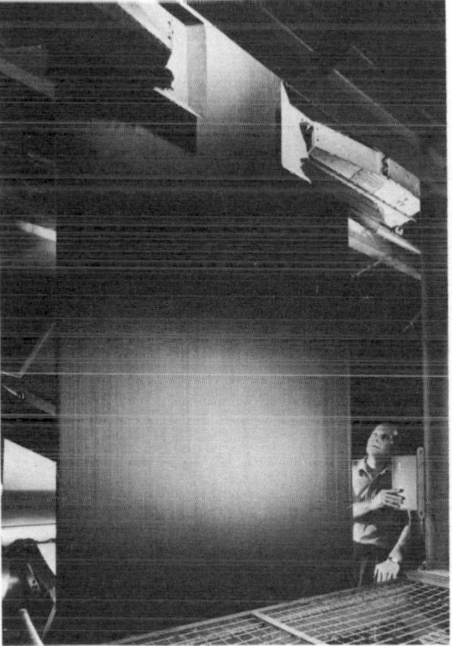
Figure 23.9 Equipment used to check the thickness of tyre cord

Industry

One example of the use of radioactive sources in industry is in the measuring of the thickness of material continuously as it is manufactured. In Figure 23.9 a radioactive strontium-90 source is placed at one side of tyre cord with a detector on the other side. If the tyre cord is too thin, the detector will indicate a higher level of radiation. Similar equipment can be used to reject automatically an article of the wrong dimensions.

Agriculture

One example of the use of radioactive sources in agriculture is in producing strains of plants with new characteristics. For example, the exposure of oats to radiation has led to the development of a new strain of oats which is particularly resistant to disease, Figure 23.10.

Other uses of radioactivity in agriculture include the sterilization of insects. Many male screw-worm flies on the island of Curaçao were bred and then sterilized to prevent further breeding. The female screw-worm only mates once, so when the sterilized males were let loose on the fly population, the numbers were subsequently reduced. This process was repeated until this insect pest was totally eliminated on the island.

Figure 23.10

Uses in art

The use of radiation in the detection of forgery in paintings and ceramics is becoming increasingly common. By bombarding a painting with radiation, it is possible to find the proportion of certain elements contained in very small amounts in the paint. It is therefore possible to assign a range of dates in history to each paint used. Art forgers have often made the mistake of using twentieth-century paints for their 'old masters'!

Carbon dating

Living plants absorb carbon dioxide from the atmosphere. Most of this carbon dioxide contains the common stable isotope carbon-12 ($^{12}_{6}C$), but some contains the radioactive isotope carbon-14 ($^{14}_{6}C$), which is continuously being formed by the effect of radiations falling on the Earth's outer atmosphere. Living plants therefore contain the same proportion of carbon-14 to carbon-12 as is in the atmosphere. We will assume that the proportion of carbon-14 to carbon-12 in the atmosphere remains constant. However, when a plant dies the carbon-14 present decays with a half-life of 5 600 years, emitting β-radiation. For example, after 5 600 years the proportion of carbon-14 to carbon-12 will have fallen to half its original level. Hence by measuring this proportion, it is possible to estimate how long ago the plant was living. This method is called **carbon dating** and has been used by archeologists to estimate the age of wood in Egyptian tombs, ancient papyrus manuscripts, and even the shawl which was said to have covered Jesus after his death. The results often confirm those obtained by other methods.

3. Nuclear fission and energy

A nucleus may contain many protons, all of which tend to repel each other. So there must be a very strong attractive nuclear force, which binds the nucleus together and prevents it from splitting up. When uranium-235 ($^{235}_{92}U$) is bombarded by neutrons, the nuclear force is overcome under certain conditions and the nucleus splits into two fragments. This process is called **nuclear fission** and is accompanied by the release of a large amount of energy and the emission of several neutrons. These neutrons can produce fission of another nucleus and the result under suitable conditions can be a **chain reaction** (Figure 23.11) with vast amounts of energy being liberated.

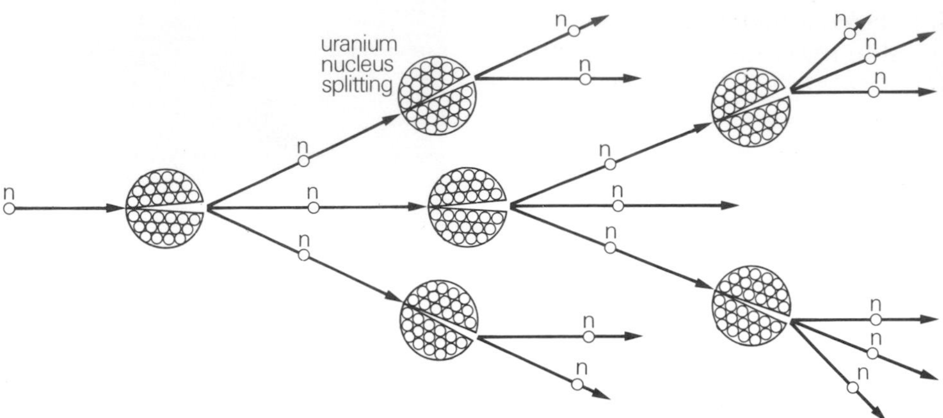

Figure 23.11 A chain reaction

Figure 23.13 Atomic bomb exploding

In a **nuclear reactor**, the chain reaction is controlled using a moderator; this is a substance which absorbs (and so reduces) the energy of the neutrons. With the reaction proceeding at a steady rate, the energy which is released in the form of heat can be used to make steam to drive turbines to produce electricity as in other power stations. Figure 23.12 shows a commercial nuclear power station which provides us with electrical energy.

If the chain reaction gets out of control, a nuclear explosion can result, Figure 23.13. This can cause great devastation as it is so much more powerful than chemical explosives and radiation lingers as an after-effect.

Figure 23.12 Nuclear power station

23.3 Landmarks in radioactivity

In conclusion, here is a brief historical development of the advances in physics relevant to our discussion on radioactivity.

1895 Professor Wilhelm Rontgen, an eminent German scientist, discovered a new form of radiation which was named X-radiation.

1896 Antoine Henri Becquerel, a French physicist, discovered what was later called radioactivity, in uranium salts.

1897 Sir Joseph J. Thomson, an English physicist and mathematician and a Professor of Physics at Cambridge from 1884 to 1919, measured the ratio of the electrical charge of an electron to its mass.

1898 Professor Pierre Curie and his wife Marie isolated radium after years of laborious work separating it from uranium ore. Madame Curie was Polish and her husband was French. Their laboratory was an old shed and although they spent practically all their waking hours in it, Madame Curie later said, 'We spent the best and happiest years of our lives entirely devoted to the task in that miserable old shed'.
In the same year, Paul Villard discovered a new form of radiation with similar properties to X-rays.

1900 Investigations into radioactivity showed three distinct types of radiation. Two types were named alpha and beta radiation by Lord Rutherford. The third was the type discovered by Villard, which he named gamma radiation.

1903 Phillip Lenard showed that most of an atom is empty space by demonstrating that electrons moving at high speed can pass through an aluminium sheet.

1909 Rutherford and Royds proved that an alpha particle is actually the nucleus of a helium atom. They detected the gas helium in a specially sealed apparatus containing a source of alpha radiation.

1910 Frederick Soddy, an English chemist and a Professor of Chemistry at Oxford from 1919 to 1936, suggested the existence of atoms with identical chemical properties but with different masses, and named them isotopes.

1911 Geiger and Marsden, under Rutherford's guidance, detected the back-scattering of alpha particles. As a result, Rutherford proposed a new 'nuclear' model of the atom.
Lord Rutherford was born in New Zealand and was a Professor of Physics at the famous Cavendish Laboratory at Cambridge.
In the same year C. T. R. Wilson, a Scottish physicist, invented the cloud chamber. The idea came as a result of studying clouds seen from the top of Ben Nevis during his summer holidays.

1913 The famous Danish physicist, Neils Bohr, took Rutherford's model of the atom and proposed that the electrons orbiting the nucleus could do so only in certain well-defined orbits. He stated that only certain jumps between energy levels were permitted.

1919 Rutherford bombarded nitrogen with alpha particles and became the first person to transform one element into another artificially. In his experiment nitrogen was changed to oxygen. The reaction can be represented by the equation,

$${}^{14}_{7}N + {}^{4}_{2}He \rightarrow {}^{17}_{8}O + {}^{1}_{1}H$$

1932 Sir James Chadwick discovered the neutron as a result of an experiment in which a beryllium target was bombarded by alpha particles,

$^{9}_{4}Be + ^{4}_{2}He \rightarrow ^{12}_{6}C + ^{1}_{0}n,$

(the term $^{1}_{0}n$ represents a neutron).

The discovery of the neutron was one of the main steps in the discovery of nuclear fission which led to the production of the atom bomb.

In the same year Sir John Cockroft and Ernest Walton used the first particle accelerator to transform the nuclei of one element to the nuclei of a different element.

Also in the same year, E. O. Lawrence used the cyclotron to split lithium nuclei. The cyclotron is a machine for accelerating charged particles to very high energies. It was devised by Lawrence in 1930 in California.

1934 The daughter of Pierre and Madame Curie, Irene, married a physicist and became Irene Joliot-Curie. Irene and her husband Frederic first produced artificial radioactivity by bombarding stable elements with alpha particles. Lawrence and an Italian, Enrico Fermi, carried out similar experiments using neutrons to bombard stable elements. Fermi's later research in U.S.A. contributed to the harnessing of atomic energy and the development of the atomic bomb.

1936 The diffusion cloud chamber was invented by Langsdorf.

1938 Nuclear fission was first observed with uranium (atomic number 92). Otto Hahn and Fritz Strassman found the element barium (atomic number 56) in what had once been pure uranium. Lise Meitner and Otto Frisch guessed that the element krypton should also be present as its atomic number is 36 (and $36 + 56 = 92$). They carried out experiments which confirmed that some uranium nuclei had indeed split into these two parts.

1940 By 1940, several hundred artificially created radioisotopes had been discovered.

1942 Fermi produced the first atomic pile in which the controlled fission of uranium could be carried out.

1945 The first atom bomb was exploded in a remote area in New Mexico.

1952 The first hydrogen bomb, which was much more powerful than an atom bomb, was exploded.

1956 The first nuclear power station was opened at Calder Hall in Cumberland.

1971 A prototype fast reactor was opened at Dounreay in the north of Scotland. In a fast reactor, no moderator is required and the plutonium that it produces as a by-product can be used as its fuel. This is why it is called a 'breeder' reactor.

Summary

Rutherford's scattering experiments showed that most of an atom is empty space and that all the positive charge and practically all the mass of an atom is concentrated in a tiny nucleus.

A tracer is a radioactive substance added in very small quantities to trace the path that it takes in a system. Tracers are used in medicine, in industry, and in agriculture.

Radioactive sources produce ionization which can lead to a living cell being destroyed. They are used in medicine, in industry, in agriculture, and for detecting art forgeries.

Carbon dating is a method of estimating the age of once-living objects containing carbon. The method depends on measuring the proportion of carbon-14 to carbon-12.

An atomic nucleus is held together by extremely strong forces, but it can be split (nuclear fission) under certain conditions. The energy released by nuclear fission can be harnessed by a nuclear reactor to produce electricity.

Problems

1 Describe Rutherford's scattering experiments and explain how they led to Rutherford developing a new model of the atom.

2 Explain what is meant by a radioactive tracer and discuss their application in
 2·1 industry,
 2.2 medicine, and
 2·3 agriculture.

3 Discuss the necessity for precautions to be taken when handling radioactive sources.

4 Workers in an establishment in which radioactive sources are used wear a sealed badge containing photographic film which is replaced and developed regularly. Explain why this is necessary.

5 Discuss briefly how gamma radiation can be used in the fight against cancer.

6 Describe how an ancient wooden bowl from an archeological site could be dated using the carbon dating technique.

7 Discuss briefly the fission of uranium and how it can be controlled productively in a nuclear reactor.

8 Explain how this apparatus can be used to check the uniformity of thickness of linoleum as it passes through the gap during its manufacture.

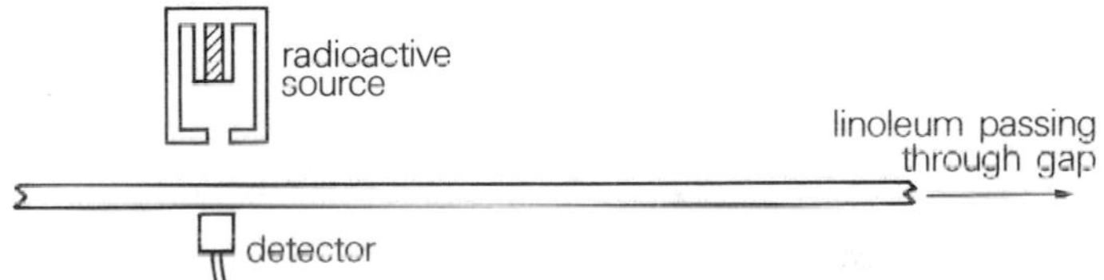

Explain which of alpha, beta or gamma sources would be the most suitable for this application. Why are the other two unsuitable? *SCEEB*

Numerical answers to problems

Chapter 1 *page 7*
1.1 a) 0·015 m b) 2·0 m c) 20 m s^{-1} d) 10 Hz
2 0·08 m s^{-1}
3 2 Hz
4 0·02 m
5.2 a) 16 mm b) 5 mm
6.1 4·0 m
6.2 1·5 Hz

Chapter 2 *page 19*
6.2 a) 0·5 cm b) 2·5 cm s^{-1} c) 5 Hz
7.1 0·01 m
7.2 0·1 m s^{-1}
7.3 0·005 m
8 2 Hz; 4 Hz
9 12 Hz
10.1 12 cm s^{-1}
10.2 2.4 cm
12.1 4·0 m
12.2 1·5 Hz

Chapter 3 *page 27*
6 3×10^{-6} s

Chapter 4 *page 40*
8 $3{\cdot}3 \times 10^{14}$ Hz
12 1 m

Chapter 5 *page 53*
2 1020 m
3 330 m s^{-1}
6.2 a) 70 m

Chapter 6 *page 63*
2 0·8 Hz
3 a) 0·10 s b) 0·20 s c) 0·06 s
5 0·2 s
6.1 1·5 s
6.2 1·0 s
7 25 Hz

Chapter 7 *page 68*
4.1 5·8 cm (329°)
4.2 5·8 cm (329°)
4.3 5·8 cm (329°)
5.1 200 m North
5.2 800 m
6.1 24·5 km
6.2 10 km (008°)
6.3 10 km (188°)

Chapter 8 *page 86*
1 2 m s^{-1}
2 30 s
3 48000 m
4 a) 0·68 m s^{-1} b) 1·03 m s^{-1} c) 0·82 m s^{-1}
6 26 m s^{-1} (023°)
7 20 m s^{-1} at 37° to the vertical; 375 m
8.1 a) 0 b) 15 m s^{-1} c) 30 m s^{-1} d) 30 m s^{-1} e) 30 m s^{-1} f) 10 m s^{-1} g) 0
8.2 a) 0-40 s b) 100-130 s c) 40-100 s
8.3 a) 0·75 m s^{-2} b) 0 c) −1·0 m s^{-2} d) 0
8.4 a) 15 m s^{-1} b) 30 m s^{-1} c) 15 m s^{-1} d) 22 m s^{-1}
9 a) 12 m s^{-2} b) 38 m s^{-2} c) −18 m s^{-2} d) −9 m s^{-2}
10.1 a) 0·12 m s^{-2} b) 0·3 m s^{-1}
11.1 9·4 m s^{-2}

Chapter 9 *page 99*
2 2400 N; 5 s
3.1 1 N; 2·6 N; 0·36 N
3.2 750 N; 1950 N; 270 N
4.1 50 N
4.2 10 N at 53° to the 6 N force
5.1 91 cm s^{-1} 105 cm s^{-1}
16 cm s^{-1} 14 cm s^{-1}
160 cm s^{-1} 140 cm s^{-1}
average acceleration = 150 cm s^{-2}
5.2 60 vibrations per second
5.4 5·5 cm, 6·7 cm, 8·1 cm, 9·5 cm
7.3 200 N
8.1 a) 0·1 s b) 150 cm s^{-1}, 240 cm s^{-1}, 350 cm s^{-1}, 440 cm s^{-1}
9.1 b) 4·0 N

Chapter 10 *page 107*
1 9·36 m s^{-2}
2.1 0·64 m s^{-1}
2.2 10 m s^{-2}
2.3 3 m s^{-1}
3.1 30 m s^{-1}
3.2 40 m s^{-1}
3.3 80 m
3.4 120 m
3.5 20 m s^{-1}
3.6 144 m s^{-1} at 34° below the horizontal
4.1 0·3 s
4.2 7 m s^{-1}

Chapter 11 *page 116*
1 6 m s^{-1}
2 0·25 m s^{-1}
3 2·4 m s^{-1}
4 0
5.1 0·2 0·6
0·06 0·06
5.2 a) 0 b) 0
7.1 18 s
7.2 6×10^{3} kg

Chapter 12 *page 133*
3 6000 J
4 400 N
5 4 m
6 50%
7 $1{\cdot}6 \times 10^{5}$ J
8 20 m s^{-1}
9 40 J
10 20 kg
11.1 20 m s^{-1}
11.2 20 J
11.3 20 J
11.4 20 m
11.5 20 m s^{-1}
12.1 4 J
12.2 4 J
12.3 3·2 J
12.4 4·5 m s^{-1}
12.5 0·8 J
13.1 2 m s^{-1}
14 100 N
15 17 W
16 12·5 s
17.1 10^{7} J
17.2 $8{\cdot}0 \times 10^{6}$ W
18.1 a) 12 m s^{-1} b) 3 m s^{-1} c) 1 kg d) 72 J e) 18 J
18.3 0·45 m
19

I	II	III	IV
12·0	72·0	12·2	37·2
9·8	48·0	9·9	16·3
20·4	104·0	20·0	50·0
23·0	132·2	22·8	86·6

20.2 0·2 kg
21.1 a) 5000 N b) 7500 J
21.2 a) 1000 J
21.3 10%
22.1 2·4 J
22.3 4 m s^{-1}
23.1 8 m s^{-1}
23.2 $1{\cdot}6 \times 10^{5}$ J
23.3 $7{\cdot}5 \times 10^{5}$ J

Chapter 13 *page 140*
2 25°C
4 90°C

Chapter 14 *page 154*
1 4500 J kg^{-1} °C^{-1}
2 1560 J
3 370 J kg^{-1} °C^{-1}
4 0·1 kg
5.1 6×10^{4} J
5.2 3×10^{3} J kg^{-1} °C^{-1}
5.3 250 s
6.1 280 W
7 20·8°C
8 46°C
9 90°C
11 $3{\cdot}34 \times 10^{4}$ J
12 $1{\cdot}49 \times 10^{5}$ J kg^{-1}
13 670 s
14 $3{\cdot}5 \times 10^{5}$ J kg^{-1}
15 $4{\cdot}3 \times 10^{5}$ J
16 0·11 kg

17 0·05 kg
18 $7{\cdot}52 \times 10^5$
20.2 a) 130°C
21.2 a) 12600 J b) 252 s
22.1 2083 J kg^{-1} °C^{-1}
23.1 360 J
23.2 289 J
24.1 400 J kg^{-1} °C^{-1}
25.1 200°C
26.1 a) 50 W e) 1250 J kg^{-1} °C^{-1}

Chapter 15 *page 170*
3 $2{\cdot}5 \times 10^5$ Pa
4 500 Pa
5.1 87°C
6.1 a) 0 K b) 123 K c) 773 K
6.2 a) −273°C b) −1°C c) 227°C
7 273°C
8 90 m^3
9 $2{\cdot}5 \times 10^5$ Pa
17.1 200 kPa
17.2 600 K or 327°C
18.2 0·14 cm^3

Chapter 16 *page 180*
5 6 units

Chapter 17 *page 194*
2 1 Ω
6.1 8 V
6.2 40 Ω
6.3 1·6 W
6.4 67%
7.1 140 C
7.2 2·3 A
7.3 0·4°C
9.1 11 Ω
10.3 7·2 Ω
11.1 6 A
11.2 24 V
12 $R_1 = 3\,\Omega$, $R_2 = 6\,\Omega$
14.1 a) 25 Ω b) 0·8 A c) 8 V, 12 V
14.2 a) 24 V b) 0·05 A

Chapter 18 *page 207*
7.1 shunt: 1·1 Ω
7.2 multiplier: 90 Ω

Chapter 19 *page 221*
1.1 400 kW
1.2 4 kW
2.3 100 turns
2.4 0·1 A
6.2 100 V a) 1 V b) 98 V, 9·8 V
c) 2 A

Chapter 20 *page 227*
1 72 W; 3 A
0·4 A; 3 A
12·5 A; 13 A
360 W; 3 A
2.1 $22\frac{1}{2}$p
2.2 2·5p
5.5 10 A

Chapter 22 *page 259*
7 0·8 μCi
8 1800 years
9.1 3·5 hours
10.1 86 protons; 136 neutrons
10.2 218; 84
11 90; 226
15.2 a) 2·5 millicuries
16.1 a) 605 counts per minute b) 8 days
c) 32 days
17.2 5 counts per minute
17.3 30 minutes
18.1 a) $5{\cdot}4 \times 10^{11}$ J b) 33%
18.2 a) 270 years

Index

Acknowledgements

The publishers and authors would like to thank the following for permission to reproduce photographs:

Lord Blackett's estate p.246 centre and bottom, p.251 bottom left and right, p.262 bottom; Paul Brierley p.8 bottom right, p.9 top, and centre; p.12 top, centre and bottom left, p.13 bottom left, p.14 (all); p.16 left and right; British Steel p.200 centre; J Allan Cash p.69 top; Cavendish Laboratory p.36 centre, p.261 centre; CEGB (Dr Alan Male) p.1 bottom right; CEGB p.212 top, p.220 left; Centre Nationale de Recherche Scientifique p.23 centre; CERN p.252 centre; Eastman Kodak p.38 centre; H Edgerton p.74 centre; Electricity Council p.224 top; EMI p.33 bottom; Ford p.31 bottom right, p.131 bottom left and right; Griffin and George Ltd. p.56 bottom, p.57 top and centre, p.58 top, p.59 centre, p.96 top, p.250 centre left and bottom right, p.251 centre, p.254 centre, p.263 top; Phillip Harris and Co. p.174 centre, p.250 centre right; Imperial War Museum p.158; International Atomic Energy Authority p.265 bottom; Keystone Press p.1 top, p.96 bottom, p.266 top; Mansell Collection p.3 centre, p.38 top, p.141 top; Middlesex Hospital (Mr Baldwin) p.31 top; MK Electrical p.225 centre; Mullard p.237, p.241 bottom; NASA p.106 top left; National Physical Laboratory p.183 centre; Nuclear Enterprises p.49 bottom right, p.263 bottom; Panax Equipment p.252 bottom; Popperfoto p.1 centre right, p.45 centre left; Rediffusion p.37 top right, Royal Institution p.209 top; Science Museum p.21 top right, p.38 centre right, p.41 top right, p.64 bottom, p.208 bottom, p.252 top, p.261 bottom; Smiths Industries p.159 top, p.160 top; Space Frontiers p.31 top left, p.42 centre right; Sparrows p.121 right; Swiss National Tourist Office p.121 left; J Tabberner p.36 top, p.218 centre, p.220 right; Tektronix p.232 top; Teltron p.229 bottom, p.230 centre, p.231 bottom; Texas Instruments p.239 bottom; Unilab p.239 top; United Kingdom Atomic Energy Authority p.38 bottom right, p.264 bottom, p.265 top, p.266 bottom.

All other photographs taken by Robb & Campbell Harper Studios and the authors.

We would also like to thank Dynamic Electronics Limited for their help with the telemetry transmitter.